C

ATMOSPHERIC
CIRCULATION SYSTEMS

Their Structure and Physical Interpretation

International Geophysics Series

Edited by

J. VAN MIEGHEM

Royal Belgian Meteorological Institute
Uccle, Belgium

ATMOSPHERIC CIRCULATION SYSTEMS

Their Structure and Physical Interpretation

E. PALMÉN

ACADEMY OF FINLAND
AND UNIVERSITY OF HELSINKI
HELSINKI, FINLAND

AND

C. W. NEWTON

NATIONAL CENTER FOR ATMOSPHERIC RESEARCH
BOULDER, COLORADO

1969

ACADEMIC PRESS New York and London

ACADEMIC PRESS, INC.
111 Fifth Avenue, New York, New York 10003

United Kingdom Edition published by
ACADEMIC PRESS, INC. (LONDON) LTD.
Berkeley Square House, London W.1

LIBRARY OF CONGRESS CATALOG CARD NUMBER: 69-12279

PRINTED IN THE UNITED STATES OF AMERICA

PREFACE

False facts are highly injurious to the progress of science, for they often endure long; but false views, if supported by some evidence, do little harm, for every one takes a salutary pleasure in proving their falseness.

Charles Darwin
The Origin of Man, Chap. 6.

The objective of this book is to describe the essential nature of the principal circulation systems in the atmosphere, as they appear in light of our present knowledge. We have limited our discussion to the "lower atmosphere," roughly below the 100-mb surface. R. A. Craig, in his volume "The Upper Atmosphere," in this same series, has given a thorough treatment of the layers higher up. However, we also adopted this limitation because we wanted to discuss the processes most directly responsible for weather and climatological phenomena in the most ordinary sense. It is not yet clear how much these phenomena are influenced by processes in the upper atmosphere; while there are certainly connections, it is established that the main driving mechanisms for the lower atmosphere are in the lower atmosphere itself.

Despite the tremendous improvement of the observational data and their treatment during the past 20 to 30 years, these data are still too inadequate to permit a complete description of the complicated nature of the diverse circulation patterns appearing in the atmosphere. It is therefore necessary to simplify and generalize by using proper methods of analysis founded partly on the principle of eliminating unessential phenomena. The limited observations available in a given system might be analyzed in different ways; it is the aim of synoptic meteorology to find the particular solution that is physically coherent. In the search for this, different hypotheses will appear, whose validity or falsity will be exposed when more observations are provided or more correct theories devised. In this book, the reader will find many such partial descriptions and interpretations, and we hope that the measure of his dissatisfaction will be his urge to improve upon them.

One of the leading objectives of meteorology is to find out, knowing the inputs of energy into the atmosphere and the properties of the fluid constituting it, what makes it operate in the way it does. Nature has devised the working parts most suited for converting energy between the various forms and for transferring the energy from one place to another as this is needed. The ultimate test of any hypothesis about a circulation system is whether in its final form the hypothesis accounts satisfactorily for these functions.

It is our purpose to describe the character of the circulation systems in light of their performance of these various functions. These systems differ in kind both according to the latitudes at which they are found, because the required functions differ in degree, and at a given latitude, because the circumstances of regional physiography require them to take different forms. Thus, to achieve any degree of completeness, we must undertake to describe a considerable variety of systems, sometimes in a very sketchy way. At the same time, we have attempted to provide enough references to enable the reader to track down a more adequate description of a particular phenomenon in which he may have a special interest.

There are practical limitations imposed upon what can be covered in a single volume. Perhaps the most discouraging aspect is that there are extensive areas which properly fall within the framework of the subject matter, but cannot be treated adequately in a few pages and which therefore must be omitted entirely. Secondly, it is disappointing to have to leave out a great deal of material as to the origins of present concepts. We have tried to strike a balance between including a fair amount of material of this sort and writing an account overloaded with references. Thus we have not given a complete historical summary, but where possible we have tried to indicate the sources of the particular aspects we have discussed. Only in Chapter 5 have we specifically included a brief historical review of the earlier development of synoptic aerology and the polar-front theory, in order to stress their special importance for our present knowledge.

We have not included anything about aerological techniques or about weather forecasting, since both subjects are treated in available texts. We have also not attempted to present the material specifically in textbook form—for example, the derivations of equations which may be found elsewhere. However, we believe this volume will be suitable for use as an advanced text by those who have some grounding in basic theory and preferably also some practice in synoptic analysis.

Only a few references are made to the results of numerical model experiments in solving general and specific atmospheric circulation problems. Obviously a new era in the study of the atmosphere and its flow patterns has been inaugurated by this technique and by the use of satellite observations. The work in these areas is so vast in scope that their summarization cannot be attempted here, and is best left for experts in these specialties. With these omissions, it is possible that some readers will consider this book somewhat too qualitative or descriptive and perhaps not up to date in every respect. Nevertheless, the basic principles have not been changed by technological advances; and the traditional methods of synoptic aerology and the recent techniques play complementary roles in investigating the atmosphere.

It must be admitted that where alternative views exist on a given subject, we have not exposed all of them, with the excuse that to do this would so encumber the discussion as to destroy its continuity. We have exercised the authors' prerogative to choose the viewpoints discussed. It will perhaps be noticed that a large proportion of the illustrations is taken form our own publications or those of our close colleagues. This is partly a matter of convenience of access and partly a natural result of the fact that we find it easiest to write about material with which we are already familiar.

Largely for these reasons, our illustrations also show a geographical bias, especially toward synoptic systems over and near North America. Readers in other areas will recognize that these systems may differ in important respects from one location to another. Since one of our main objectives is to stress the physical principles, it is hoped that these will not suffer from the illustration of circulation systems of special types; the same basic principles apply regardless of locality. Nevertheless, we must apologize for not having extended the description to certain broad regions.

Thus, in many respects, this book is not what we had hoped it would be, but what we were able to make it. We trust that we have given an accurate account of the work of others. Such an aspiration is difficult to realize, since the meaning of what one reads is likely to be colored by his own thoughts on the subject, so that in mentioning an author's work, one may either confound the original message or put the emphasis where the author does not intend it. At most, we pray that we have not obscured what is fact and what is hypothesis, and thereby violated Darwin's maxim used as a motto for this book.

In the preparation of this volume, we have been materially assisted by many persons as well as several institutions. We would specifically like to thank The University of Chicago, The University of Stockholm, The University of Helsinki, the United States Weather Bureau, and the National Center for Atmospheric Research for making available support and facilities that made it possible to carry out this work. We are personally indebted to Professors Sverre Petterssen, Bert Bolin, Bo Döös, and Lauri A. Vuorela, and Dr. Robert H. Simpson for their kindness in making the necessary arrangements. A number of our colleagues have given comments on parts of the manuscript or offered advice useful in its preparation. Ralph L. Coleman has been especially helpful in analyzing some of the figures and in drafting most of the new illustrations. We are grateful to Miss Margaret Johnstone, Mrs. Nancy Seiller, Mrs. Lois Gries, and Mrs. Eileen Workman, who cheerfully suffered through many drafts, for their expert typing and checking of the manuscript. We are particularly indebted to Professor Carl W. Kreitzberg for reading the manuscript and calling attention to a number of errors we would otherwise have overlooked.

March 1969

E.P.

C.W.N.

CONTENTS

3. SEASONAL AND ZONAL VARIATIONS OF THE MEAN ATMOSPHERIC STRUCTURE AND FLOW PATTERNS

4. PRINCIPAL AIR MASSES AND FRONTS, JET STREAMS, AND TROPO-PAUSES

5. THE POLAR-FRONT THEORY AND THE BEGINNINGS OF SYNOPTIC AEROLOGY

6. EXTRATROPICAL DISTURBANCES IN RELATION TO THE UPPER WAVES

7. THERMAL STRUCTURE OF FRONTS AND CORRESPONDING WIND FIELD

8. PRINCIPAL TROPOSPHERIC JET STREAMS

9. FRONTOGENESIS AND RELATED CIRCULATIONS

14. Circulation and Disturbances of the Tropics

15. Tropical Cyclones, Hurricanes, and Typhoons

16. Energy Conversions in Atmospheric Circulation Systems

17. Summary of the Atmospheric Circulation Processes

Contents

LIST OF SYMBOLS

Special Conventions

$C\,(E_1, E_2)$	conversion of energy from form E_1 to form E_2, where the E's may be either A_m, A_e, K_m, or K_e
$\dot{Q}$	dQ/dt
$\bar{Q}$	mean value around a latitude circle
$\hat{Q}$	mean value around an arbitrary boundary
$[Q]$	mean value over an area
Q'	local deviation from mean value
Q''	local deviation from area mean $[Q]$, where deviations (Q') from boundary means are also used
Q_0	value of Q at earth's surface (unless otherwise specified)
Q_g	geostrophic value
Q_x, Q_{xy}	component of Q in x-direction or in xy-plane, etc.

Common Symbols

Only symbols used repeatedly are listed. Others are defined as needed in text, sometimes using the same letters. Boldface letters indicate vector quantities.

a	radius of earth
c	speed of movement of a circulation system, such as a wave
c_p	specific heat of air at constant pressure
c_v	specific heat at constant volume
f	coriolis parameter, $2\,\Omega \sin \varphi$
g	acceleration of gravity
h	(1) thickness between two isobaric surfaces; (2) sensible heat content per unit mass

List of Symbols

k	(1) kinetic energy per unit mass; (2) trajectory curvature
k_m	kinetic energy of zonally averaged motions, per unit mass
k_λ, k_φ	kinetic energies of mean zonal (λ) and meridional (φ) motion, per unit mass
k_s	streamline curvature
n	distance normal to wind direction or to a curve as specified
p	pressure
q	(1) specific humidity; (2) heat content per unit mass
r	(1) radius in cylindrical coordinate system; (2) radius of flow curvature
s	distance downwind along a streamline
t	time
u, v, w	eastward, northward, and upward components of wind velocity
x, y, z	distance eastward or northward, and height above sea level
A	(1) area; (2) wave amplitude
A_e, A_m	integrated eddy or mean available potential energy within a specified volume
A_s	streamline amplitude
C	horizontal movement of an isopleth
D	(1) divergence (horizontal, unless otherwise specified by subscripts); (2) rate of frictional dissipation of kinetic energy within a specified volume
E	(1) mass of water evaporated from earth's surface, per unit time and area; (2) static energy ($c_p T + gz + Lq$)
F_{xy}, F_{xz}	convergence, in the poleward or vertical direction, of the eddy flux of eastward linear momentum
I	total energy within a specified volume
J	eddy flux of absolute angular momentum, in direction specified by subscript
K	total kinetic energy within a specified volume
K_e, K_m	integrated kinetic energy of the eddy or mean motions within a specified volume
K_φ	transfer of kinetic energy across latitude φ
K_y	mean kinetic energy of the meridional motions, integrated within a volume bounded by latitude circles and pressure surfaces
L	(1) wavelength (L_s, stationary wavelength); (2) length of a boundary; (3) latent heat of vaporization
M	total absolute angular momentum in ring bounded by two latitudes and pressure surfaces
$(M)_p$	downward mass flux through an isobaric surface, in specified area of cold air
M_r	absolute angular momentum about the central axis of a cyclone, per unit mass, at distance r from center
$(M_r)_r$	radial influx of absolute angular momentum through a cylinder at distance r from the center of a revolving system
$(M_r)_z$	upward flux of absolute angular momentum through an isobaric surface between two radii of a circulation system
M_φ	absolute angular momentum transferred poleward across latitude φ between two pressure surfaces

M_z	total vertical flux of absolute angular momentum through an isobaric surface, over a specified area
P	(1) mass of precipitation reaching earth's surface, per unit time and area; (2) total potential energy within a specified volume
Q_a, Q_e	heat content of atmosphere or earth
R	gas constant for air
R_a, R_e, R_{ae}	latitudinal mean net (excess of absorbed over emitted) radiation per unit time and area received by atmosphere, earth, or earth-atmosphere system
Q_s	transfer of sensible heat from earth to atmosphere, per unit time and area
T	temperature (in some contexts, virtual temperature)
ΔT	temperature excess of an air parcel in a convective cloud, relative to the undisturbed environment temperature
V	total wind speed
V_a	scalar value of ageostrophic wind velocity $\mathbf{V}_a = (\mathbf{V} - \mathbf{V}_g)$
W	energy per unit area (excluding latent heat)
$\overline{W}_\varphi$	mean poleward flux of energy (other than latent heat) per unit length of a latitude parallel, in atmosphere or earth as designated by secondary subscript a or e; $\overline{W}'_{\varphi a}$ includes latent heat flux
α	specific volume
β	df/dy, the variation of coriolis parameter poleward
γ	lapse rate $(-\partial T/\partial z)$ of temperature with height
γ_a	adiabatic lapse rate $(-dT/dz)$, dry or moist, depending on saturation
ζ	relative vorticity about a vertical axis
ζ_a	absolute vorticity $(f + \zeta)$
ζ_θ	relative vorticity in an isentropic surface
η	absolute vorticity $(f + \bar{v}_\theta/r + \partial \bar{v}_\theta/\partial r)$ in symmetrical cyclone
θ	(1) potential temperature; (2) angular measure in cylindrical coordinate system, increasing in cyclonic sense
θ_e, θ_w	equivalent and wet-bulb potential temperature
λ	longitude
μ	coefficient of eddy viscosity
ϱ	density
σ	(1) vertical shear $\partial u/\partial z$; (2) horizontal wind speed inside a convective cloud; (3) area of a surface bounding a specified volume
τ	horizontal stress
φ	latitude
ψ	(1) Montgomery stream function $(c_p T + gz)$; (2) angle between a frontal surface and the horizontal; (3) slope of a streamline in the vertical plane
ω	vertical wind component (dp/dt) in pressure coordinate system
Γ	lapse rate $(\partial T/\partial p)$ in pressure coordinate system
Γ_a	adiabatic lapse rate (dT/dp), dry or moist process, depending on saturation
Φ	geopotential (gz)
Ω	angular velocity of earth's rotation
$\mathbf{F}$	(1) frontogenetic function, with components F_x, F_y, F_z; (2) frictional force per unit mass
$\mathscr{T}$	tensor of the Navier-Stokes viscosity stresses

1

THE MEAN STRUCTURE OF THE ATMOSPHERE,
AND THE MAINTENANCE OF
THE GENERAL CIRCULATION
IN THE NORTHERN HEMISPHERE

It is appropriate to begin our description of atmospheric circulation processes with a broad review of the requirements for momentum and energy balance. The circulation systems adapt themselves to the forms that are peculiarly effective in generating and transferring these properties and, as will be seen, the most characteristic forms vary according to the functions required in different geographical locations and seasons. Because of these diversities in temperature, pressure and wind patterns, it is useful to examine the mean state of the atmosphere before dealing with the natures of its disturbances.

The most concise way of describing the atmosphere is to present the mean structure in a single meridional cross section, averaged around the earth, and extending from the North Pole to the South Pole. Although this will provide a frame of reference, it has been convincingly demonstrated that the mean conditions in such a section cannot satisfactorily be explained by the *average* air motions in connection with the existing heat and momentum sources and sinks. This was emphasized by V. Bjerknes *et al.* (1933) in "Physikalische Hydrodynamik." There the proposition was elaborated that it would be impossible to explain the general circulation in terms of an atmospheric flow that is axially symmetrical about the globe.

For example, a simple toroidal circulation with rising motion in low latitudes and sinking in high latitudes would, on the rotating earth, generate excessively large zonal wind speeds in the connecting meridional branches of the flow. The observed breakdown of the subtropical and higher-latitude circulation into cyclonic and anticyclonic eddies was viewed as a condition necessary to avoid this difficulty. It was pointed out that, at the same time, these eddies accomplish a large part of the needed meridional heat exchange, owing to their asymmetry in thermal properties. These authors also introduced a scheme of organized vertical motions in the cellular circulations such that a continuous transformation of potential energy into kinetic energy is possible.

Thus, it is essential to take account of the fluxes of given properties due to *eddies*, which exist in a wide range of sizes and kinds. These eddies, being partly transient and mobile, disappear when the atmospheric properties are averaged over a period of time and around all longitudes. It will also become evident that consideration of the eddies alone would provide only a partial conception of the processes at work. Both the mean circulation and the disturbances contribute in important ways, and they cooperate in a physically harmonious fashion to balance the energy and momentum budgets.

1.1 MEAN MERIDIONAL DISTRIBUTION OF TEMPERATURE AND WIND

In both hemispheres (Fig. 1.1) the mean temperature distribution is characterized by a horizontal (isobaric) decrease poleward at all levels of the troposphere except above about 200 mb in the tropics. The overall pole-to-equator temperature contrast in the troposphere is much stronger during the cold season than during the warm season, in both hemispheres. The most characteristic features in these mean cross sections are the strong baroclinity in middle latitudes and the clear distinction between the tropopause in low latitudes and that in middle and high latitudes.

Considering the variation of the coriolis parameter with latitude, the mean zonal wind field is approximately consistent with the meridional temperature distribution. This is conveniently illustrated by Fig. 1.2 (Mintz, 1954), which shows the average zonal wind component in both hemispheres for winter and summer. The essential features are the low-latitude belt of easterlies and the middle- and high-latitude westerlies. Tropical easterlies occupy a zone whose width is somewhat more than 60° latitude in low levels but narrows upward. With changes in season the centers of the easterly wind belts shift slightly northward and southward, as does the equatorial trough.

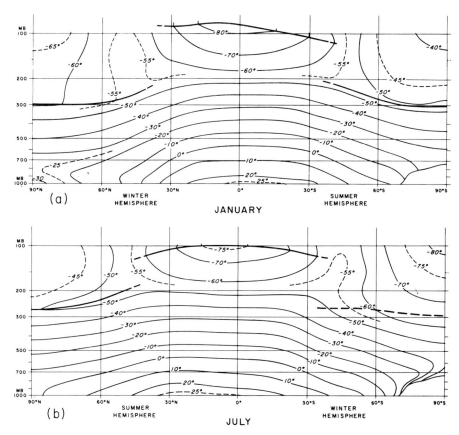

Fig. 1.1 Mean temperature (°C) averaged around latitude circles for (a) January and (b) July. Heavier lines show approximate mean tropopauses. Data were taken mostly from Goldie *et al.* (1958). South of 50°S, mean soundings were provided by H. van Loon; surface temperatures in Southern Hemisphere from van Loon (1966) and (over Antarctica) from Phillpot (1962). In Northern Hemisphere, temperatures in stratosphere are a compromise between data taken from Muench (1962) and from Reed and Mercer (1962). (Constructed by Ralph L. Coleman.)

In winter, circumpolar west-wind maxima near the 200-mb level are especially pronounced, slightly equatorward from latitudes 30°N and 30°S. These maxima shift poleward in summer by about 15° latitude in the Northern Hemisphere but by a smaller amount in the Southern Hemisphere. Especially in the Northern Hemisphere, the summer wind maximum is very much weaker than the corresponding winter maximum.[1] These west-wind

---

[1] At the 500-mb level, the middle-latitude westerlies of the Southern Hemisphere are

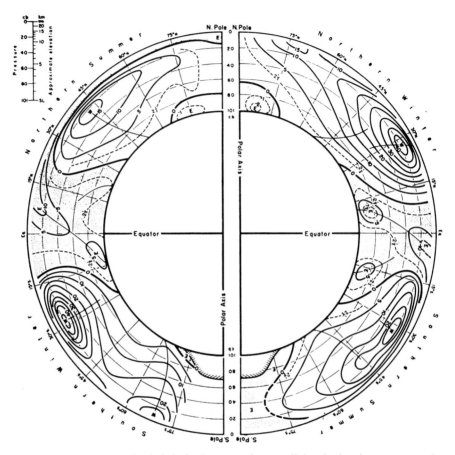

Fɪɢ. 1.2   Mean zonal wind (m/sec) averaged over all longitudes, in summer and winter. Negative values denote mean easterly wind. (After Mintz, 1954.)

maxima are situated almost vertically above the average positions of the subtropical high-pressure belts at sea level.

While the zonal mean circulation is generally in approximate geostrophic balance with the meridional pressure gradients, the *meridional* wind averaged around a latitude circle represents an ageostrophic wind. Hence it must be evaluated from direct upper-wind measurements, and it is difficult to arrive at accurate results because the mean meridional wind is weak. Nevertheless several attempts have been made to compute the mean south-north winds in

---

stronger in summer (Section 3.4). Abundant evidence shows that the subtropical wind maximum higher up is stronger in winter, as in Fig. 1.2 (Gabites, 1953; Phillpot, 1959; Hofmeyr, 1961; U.S. Weather Bureau, 1961; Schwerdtfeger and Martin, 1964).

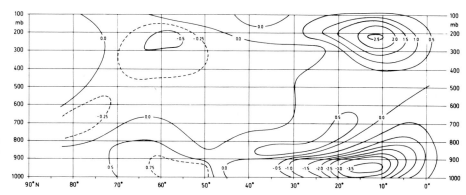

FIG. 1.3 Mean meridional wind components (m/sec, positive for south wind) in the Northern Hemisphere during the winter season, December-February. (After Palmén and Vuorela, 1963.)

the Northern Hemisphere. Data are sparse over the extensive oceanic regions between the Equator and 30°N, and the problem is further complicated by the strong seasonal variation of the position of the equatorial trough.

Utilizing mostly the data compiled by Crutcher (1961), Palmén and Vuorela (1963) and Vuorela and Tuominen (1964) derived the hemispheric mean meridional wind distributions shown in Figs. 1.3 and 1.4 for winter and summer. These results are at least partly supported by related computations by Riehl and Yeh (1950), Mintz and Lang (1955), Palmén (1955), Tucker (1959), and Defant and van de Boogaard (1963). With better data in the future we may expect some changes to the values in Figs. 1.3 and 1.4, but their essential features appear to be correct.

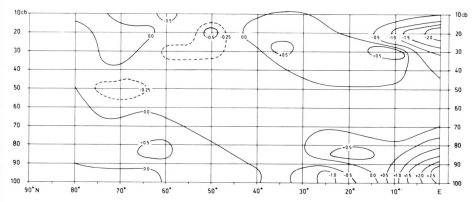

FIG. 1.4 Same as Fig. 1.3, for the summer season, June–August. (After Vuorela and Tuominen, 1964.)

According to Fig. 1.3, an appreciable mean northerly wind prevails in the lowest troposphere in the tropics during winter (maximum speed about 3.5 m/sec), with a pronounced southerly component in the upper troposphere (maximum about 2.5 m/sec). Through the deep layer 750 to 350 mb, the meridional wind is very weak. In higher latitudes, low-level southerly and high-level northerly components prevail, but the mean speeds are very modest. During the summer (Fig. 1.4) moderate low-level northerlies, up to about 1 m/sec, prevail between 40°N and 13°N, with a somewhat weaker south wind through a deeper layer in upper levels. Close to the Equator, mean winds from the south reach speeds of 2.5 m/sec in low levels, with northerlies up to 2 m/sec in upper levels.

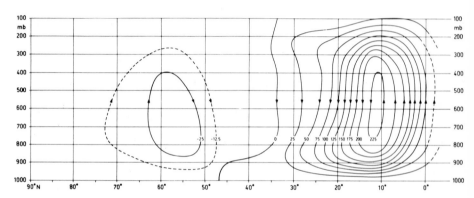

FIG. 1.5   Total mean meridional circulation in the Northern Hemisphere during the winter season. The transport capacity of each streamline channel is 25×10^6 ton/sec, in the direction indicated by arrows. (After Palmén and Vuorela, 1963.)

Figures 1.5 and 1.6 show the mass circulations derived from Figs. 1.3 and 1.4. In winter the mean mass circulation is very intense between the equatorial belt and 30°N, corresponding to the well-known "Hadley cell" of the tropics. This cell, with ascending warm air in the south and descending cooler air in the north, has a character of a "direct," kinetic-energy-producing circulation. As its counterpart, a much weaker mean meridional circulation in the opposite sense, the "Ferrel cell," can be observed in middle latitudes.[2] In winter, the total mass circulation in the Hadley cell amounts to about 230×10^6 ton/sec, whereas the Ferrel cell mass circulation

[2] Here the sense of the circulation would be opposite if averaging were performed relative to the polar-front zone rather than latitude circles. This must be considered in interpreting the overall production of kinetic energy in the latitudes of the Ferrel cells (Section 16.3).

is only 30 × 10⁶ ton/sec. In summer, the Northern Hemisphere Hadley cell, shifted northward about 20° latitude, has a feeble circulation of about 30 × 10⁶ ton/sec, while the Ferrel cell has about the same strength as in winter. The strongest meridional circulation in summer is associated with the extension, north of the Equator, of the intense Hadley cell of the Southern Hemisphere winter.[3]

As noted in Section 2.2, it is quite possible that the value given above, for the winter mass circulation of the Hadley cell, should be somewhat reduced. Weaker mass circulations are shown, e.g., by Guterman (1965) and Lorenz (1967).

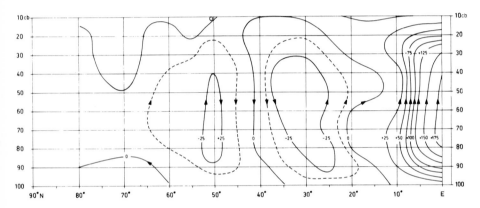

FIG. 1.6   Same as Fig. 1.5, for the summer season. (After Vuorela and Tuominen, 1964.)

Estimates of the mean meridional circulation based on Figs. 1.5 and 1.6 will be used in the ensuing discussion of the generation and fluxes of angular momentum (and, in Chapter 2, of energy). Later, the Hadley circulation will be invoked as a mechanism for maintenance of the subtropical jet stream and of the characteristic temperature field in the subtropics associated with this important global phenomenon.

[3] As part of a more general study based on observed winds in the Southern Hemisphere, Obasi (1963) evaluated the mean meridional circulations, but acknowledged their uncertainty in this region where observations are sparse and geographically biased. Using Obasi's calculations of the mean zonal flow and of the momentum transfer by eddies, Gilman (1965) derived the meridional circulation field by an indirect method based on momentum transfer requirements. The results indicated a winter Hadley cell corresponding in intensity to that calculated in a similar way by Mintz and Lang (1955) for the Northern Hemisphere. A much weaker, but still appreciable, Hadley circulation was deduced for summer. The intensity of the Ferrel cell, nearly the same in winter and summer, was found to be comparable with that of the Northern Hemisphere.

1.2   Maintenance of the Angular Momentum in the Atmosphere

The question of maintenance of the mean atmospheric circulation has been the subject of a large number of investigations during the past 20 years. Only a few of the salient results will be repeated here. These are founded upon the principle of meridional and vertical exchange of angular momentum.

Essentially, the global wind system is characterized by surface easterlies in low latitudes and surface westerlies in higher latitudes. Since the torque exerted by the earth on the atmosphere is proportional to the distance from the axis of rotation, the existence of small regions of surface easterlies in high latitudes can be mainly disregarded in a discussion of the global angular-momentum budget.

In the belt of easterlies that occupies the tropical half of the earth's surface (Fig. 1.2), the earth exerts an eastward torque upon the atmosphere above, owing to surface friction. This is equivalent to a flux of westerly angular momentum from earth to atmosphere. The opposite frictional torque in the extratropical belts of westerlies of both hemispheres is then equivalent to a flux of angular momentum from atmosphere to earth. The mean zonal surface stresses over the oceans at various latitudes have been computed by Priestley (1951), and are shown in Fig. 1.7. Since the torque

$$T = 2\pi a^3 \int \bar{\tau}_x \cos^2 \varphi \, d\varphi$$

where a is earth radius and $\bar{\tau}_x$ is the mean eastward surface stress at a given latitude, Priestley plotted the stress against an ordinate scale proportional to $(2\varphi + \sin 2\varphi)$, which makes the area between the curve and the y-axis proportional to the torque.

For sufficiently long time periods the total angular momentum of the whole atmosphere must remain unchanged, so *the upward flux in low latitudes must equal the downward flux in middle and high latitudes.*[4] Angular momentum must therefore be transferred poleward by processes within the atmosphere, from source to sink regions according to the scheme in Fig.

[4] In Fig. 1.7, there appears an excess of eastward over westward torque in all seasons. The use of the same surface drag coefficient in all latitudes may have resulted in a relative underestimate of the stresses in the westerlies, where (on the whole) winds are stronger and the ocean surface rougher. In addition, however, pressure differences between the west and east sides of mountain massifs exert a torque. According to White (1949), in most latitudes of the Northern Hemisphere the mountain torque acts in the same direction as the frictional torque, and in winter amounts to about 40% of the total in the latitude of strongest westerlies.

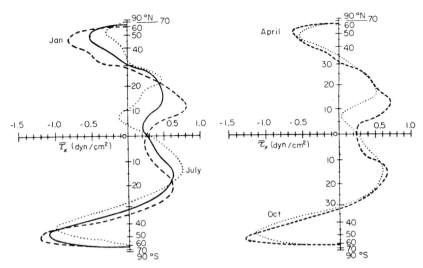

FIG. 1.7   (a) Mean zonal surface stress distribution over all oceans for January (dashed), July (dotted), and year (full line);   (b) the same for April (dashed) and October (dotted). (After Priestley, 1951.)

1.8. Since angular momentum is imparted to the atmosphere throughout the belt of easterlies, and removed within the belt of westerlies, *the maximum poleward flux of angular momentum must occur at those latitudes in both hemispheres that separate the zones of surface easterlies and westerlies.* This means roughly the latitudes of the subtropical high-pressure belts.

Jeffreys (1926) pointed out that in treating the atmospheric flux of angular momentum, two processes have to be considered: meridional mass circulations of the type proposed by Hadley (1735) and eddy fluxes of the kind discussed originally by Reynolds (1894). For discussions of the principles, the reader is referred to the formulation by J. Bjerknes (1948) and to an essay by Starr (1948), which provides an elegant and concise statement of the general circulation problem without resort to mathematics.

Jeffreys made the observation that mean mass circulations in a meridional direction do not adequately account for the necessary poleward flux of angular momentum, at least in middle and high latitudes. Extensive investigations have since shown that *except in low latitudes, the poleward momentum transfer essentially takes place through the agency of eddy fluxes.* When upper-level troughs and ridges are tilted as in Fig. 1.9a, there is a transport of angular momentum poleward because northward-moving air particles are associated with a greater west-wind component than are southward-moving particles (see Machta, 1949, for a theoretical discussion of

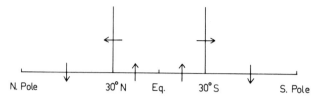

FIG. 1.8 Scheme showing the directions of angular momentum flux between the atmosphere and the earth's surface, and of the flux within the atmosphere across subtropical latitudes.

the properties of such a wave). Thus, *at a given level, relative momentum can be transferred poleward by eddies without the necessity of a poleward mass flux of air.* The same is true of the closed circulation pattern in Fig. 1.9b. Starr concludes that "the separation of the high-pressure belts of the tropics into individual cells is a necessary automatic adjustment in the atmosphere which provides for a poleward transfer of absolute angular momentum from the low-latitude easterlies."

For the global budget of atmospheric angular momentum, *it is necessary also to consider meridional mass circulations of the types in Figs. 1.5 and 1.6, particularly in low latitudes where these circulations are pronounced.*

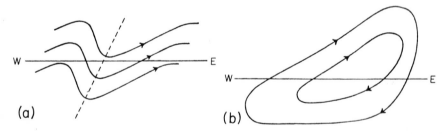

(a)

(b)

FIG. 1.9   (a) Schematic picture of horizontal streamlines in a typical wave in the upper westerlies; (b) schematic horizontal streamlines in a subtropical anticyclone. (After Starr, 1948.)

If we consider a latitude in the middle of the Hadley cell, the dominant association expressing the poleward transfer of angular momentum is that between southward and westward wind components in the low-level trades and between northward and eastward motion in the upper branch of the circulation. This correlation, as discussed later, results from the conversion of earth-angular momentum to the relative angular momentum of the wind field.

1.3 QUANTITATIVE EVALUATIONS OF ANGULAR MOMENTUM FLUX

If we limit our discussion to the Northern Hemisphere and assume that the exchange of angular momentum across the Equator is negligible,[5] we can state that the maximum poleward flux occurs near latitude 30°N in winter and somewhat farther north in summer. This flux then equals the total torque about the earth's axis, owing to surface interactions in the belt of easterlies north of the Equator. Based on the frictional stresses in Fig. 1.7, the total torque in the trade-wind belt of the Northern Hemisphere would be 40×10^{25} gm cm² sec^{-2} in winter and 12×10^{25} gm cm² sec^{-2} in summer. Palmén (1955) made an attempt to estimate the surface torque for the same region in winter, using the southward movement of air between the earth's surface and the level of strongest easterlies and assuming the zonal stress to vanish at that level, according to the principle expressed by Sheppard and Omar (1952). By this method, which does not involve the assumption of a surface drag coefficient, the surface torque in the northern trade-wind region was estimated to be 50×10^{25} gm cm² sec^{-2}. This latter value will be adopted here for winter; for summer, the assumed torque is 13×10^{25} gm cm² sec^{-2}. The total poleward flux of angular momentum across about latitude 30°N should then correspond to these numerical values.

The total angular momentum of an entire ring of the atmosphere, of unit width in the meridional direction, at latitude φ, and bounded by the isobaric surfaces p_1 and p_2 ($p_1 > p_2$), is given by the expression

$$M = \frac{a^2 \cos^2 \varphi}{g} \int_{p_2}^{p_1} \int_0^{2\pi} (\Omega a \cos \varphi + u) \, d\lambda \, dp \qquad (1.1)$$

Here a is the earth's radius, g the acceleration of gravity, Ω the angular velocity of earth's rotation, u the zonal wind component, and λ the longitude. The corresponding poleward transport of angular momentum across the same ring is then determined by

$$M_\varphi = \frac{2\pi a^2 \cos^2 \varphi}{g} \int_{p_2}^{p_1} (\Omega a \cos \varphi \bar{v} + \bar{u}\bar{v} + \overline{u'v'}) \, dp \qquad (1.2)$$

where v denotes the meridional wind component, a bar over the letter

---

[5] According to Tucker (1965), this assumption is not completely justified; he assessed a southward transfer across the Equator in January equivalent to about 10% of the total westerly angular momentum source in the Northern Hemisphere.

indicates the mean value along a latitude circle, and a prime designates the local deviation from the mean value.

If the integration is extended through the *whole depth* of the atmosphere and the slow seasonal changes of the atmospheric mass distribution are disregarded, Eq. (1.2) becomes

$$M_\varphi = \frac{2\pi a^2 \cos^2 \varphi}{g} \int_0^{p_0} (\bar{u}\bar{v} + \overline{u'v'})\, dp \qquad (1.3)$$

In this expression, $\bar{u}\bar{v}$ represents the *circulation flux* resulting from the mean meridional motion $\bar{v}$, whereas $\overline{u'v'}$ represents an *eddy flux*. The contribution of the circulation flux is generally minor compared with that of the eddy flux, except in the low latitudes with their strong Hadley circulations. It should be stressed that if a limited depth is considered, Eq. (1.2) should be used, since then the term containing Ω can be very large.

It should be noted that Eq. (1.3) gives the flux at a particular instant. For an extended time period, the flux is given (Starr and White, 1952) by

$$\frac{2\pi a^2 \cos^2 \varphi}{g} \int_0^{p_0} (\widetilde{\bar{u}\bar{v}} + \widetilde{\overline{u'v'}})\, dp = \frac{2\pi a^2 \cos^2 \varphi}{g} \int_0^{p_0} (\tilde{\bar{u}}\tilde{\bar{v}} + \overline{\tilde{u}^*\tilde{v}^*} + \widetilde{\overline{u'v'}})\, dp \qquad (1.3')$$

Here the tilde indicates a mean over a period of time. The first right-hand-side (r.h.s.) term expresses the influence of the mean meridional circulation averaged over a period of time, and the second term (where the asterisks indicate a deviation of the longitudinally averaged u and v from the longer-term means) represents the contribution of the time variation of the mean meridional circulation. This is probably small; hence we assume in the following discussion that $\bar{u}\bar{v} = \tilde{\bar{u}}\tilde{\bar{v}}$ and $\overline{u'v'} = \widetilde{\overline{u'v'}}$.

Several attempts have been made to compute the eddy flux of momentum, using principles formulated by Priestley (1949) and others. We shall here use the values computed by Mintz (1955), which are based on geostrophic rather than real winds, an approximation that seems satisfactory in dealing with eddy flux. According to Mintz, the eddy flux for the winter months January and February, 1949, amounted to 45×10^{25} gm cm^2 sec^{-2}, and for the summer months July and August, 1949, to 13×10^{25} gm cm^2 sec^{-2}, at latitude 30°N. At this approximate latitude where the required transfer is greatest, Starr and White (1951, 1954) demonstrated (using real wind data) that this is essentially accomplished entirely by the eddies.[6] However, Pal-

---

[6] A similar result was obtained by Obasi (1963) for the Southern Hemisphere.

mén (1955) found that farther south, around 15°N to 20°N, the circulation flux is of almost the same magnitude as the eddy flux. According to Fig. 1.3, at latitude 30°N there still exists a northward mean-wind component in upper levels where the zonal wind is strong, so the product $\bar{u}\bar{v}$ is not completely negligible there. To establish a balance with the surface torque of 50×10^{25} gm cm² sec⁻², quoted earlier, we hence assume that the sum of eddy and circulation fluxes at latitude 30°N has this same value in winter.

The poleward flux of angular momentum in the 1000 to 100-mb layer is graphed in Fig. 1.10. This is essentially after Mintz, but with consideration of the transport by the Hadley cell in winter. For the flux in summer, the

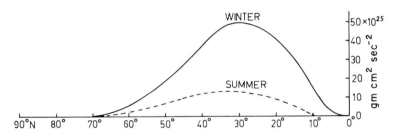

FIG. 1.10 Total northward flux of angular momentum in winter and summer in the Northern Hemisphere. Units are 10^{25} gm cm² sec⁻². Based upon geostrophic eddy fluxes computed by Mintz (1955) at latitudes 20° to 75°N, with flux by Hadley circulation added for winter.

relatively weak Hadley cell of the Northern Hemisphere is neglected, and because of lack of data close to the Equator, the curve is not extended to the equatorial region of the Southern Hemisphere Hadley cell (cf. Fig. 1.6). Table 1.1 shows the mean meridional flux of angular momentum and the corresponding surface torques in different latitude belts.[7]

For a deeper understanding of the maintenance of the zonal wind field, it is necessary to study not only the mean meridional exchange of momentum, but also the exchange in the vertical, and in addition the conversion between the Ω-angular momentum and the relative angular momentum represented by the two terms in Eq. (1.1).

The simplest way to describe the momentum flux would be to draw streamlines of momentum transport following the arrows in Fig. 1.8, these

[7] Note that, owing to the variation in areas of these belts and of the distances to the earth's axis, no immediate conclusions about the mean surface wind stress should be drawn from this table. For example, the mean surface stresses are almost equal in winter in the belts 10° to 20°N and 40° to 50°N, where the torques are very different.

TABLE 1.1

Northward Flux of Angular Momentum in the Northern Hemisphere, and
Surface Torque in 10°-Latitude Belts, in Units of 10^{25} gm cm^2 sec^{-2}

Season Latitude	0°	10°	20°	30°	40°	50°	60°	70°	80°	90°
Winter										
Flux	0	14	41	50	37	19	6	0.5	0	0
Torque		14	27	9	−13	−18	−13	−5.5	−0.5	0
Summer										
Flux	0	4	9	13	12	7	3	0.5	0	0
Torque		4	5	4	−1	−5	−4	−2.5	−0.5	0

being directed upward from the tropical source, poleward with a concentration in the upper troposphere (where, as established first by Priestley, the greatest flux occurs), and downward to the extratropical sink region. In the tropical zone the vertical momentum flux would be directed from lower to higher angular momentum, since the west wind generally increases with height to around 200 mb. Streamline pictures of this type have been presented by Widger (1949) and Smagorinsky (1963). Such pictures result if the conversion between Ω-angular momentum and u-momentum is disregarded. The essential mechanism of angular momentum flux cannot, however, be appreciated without considering this conversion.

Let A denote the total area between latitudes 5°S (the mean intertropical convergence zone) and 30°N, this area representing the principal region of the Hadley cell in Northern Hemisphere winter. The total vertical flux of absolute angular momentum resulting from the mean meridional circulation can be expressed by

$$(M_z)_c = - \int_A (\Omega a^2 \cos^2 \varphi + \bar{u} a \cos \varphi) \frac{\bar{\omega}}{g} \, dA \qquad (1.4)$$

where $\bar{\omega} = \overline{dp/dt}$ is the mean vertical wind component in pressure coordinates.

From the mass-transport streamlines of Fig. 1.5, the distribution of $\bar{\omega}$ can be determined. Between 5°S and 11°N, Ω-momentum is transported upward, while north of this belt Ω-momentum is carried downward. Since the upward-moving mass equals the downward-moving mass, but the distance from the earth's axis decreases poleward, less Ω-momentum is brought down than is carried up. On the other hand, since $\bar{u}$ in the middle and upper troposphere increases poleward, more $\bar{u}$-momentum is brought down than is carried up. Applying the same principles to extratropical latitudes (where $\bar{u}$ decreases poleward), the Ferrel cell brings down more of both Ω- and $\bar{u}$-momentum than is carried upward. This viewpoint was advanced by Palmén (1954), and has been emphasized by Yeh and Yang (1955).

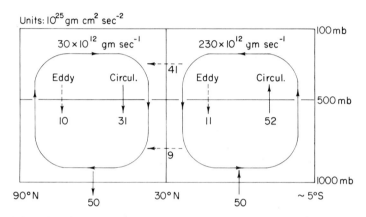

FIG. 1.11   The winter total mass circulations north and south of latitude 30° in the Northern Hemisphere. Arrows denote angular momentum fluxes between atmosphere and earth, through the 500-mb surface, and across 30°N above and below 500 mb. Dashed arrows signify portions of fluxes due to eddies. Units are 10^{25} gm cm² sec⁻².

In Fig. 1.11, a budget of the meridional and vertical fluxes in winter is presented. For simplicity, only the fluxes through a latitudinal wall at 30°N and through the 500-mb surface are represented. The scheme could, how-ever, be given in more detail for fluxes through other surfaces. This budget is based on the following:

(1) Surface torques of 50 × 10^{25} gm cm² sec⁻², in accord with the earlier discussion.[8]
(2) Northward fluxes across 30°N above and below 500 mb, after Mintz, with small adjustments to make the total balance (step 1).
(3) From the mass circulation of Fig. 1.5, the mean wind at 500 mb, and Eq. (1.4), an upward net transport of Ω-momentum in the Hadley cell, amounting to 68 units, and a downward net transport of $\bar{u}$-mo-

[8] This should perhaps be somewhat larger because the zone 0° to 5°S has been added. However, the value 50 units was partly assumed to give balance with the flux across 30°N. As noted earlier, Priestley's computation gave only 40 units.

mentum of 16 units, resulting in an overall upward circulation flux of 52×10^{25} gm cm² sec^{-2} of angular momentum.

(4) As a residual to balance the budget in each box, a downward eddy flux of 11 units through the 500-mb surface, in the tropical cell.

(5) Fluxes through the 500-mb surface in the extratropical region, arrived at in a manner similar to steps (3) and (4).

The downward-directed eddy fluxes correspond to an equivalent stress of about 0.14 dyn/cm² in the tropical region and about 0.26 dyn/cm² in the extratropical region. At the boundary of the tropics the horizontal transfer is, as noted earlier, predominantly accomplished by an eddy flux of the type emphasized by Jeffreys (1926), J. Bjerknes (1948), and Starr (1948). No corresponding scheme will be presented here for the summer season, when the Northern Hemisphere circulation fluxes are weak, as are also the eddy fluxes.

1.4   Maintenance of the Zonal Circulation

We shall now turn to the main question of the maintenance of the mean atmospheric circulation. The change per unit time of the angular momentum M of a fixed zonal ring of air, of mean density ϱ and with unit width and depth, can be written as

$$\frac{\partial M}{\partial t} = -\frac{\partial}{\partial y}(C_\varphi + J_\varphi) - \frac{\partial}{\partial z}(C_z + J_z) \tag{1.5}$$

Here C_φ and C_z represent the meridional and vertical circulation fluxes, and J_φ and J_z the corresponding eddy fluxes. These are (if fluctuations of the density ϱ are neglected) given by

$$\begin{aligned}
C_\varphi &= 2\pi a^2 \cos^2 \varphi (\Omega a \cos \varphi + \bar{u})\varrho\bar{v} \\
J_\varphi &= 2\pi a^2 \cos^2 \varphi \varrho \overline{u'v'} \\
C_z &= 2\pi a^2 \cos^2 \varphi (\Omega a \cos \varphi + \bar{u})\varrho\bar{w} \\
J_z &= 2\pi a^2 \cos^2 \varphi \varrho \overline{u'w'}
\end{aligned} \tag{1.6}$$

Inserting Eqs. (1.6) into Eq. (1.5) and utilizing the equation of continuity with the assumption that the mass of air within a zonal ring is conserved, Palmén (1955) arrived at the following expression for the maintenance of the zonal circulation:

$$\frac{\partial(\varrho\bar{u})}{\partial t} = \varrho\bar{v}\left(f + \frac{\bar{u}}{a}\tan \varphi - \frac{\partial\bar{u}}{\partial y}\right) - \varrho\bar{w}\frac{\partial\bar{u}}{\partial z} + F_{xy} + F_{xz} \tag{1.7}$$

Here f is the coriolis parameter, w is the vertical velocity, and the last two terms are

$$F_{xy} = -\frac{1}{\cos^2\varphi}\,\frac{\partial}{\partial y}\,(\overline{\varrho u'v'}\cos^2\varphi) \tag{1.7a}$$

$$F_{xz} = -\frac{\partial}{\partial z}\,(\overline{\varrho u'w'}) \tag{1.7b}$$

These terms represent the contributions of the meridional and vertical divergences of the eddy fluxes of angular momentum, and can be considered analogous to "frictional" terms caused by large- and small-scale eddy (Reynolds) stresses. In F_{xy}, the large-scale stresses are the dominating ones, whereas in F_{xz}, both large- and small-scale processes have to be considered. The second r.h.s. term of Eq. (1.7) represents the change of $\bar{u}$ associated with the mean vertical wind $\bar{w}$.

The first term of Eq. (1.7) determines the local change of $\bar{u}$ due to meridional movement of the whole ring of air. It is important where $\bar{v}$ is large, but is also proportional to the absolute vorticity of the mean zonal flow which is the coefficient in parentheses. In this coefficient, f generally is appreciably larger than $(\bar{u}/a)\tan\varphi$ and $\partial\bar{u}/\partial y$ except for regions close to the Equator. In the Hadley cell,

$$\bar{v}\left(f + \frac{\bar{u}}{a}\tan\varphi - \frac{\partial\bar{u}}{\partial y}\right)$$

transforms the Ω-angular part of the total momentum that is carried upward in the southern limb of the cell into u-momentum, resulting in a poleward increase of the mean zonal circulation. As we shall see later, this transformation is essential for the existence of the subtropical jet stream.

The general principle of the maintenance of the mean zonal circulation can be expressed in the following way: At a fixed level and latitude the zonal circulation tends to change by the effect of the mean meridional and vertical drift of air expressed by the first two r.h.s. terms in Eq. (1.7). This tendency is balanced by the combined effect of the eddy stresses in Eqs. (1.7a) and (1.7b). As emphasized earlier, F_{xy} is dominated by, and can be computed from, the divergence of the large-scale eddy flux of momentum.

The influence of F_{xz} is more difficult to evaluate. At least in some latitudes, the effect of small-scale eddies may be opposite to that of large eddies. For small-scale turbulence we may write

$$F_{xz} = \frac{\partial}{\partial z}\left(\mu\,\frac{\partial\bar{u}}{\partial z}\right) \tag{1.7b'}$$

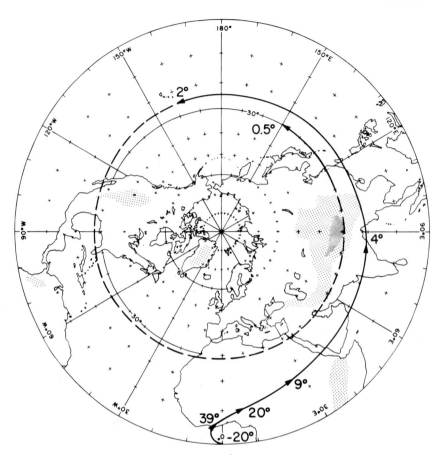

FIG. 1.12   The average motion of air, starting at the Equator and ending at 30°N, at 200 mb in winter. The angles between the mean air trajectory and the latitude parallels are indicated at every 5° latitude. (From Palmén, 1964.)

where μ is the coefficient of eddy viscosity. If μ were constant with height, the effect of small-scale turbulence could be determined from the vertical wind profile, but since we know very little about the variation of μ, no quantitative estimates can be made. It is equally difficult to evaluate the frictional effect of large-scale eddies on the vertical transfer. From the earlier discussion of the momentum budget (Fig. 1.11), we concluded that the combined effect of all types of eddies in the middle troposphere corresponds to a stress of the order of magnitude 0.15 to 0.25 dyn/cm².[9]

---

[9] These are only rough magnitudes, assuming that the stresses are uniform over the latitude belts concerned.

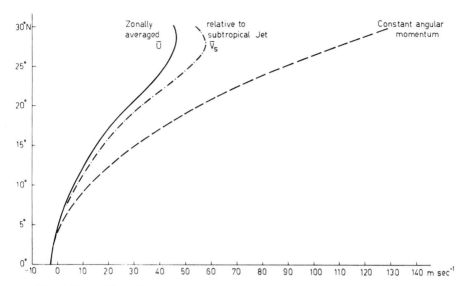

FIG. 1.13   Profiles of the 200-mb mean zonal wind between the Equator and 30°N in winter, compared to the profile corresponding to constant angular momentum. Curve $\bar{U}$ represents the zonal wind averaged around latitude circles (Mintz, 1954), while in the case of profile $\bar{V}_s$, the winds were averaged relative to the core of strongest wind in the subtropical jet stream (Krishnamurti, 1961), which meanders north and south about a winter mean latitude of 27.5°N. (From Palmén, 1964.)

From Fig. 1.3 we can compute that at the 200-mb level, a time of about 15 days would be required for an air parcel to move from 5°N to 25°N in the mean circulation, or about 30 days to move from the Equator to 30°N. The corresponding mean air trajectory is shown in Fig. 1.12. During the same times, the eastward displacement of the air would be 21,000 and 52,000 km, respectively.[10]

From Eq. (1.7), at the level of mean maximum wind ($\partial \bar{u}/\partial z = 0$) the mean meridional profile of wind is determined by

$$\frac{\partial \bar{u}}{\partial y} = f + \frac{\bar{u}}{a}\tan\varphi + \frac{F_{xy} + F_{xz}}{\varrho\bar{v}} \tag{1.8}$$

in which the first two r.h.s. terms would correspond to the profile of constant angular momentum. This profile is shown in Fig. 1.13 along with

[10] It would be interesting to verify such a trajectory by releasing constant-level balloons near the Equator and following their mean motions over periods of many days. Depending on their latitude of release, the balloons should tend to accumulate in the belt of the subtropical jet stream after 15 to 30 days, having moved $\frac{1}{2}$ to $1\frac{1}{2}$ times around the globe.

the profile of observed winter zonal wind from Fig. 1.2 and of the mean wind computed relative to distance from the subtropical jet-stream core (Krishnamurti, 1961). Although the influences of F_{xy} and F_{xz} cannot be separated, the deviation of the real profiles from the one for conserved angular momentum shows that these influences are very significant. According to estimates based on Figs. 1.3 and 1.13 (Palmén, 1964), a northward-moving air parcel south of the subtropical jet in latitudes 10° to 25° has its speed retarded by eddy stresses at the rate of 3 to 5 m/sec per day, although this is more than overcome by the influence of the first two terms of Eq. (1.8).

1.5 GENERATION AND TRANSFER OF KINETIC ENERGY

The preceding discussion relating to angular momentum naturally carries implications for the kinetic energy of the mean zonal motion. Just as the meridional transfer of angular momentum is dominated in lower subtropical latitudes by the mean circulation (vertical correlation between $\bar{v}$ and $\bar{u}$), and in other latitudes by the eddies (horizontal correlation between v' and u'), the same is true of the transfer of kinetic energy. Kuo (1951) has derived an equation for the local change of zonal kinetic energy, which has been interpreted by Starr (1953) in terms of observational data. An analogous expression may be obtained through multiplication of Eq. (1.7) by $\bar{u}$:

$$\frac{\partial(\varrho\bar{k})}{\partial t} = \varrho\bar{u}\bar{v}\left(f + \frac{\bar{u}}{a}\tan\varphi - \frac{\partial\bar{u}}{\partial y}\right) - \frac{1}{\cos^2\varphi}\,\bar{u}\,\frac{\partial}{\partial y}\left(\overline{\varrho u'v'}\cos^2\varphi\right)$$

$$- \varrho\bar{w}\bar{u}\frac{\partial\bar{u}}{\partial z} - \bar{u}\,\frac{\partial}{\partial z}\left(\overline{\varrho u'w'}\right) \tag{1.9}$$

where $\bar{k} = \bar{u}^2/2$.

The interpretation may be most readily seen by application at the level of strongest mean westerly wind. In that case the next last term, representing vertical advection of mean kinetic energy, vanishes. The last term, which is $-\bar{u}F_{xz}$, represents generally a frictional dissipation at this level. For a steady mean condition, the effect of this must be balanced by the other two r.h.s. terms.

The first of these expresses the production of zonal kinetic energy by mean meridional circulations (see Section 16.2), modified by the meridional advection of kinetic energy. The last part of this term, equivalent to $-\varrho\bar{v}\,\partial\bar{k}/\partial y$, contributes to a local decrease of $\bar{k}$ if $\partial\bar{u}/\partial y > 0$ and thus partially offsets the production equatorward of the latitude of strongest wind. Since the terms in the parentheses always have collectively a positive sign, the first r.h.s. term of Eq. (1.9) contributes to a local increase of kinetic energy if $\bar{v} > 0$ and to a decrease if $\bar{v} < 0$. Hence the mean meridional flow augments the kinetic energy of the westerlies in the upper troposphere of the Hadley cell, and diminishes it in upper levels of the Ferrel cell.

Considering the extratropical westerlies, where there is also a frictional loss, it follows that the mean zonal kinetic energy can be maintained only by the action of the eddy transfers, expressed by the second r.h.s. term of Eq. (1.9). This term can be interpreted as the contribution of the meridional convergence of the eddy transfer of westerly momentum, multiplied by the existing value of $\bar{u}$, to give the increase of $\bar{k}$. Alternatively, the kinetic energy of the westerlies within a zonal ring is augmented if a net eastward torque is exerted upon it, i.e., if the integrated eastward eddy stress acting across a vertical wall at its equatorward boundary exceeds the eastward stress imparted by the air within this ring to the air poleward of the other boundary.

Since (Fig. 1.10) the eddy momentum flux is greatest at about 30°N, near which latitude $\bar{u}$ is also greatest, it is clear that the second r.h.s. term of Eq. (1.9) is positive in the latitudes of the Ferrel westerlies. Thus, as advocated by Starr (1953), *the mean westerly winds in extratropical latitudes are largely maintained by the eddy transfer of zonal kinetic energy poleward across the subtropical latitudes.*

It follows also from Fig. 1.10 that the second r.h.s. term of Eq. (1.9) contributes to a local decrease of kinetic energy in the upper westerlies equatorward of 30°N. Considering also the dissipation represented by the last term, balance must be maintained by the first term. It can be concluded, as stressed by Palmén (1951) with regard to the zonal momentum, that *the kinetic energy of the upper westerlies in the tropical latitudes is generated by the mean meridional motions, and the portion of this that is not dissipated within the tropics is exported by eddy transfers, contributing to maintain the mean westerlies of the extratropical latitudes.*

While the available evidence indicates that the meridional *transfer* of kinetic energy by the eddies can be expressed quite well by the geostrophic motions and that this eddy transfer dominates except in low latitudes, the *generation* is a result of ageostrophic motions. In the tropics, these take the form mainly of the mean meridional motions of the Hadley cells. In higher latitudes, the interpretation of the ageostrophic motions is more complicated and is discussed in Section 16.3.

The meridional transport of kinetic energy by eddies, of which the second r.h.s. term of Eq. (1.9) is an expression, has been evaluated for different levels and latitudes by Kao (1954) and by Pisharoty (1955). Based on data by Mintz (1951), Kao found a vertically integrated transport across latitude 30°N amounting to 24×10^{10} kJ/sec in winter, and Pisharoty's determination was 20×10^{10} kJ/sec. These values correspond to 67 to 80% of the estimated generation in the Hadley cell (Table 16.1).

1.6 SUMMARY OF CONCLUSIONS CONCERNING MERIDIONAL AND VERTICAL EXCHANGE OF ANGULAR MOMENTUM

It has been shown that it is possible to achieve a realistic picture of atmospheric processes involved in the meridional and vertical exchange of angular momentum, *only if the influences of both the meridional mean circulations and eddy processes are considered.*

These meridional circulations are essential for the *vertical* exchange of angular momentum as well as for the conversion of Ω-angular momentum into *u*-momentum in the Hadley cell, and vice versa in the Ferrel cell. Except in roughly the equatorward halves of the Hadley cells, however, eddy processes dominate the *meridional* exchange of momentum. This exchange is adequate to transfer angular momentum generated by the eastward torques of the tradewind belts to the regions of dissipation of westerly momentum in higher latitudes. It seems likely that eddies carry momentum downward in general. The required vertical flux of momentum, equivalent to a mean stress of about 0.15 to 0.25 dyn/cm² at the 500-mb level, must in the lowest levels be accomplished by small-scale turbulence. This influence probably decreases upward, and in the middle troposphere the momentum transfer is probably in higher latitudes dominated by synoptic-scale eddies. Westerly momentum is evidently transferred upward in some types of synoptic disturbances, or in certain stages of their development, and downward in others (see Section 10.3). However, very little is known about the statistical result of this type of momentum flux and its interaction with the momentum flux due to small-scale turbulence. The obvious reason for this uncertainty (in contrast with the situation regarding meridional transfer) is that a detailed knowledge of the correlation between horizontal and vertical motions would be required to arrive at an evaluation, and that vertical motions are not directly measured and are difficult to evaluate with confidence.

In Fig. 1.11, the inferred vertical eddy flux is by all kinds of eddies. The influence of medium-scale eddies has also never been evaluated. Computations in Section 13.6 suggest that the vertical transport of angular momentum by cumulonimbus clouds may be a dominant mechanism in middle latitudes during the warmer months.

In this very brief review of the main factors in the maintenance of the mean atmospheric circulation, many problems have been omitted. For instance, it might have been desirable to treat the seasonal variation of the meridional and vertical momentum flux and combine this with the corresponding variation of the surface torque. However, our purpose was mainly to stress the general principles. For this purpose the conditions in winter were more suitable because of the much stronger circulation during this season. The general principles discussed above are also valid for the Southern Hemisphere; however, quantitative evaluations are more uncertain there because of sparse observations. But also for the Northern Hemiphere considerable corrections to the quantities quoted above will be necessary when more complete data become available, especially in the tropical zone.

For a thorough discussion of the general circulation, we refer especially to the recent monograph by Lorenz (1967). This contains, in addition to a valuable summary of the theory and observed conditions, an interesting account of the historical development of general circulation concepts.

REFERENCES

Bjerknes, J. (1948). Practical application of H. Jeffreys' theory of the general circulation. Résumé des mém. (Los Angeles), pp. 13–14. Réunion de Oslo, Assoc. de Météorol., U.G.G.I.

Bjerknes, V. Bjerknes, J., Solberg, H., and Bergeron, T. (1933). "Physikalische Hydrodynamik." Springer, Berlin.

Crutcher, H. L. (1961). Meridional cross sections. Upper winds over the Northern Hemisphere. Tech. Paper No. 41, 307 pp. U.S. Weather Bur., Washington, D.C.

Defant, F., and van de Boogaard, H. M. E. (1963). The global circulation features of the troposphere between the Equator and 40°N, based on a single day's data (I). *Tellus* **15**, 251–260.

Gabites, J. F. (1953). Mean westerly wind flow in the upper levels over the New Zealand region. *New Zealand J. Sci. Technol.* **B 34**, 384–390.

Gilman, P. A. (1965). The mean meridional circulation of the southern hemisphere inferred from momentum and mass balance. *Tellus* **17**, 277–284.

Goldie, N., Moore, J. G., and Austin, E. E. (1958). Upper air temperature over the world. *Geophys. Mem.* **13**, No. 101, 1–228.

Guterman, I. G. (1965). "Raspredelenie Vetra nad Severnim Palushariem." (Wind Distribution in the Northern Hemisphere). Gidrometeorologicheskoe Izdatel'stvo, Leningrad.

Hadley, G. (1735). Concerning the cause of the general trade winds. *Phil. Trans. Roy. Soc. London* **39**, 58; reprinted in *Smithsonian Inst. Misc. Collections* **51**, 5–7 (1910).

Hofmeyr, W. L. (1961). Statistical analysis of upper air temperatures and winds over tropical and subtropical Africa. *Notos* **10**, 123–149.

Jeffreys, H. (1926). On the dynamics of geostrophic winds. *Quart. J. Roy. Meteorol. Soc.* **52**, 85–104.

Kao, S.-K. (1954). The meridional transport of kinetic energy in the atmosphere. *J. Meteorol.* **11**, 352–361.

Krishnamurti, T. N. (1961). The subtropical jet stream of winter. *J. Meteorol.* **18**, 172–191.

Kuo, H. L. (1951). A note on the kinetic energy balance of the zonal wind systems. *Tellus* **3**, 205–207.

Lorenz, E. N. (1967). "The Nature and Theory of the General Circulation of the Atmosphere." World Meteorological Organization, Geneva.

Machta, L. (1949). Dynamic characteristics of a tilted-trough model. *J. Meteorol.* **6**, 261–265.

Mintz, Y. (1951). The geostrophic poleward flux of angular momentum in the month January 1949. *Tellus* **3**, 195–200.

Mintz, Y. (1954). The observed zonal circulation of the atmosphere. *Bull. Am. Meteorol. Soc.* **35**, 208–214.

Mintz, Y. (1955). Final computation of the mean geostrophic poleward flux of angular momentum and of sensible heat in the winter and summer of 1949 (Article V). Final Rept., Gen. Circ. Proj., Contr. AF 19(122)-48. Dept. Meteorol., Univ. California, Los Angeles.

Mintz, Y., and Lang, J. (1955). A model of the mean meridional circulation (Article VI). Final Rept., Gen. Circ. Proj., Contr. AF 19(122)-48. Dept. Meteorol., Univ. California, Los Angeles.

Muench, H. S. (1962). Atlas of monthly mean stratosphere charts, 1955-1959. Part I, January-June; Part II (with T. R. Borden, Jr.), July-December. *Air Force Surveys in Geophysics*, No. 141, AFCRL-62-494. Air Force Cambridge Res. Lab., Bedford, Massachusetts.

Obasi, G. O. P. (1963). Poleward flux of angular momentum in the southern hemisphere. *J. Atmospheric Sci.* **20**, 516–528.

Palmén, E. (1951). The rôle of atmospheric disturbances in the general circulation. *Quart. J. Roy. Meteorol. Soc.* **77**, 337–354.

Palmén, E. (1954). On the relationship between meridional eddy transfer of angular momentum and meridional circulations in the earth's atmosphere. *Arch. Meteorol., Geophys. Bioklimatol.* **A7**, 80–84.

Palmén, E. (1955). On the mean meridional circulation in low latitudes in the Northern Hemisphere in winter and the associated meridional flux of angular momentum. *Soc. Sci. Fennica, Commentationes Phys.-Math.* **17**, No. 8, 1–33.

Palmén, E. (1964). General circulation of the tropics. *In* "Proceedings of the Symposium on Tropical Meteorology" (J. W. Hutchings, ed.), pp. 1–30. New Zealand Meteorol. Serv., Wellington, New Zealand.

Palmén, E., and Vuorela, L. A. (1963). On the mean meridional circulations in the Northern Hemisphere during the winter season. *Quart. J. Roy. Meteorol. Soc.* **89**, 131–138.

Phillpot, H. R. (1959). Winds at 30,000 and 40,000 feet in the Australia-New Zealand-Fiji area. Proj. Rept. 59/2645. Australian Bur. Meteorol., Melbourne.

Phillpot, H. R. (1962). Some aspects of the climate of the Antarctic Continent. Working Paper 62/707, 19 pp., plus 9 tables and 8 figures. Bur. Meteorol., Commonwealth of Australia, Melbourne.

Pisharoty, P. R. (1955). The kinetic energy of the atmosphere (Article XIV). Final Rept., Gen. Circ. Proj., Contr. AF 19(122)-48. Dept. Meteorol., Univ. California, Los Angeles.

Priestley, C. H. B. (1949). Heat transport and zonal stress between latitudes. *Quart. J. Roy. Meteorol. Soc.* **75**, 28–40.

Priestley, C. H. B. (1951). A survey of the stress between the ocean and the atmosphere. *Australian J. Sci. Res.* **A4**, 315–328.

Reed, R. J., and Mercer, J. M. (1962). Arctic forecast guide. NWRF 16-0462-058, 107 pp. U. S. Naval Weather Research Facility, Norfolk, Virginia.

Reynolds, O. (1894). On the dynamical theory of incompressible viscous fluids and the determination of the criterion. *Phil. Trans. Roy. Soc. London* **A196**, 123–164.

Riehl, H., and Yeh, T. C. (1950). The intensity of the net meridional circulation. *Quart. J. Roy. Meteorol. Soc.* **76**, 182–188.

Schwerdtfeger, W., and Martin, D. W. (1964). The zonal flow in the free atmosphere between 10N and 80S, in the South American sector. *J. Appl. Meteorol.* **3**, 726–733.

Sheppard, P. A., and Omar, M. H. (1952). The wind stress over the ocean from observations in the Trades. *Quart. J. Roy. Meteorol. Soc.* **78**, 583–589.

Smagorinsky, J. (1963). General circulation experiments with the primitive equations: I. The basic experiment. *Monthly Weather Rev.* **91**, 99–164.

Starr, V. P. (1948). An essay on the general circulation of the earth's atmosphere. *J. Meteorol.* **5**, 39–48.

Starr, V. P. (1953). Note concerning the nature of the large-scale eddies in the atmosphere. *Tellus* **5**, 494–498.

Starr, V. P., and White, R. M. (1951). A hemispherical study of the atmospheric angular-momentum balance. *Quart. J. Roy. Meteorol. Soc.* **77** 215–225.

Starr, V. P., and White, R. M. (1952). Note on the seasonal variation of the meridional flux of angular momentum. *Quart. J. Roy. Meteorol. Soc.* **78**, 62–69.

Starr, V. P., and White, R. M. (1954). Balance requirements of the general circulation. *Geophys. Res. Papers (U.S.)* **35**, 1–57.

Tucker, G. B. (1959). Mean meridional circulations in the atmosphere. *Quart. J. Roy. Meteorol. Soc.* **85**, 209–224.

Tucker, G. B. (1965). The equatorial tropospheric wind regime. *Quart J. Roy. Meteorol. Soc.* **91**, 140–150.

U. S. Weather Bureau. (1961). Monthly mean aerological cross sections pole to pole along meridian 75ºW for the IGY period. U. S. Weather Bur., Washington, D. C.

van Loon, H. (1966). On the annual temperature range over the southern oceans. *Geograph. Rev.* **56**, 497–515.

Vuorela, L. A., and Tuominen, I. (1964). On the mean zonal and meridional circulations and the flux of moisture in the Northern Hemisphere during the summer season. *Pure Appl. Geophys.* **57**, 167–180.

White, R. M. (1949). The role of the mountains in the angular momentum balance of the atmosphere. *J. Meteorol.* **6**, 353–355.

Widger, W. K. (1949). A study of the flow of angular momentum in the atmosphere. *J. Meteorol.* **6**, 291–299.

Yeh, T. C., and Yang, T. S. (1955). The annual variation of the atmospheric angular momentum of the Northern Hemisphere and the mechanism of its transfer. *Acta Meteorol. Sinica* **26**, 281–294.

2

HEAT BALANCE OF THE EARTH'S ATMOSPHERE, AND THE MERIDIONAL AND VERTICAL TRANSFER OF ENERGY

In Chapter 1 it was emphasized that the generation and redistribution of absolute angular momentum is accomplished both by meridional circulations and by eddies, whose relative contributions are different in tropical and extratropical latitudes. The same is true of the redistribution of heat and related forms of energy, the mechanisms for which adapt themselves to the most effective forms to meet the requirements, as prescribed by the sources and sinks of energy in the atmosphere and in a shallow layer beneath the earth's surface.

As in the earlier discussion, our intent is only to touch on some general features fundamental to an understanding of the roles of circulation systems of different scales in the processes of heat transfer. Again, the more detailed parts of the discussion will be confined to the Northern Hemisphere; it is applicable with modifications to the Southern Hemisphere, for which only the broad energy-budget requirements will be indicated. For a more comprehensive treatment of many of the aspects, the reader is referred especially to a summary article by Malkus (1962). The heat and momentum exchange processes at and near the air-sea interface are dealt with in detail in the book by Roll (1965), and Kraus (1967) summarizes the theory of surface stresses.

2.1 GENERAL PRINCIPLES CONCERNING FLUX AND STORAGE OF ENERGY

The difference between the radiation absorbed and emitted by the atmosphere, in combination with the heat exchange between earth and atmosphere and the transfer of energy by motions within the atmosphere and the oceans, determines the heat budget.

Most of the atmosphere is characterized by a strong radiative deficit, which in itself would lead to a continual decrease of the tropospheric temperature. This deficit must be compensated by a net transfer of energy from the earth's surface, where there is a surplus of radiative heating. Sensible and latent heat transferred by conduction and evaporation at the earth's surface are carried upward by air motions, in the form of small eddies in lower levels and, increasingly at higher elevations, by organized circulation systems on scales ranging from cumulonimbus clouds to the Hadley circulation. Latent heat is realized on condensation in the ascending branches of these. Thus the overall energy balance in the vertical results from both radiative processes and the air motions.

Likewise, the meridional temperature distribution must be viewed as the result of radiative processes combined with the meridional exchange of energy by fluid motions. Simpson (1928, 1929) was the first to provide a quantitative evaluation of the global radiation distribution. Later a comprehensive investigation of the Northern Hemisphere energy budget, including not only radiation but also an assessment of transfer requirements, was carried out by Baur and Philipps (1934, 1935). Godske *et al.* (1957) give an overall summary of their results. Simpson's values for the incoming and outgoing radiation at different latitudes have been recalculated by Houghton (1954), in light of later physical data.[1]

Figure 2.1a shows the seasonal variation of effective insolation and outgoing terrestrial radiation, determined from satellite measurements between 60°N and 60°S by Rasool and Prabhakara (1966). The values shown may, of course, be somewhat changed when a longer period of observations is analyzed; however, the main features are representative. Insolation varies strongly, especially in higher latitudes, in accord with differences of solar elevation. By comparison, the variability of outgoing radiation, both latitudinally and seasonally, is weak. Consequently, the isopleths of net radiation (Fig. 2.1b) strongly resemble those of incoming radiation. Both Simpson and Houghton found that, considering the year as a whole, there

---

[1] Houghton's calculations indicate that Simpson underestimated both incoming and outgoing radiation. As noted by Benton (1954), the net radiation is nearly identical for the two sets of computations.

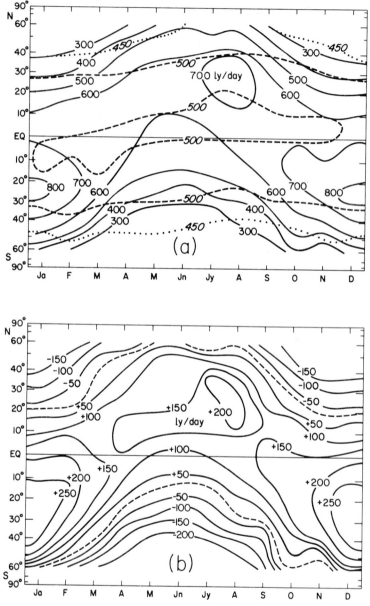

FIG. 2.1 (a) Isopleths of effective insolation (solid) and outgoing radiation from earth-atmosphere system (dashed and dotted) in langleys per day, determined by Rasool and Prabhakara (1966) from a year of TIROS satellite measurements. (b) The corresponding net radiation. (Reprinted, redrawn, from *Problems of Atmospheric Circulation*, R. Garcia and T. Malone (eds.), Spartan Books, New York, 1966.)

is a surplus of radiative heating equatorward of about latitude 37° and a corresponding deficit at higher latitudes. Thus the flux of energy from the tropics is strongest at a latitude slightly poleward of the annual average latitude of the subtropical highs.

Although on a long-term annual basis the global insolation (0.5 ly/min, reduced by reflection to a mean of about 0.3 ly/min) and eradiation must balance, this is not so when a given hemisphere is considered for a shorter period. As shown by Fig. 2.1b, in midwinter there is a radiative deficit at all latitudes except roughly the equatorward two-thirds of the tropics, while in midsummer there is a radiative surplus at all latitudes up to somewhat above 60°.

For the earth-atmosphere system of the Northern Hemisphere as a whole, London (1957) calculated that the excess of insolation over eradiation during spring amounts to 0.052 ly/min on the average.[2] In summer there is a mean excess of 0.072 ly/min, in autumn a deficit of 0.044 ly/min, and in winter a deficit of 0.085 ly/min. If the heat exchange across the Equator is considered relatively small (as justified later), the heat surplus must be stored within a hemisphere during spring and summer and given up from storage during autumn and winter. Considering these values, during spring and summer the total required change in storage amounts to an average of 1.63×10^4 cal/cm^2 for the whole hemisphere. Spar (1949) computed an increase of potential and internal energy of the atmosphere, from winter to summer, of 19×10^{28} ergs, or 4.54×10^{21} cal, for the hemisphere, which is equivalent to a change in atmospheric energy storage of 1.78×10^3 cal/cm^2. This is only 11% of the total change in storage; the remaining 89%, or 1.45×10^4 cal/cm^2, must be stored in the earth if there is no trans-equatorial transfer. Since (see, e.g., Godske *et al.*, 1957) the heat storage in solid earth is minor by comparison with that in water, if we disregard the relatively small ice-covered regions the surplus heat must be stored mainly in the oceans and withdrawn from them during the period of radiative deficit. Thus the essential effect is to raise the mean ocean temperature during spring and summer and to lower it during autumn and winter.

Since 61% of the hemisphere is occupied by ocean, this heat excess implies an increase in heat storage, averaged over the oceanic area, of about 2.4×10^4 cal/cm^2 from the end of winter to the end of summer. This corresponds to a mean temperature increase of 1.2°C in the upper 200 m, the

---

[2] The langley (ly) is 1 cal/cm^2. Other units, often used interchangeably in energy budget discussions, are the joule (J), the erg, and the (gram) calorie. 1 joule (J) $= 10^7$ ergs ≈ 0.24 cal. With a wind of 10 m/sec, the kinetic energy of a gram of air is 0.5×10^6 ergs, or 0.05 joule. 1 watt (W) $= 1$ joule/sec.

average depth to which significant seasonal changes penetrate (Defant, 1961). This difference should be counted from about March 1 to September 1. Although the actual change of oceanic heat content cannot yet be given exactly, according to Defant the mean annual temperature change of the surface water of the Atlantic Ocean is about 4°C. Considering the characteristic variation with depth, the change in the 0 to 100-m layer may be estimated at 2°C, and in the 100- to 200-m layer at 0.4°C. *The average of* 1.2°C *for the* 200-m *deep layer corresponds to the storage required by the spring and summer radiation surplus.*

Let the mean excess of absorbed over emitted radiation per unit time and unit area of the earth at a fixed latitude be $\bar{R}_a$ for the atmosphere and $\bar{R}_e$ for the earth, the excess of the total system being $\bar{R}_{ae} = \bar{R}_a + \bar{R}_e$. Denote also the transfer of sensible heat from earth to atmosphere by $\bar{Q}_s$ and the transfer of latent heat through evapotranspiration by $L\bar{E}$, where L is heat of vaporization and $\bar{E}$ is the mass of water evaporated per unit time and area. Then a heating function for the earth may be defined by

$$\left(\frac{d\bar{Q}}{dt}\right)_e = \bar{R}_e - \bar{Q}_s - L\bar{E} \tag{2.1}$$

In general, the water evaporated will at least be carried partially in vapor form to other regions, where on condensation the latent heat will be given up to the atmosphere. If the rate of condensation is expressed by the rate of precipitation $\bar{P}$, the heating function $(d\bar{Q}/dt)_a$ of the atmosphere is determined by

$$\left(\frac{d\bar{Q}}{dt}\right)_a = \bar{R}_a + \bar{Q}_s + L\bar{P} \tag{2.2}$$

By combining Eqs. (2.1) and (2.2) we get

$$\left(\frac{d\bar{Q}}{dt}\right)_a = \bar{R}_{ae} - \left(\frac{d\bar{Q}}{dt}\right)_e + L(\bar{P} - \bar{E}) \tag{2.3}$$

as an expression for the atmospheric heating function.[3]

---

[3] It should be remarked that in Eqs. (2.2) and (2.3) only the heat of vaporization is considered. In regions where precipitation occurs partly in solid form, the heat of fusion should be added. Its inclusion would increase the value of LP by no more than about 13%; considering also that solid forms occur generally in regions with relatively small precipitation amounts, their neglect is considered justifiable. Also, we have used in all computations a constant value of L, 600 cal/gm deg. The errors involved in neglecting the variation of L with temperature are small compared with those due to observational uncertainties.

The effect of heat gain or loss is expressed partly as a local change of energy and is redistributed partly through exchange with neighboring parts of the atmosphere and earth. Hence, the heating functions can be written in the forms

$$\left(\frac{d\bar{Q}}{dt}\right)_e = \frac{\partial \bar{W}_e}{\partial t} + \mathrm{div}(\bar{W}_{\varphi e})$$

$$\left(\frac{d\bar{Q}}{dt}\right)_a = \frac{\partial \bar{W}_a}{\partial t} + \mathrm{div}(\bar{W}_{\varphi a})$$

$$(2.4)$$

where $\bar{W}_e$ and $\bar{W}_a$ denote the zonally averaged energies per unit area at a fixed latitude. In the present context, $\bar{W}_e$ corresponds to the heat stored in the hydrosphere and $\bar{W}_a$ to the sum of enthalpy and geopotential in a whole atmospheric column. Under the hydrostatic assumption these are interchangeable in a dry-adiabatic process; latent heat is not included in $\bar{W}_a$. In the following discussion, the unqualified term "energy" refers to *realized* energy ($c_p T + \Phi$). In Eqs. (2.4), $\bar{W}_{\varphi e}$ and $\bar{W}_{\varphi a}$ denote the *northward fluxes, per unit length of a fixed parallel,* of $\bar{W}_e$ and $\bar{W}_a$, and $\mathrm{div}(\bar{W})$ is the corresponding meridional divergence of the flux.[4]

By substitution from Eqs. (2.4) into Eqs. (2.1), (2.2), and (2.3), respectively, we get

$$\mathrm{div}(\bar{W}_{\varphi e}) = \bar{R}_e - \bar{Q}_s - L\bar{E} - \frac{\partial \bar{W}_e}{\partial t} \tag{2.5a}$$

$$\mathrm{div}(\bar{W}_{\varphi a}) = \bar{R}_a + \bar{Q}_s + L\bar{P} - \frac{\partial \bar{W}_a}{\partial t} \tag{2.5b}$$

$$\mathrm{div}(\bar{W}_{\varphi a}) = \bar{R}_{ae} + L(\bar{P} - \bar{E}) - \frac{\partial \bar{W}_e}{\partial t} - \frac{\partial \bar{W}_a}{\partial t} - \mathrm{div}(\bar{W}_{\varphi e}) \tag{2.5c}$$

From these equations the northward energy flux in the atmosphere and hydrosphere may be computed. If only mean *annual* values are considered, $\partial \bar{W}_e / \partial t$ and $\partial \bar{W}_a / \partial t$ vanish. However, for computations of the energy fluxes for shorter periods such as seasons, the changes in storage of energy may be very important, especially in the oceans with their large heat capacity. By areal integration of Eq. (2.5c) over the region north of any latitude φ, and application of Gauss' theorem at this boundary, the flux of energy

---

[4] In this kind of expression, as applied to a mean poleward flux $\bar{F}$ of any quantity, $\mathrm{div}\,\bar{F} = (\partial \bar{F}/\partial y - \bar{F}a^{-1} \tan \varphi)$, where a is earth radius. This takes account of the convergence of meridians.

(in both atmosphere and oceans) across φ is

$$2\pi a \cos \varphi(\bar{W}_{\varphi a} + \bar{W}_{\varphi e}) = -2\pi a^2 \int_{\varphi}^{\pi/2} \left\{ \bar{R}_{ae} + L(\bar{P} - \bar{E}) \right.$$

$$\left. - \frac{\partial}{\partial t} (\bar{W}_a + \bar{W}_e) \right\} \cos \varphi \, d\varphi \qquad (2.6)$$

We can also include the latent heat in the atmospheric energy flux. By release upon condensation of the vapor, this energy form goes over into internal and potential energy. It can be shown (see, e.g., Riehl and Malkus, 1958) that

$$\bar{P} - \bar{E} = -\operatorname{div} \int_0^{p_0} \overline{qv} \, \frac{dp}{g} \qquad (2.7)$$

where q is specific humidity. The formula is strictly valid if no change in storage of water vapor occurs and all water vapor condensed immediately precipitates. If the latent heat is included in the atmospheric energy flux, Eq. (2.6) may be replaced by

$$2\pi a \cos \varphi(\bar{W}'_{\varphi a} + \bar{W}_{\varphi e}) = -2\pi a^2 \int_{\varphi}^{\pi/2} \left\{ \bar{R}_{ae} \right.$$

$$\left. - \frac{\partial}{\partial t} (\bar{W}_a + \bar{W}_e) \right\} \cos \varphi \, d\varphi \qquad (2.8)$$

where $\bar{W}'_{\varphi a}$ represents the atmospheric flux of $(c_p T + \Phi + Lq)$, which C. W. Kreitzberg calls "static energy." This includes all forms of atmospheric energy excepting the kinetic energy associated with motion.

In the form of Eq. (2.8), the total northward flux of energy in atmosphere and oceans is determined by the distribution of $\bar{R}_{ae}$ and the total change of energy storage in the atmosphere and earth north of latitude φ. Figure 2.2, derived from the radiation computations by London, shows the values of $-\bar{R}_{ae}$ integrated from the North Pole in winter and summer. If there were no storage, corresponding meridional energy fluxes would be required. As indicated in Section 2.3, the seasonal radiative imbalance, for the hemisphere as a whole, is essentially balanced by storage.

The flux quantities $\bar{W}_{\varphi a}$ and $\bar{W}'_{\varphi a}$ in Eqs. (2.5), (2.6), and (2.8) can also be evaluated from aerological observations by using the equations

$$\bar{W}_{\varphi a} = \frac{1}{g} \int_0^{p_0} (c_p \overline{Tv} + \overline{\Phi v} + \overline{kv}) \, dp \qquad (2.9)$$

$$\bar{W}'_{\varphi a} = \frac{1}{g} \int_0^{p_0} (c_p \overline{Tv} + \overline{\Phi v} + \overline{kv} + L\overline{qv}) \, dp \qquad (2.10)$$

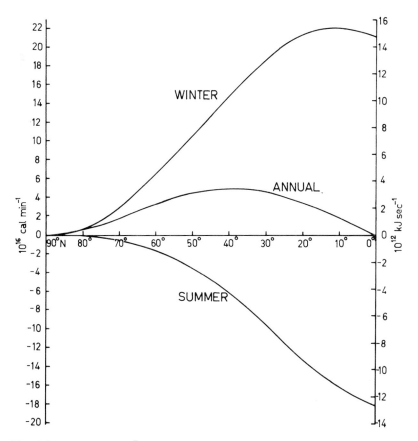

FIG. 2.2   Net radiation $\bar{R}_{ae}$ of the earth-atmosphere system, integrated from North Pole to give the total amount of heat lost or gained by radiation over the portion of the hemisphere north of a given latitude. Curve shows mean values for winter (December–February), summer (June–August), and the year. (After London, 1957.)

where c_p is the specific heat at constant pressure, T is absolute temperature, Φ is geopotential, and k is kinetic energy per unit mass.

Analogous with the angular momentum flux (Chapter 1), the energy flux can be divided into two parts: the flux due to mean meridional circulations and the eddy flux. Equation (2.9) can then be written as

$$\bar{W}_{\varphi a} = \frac{1}{g} \int_0^{p_0} (c_p \bar{T} + \Phi + \bar{k}) \bar{v} \, dp$$

$$+ \frac{1}{g} \int_0^{p_0} (c_p \overline{T'v'} + \overline{\Phi'v'} + \overline{k'v'}) \, dp \qquad (2.11)$$

The circulation flux can be computed if $\bar{T}$, $\bar{\Phi}$, $\bar{k}$, and $\bar{v}$ are known for selected periods of time. Mintz (1955) computed the geostrophic eddy flux, in which the term $\overline{\Phi'v_g'}$ vanishes.[5] If we further disregard the flux of kinetic energy that is small compared with the total flux of enthalpy and potential energy,[6] the approximate formula

$$\bar{W}_{\varphi a} = \frac{1}{g} \int_0^{p_0} (c_p\bar{T} + \bar{\Phi})\bar{v}\, dp + \frac{1}{g} \int_0^{p_0} c_p\overline{T'v'}_g\, dp \qquad (2.11')$$

can be used (see Section 2.3). A corresponding simplification is also applicable to Eq. (2.10), but will not be used here.

2.2 Mean Annual Energy Budget over the Globe

If only annual means are considered, the storage terms in Eqs. (2.5), (2.6), and (2.8) vanish, and the meridional energy flux can be evaluated from the annual mean values of R_e, R_a, Q_s, E, and P. Since the values of these quantities are still not very exactly known, the result may in the future be modified in several respects. The problem has recently been treated very thoroughly by Budyko (1956, 1963), Smagorinsky et al. (1965), and Sellers (1965), and the discussion here follows closely the principles outlined by these authors. We call attention also to the analysis by Davis (1963) of various components of the heat budget in the region 20° to 70°N, up to the 25-mb level in different seasons.

Water vapor plays a very important role in the atmospheric heat budget. For the whole globe the annual mean of $(P - E)$ must vanish; this gives a check on the accuracy of the estimates of P and E. Already in 1934 Meinardus published data for the global distribution of P. If these are compared with Budyko's values of E, the global balance is nearly achieved. Later, Möller (1951) presented revised precipitation values which, while they are not compatible with Budyko's estimates of E, can be used to estimate the seasonal apportionment.

In Table 2.1 are presented the mean annual precipitation and evaporation,

---

[5] At present it is not possible to estimate the errors following from the geostrophic assumption. These are probably not considerable in the eddy flux of enthalpy, but should be kept in mind when judging the results of the total balance requirement. Priestley (1949), Starr and White (1954), and others have computed the fluxes of heat and other quantities, using real winds for selected stations.

[6] At 30°N where it is strongest, the poleward flux of kinetic energy is about 2×10^{11} kJ/sec in winter (Section 1.5), or 7% of the internal plus potential energy flux in Fig. 2.7.

TABLE 2.1

ZONAL MEAN VALUES OF ANNUAL PRECIPITATION AND EVAPORATION[a]

Latitude zone	P (mm/year)	E (mm/year)	$P - E$ (mm/year)	$A(P - E)$ (10^{13} kg/year)
90–70°N	169	119	50	77
70–60°N	415	333	82	155
60–50°N	789	469	320	820
50–40°N	907	641	266	838
40–30°N	872	1002	−130	− 473
30–20°N	790	1246	−456	−1833
20–10°N	1151	1389	−238	−1018
10–0°N	1934	1235	699	3081
90–0°N	1009	944	65	1647
0–10°S	1445	1304	141	622
10–20°S	1132	1541	−409	−1748
20–30°S	857	1416	−559	−2247
30–40°S	932	1256	−324	−1179
40–50°S	1226	895	331	1043
50–60°S	1046	520	526	1347
60–70°S	418	174	244	461
70–90°S	69	34	35	54
0–90°S	1000	1065	− 65	−1647

[a] After Sellers (1965).

and the mass excess of P over E, for different latitude belts. The values of P are essentially those given by Meinardus, and the values of E correspond closely to Budyko's values, both having been slightly modified by Sellers. This table shows an excess of precipitation over evaporation in high latitudes of both hemispheres and also in the equatorial belt where, especially north of the Equator, the excess is extremely large. In the belts 10° to 40° of both hemispheres, on the contrary, evaporation exceeds precipitation. On the whole $(P - E)$ is positive in the Northern Hemisphere and negative in the Southern Hemisphere. To achieve balance, 1647×10^{13} kg/year of water vapor must hence be transported northward across the Equator. This corresponds to an annual transport of about 10^{22} cal of latent heat.

The values of Table 2.1 are presented graphically in Fig. 2.3. For the Northern Hemisphere we have also added a curve showing $(E - P)$ computed from a hemispheric map of the divergence of the water vapor flux

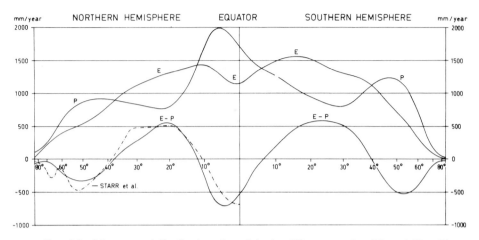

FIG. 2.3   Mean annual distribution of precipitation (*P*), evaporation (*E*), and ($E - P$) over the globe, in millimeters per year (after Sellers, 1965). The dash-dotted curve represents ($E - P$) for one year, as computed by Starr *et al.* (1965) from the atmospheric flux of water vapor in the Northern Hemisphere. (Reprinted, redrawn, from W. D. Sellers, *Physical Climatology*, by permission of The University of Chicago Press, copyright 1965 by The University of Chicago.)

according to Starr, Peixoto, and Crisi (1965). In contrast with Sellers' data, the curve indicates an approximate balance between evaporation and precipitation over the Northern Hemisphere as a whole. The general agreement in other respects is of interest, since Sellers' curve is based on the left side of Eq. (2.7), while the computations by Starr *et al.* (for the year 1958 only) were based on the atmospheric fluxes expressed by the right side. From the last column of Table 2.1, the annual northward flux of water vapor has been computed and is given in the upper row for each hemisphere in Table 2.2. Corresponding values based on aerological data, by Peixoto and Crisi (1965) and by Peixoto (1967), are given in the lower rows.[7] In both types of computation, there are considerable uncertainties in the data over extensive oceanic areas. Also, the two types incorporate data from different time periods. Nevertheless, a broad agreement is evident.

In both hemispheres the water-vapor flux is directed poleward in middle and high latitudes, whereas in low latitudes the flux is directed toward the

---

[7] The atmospheric flux values shown here for the Northern Hemisphere were computed from profiles, plotted for each latitude, of water-vapor transfers, given by Peixoto and Crisi (their Table 4) for the 1000-, 850-, 700-, and 500-mb levels. These are considered more reliable than the integrated values in their Table 5, which were arrived at in a different way.

mean "heat equator," the average position of which is near 5°N (Riehl, 1954). The water-vapor convergence around the equatorial trough is a result of the mass transport in the Hadley cells of both hemispheres. The boundary between these circulation cells shows an average meridional displacement from an extreme southern position near 5°S during northern winter to an extreme northern position near 13°N during northern summer. During the former season, water vapor is transported southward across the Equator, but during the latter, a considerably stronger transport takes place northward across the Equator (cf. Figs. 1.5 and 1.6).

TABLE 2.2

MEAN ANNUAL NORTHWARD FLUX OF WATER VAPOR (10^{14} kg/year) DERIVED FROM $P - E$ BALANCE AND FROM ATMOSPHERIC FLUX COMPUTATIONS

Hemisphere	Latitude	70°	60°	50°	40°	30°	20°	10°	0°
Northern	$P - E$ balance[a]	8	23	105	189	142	−42	−143	165
	Atmos. flux[b]	14	55	137	254	194	87	−105	103
Southern	$P - E$ balance[a]	−5	−52	−186	−291	−173	52	227	165
	Atmos. flux[c]	−1	−42	−175	−229	−163	−43	225	141

[a] Based on Table 2.1.
[b] Based on computations by Peixoto and Crisi (1965) for year 1958.
[c] Unpublished computations by Peixoto (1967), by permission.

In Table 2.3 the quantities needed for computing the different components of the meridional energy flux are presented. These are essentially the same as used by Sellers. By considering the areas of the different belts of latitude, the components of the energy flux across standard latitude circles were evaluated (Table 2.4) according to Eqs. (2.5), (2.6), (2.8), and (2.9), in which, on an annual basis, the local change terms are not relevant.

Table 2.4 shows many interesting features, among which is the relative importance of the poleward flux of heat in the oceans ($\overline{W}_{\varphi e}$), especially in lower latitudes. Although at latitudes 60° this accounts for only 10% of the total flux in the last column, at latitudes 30° it amounts to about 30%, and at latitudes 10° to about 60% of the total flux. At the Equator, virtually the entire flux toward the Southern Hemisphere is due to the oceanic heat exchange. Here the northward flux of latent heat ($\overline{W}_{\varphi L}$) is nearly compensated by an almost equally large southward transport of realized energy ($\overline{W}_{\varphi a}$) in the atmosphere.

TABLE 2.3

QUANTITIES NEEDED FOR COMPUTATION OF ENERGY FLUXES FROM EQS. (2.6) AND (2.8)[a]

Latitude belt	$\bar{R}_e$	$\bar{R}_a$	$\bar{R}_{ae}$	$L\bar{P}$	$L\bar{E}$	$L(\bar{P} - \bar{E})$	$\bar{Q}_s$
90–70°N	−0.003	−0.151	−0.154	0.019	0.014	0.005	−0.006
70–60°N	0.040	−0.153	−0.113	0.047	0.038	0.009	0.019
60–50°N	0.057	−0.131	−0.074	0.090	0.053	0.037	0.027
50–40°N	0.091	−0.121	−0.030	0.102	0.072	0.030	0.032
40–30°N	0.139	−0.121	0.018	0.097	0.112	−0.015	0.046
30–20°N	0.183	−0.135	0.048	0.088	0.139	−0.051	0.045
20–10°N	0.202	−0.143	0.059	0.128	0.154	−0.026	0.030
10–0°N	0.200	−0.137	0.063	0.215	0.137	0.078	0.021
0–10°S	0.200	−0.138	0.062	0.161	0.145	0.016	0.019
10–20°S	0.198	−0.137	0.061	0.125	0.171	−0.046	0.021
20–30°S	0.179	−0.138	0.041	0.095	0.158	−0.063	0.030
30–40°S	0.152	−0.143	0.009	0.104	0.141	−0.037	0.021
40–50°S	0.106	−0.139	−0.033	0.138	0.101	0.037	0.019
50–60°S	0.053	−0.133	−0.080	0.118	0.059	0.060	0.021
60–70°S	0.025	−0.147	−0.122	0.048	0.019	0.029	0.021
70–90°S	−0.008	−0.146	−0.154	0.008	0.004	0.004	−0.011

[a] Units: ly/min. Values are essentially from Sellers (1965), following Budyko (1963).

The values in Table 2.4 are presented graphically in Fig. 2.4, which is almost identical to a similar presentation by Sellers (1965). Although the reliability of the basic quantities in Table 2.3 is still somewhat doubtful, computations of diverse kinds made in recent years, such as those in Section 2.3, suggest that the various components of the energy flux derived from these values are realistic.

Figure 2.4 indicates that in each hemisphere the atmospheric flux (enthalpy plus potential energy) reaches two maxima, near 20° and near 50° to 55° latitude. Near latitudes 10°, this poleward energy flux is to a large extent counteracted by the equatorward flux of latent heat in the trade winds. *Somewhat poleward of the subtropical high belts, the poleward flux of latent heat nearly equals the flux of realized energy in the same direction, so that the total atmospheric flux reaches maximum values.* Still farther poleward, the relative contribution of latent heat flux diminishes markedly.

Figure 2.5 presents the annual heat budget of the atmosphere by zones of latitude. This figure contains the values of $\bar{R}_a$, $L\bar{P}$, and $\bar{Q}_s$ from Table 2.3, the atmospheric energy fluxes ($\bar{W}_{\varphi a}$) derived from these (Table 2.4),

TABLE 2.4

MEAN ANNUAL NORTHWARD ENERGY FLUX IN ATMOSPHERE AND OCEANS (10^{16} cal/min)

Latitude	Oceans Heat flux $(\bar{W}_{\varphi e})$	Flux of $(c_p T + \Phi)$ $(\bar{W}_{\varphi a})$	Latent heat flux $(\bar{W}_{\varphi L})$	Total energy flux $(\bar{W}_{\varphi a} + \bar{W}_{\varphi L})$	Atmospheric plus oceanic energy flux $(\bar{W}_{\varphi a} + \bar{W}_{\varphi L} + \bar{W}_{\varphi e})$
70°N	0.17	2.16	0.09	2.25	2.42
60°N	0.49	3.81	0.26	4.07	4.56
50°N	1.08	4.17	1.20	5.37	6.45
40°N	1.49	3.76	2.15	5.91	7.40
30°N	2.09	2.94	1.62	4.56	6.65
20°N	2.17	3.01	−0.46	2.55	4.72
10°N	1.40	2.40	−1.60	0.80	2.20
0°	−0.45	−2.05	1.92	−0.13	−0.58
10°S	−2.04	−3.90	2.63	−1.27	−3.31
20°S	−2.30	−4.27	0.65	−3.62	−5.92
30°S	−1.94	−3.73	−1.90	−5.63	−7.57
40°S	−1.58	−3.11	−3.21	−6.32	−7.90
50°S	−1.14	−3.60	−2.12	−5.72	−6.86
60°S	−0.45	−3.78	−0.58	−4.36	−4.81
70°S	−0.16	−2.29	−0.06	−2.35	−2.51

and the heating function $(dQ/dt)_a$ according to Eq. (2.2). The latter, which is also the divergence of the flux of realized energy according to Eq. (2.5b), is greatest in the region of the heat equator, around 5°N. This is essentially a result of the excessive release of latent heat in the average convergence belt between the Hadley circulations of the two hemispheres.

The whole equatorial zone, roughly between 20°N and 20°S, hence represents an intense heat source for the atmosphere, strongest somewhat north of the geographical Equator. Other heat sources appear in the belts 30°N to 50°N and 40°S to 55°S, roughly corresponding to the average belts of the polar-front disturbances with their excess of P over E. The asymmetry of the heat sources with respect to the geographical Equator is obviously caused by the unlike distributions of continents and oceans in the hemispheres.

The global meridional flux of energy could also be determined from Eq. (2.9). However, for the tropics of the Northern Hemisphere and for the whole Southern Hemisphere the aerological data available for this purpose

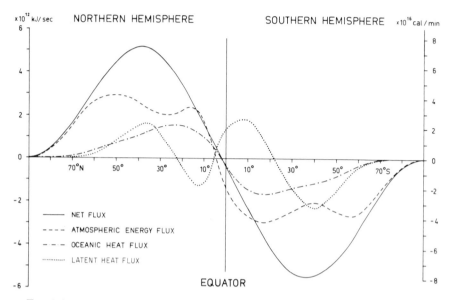

Fɪɢ. 2.4 Solid curves show the mean annual northward flux of net energy in the atmosphere-ocean system. Other curves show atmospheric flux of realized energy (enthalpy + potential energy), atmospheric flux of latent heat, and oceanic flux of heat. Units are (right scale) 10^{16} cal/min or (left scale) 10^{12} kJ/sec. (Reprinted, redrawn, from W. D. Sellers, *Physical Climatology*, by permission of The University of Chicago Press, copyright 1965 by The University of Chicago.)

are scanty. Figure 2.6 shows a comparison of the atmospheric flux deduced from energy balance requirements (from $\bar{R}_a$, $\bar{Q}_s$, and $L\bar{P}$), with the mean of the winter and summer fluxes computed from aerological data (Section 2.3). The curves are fairly similar, and the differences are difficult to interpret. Holopainen (1965) suggests that the exaggerated peak in the aerological values in the tropics may be due to an overestimate of the Hadley-cell mass circulation in winter, and that this should probably be reduced 20 to 30% from the 230×10^6 ton/sec shown in Fig. 1.5, to about 160 − 180×10^6 ton/sec. However, there are other possible sources of error. For example, since the quantity $(c_p T + \Phi)$ increases rapidly with height, the computed net meridional flux of this quantity is sensitive not only to the amount of the mean mass circulation, but also to the precise distribution of the mean meridional wind component with height. The same is true of the water-vapor flux. Considering also that the winter-summer mean may not represent a true annual average, it is difficult to judge which of the curves in Fig. 2.6 better represents the real northward flux of energy in the atmosphere.

Atmospheric Mean Annual Heat Budget (ly/min)
and Poleward Flux of Energy (Unit: 10^{16} cal/min)

Latitude zone	R_a	$(dQ/dt)_a$	LP	Q_s	Energy Flux (at lower boundary)
90°–70°N	−0.151	−0.138	0.019	−0.006	2.2
70°–60°	−0.153	−0.087	0.047	0.019	3.8
60°–50°	−0.131	−0.014	0.090	0.027	4.2
50°–40°	−0.121	0.013	0.102	0.032	3.8
40°–30°	−0.121	0.022	0.097	0.046	2.9
30°–20°	−0.135	−0.002	0.088	0.045	3.0
20°–10°	−0.143	0.015	0.128	0.030	2.4
10°–0°	−0.137	0.099	0.215	0.021	2.0
0°–10°S	−0.138	0.042	0.161	0.019	3.9
10°–20°	−0.137	0.009	0.125	0.021	4.3
20°–30°	−0.138	−0.013	0.095	0.030	3.7
30°–40°	−0.143	−0.018	0.104	0.021	3.1
40°–50°	−0.139	0.018	0.138	0.019	3.6
50°–60°	−0.133	0.006	0.118	0.021	3.8
60°–70°	−0.147	−0.078	0.048	0.021	2.3
70°–90°S	−0.146	−0.149	0.008	−0.011	

Northern Hemisphere   Thermal Equator 5°N   Southern Hemisphere

FIG. 2.5 Tentative annual heat budget of the atmosphere, showing the heating function $(dQ/dt)_a = R_a + Q_s + LP$ and the contributions of the individual terms (ly/min) in latitudinal zones. Numbers over arrows (which show direction of flux) indicate the resulting energy fluxes across fixed latitudes, in units of 10^{16} cal/min.

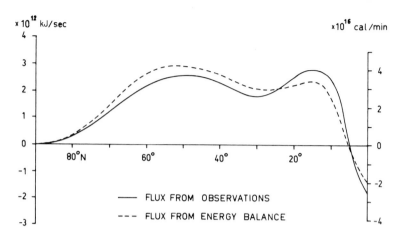

FIG. 2.6 Comparison between the annual northward energy flux in the Northern Hemisphere, as computed from aerological observations for three winter and three summer months, and the flux required for energy balance.

2.3 ENERGY BUDGET OF THE NORTHERN HEMISPHERE
IN WINTER AND SUMMER

Because of the strong seasonal variation of the radiation budget, and correspondingly of the requirements for energy transfer, the annual mean values are not very satisfactory for an aerological interpretation of the whole mechanism of the meridional and vertical energy fluxes. We shall therefore attempt to study the energy budgets separately for the winter and summer seasons, during which (according to the discussion in Section 2.1) the changes of energy stored in the oceans are considerable. Because of the lack of adequate aerological data from the Southern Hemisphere, the following discussion will be entirely confined to the processes in the Northern Hemisphere.

Holopainen (1965) has evaluated separate values for $\overline{W}_{\varphi a}$ from Eq. (2.11) for the winter and summer seasons. For the circulation flux (first r.h.s. integral) he considered the mean mass circulations in the Hadley cell shown in Figs. 1.5 and 1.6. North of about 30°N the mass circulation of the Ferrel cell after Mintz and Lang (1955) was utilized. For the eddy flux he used Mintz' values (1955) of the geostrophic fluxes. The computed values of the poleward energy flux are presented in Table 2.5 for winter and in Table 2.6 for summer.

With use of these data, we are now in a position to appraise the disposition of the large radiative deficit or excess within a hemisphere during

TABLE 2.5

NORTHWARD ENERGY FLUX ($c_p T + \Phi$) ACROSS DIFFERENT LATITUDES IN WINTER AND DIVERGENCE OF FLUX, IN NORTHERN HEMISPHERE[a]

Latitude	Circulation flux	Eddy flux	Total flux	Flux diff.
0°	2.90	0.00	2.90	
				4.18
10°	7.08	0.00	7.08	
				−2.22
20°	4.40	0.46	4.86	
				−2.02
30°	0.79	2.05	2.84	
				1.00
40°	−0.65	4.49	3.84	
				0.38
50°	−0.73	4.95	4.22	
				−0.52
60°	−0.56	4.26	3.70	
				−1.59
70°	0.03	2.08	2.11	
				−2.11
90°	0.00	0.00	0.00	
0–90°	—	—	—	−2.90

[a] Units: 10^{12} kJ/sec.

a given season. As noted earlier, the imbalance of the integrated R_{ae} values (Fig. 2.2) must be compensated either by a meridional energy flux or change in energy storage, or both, according to Eq. (2.8). As regards the oceanic flux, no final seasonal values are available. According to Bjerknes (1964) the mean annual flux of heat across the Equator, considering both the Atlantic and Pacific oceans, is northward about 0.4×10^{16} cal/min, while Sellers (1965) cites a corresponding southward flux for all oceans (Table 2.4). In any event, it is probably safe to assume that the oceanic transport is small compared with the large values at the Equator in Fig. 2.2.

Table 2.5 indicates for winter a northward transequatorial flux of atmospheric energy ($c_p T + \Phi$) of 2.9×10^{12} kJ/sec, or 4.2×10^{16} cal/min. However, the effect of the Hadley circulation (Fig. 1.5) is to carry large quantities of water vapor southward in low levels and small quantities

TABLE 2.6

NORTHWARD ENERGY FLUX ($c_pT + \Phi$) ACROSS DIFFERENT LATITUDES IN SUMMER AND
DIVERGENCE OF FLUX, IN NORTHERN HEMISPHERE[a]

Latitude	Circulation flux	Eddy flux	Total flux	Flux diff.
0°	−6.48	0.00	−6.48	
				4.34
10°	−2.14	0.00	−2.14	
				2.23
20°	0.30	−0.21	0.09	
				0.62
30°	0.57	0.14	0.71	
				0.09
40°	0.00	0.80	0.80	
				0.16
50°	−0.12	1.08	0.96	
				−0.26
60°	−0.10	0.80	0.70	
				−0.35
70°	0.01	0.34	0.35	
				−0.35
90°	0.00	0.00	0.00	
0–90°	—	—	—	6.48

[a] Units: 10^{12} kJ/sec.

northward in upper levels. The resulting net transfer of latent heat southward, given by Palmén and Vuorela (1963) as 3.6×10^{16} cal/min, almost
balances the northward transfer of enthalpy and potential energy. Hence
the total atmospheric flux across the Equator is northward in the amount
0.6×10^{16} cal/min. The lowest atmospheric temperature is reached in mid-
January, so that the change of atmospheric energy storage during the winter
season can be neglected as a small quantity. If we neglect the small transequatorial transport by ocean currents, but consider the atmospheric flux,
the total radiative deficit of 21×10^{16} cal/min (Fig. 2.2) is diminished only
to 20.4×10^{16} cal/min. This amount of heat given up from storage by the
oceans during the winter months December to February corresponds to a
mean cooling of the upper 200-m layer of 0.86°C. (Again, the change of
heat storage by continental surfaces has been neglected).

In Section 2.1 it was estimated that the total change in heat storage during the autumn and winter months combined should correspond to about 1.2°C temperature change in this layer. The implication from the preceding estimate is that about 70% of the total change takes place in the winter season. This seems plausible, since during that season the air masses overlying the oceans are coldest and the synoptic systems most vigorous, so that the sea-air exchange should be greater than during autumn. The actual change in oceanic heat storage by seasons has not been established and therefore we have no direct check on the estimate.

In summer, Table 2.6 indicates that both the circulation and eddy fluxes of the Northern Hemisphere circulation systems are generally weak compared with those in winter, but in the latitudes where the Southern Hemisphere Hadley cell encroaches north of the Equator (cf. Fig. 1.6), the southward flux of realized energy is strong. During this season the mean flux of $(c_pT+\Phi)$ across the Equator amounts to -6.5×10^{12} kJ/sec, or -9.3×10^{16} cal/min. Considering the opposing latent heat flux of 9.1×10^{16} cal/min (Vuorela and Tuominen, 1964), the total atmospheric flux across the Equator is southward, 0.2×10^{16} cal/min. On the same basis as used above, the amount of excess radiation (Fig. 2.2) not accounted for by atmospheric storage and transequatorial transfer is about 17.8×10^{16} cal/min. The implied increase in mean temperature of the upper 200-m layer of the oceans during the summer months June to August is 0.75°C.

These results indicate that *the transfer of energy across the Equator by atmospheric and oceanic motions is small*; *thus the seasonal radiative surplus or deficit within a hemisphere must be essentially stored in, or given up by, the oceans*. The small net transfer by the atmosphere expresses the circumstance that although both the upper and lower branches of the Hadley circulation carry considerably large amounts of static energy across the Equator, these transports are in opposite directions and the difference in their magnitudes is small (Section 17.3).

Baur and Philipps (1935) recognized the importance of heat storage in the oceans, but with the information available they also estimated a rather large transport of heat northward across the Equator in northern winter, and southward in summer. Gabites (1950) estimated the oceanic storage and emphasized its great significance in the context of the overall energy balance. Newer appraisals have been made by Fritz (1958) on the basis of oceanographic data, and by Rasool and Prabhakara (1966) in their study of the global heat budget incorporating satellite radiation data. Considering the Northern Hemisphere as a whole, the rates of storage they computed are considerably below our estimates, and would, as indicated by Rasool and Prabhakara, require large transequatorial heat fluxes during

the seasons (but not on an annual average). The studies mentioned above
show that the changes in oceanic heat storage are greatest near latitudes
30° to 40°, as would be expected from the large seasonal variations of the
air masses sweeping over the oceans in these latitudes, as well as from the
variations of insolation.

The fluxes presented in Tables 2.5 and 2.6 are shown graphically in
Figs. 2.7 and 2.8. In the last columns of Tables 2.5 and 2.6 are tabulated
the differences between the energy flux through neighboring latitudes. These
may be converted to average values of the flux divergence for the latitude

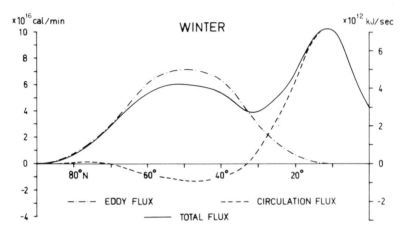

FIG. 2.7   Mean northward atmospheric energy flux in winter (December–February).
(After Holopainen, 1965.)

belts, in langleys per minute, by dividing by the area of each latitude belt
and by the mechanical equivalent of heat. According to Eq. (2.5b), if we
neglect the small change of atmospheric energy storage during a season,
this flux divergence must be balanced by the sum of heat sources and sinks
within the latitude belt, and is thus equivalent to the atmospheric heating
function $(dQ/dt)_a$ in Eq. (2.2). Using this and other information, an attempt will be made here to use the heating function derived in this way
to obtain seasonal energy budgets by zones of latitude for the Northern
Hemisphere.

Unfortunately, not all seasonal values of the annually averaged quantities
in Table 2.3 are readily available in a form that can directly be used for
this purpose. For the winter and summer seasons, $\bar{R}_e$, $\bar{R}_a$, and $\bar{R}_{ae}$ were
computed by London (1957), and values of $\bar{P}$ for the same seasons have
been presented by Möller (1951). As noted earlier, comparison of Möller's

valu_s with Budyko's estimates of the evaporation do not result in a complete water balance of the whole atmosphere. To get results comparable with those used for the annual heat budget, some adjustments have been made, taking into account Meinardus' data for the mean annual precipitation. We have tentatively used London's $\bar{R}_a$ values for the atmosphere below 100 mb only (essentially the troposphere), since the energy flux from which $(dQ/dt)_a$ is derived was computed for this layer of the atmosphere. It should be pointed out that the mean of these winter and summer R_a values is not compatible with the annual values (from a different source)

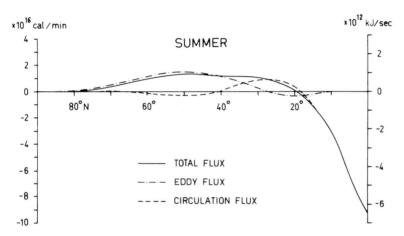

FIG. 2.8   Mean northward atmospheric energy flux in summer (June–August). (After Holopainen, 1965.)

in Fig. 2.5. The values in Fig. 2.5 are smaller in latitudes 10°N to 50°N, and larger above latitude 60°N. The method is not completely satisfactory, but in any case it gives a possible way of judging the accuracy of the computed atmospheric energy fluxes.

The results of these estimates are shown in Fig. 2.9. For each latitude belt the heating function computed from the aerologically determined divergence of the northward energy flux is shown in a rectangle. At the tops are shown the radiation deficits of the troposphere, and in the lower parts are the values of $L\bar{P}$. The transfer of sensible heat from earth to atmosphere ($\bar{Q}_s$) was computed as a residual, according to Eq. (2.2), and is shown at the bottoms of Figs. 2.9a and 2.9b.

In winter (Fig. 2.9a) the value of $\bar{Q}_s$ in the southernmost belt appears to be much too large. This could plausibly result from an underestimate of the precipitation or an overestimate of the atmospheric energy flux

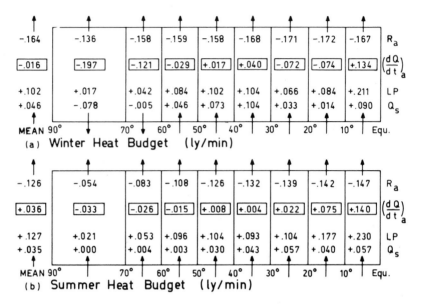

Fig. 2.9   The heat budget in latitude belts of the Northern Hemisphere in (a) winter and (b) summer. Arrows at top (100 mb) indicate radiative loss R_a from atmosphere; $(dQ/dt)_a$ represents the divergence of the latitudinal heat flux in the atmosphere ($c_p T + \Phi$); LP is the release of latent heat due to precipitation within the latitude belt; and Q_s, computed as a residual, is the transfer of sensible heat from earth to atmosphere. All quantities are expressed in langleys per minute.

divergence $(d\bar{Q}/dt)_a$. It is impossible to appraise the former possibility,[8] but the latter seems at least an equally likely source of error. This would suggest either that the northward energy flux across the Equator (Table 2.5) is too small or that the Hadley circulation is somewhat too strong, resulting in an export of heat across 10°N, which is too great. For example, if the flux of realized energy across the Equator were increased from the computed 2.9 to 4.9 × 10¹² kJ/sec, $(d\bar{Q}/dt)_a$ in the 0°N to 10°N belt would decrease from 0.134 to 0.070 ly/min, with a resulting change of $\bar{Q}_s$ to 0.026 ly/min, which is more in line with the other low-latitude belts. It should again be stressed that the inadequacy of sounding stations in low latitudes makes the computed fluxes there very uncertain.

The second maximum of $\bar{Q}_s$, between 30°N and 40°N, also appears somewhat large, as does the value in the neighboring belt 40°N to 50°N.

[8] The estimates vary widely, as shown in the comparisons made by Brooks and Hunt (1930) of estimates by different authors. The estimate of Brooks and Hunt gives a smaller value for $L\bar{P}$ in this latitude belt.

However, in these latitudes, vigorous heat transfer in winter is favored by the numerous strong outbreaks of cold air over the relatively warm oceans, especially in the regions of the Gulf Stream and the Kuroshio Current (Jacobs, 1942; Petterssen et al., 1962; Budyko, 1963), and in the Mediterranean, Black, and Caspian seas.

In the subarctic and arctic regions, the negative values of $\bar{Q}_s$ may be associated with the regular presence of surface inversions that favor the downward eddy transfer of heat. The value in the northernmost zone is, however, obviously too large. For this region the mean radiational cooling of the earth's surface layer is 0.068 ly/min, according to London. A downward heat flux of 0.078 ly/min, as in Fig. 2.9a, would more than counteract the radiative loss (especially since practically none of the latent heat released in this region is likely to be derived from local evaporation) and would not permit the freezing of surface ice, whose thickness is estimated to increase by about 1 m during winter. The freezing of 1 m of ice (of density 0.8) during a three-month period would release heat at the mean rate of 0.050 ly/min. If the radiative loss is 0.068 ly/min and we consider that the arctic ice does not cover the entire region, the $\bar{Q}_s$ value may hardly be more than about −0.02 ly/min.[9] To reduce $\bar{Q}_s$ from −0.078 to this value, the total atmospheric heat flux across latitude 70°N should be reduced in Table 2.5 from 2.11×10^{12} kJ/sec to about 1.5×10^{12} kJ/sec. Such an adjustment would, however, also influence the values in the next zone south (60°N to 70°N), where $(d\bar{Q}/dt)_a$ would correspondingly increase to −0.168 ly/min and $\bar{Q}_s$ from −0.005 to −0.052 ly/min. The further possibility must be admitted that $\bar{R}_a$ is too small in the polar cap (as suggested by comparison with the value in Fig. 2.5). If it were increased, the required air-to-ground transfer $\bar{Q}_s$ would be correspondingly diminished.

The preceding remarks show how sensitive the different components of the heat budget in Fig. 2.9 are to relatively modest corrections of the northward energy flux in Tables 2.5 and 2.6 or of the values of $\bar{R}_a$ or $L\bar{P}$. The values of $\bar{Q}_s$, which are here derived as residuals, obviously cannot be

[9] Fletcher (1965) gives a thorough review of the literature in his appraisal of the heat budget of the arctic region, which contains pertinent tabulated data. He quotes Doronin (1963) as estimating the turbulent heat flux from the atmosphere to the ice-covered surface of the central Arctic Ocean at 2735 cal/cm² during the season December to February. This gives an average value for Q_s of −0.021 ly/min, which corresponds well with the above rough estimate. For the region around the South Pole, Hanson and Rubin (1962) find the annual average loss of heat by the atmosphere to the snow surface as $(Q_s + Q_e)$ = −0.024 ly/min. The atmospheric energy fluxes across the Antarctic boundary have been computed by Rubin and Weyant (1963).

accepted as more than first estimates. However, the general magnitudes are probably for the most part realistic except in the northernmost latitudes and near the Equator. It should be stressed that there are sources of uncertainty in the assumptions or observations, or both, utilized in all estimates of $\bar{Q}_s$, a quantity that is not measured directly, so that the estimates by different authors vary widely.

In summer (Fig. 2.9b) the computed values of $\bar{Q}_s$ seem more reliable than the corresponding winter values. Also in this case, however, the value 0.057 ly/min in the southernmost zone is probably too large. The high values between 20°N and 40°N may be explained by the large arid continental regions, where evaporation is strongly suppressed and high surface temperatures result. The low $\bar{Q}_s$ values north of 50°N, despite the expected strong upward flux of sensible heat over the continents, may reasonably result from the circumstance that sensible heat is transferred from atmosphere to earth over large portions of the oceanic parts of this belt.

In the arctic region at 70°N to 90°N, the net absorption of radiation by the earth's surface is 0.089 ly/min on the average, according to London. If a part of the latent heat released, 0.021 ly/min, is assumed to be derived from water vapor evaporated from the surface within the region, this would reduce the heat available to about 0.075 ly/min during the summer season. This amount of heating would suffice to account for the estimated melting of the arctic ice and for some heat storage in the ocean on the fringes of the ice region. Hence, practically no flux of sensible heat into the atmosphere, or vice versa, may be expected.[10] In this respect the heat budget of the arctic belt in Fig. 2.9 is more satisfactory in summer than in winter.

The hemispheric mean ratio Q_s/LP is, according to the values at the left side of Fig. 2.9, 0.45 in winter and 0.28 in summer. The greater value in winter may be explained by the dominance of the oceans, where the transfer of sensible heat is on the whole large in winter and small, or even negative, in summer. The mean ratio Q_s/LP is not the same as the "Bowen ratio" Q_s/LE, which is a measure of the partition of the energy given off locally by the earth's surface (other than by radiation) between sensible and latent heat. Various authors (e.g., Jacobs, 1942; Sverdrup, 1951) give different values for this ratio, which varies widely with location and season. The mean annual values by latitude belts, according to Budyko (1956, 1963), are presented in Table 2.7. This shows an increase of $\bar{Q}_s/L\bar{E}$ from quite low values in the tropics to moderate values in middle and high latitudes.

---

[10] For this season, Doronin (1963) estimated the turbulent heat flux into the atmosphere in the central arctic to be 1092 cal/cm², corresponding to a $\bar{Q}_s$ value of $+0.008$ ly/min.

TABLE 2.7

ANNUAL MEAN VALUES OF $\bar{Q}_s/L\bar{E}$ FOR DIFFERENT LATITUDE BELTS[a]

Lat. Belt	0–10	10–20	20–30	30–40	40–50	50–60	60–70
Q_s/LE	0.13	0.19	0.33	0.39	0.45	0.46	0.45

[a] According to Budyko (1963).

There are, however, also substantial seasonal variations and very large geographical variations other than with latitude. Over the oceans the ratio is generally positive in winter, but it is negative over extensive regions in summer. In subtropical regions of the continents the ratio may be very high in summer, but it is extremely variable, depending on the character of the earth's surface—its wetness, vegetation, etc.[11]

For the whole Northern Hemisphere, the annual mean of Q_s/LE, based on an area weighting of the values in Table 2.7, is 0.32. This is somewhat smaller than the mean of the winter and summer values of Q_s/LP in Fig. 2.9, which is 0.35. To convert to the ratio Q_s/LE, this latter value should be adjusted to take account of the net annual transfer of latent heat from the Southern Hemisphere, which according to Table 2.4 amounts to 1.92 $\times 10^{16}$ cal/min or, when distributed over the area of the hemisphere, to 0.0075 ly/min. After this adjustment our mean value of Q_s/LE would be slightly increased to 0.37. Since this latter value is derived from only three

[11] Part of the systematic variation with latitude is connected with the temperature dependence of saturation specific humidity. At temperatures of 20° to 30°C, characteristic of the tropics, this increases 1.0 to 1.4 gm/kg for each degree of temperature rise at 1000 mb. If the air is heated and kept at 80% relative humidity, a common value over the oceans, the ratio $c_p \Delta T/L \Delta q$ shows that increasing the moisture content requires about three times as much energy as increasing the sensible heat content. At 0°C, a corresponding calculation shows that increasing the sensible heat content requires 1.8 times as much energy as increasing the water vapor. The ratio Q_s/LE is, of course, influenced by additional processes, but this temperature dependence partly accounts for its lower values in the tropics than in higher latitudes. Also (Fig. 2.9a), in winter the sensible heat transfer takes place predominantly in middle latitudes, into cold air. This would contribute to a larger hemispheric average (of Q_s/LE) in winter than in summer, when the sensible heat transfer occurs mainly at lower latitudes (Fig. 2.9b) with higher temperatures. Manabe (1957) found that during a typical continuous cold outburst in winter over the Japan Sea, in which the air-sea temperature difference exceeded 10°C, the heat supplied from the sea surface was 1030 ly/day compared with a latent heat supply of 450 ly/day, giving a value of $Q_s/LE = 2.3$.

winter and three summer months, it is not strictly comparable with Budyko's values. Furthermore, in our seasonal heat budget we used London's R_a values, which give a somewhat stronger cooling of the atmosphere than the radiational data used in Table 2.3, according to Sellers. This stronger cooling is reflected, owing to our method, in a correspondingly greater value of the transfer of sensible heat from the earth's surface.

The heating function and the poleward flux of energy according to Fig. 2.9, where the fluxes were computed aerologically, was in Fig. 2.6 compared with the corresponding mean annual values derived from energy balance requirements. The evaluation above of the heat budget of the Northern Hemisphere by use of the aerologically computed meridional energy flux is on the whole quite satisfactory, but considerable improvement in details is still needed. Especially in the tropics the inadequate observations make it at the present time impossible to determine the mass circulation and the energy transport of the Hadley cell with satisfactory accuracy. However, the similarity between the curve showing the poleward energy flux computed from aerological observations and that evaluated from radiation and heat exchange at the earth-atmosphere interface is very encouraging.

2.4 VERTICAL HEAT FLUX IN NORTHERN HEMISPHERE WINTER

For a better understanding of the atmospheric processes on different scales, it is necessary to consider their influence not only on the mean meridional heat exchange, but also on the vertical flux. The mean upward flux of energy $[W_p]$ per unit area through an isobaric surface p in the belt between latitudes φ_1 and φ_2 is, if the latent heat is excluded, determined by

$$[W_p] = - \frac{1}{g(\sin \varphi_2 - \sin \varphi_1)} \int_{\varphi_1}^{\varphi_2} \left\{ (c_p \bar{T} + \bar{\Phi})\bar{\omega} + c_p \overline{T'\omega'} + \overline{\Phi'\omega'} \right\} \cos \varphi \, d\varphi \tag{2.12}$$

In this expression the first and third integral terms represent the vertical flux of sensible heat, and the second and fourth terms the flux of potential energy. The fluctuations of T, Φ, and ω are measured on a constant-pressure surface. If the computation were carried out on a horizontal surface corresponding to the mean height of the isobaric surface, the fluctuation of Φ would vanish, but then the variations of T would change correspondingly. In that case,

$$[W_z] \approx - \frac{c_p}{g(\sin \varphi_2 - \sin \varphi_1)} \int_{\varphi_1}^{\varphi_2} (\bar{T}_H \bar{\omega}_H + \overline{T_H'\omega_H'}) \cos \varphi \, d\varphi \tag{2.13}$$

The first term represents the heat flux due to a mean vertical motion at the fixed latitude; the second term, the eddy flux resulting from the correlation between $\omega_H{}'$ and $T_H{}'$ at the fixed height. If a mean meridional circulation exists, the contribution by the first term represents the average vertical heat flux due to this circulation, whereas the second term gives the average vertical eddy heat flux.

As a basis for discussion of the regional heat budget in Northern Hemisphere winter, Fig. 2.10 is presented.[12] A corresponding heat budget could, of course, be derived for the summer season or the whole year. However,

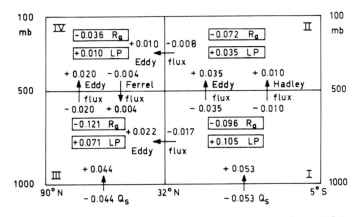

FIG. 2.10   Winter heat budget in the belts 5°S to 32°N and 32°N to 90°N, and the vertical heat fluxes through the 500-mb surface and from the earth's surface. All quantities are expressed in langleys per minute averaged over the areas concerned.

since our purpose is mainly to investigate the physical processes involved in the vertical heat flux, which on the synoptic scales are more vigorous during winter, we shall limit our presentation to this season.

For the simplest discussion, the atmosphere is shown in Fig. 2.10 divided into four blocks, separated by the 500-mb surface, the southern blocks being bounded by latitudes 5°S and 32°N, corresponding to the Hadley cell. At the northern boundary of the latter, the circulation flux of heat can be disregarded. The budgets of the various blocks were arrived at in the following way:

[12] The discussion in the rest of this chapter in principle follows that by Palmén (1966). The numerical values, however, are newer estimates. The difference arises from changes of the LP values; Palmén used essentially Möller's data, but here these data have been adjusted as outlined in Section 2.3.

(1) In each, the mean radiation deficit R_a (from London's data) is shown in the upper rectangle. This must be balanced by horizontal and vertical fluxes through the boundaries, together with the latent heat released within the block, to maintain a steady heat content (the slow seasonal change is disregarded).

(2) The heat transfers across 32°N are marked above the horizontal arrows, according to the northward fluxes of heat above and below 500 mb computed by Mintz (1955). The values on the two sides are consistent but numerically different because they represent the total heat transports across this latitude, converted to langleys per minute heat loss or gain for the two regions by dividing by their horizontal areas, which are not the same.

(3) The mean sensible heat transfer Q_s from earth's surface to atmosphere is indicated by the arrows at the bottoms of blocks I and III. The mean values of Q_s were computed from the corresponding values in Fig. 2.9a, extrapolated to 5°S.

(4) The latent heat LP released in condensation (lower rectangles in each box) is apportioned to the layers above and below 500 mb, according to the moist-adiabatic process, assuming the water to fall out upon condensation and taking representative temperature regimes for the two latitude zones (at higher temperatures, a smaller proportion of the water vapor condenses below 500 mb).

(5) The vertical flux of sensible heat and geopotential by the mean circulations in the southern and northern regions are computed from Eq. (2.12), using the actual distributions of mean temperature and height at the 500-mb surface and the mean vertical velocities derived from Fig. 1.5. Latent heat flux is omitted because this is included in LP in the upper blocks. The upward transfer of heat in the Hadley circulation, 0.010 ly/min, is roughly equivalent to that provided by a mass circulation of 230×10^{12} gm/sec (Fig. 1.5) if there is an average temperature difference of 4.5°C on a level surface between the ascending and descending branches. In the Ferrel cell, the mass circulation, roughly 30×10^{12} gm/sec, represents a downward energy transport of 0.004 ly/min if the corresponding temperature difference is 11°C.

(6) For balance, an overall upward transfer of 0.045 ly/min is required from block I to II, and 0.016 ly/min from block III to IV. The required eddy fluxes are obtained as residuals between these values and the circulation fluxes in step (5).

Although the procedure above fixes the approximate magnitude of the vertical eddy-heat fluxes, it does not specify the mechanisms by which they are accomplished. Depending on the level considered, these mechanisms can be very different. For instance, in the friction layer (roughly 500 to 1000 m deep) the principal means of transferring heat vertically is through very small-scale eddies. Beginning at the average condensation level, small- to medium-scale eddies of the cumulus type and large-scale eddies of the cyclone type take over a considerable part of the transfer mechanism. A general review of the small- and medium-scale transfer processes has been given by Sheppard (1958); these will be discussed here only in terms of their gross effects.

In middle and higher latitudes, cumulus activity in winter is mainly confined to oceanic regions, which can supply significant quantities of heat; and this activity is mostly limited to relatively low levels (except in special regions of intense outbreaks of cold air where subsidence is not strong).[13] Consequently, *in extratropical regions the relative importance of synoptic-scale disturbances for the vertical transfer of energy increases upward and dominates in the middle and upper troposphere.* Because of the increase of potential temperature with height, the flux of heat due to microscale eddies is in general directed downward, and only in limited regions is this overcompensated by upward heat flux in cumuli at the 500-mb level. For this reason, the upward eddy flux due to large-scale disturbances (mainly extratropical cyclones) can be expected to be somewhat larger than the 0.020 ly/min required for the total eddy flux.

2.5 VERTICAL HEAT FLUX IN THE TROPICS

It is doubtful that any appreciable part of the large vertical eddy-heat transport in the tropics can be accounted for by organized circulations on the synoptic scale, although embedded convective clouds are significant for this transport. Only in well-developed tropical storms and "monsoon disturbances" (see Chapters 14 and 15) is there a clear-cut negative correlation between T' and ω'; however, no data are available for a quantitative estimate of this heat flux. While due to the high content of water vapor in lower levels and its rapid decrease with height there is a strong correlation between $-\omega'$ and q', the available evidence suggests that in some types of

---

[13] See Section 12.2. Even over the Japan Sea where the air in winter is often very much colder than the water, cumulus convection is generally confined to the lower troposphere (Manabe, 1957; Asai, 1965).

common disturbances there may be a positive correlation between ω' and T'.

One might examine the possibility that the transfer and liberation of latent heat accounts for the warming of block II, required to offset radiative cooling. This could be accomplished in principle by a reasonable correlation between ω' and q'. Another difficulty would, however, arise. Since the upward transfer of sensible heat by the Hadley circulation is only 0.010 ly/min, this just about compensates the northward flux of heat so that a latent heat release of 0.070 ly/min would be needed in block II. Based on this supposition, a release of 50% of the total (0.140 ly/min) latent heat available would have to take place above 500 mb. If we consider a moist-adiabatic (convective cloud) process, the release above that level could hardly exceed 30%, as indicated in Fig. 2.10. The only alternative is that the bulk of the vertical flux is in the form of sensible heat. This would largely originate from latent heat that was released below 500 mb, but which was carried upward in cumulus chimneys.

Riehl and Malkus (1958) have provided strong arguments that *virtually all heat and mass transfer upward in the equatorial zone takes place in tall cumulonimbus clouds.* According to their reasoning, a uniformly slow upward motion in the equatorial branch of the Hadley cell would be impossible. Instead, the slow mean circulation indicated in Fig. 1.5 should be interpreted as the statistical result of vigorous vertical motions in convective clouds that penetrate to the upper troposphere only in a very small portion of the total region. This concept is elaborated upon in Sections 14.4 and 17.2. In the subtropical descending branch of the Hadley cells, the situation is quite different. We may, following Vuorela (1957), examine whether a uniformly slow sinking is realistic in this branch.

In Fig. 1.5, the total upward transport at 500 mb between the circulation center (11°N) and 5°S is 230×10^{12} gm/sec; this is the same as the downward mass transport between 11°N and 32°N. Dividing by the areas concerned, the mean vertical motion in the southern branch is found to be about $-0.32\ \mu$b/sec or -28 mb/day, while that in the northern branch is $+0.27\ \mu$b/sec or $+24$ mb/day. Considering that at 500 mb in the descending branch, the horizontal advection and the local change are probably small, we may write

$$\left(\frac{d\bar{T}}{dt}\right)_D \approx \bar{\omega}(\bar{\Gamma} - \bar{\Gamma}_d) \tag{2.14}$$

where subscript D denotes diabatic temperature change from any cause Γ is the existing lapse rate $\partial T/\partial p$, and Γ_d is the dry-adiabatic lapse rate

From Jordan's (1958) mean West Indies sounding for January, $(\bar{\Gamma} - \bar{\Gamma}_d)$ is about $-4.7°C/(100 \text{ mb})$ at 500 mb. Inserting this and $\bar{\omega} = 24 \text{ mb/day}$ into Eq. (2.14), we find $(d\bar{T}/dt)_D = -1.13°C/\text{day}$. Since some latent heat is released and contributes in a positive sense to this quantity, the radiative cooling required for the air to sink would be somewhat larger than this value. The cooling indicated at 500 mb by London's data (Fig. 17.1) is about 1.4°C/day. Thus, *it is consistent with observations to regard this branch, over the trade-wind belt, as dominated by a broad-scale gentle descent* wherein radiative loss of heat permits the air to settle toward lower levels where the potential temperature is lower. At levels closer to the earth's surface, it would, of course, be essential to consider the influences of vertical heat transfer by eddies and of the divergence of the heat flux (Section 17.2).

2.6   VERTICAL HEAT FLUX IN EXTRATROPICAL REGIONS AND CORRESPONDING GENERATION OF KINETIC ENERGY

It is of special interest to study the mechanism of the vertical heat flux in the extratropical region defined as the region of blocks III and IV in Fig. 2.10. For this purpose we present Fig. 2.11, which gives the eddy flux of heat in winter across latitude 32°N in 100-mb layers after Mintz (1955) and also the vertical heat flux computed from balance requirements according to the principles used in Fig. 2.10. These vertical heat fluxes across isobaric surfaces up to 100 mb (where the heat flux is assumed to stop) represent the combined effects of the mean meridional circulation (Ferrel circulation) and of eddy fluxes by all kinds of disturbances, ranging from synoptic systems down to the smallest-scale turbulence. Since the Ferrel flux is small and opposes the eddy flux, the vertical fluxes shown in Fig. 2.11 are dominated by the eddies. At the 500-mb level, the eddy flux is (from Fig. 2.10) about 25% greater than the net flux shown in Fig. 2.11.

Our aim is to separate the contribution of synoptic disturbances from the influence of other disturbances that cannot be studied in detail on regular synoptic charts. This separation is important for an estimate of the generation of large-scale kinetic energy (Chapter 16), but cannot be carried out in a completely satisfactory fashion by aid of available data.

For our purpose we shall now make some general assumptions concerning the probable contribution of small-scale turbulence (including the cu type of convection) and that of large-scale processes (including the Ferrel circulation). The large-scale processes give zero contribution to the vertical heat flux at the earth's surface (1000 mb). The vertical heat flux due to

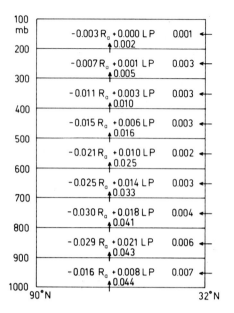

FIG. 2.11   Heat balance in 100-mb deep layers north of latitude 32°N in winter. Horizontal arrows give the eddy flux of heat across 32°N (ly/min) in the different layers, and vertical arrows represent the vertical heat flux necessary for balance. R_a and LP have the same meaning as in Figs. 2.9 and 2.10.

these processes should also approach zero at the level where the thermal field of the troposphere changes to the characteristic stratospheric field. We put that level around 200 mb for the whole belt from 32°N to 90°N.

Numerous investigations have shown that ω (in synoptic-scale disturbances) generally has its maximum absolute values around 600 to 500 mb. This is also a level where the horizontal temperature gradient is still large. Furthermore, it follows from the vertical heat flux that ω and T are, on the average, negatively correlated and that the vertical heat flux due to large-scale processes should reach quite high positive values in the middle troposphere.

In the lowest layer of the troposphere, below the average condensation level, very small-scale turbulence carries heat upward. Above the condensation level a part of this heat flux is taken over by cu-convection. Hence the turbulent vertical heat flux, at least in the lower atmosphere, is directed upward also in an atmosphere that, on the average, is stable. This type of turbulent heat flux in which warm elements selectively rise, referred to as convective turbulence, has especially been stressed by Ertel (1942), Priestley and Swinbank (1947), and Priestley and Sheppard (1952). In the extratropical region in winter, convective clouds mostly appear over relatively warm ocean surfaces and generally do not reach very high except in limited portions of active cyclonic systems (cf. Figs. 12.4, 12.5). On the other hand, the regular dynamic (mechanically forced) turbulence should tend to transport heat downward in a stable atmosphere, and hence act against the upward heat flux by convective turbulence, that dominates in the lowest layers of the atmosphere and in areas of convection. According to this assumption we may expect the small-scale heat flux to decrease upward and become negative at the level where the influence of cu-convection is negligible. As zero level for the small-scale turbulent heat flux we tentatively select 600 mb.

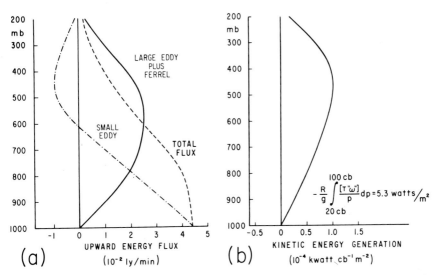

Fig. 2.12   (a) Mean total vertical heat flux in winter north of 32°N (dashed curve) taken from Fig. 2.11. Based on considerations in text, the total heat flux is partitioned between the small-scale eddy flux (dash-dotted curve) and the large-scale eddy flux plus the Ferrel flux (solid curve). (b) Total mean rate of generation of kinetic energy in winter north of 32°N, computed from the solid curve in (a). Rate of generation expressed in watt per square meter.

In Fig. 2.12a the dashed curve represents the total mean vertical energy flux per square centimeter; the solid curve, the large-scale energy flux; and the dash-dotted curve, the small-scale flux; all as functions of pressure. The positions of these curves are fixed at 1000 and 600 mb by the assumptions above. Of course they are to some extent subject to criticism, but the curve representing the large-scale flux is in satisfactory agreement with synoptic experience (Chapters 10 and 16), according to which extratropical disturbances are effective in carrying heat upward in the troposphere.

The vertical heat flux representing large-scale processes, per unit area and through a fixed level L, can be expressed by

$$[W_{zL}] = c_p[\varrho wT] \tag{2.15}$$

where the brackets denote an areal mean for the whole region 32°N to 90°N, and the bracketed quantities refer to the large-scale processes. This may also be written as

$$[W_{zL}] = c_p[\varrho w][T] + c_p[(\varrho w)''T''] \tag{2.16}$$

where the double primes denote a local deviation from the areal mean. Considering that $[\varrho w] = 0$ because of the selection of the southern boundary, and that $\varrho w = -\omega/g$, Eq. (2.16) is reduced to

$$[W_{zL}] = -\frac{c_p}{g}[T''\omega''] \tag{2.17}$$

In this expression, T'' and ω'' denote the variations at a constant level, as in Eq. (2.13). If $[W_{zL}]$ is multiplied by $(R/c_p p)$, where R is the gas constant, Eq. (2.17) becomes

$$\frac{R}{c_p p}[W_{zL}] = -\frac{R}{g}\frac{[T''\omega'']}{p} \tag{2.18}$$

This expression represents the production of kinetic energy of the horizontal motions per unit area and pressure (White and Saltzman, 1956), as discussed in Chapter 16.

Alternatively, we could write, for the flux through a fixed isobaric surface at a corresponding mean level,

$$[W_{pL}] = -\frac{1}{g}(c_p[T][\omega] + [\Phi][\omega] + c_p[T''\omega''] + [\Phi''\omega'']) \tag{2.19}$$

Since, over the area specified, $[\omega] = 0$,

$$[W_{pL}] = -\frac{1}{g}(c_p[T''\omega''] + [\Phi''\omega'']) \tag{2.20}$$

It may be assumed that the last term is numerically small compared with the first one. In the case shown in Fig. 16.2a for example, the 500-mb height difference is 200 m and the temperature difference is about 15°C between the regions of strongest descent and ascent (Fig. 16.2c). This corresponds to a difference of 2 J/g for Φ, compared with 15 J/g for $c_p T$, in this typical disturbance. Hence

$$[W_{pL}] \approx -\frac{c_p}{g} [T'' \omega''] \qquad (2.21)$$

which is analogous to Eq. (2.17).

In Fig. 2.12b the quantity on the right of Eq. (2.18) is presented as a function of pressure. When integrated from 1000 to 200 mb, it gives *a total mean generation of kinetic energy connected with the large-scale vertical heat flux, in extratropical latitudes in Northern Hemisphere winter, amounting to 5.3 W/m².* In this the negative production of the Ferrel cell is also included. As we shall see in Chapter 16, this estimate is consistent with energy production values deduced in other ways.

In this computation of generation of kinetic energy the small-scale (convective cloud) turbulence was disregarded. These also generate kinetic energy of smaller-scale turbulent motion, which cannot be evaluated from regular synoptic data and therefore is not included (as such) in studies of the atmospheric kinetic energy. In Chapters 13 and 14, it is brought out that cumulus convection is a very effective means of transferring energy vertically, essentially by the correlation expressed in Eq. (2.18). As mentioned in Section 2.5, the mass flux in the *ascending* branch of the Hadley circulation evidently takes place predominantly in the cumulonimbus clouds. In some cyclones (Section 13.4), virtually all mass transfer from the lower to the upper troposphere also appears to take place in convective cloud systems. Thus, although the details of the convective systems are not specifically presented, their influence, as expressed by energy liberated and subsequently transferred from one place to another by synoptic-scale motions, is in effect at least partly taken into account.[14]

---

[14] The motivation for subtracting the heat flux by small-scale eddies from the total may be seen from the example of an area with uniformly distributed cumulus clouds. These would transfer heat upward, and although this would (if not countered by radiation) increase the total potential energy, it would have no effect in creating available potential energy from which the large-scale circulations derive their kinetic energy. On the scale of the clouds, available potential energy would be generated, and lateral motions would result, but in all directions. Only to the extent that convective clouds are systematically distributed, as in a favored portion of a cyclone or in the equatorial trough, can they contribute to generate kinetic energy in an organized fashion.

2.7   SUMMARY OF CONCLUSIONS CONCERNING THE MERIDIONAL AND VERTICAL EXCHANGE OF ENERGY

Because of the radiative deficit of the atmosphere, its mean temperature can be maintained only through a strong transfer of heat from the earth where there is a radiative heat surplus. On an annual average, the greater part of this heat flux from the earth's surface occurs in the form of latent heat, the remainder being in the form of sensible heat. These heat fluxes and their apportionment between latent and sensible heat (Fig. 2.9 and Table 2.7) show strong seasonal and geographical variations. It is essential also to consider that latent heat is not necessarily converted into other energy forms at the same latitudes where vapor was transferred from earth to atmosphere.

The process of redistribution of energy within the atmosphere hinges on a variety of motions of different characters, ranging from small-scale turbulence and cumulus convection through synoptic-scale disturbances to the large-scale mean meridional circulations. *In low latitudes the Hadley circulation is decisive for the poleward flux of energy, but in middle and high latitudes the poleward energy flux depends almost entirely on large-scale eddies.*

An estimated 22% of the upward flux of energy at 500 mb, in the tropics in Northern Hemisphere winter, is due to the Hadley mass circulation. Above the condensation level, the largest part is associated with cumulus activity, confined in winter mostly to the tropical rain belt. Disturbances such as easterly waves are apparently quite ineffective in carrying sensible heat upward, but owing to a positive correlation between vertical velocity and water-vapor content, they carry latent heat upward. They are, however, associated with convective systems in which the vertical mass exchange largely occurs in "hot towers," wherein the temperature can be considerably higher than in the surroundings. Still larger disturbances, e.g., of the monsoon type, also contribute to the eddy flux as it is defined here, but no estimates of this contribution are available.

In middle and high latitudes, *a pronounced positive correlation between temperature, moisture content, and vertical velocity in cyclones and anticyclones shows that these disturbances are very effective in transporting sensible and latent heat upward.* In these latitudes, cumulus activity is generally of minor importance in carrying heat upward in winter, except in the lower troposphere, whereas in the warm season it has considerable influence in the heat transfer over continents (Chapter 13). The Ferrel circulation transports heat downward, but its influence is small compared with that of synoptic disturbances. In Chapters 10, 15, and 16, examples will be presented to demonstrate the heat flux and the generation of kinetic energy in connection with extratropical and tropical cyclones. This complex problem can be illustrated only by real synoptic case studies.

It is probable that, except in cumulus convection, small-scale turbulence carries heat downward in the free atmosphere but upward in the lowest parts of the atmosphere. Only in arctic and subarctic regions does small-scale turbulence result in an average downward flux at the surface.

The vertical flux of heat, owing to both large-scale eddies and the mean meridional circulations, is closely related to the conversion between potential and internal energy on the one hand and kinetic energy on the other. The vertically integrated flux of sensible heat can be taken as a measure of this energy conversion. The Hadley cell in low latitudes generates large amounts of kinetic energy, while the Ferrel cell, during the cold season, consumes a much smaller amount of kinetic energy. In extratropical regions, synoptic disturbances are the essential source of kinetic energy, as will be demonstrated in Chapter 16.

In the warm season, the importance of mean meridional circulations for the heat fluxes and generation of kinetic energy is much smaller than during the cold season. This difference is associated with the pronounced change in the distribution of insolation, there being in summer much less need for meridional exchange of heat. During this season a much larger part of the vertical flux of heat results from small-scale processes, including convective clouds. These are very important over continents, but strongly suppressed over most of the ocean regions in middle and high latitudes, where a large part of the heat goes into storage in the upper layers of the oceans.

Finally, it is instructive to compare the broad aspects of the energy budget (Fig. 2.10) with the corresponding features of the angular momentum budget in Fig. 1.11. As discussed by Yeh and Chu (1958), the requirements for transfer of these quantities are connected in an intimate fashion, but while there are obviously similarities, there are also important differences.

The mean meridional circulations carry both heat and absolute angular momentum upward in the tropics and downward in extratropical latitudes, considering integrated effects over the entire region in each case. In both latitude belts, the overall influence of eddies is to transfer heat upward and angular momentum downward in the middle troposphere. In the case of angular momentum transfer, the mean circulation flux dominates, while in the case of heat transfer, the effect of eddies is predominant. The overall consequence is thus to move both heat and angular momentum upward in the tropics, where the earth's surface represents a source for both, and heat upward and angular momentum downward in higher latitudes in accord with the respective source and sink at the earth's surface.

Whereas at latitude 30° the poleward transfer of heat (latent as well as sensible) takes place mostly in the lower troposphere, the transfer of angular momentum reaches a maximum near the tropopause level. The latter is a necessary consequence of the Hadley circulation, with its very large transfer of earth-angular momentum into the upper troposphere whence, upon conversion to relative angular momentum, it must be exported to higher latitudes. The poleward transfer is rendered effective by the presence of very strong westerlies in upper levels, requiring relatively small asymmetries of the wave systems for its accomplishment.

By contrast, the flux of sensible and latent heat upward into the higher troposphere of the tropics barely suffices to overcome the radiative loss, and comparatively little energy remains for export in upper levels. Also, it is characteristic of synoptic systems in general that the isotherms and contours tend to be strongly out of phase in the lower troposphere and more nearly in phase in the middle and upper troposphere (see, e.g., Fig. 7.8). This condition and a similar correlation between wind and specific humidity, which has its greatest concentration in the lower troposphere, make these systems most effective in transferring energy poleward in lower levels.

REFERENCES

Asai, T. (1965). A numerical study of the air-mass transformation over the Japan Sea in winter. *J. Meteorol. Soc. Japan* **43**, 1–15.

Baur, F., and Philipps, H. (1934, 1935). Der Wärmehaushalt der Lufthülle der Nordhalbkugel im Januar und Juli und zur Zeit der Äquinoktien und Solstitien. *Beitr. Geophys.* **42**, 160–207; **45**, 82–132.

Benton, G. S. (1954). Comments on the heat balance of the northern hemisphere. *J. Meteorol.* **11**, 517–518.

Bjerknes, J. (1964). Atlantic air-sea interaction. *Advanc. Geophys.* **10**, 1–82.

Brooks, C. E. P., and Hunt, T. M. (1930). The zonal distribution of rainfall over the earth. *Mem. Roy. Meteorol. Soc.* **3**, 139–158.

Budyko, M. I. (1956). "Teplovoi Balans Zemnoi Poverkhnosti," 255 pp. Gidrometeorologicheskoe Izdatel'stvo, Leningrad. (English transl., N. A. Stepanova, "The Heat Balance of the Earth's Surface." Office Tech. Serv., U. S. Dept. Commerce, Washington, D. C., 1958.)

Budyko, M. I. (1963). "Atlas Teplovogo Balansa Zemnogo Shara" ("Atlas of Heat Balance of the Earth's Surface"), 69 pp. U.S.S.R. Glavnaia Geofizicheskaia Observatoriia, Moscow.

Davis, P. A. (1963). An analysis of the atmospheric heat budget. *J. Atmospheric Sci.* **20**, 5–22.

Defant, A. (1961). "Physical Oceanography," Vol. 1, 113–114. Pergamon Press, Oxford.

Doronin, Yu. P. (1963). On the heat balance of the central Arctic. (In Russian.) *Proc. Arctic Antarctic Sci. Res. Inst.* **253**.

Ertel, H. (1942). Der vertikale Turbulenz-Wärmestrom in der Atmosphäre. *Meteorol. Z.* **59**, 250–253.

Fletcher, J. O. (1965). "The Heat Budget of the Arctic Basin and its Relation to Climate," 179 pp. Rand Corp., Santa Monica, California.

Fritz, S. (1958). Seasonal heat storage in the ocean and heating of the atmosphere. *Arch. Meteorol., Geophys. Bioklimatol.* **A10**, 291–300.

Gabites, J. F. (1950). Seasonal variations in the atmospheric heat balance. Unpublished Sc.D. Thesis, M.I.T., Cambridge, Massachussetts.

Godske, C. L., Bergeron, T., Bjerknes, J., and Bundgaard, R. C. (1957). "Dynamic Meteorology and Weather Forecasting," Chapter 5. Am. Meteorol. Soc., Boston, Massachusetts, and Carnegie Inst., Washington, D.C.

Hanson, K. J., and Rubin, M. J. (1962). Heat exchange at the snow-air interface at the South Pole. *J. Geophys. Res.* **67**, 3415–3424.

Holopainen, E. O. (1965). On the role of mean meridional circulations in the energy balance of the atmosphere. *Tellus* **17**, 285–294.

Houghton, H. G. (1954). On the annual heat balance of the Northern Hemisphere. *J. Meteorol.* **11**, 1–9.

Jacobs, W. C. (1942). On the energy exchange between sea and atmosphere. *J. Marine Res.* **5**, 37–66.

Jordan, C. L. (1958). Mean soundings for the West Indies area. *J. Meteorol.* **15**, 91–97.

Kraus, E. B. (1967). Wind stress along the sea surface. *Advan. Geophys.* **12**, 213–255.

London, J. (1957). A study of the atmospheric heat balance. Final Rept., Contr. AF 19(122)-165 (AFCRC-TR-57-287), 99 pp. Dept. Meteorol. and Oceanog., New York University.

Malkus, J. S. (1962). Large-scale interactions. *In* "The Sea" (M. N. Hill, ed.), Vol. 1, pp. 88–294. Wiley (Interscience), New York.

Manabe, S. (1957). On the modification of air-mass over the Japan Sea when the outburst of cold air predominates. *J. Meteorol. Soc. Japan* **35**, 311–326.

Meinardus, W. (1934). Die Niederschlagsverteilung auf der Erde. *Meteorol. Z.* **51**, 345–350.

Mintz, Y. (1955). Final computation of the mean geostrophic poleward flux of angular momentum and of sensible heat in the winter and summer of 1949 (Article V); The total energy budget of the atmosphere (Article XIII). Final Rept., Gen. Circ. Proj., Contr. AF 19(122)-48. Dept Meteorol., Univ. California, Los Angeles.

Mintz, Y., and Lang, J. (1955). A model of the mean meridional circulation (Article VI). Final Rept., Gen. Circ. Proj., Contr. AF 19(122)-48. Dept. Meteorol., Univ. California, Los Angeles.

Möller, F. (1951). Vierteljahrskarten des Niederschlages für die ganze Erde. *Petermanns Geograph. Mitt.* **95**, 1–7.

Palmén, E. (1966). On the mechanism of the vertical heat flux and generation of kinetic energy in the atmosphere. *Tellus* **18**, 838–845.

Palmén, E., and Vuorela, L. A. (1963). On the mean meridional circulations in the Northern Hemisphere during the winter season. *Quart. J. Roy. Meteorol. Soc.* **89**, 131–138.

Peixoto, J. P. (1967). Unpublished computations of water-vapor transfer in the Southern Hemisphere, Serviço Meteorológico Nacional, Lisbon, Portugal.

Peixoto, J. P., and Crisi, A. R. (1965). Hemispheric humidity conditions during the IGY.

Sci. Rept. No. 6, 166 pp. Planetary Circ. Proj., Dept. Meteorol., M.I.T., Cambridge, Massachusetts.

Petterssen, S., Bradbury, D. L., and Pedersen, K. (1962). The Norwegian cyclone models in relation to heat and cold sources. *Geofys. Publikasjoner, Norske Videnskaps–Akad. Oslo* **24**, 243–280.

Priestley, C. H. B. (1949). Heat transport and zonal stress between latitudes. *Quart. J. Roy. Meteorol. Soc.* **75**, 28–40.

Priestley, C. H. B., and Sheppard, P. A. (1952). Turbulence and transfer processes in the atmosphere. *Quart. J. Roy. Meteorol. Soc.* **78**, 489–529.

Priestley, C. H. B., and Swinbank, W. C. (1947). Vertical transport of heat by turbulence in the atmosphere. *Proc. Roy. Soc.* **A189**, 543–561.

Rasool, S. I., and Prabhakara, C. (1966). Heat budget of the Southern Hemisphere. *In* "Problems of Atmospheric Circulation" (R. V. Garcia and T. F. Malone, eds.), pp. 76–92. Spartan Books, New York.

Riehl, H. (1954). "Tropical Meteorology," 392 pp. McGraw-Hill, New York.

Riehl, H., and Malkus, J. S. (1958). On the heat balance in the equatorial trough zone. *Geophysica (Helsinki)* **6**, 503–537.

Roll, H. U. (1965). "Physics of the Marine Atmosphere," 426 pp. Academic Press, New York.

Rubin, M. J., and Weyant, W. S. (1963). The mass and heat budget of the Antarctic atmosphere. *Monthly Weather Rev.* **91**, 487–493.

Sellers, W. D. (1965). "Physical Climatology," 272 pp. Univ. of Chicago Press, Chicago, Illinois.

Sheppard, P. A. (1958). Transfer across the earth's surface and through the air above. *Quart. J. Roy. Meteorol. Soc.* **84**, 205–224.

Simpson, G. C. (1928). Further studies in terrestrial radiation. *Mem. Roy. Meteorol. Soc.* **3**, No. 21, 1–26.

Simpson, G. C. (1929). The distribution of terrestrial radiation. *Mem. Roy. Meteorol. Soc.* **3**, No. 23, 53–78.

Smagorinsky, J., Manabe, S., and Holloway, J. L., Jr. (1965). Numerical results from a nine-level general circulation model. *Monthly Weather Rev.* **93**, 727–768.

Spar, J. (1949). Energy changes in the mean atmosphere. *J. Meteorol.* **6**, 411–415.

Starr, V. P., and White, R. M. (1954). Balance requirements of the general circulation. *Geophys. Res. Papers (U.S.)* **35**, 1–57.

Starr, V. P., Peixoto, J. P., and Crisi, A. R. (1965). Hemispheric water balance for the IGY. *Tellus* **17**, 463–472.

Sverdrup, H. U. (1951). Evaporation from the oceans. *In* "Compendium of Meteorology" (T. F. Malone, ed.), pp. 1071–1081. Am. Meteorol. Soc., Boston, Massachusetts.

Vuorela, L. A. (1957). On the observed zonal and meridional circulations at latitudes 15°N and 30°N in winter. *Geophysica (Helsinki)* **6**, 106–120.

Vuorela, L. A., and Tuominen, I. (1964). On the mean zonal and meridional circulations and the flux of moisture in the Northern Hemisphere during the summer season. *Pure Appl. Geophys.* **57**, 167–180.

White, R. M., and Saltzman, B. (1956). On conversion between potential and kinetic energy in the atmosphere. *Tellus* **8**, 357–363.

Yeh, T.-C., and Chu, P.-C. (1958). "Some Fundamental Problems of the General Circulation of the Atmosphere," 159 pp. Inst. Geophys. Meteorol., Academia Sinica, Peking. (In Chinese, extended English abstract, pp. 147–156).

3

SEASONAL AND ZONAL VARIATIONS
OF THE MEAN ATMOSPHERIC
STRUCTURE AND FLOW PATTERNS

The wind and temperature distributions, and the distributions of the various derived quantities, discussed in Chapters 1 and 2, represent conditions averaged zonally around the globe. In this chapter we describe some of the salient variations with season and longitude, in both the Northern and Southern Hemispheres.[1] Also, although we shall elsewhere confine the discussion to the synoptic systems of the "lower atmosphere," we will here briefly summarize the main features of the upper stratosphere, mesosphere, and lower thermosphere.

3.1 MID-TROPOSPHERIC MEAN FLOW PATTERNS

The existence of significant zonal variations in the flow can be clearly seen in 500-mb mean charts of the Northern Hemisphere for January and July; see Figs. 3.1 and 3.2. In January, the most prominent features are the pronounced troughs near 80°W and 140°E, with a weaker third trough at 10°E to 60°E. In July the corresponding major troughs are less marked,

---

[1] For mean data at different levels, longitudes, and seasons, the reader is referred to U.S. Weather Bureau (1952), Goldie *et al.* (1958), I. Jacobs (1958), Wege *et al.* (1958), Lahey *et al.* (1958,1960), U. S. Navy (1959), Heastie and Stephenson (1960), and Tucker (1960).

and there is a suggestion of a greater number of waves (more evident in the mean charts by Lahey *et al.*, 1958).

The deviations from purely zonal flow are obviously due to features of the earth's surface. Computations by Charney and Eliassen (1949) and by Bolin (1950) of orographic influences, mainly by the Rocky Mountains

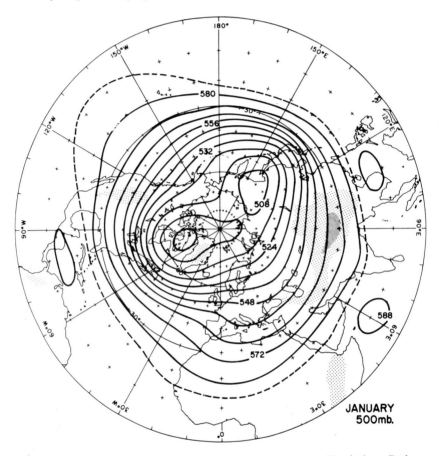

Fɪɢ. 3.1   Mean 500-mb contours in January (winter), Northern Hemisphere. Redrawn at 80-m intervals from I. Jacobs (1958). Light and heavier stippling show regions where elevations are above 1.5 km and 5 km (smoothed over 5° latitude-longitude tessera), from Berkofsky and Bertoni (1955).

and the Himalaya complex, gave encouragingly realistic approximations to the winter mean wave patterns. Sutcliffe (1951) drew attention to significant differences between the summer and winter flow patterns, challenging the dominance of mechanical orographic influences. He maintained rather that the direct thermal effects of land and sea, modified by baroclinic synoptic disturbances, was of greatest significance. The great influence of the oceans as heat sources in autumn and winter, and as relative heat sinks in spring and summer when most of the insolation goes into oceanic storage and land is heated more, was emphasized in Chapter 2.

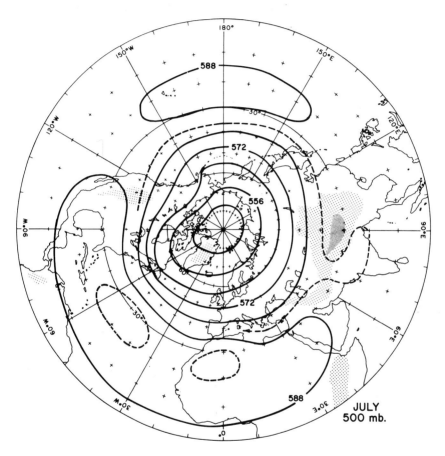

FIG. 3.2 Mean 500-mb contours in July (summer), Northern Hemisphere. (Redrawn, from I. Jacobs, 1958.)

The strong seasonal and local variations in the total heat flux from oceans and continents, demonstrated by W. C. Jacobs (1951), Budyko (1963), and others, indicate the importance of thermal influences on the deviations from the longitudinal mean flow pattern.

Actually, as discussed by Smagorinsky (1953), both orographic and thermal influences are of great significance. Yeh and Chu (1958) reproduced charts showing their computations of the separate and combined effects at 500 mb in winter. Although the effects of heat sources and orography differ, especially over Asia, these influences cooperate in maintaining mean troughs near the east coasts of Asia and North America in winter.

Orographic influences are partly mechanical and partly thermal in nature. In addition to the dynamical computations mentioned above, the character of the mechanical influence has been demonstrated in laboratory experiments by Fultz and Long (1951), involving obstacles in a rotating hemispherical shell. With "westerly" flow a train of waves was generated, dynamically similar to the long Rossby-Haurwitz waves of the atmosphere; the same characteristic resulted in Kasahara's (1966) numerical simulation of the Fultz-Long experiment. An illuminating feature of Kasahara's experiment is that these waves were well reproduced when the variation of coriolis parameter was included. However, when this parameter was held constant, only a single disturbance appeared in the "hemisphere," and this traveled eastward with nearly the speed of the basic current. This demonstrates the importance of the retarding influence of the latitudinal variation of coriolis parameter (Section 6.1) in holding the orographically generated long waves in favored geographic locations.

An especially interesting aspect, discussed by Frenzen (1955), is the dependence of fluid behavior upon the latitude at which a cylindrical obstacle is placed in the hemispherical shell. With the obstacle at 45° latitude, a steady wave pattern resulted; when the obstacle was placed at 30°, the amplitude of this increased; at 60° no steady wave pattern resulted, but traveling cyclonic vortices formed periodically. Frenzen points to the analogy with the Himalayas and the Andes, which lie astride the 30° parallels and are thus situated where they can be especially effective in generating standing waves in the westerlies; Greenland, at a high latitude, has a different effect.

Thermal effects on the formation of upper high-pressure systems over or near large mountain complexes during the warm season have especially been emphasized by Flohn (1960). Whereas Flohn considers the primary process to be the transfer of sensible heat from elevated terrain, Gutman and Schwerdtfeger (1965) also stress the importance of latent heat release in connection with the occurrence of numerous thunderstorms and showers. These influences, for the southeast Asia region, are discussed in Sections 8.7 and 14.5.

Both over warm oceans in winter and warm continents in summer, the sensible heat flux from the earth's surface tends to initiate circulations that increase condensation and add heat to the systems. These processes tend to build up (in the upper troposphere over the corresponding regions of heat sources) systems with pressures in excess of the longitudinal mean. Where these processes occur in a belt of strong westerlies, their effect is spread eastward from the regions of excessive heating. The orographic influence on the release of latent heat is especially strong over mountain ranges on the west coast of a continent. This makes it difficult to distinguish between the disturbances of the westerly flow induced directly by orography and the more indirect influence of a thermal nature, which is also caused by the orography.

Evidently the only way to appraise the relative contributions of these mixed effects is through laboratory experiments or numerical simulation in which orographic, thermal, and frictional influences can be incorporated or omitted. An interesting example is given by Mintz. Sutcliffe (1951) has stressed the well-known climatological influence of the long west-east mountain chain across southern Eurasia, which by preventing the free advection of lower-tropospheric air introduces a large meridional contrast in thermal and moisture conditions. Mintz (1965) found that when mountains were omitted from his long-term numerical integrations, the wintertime Siberian High did not appear on the mean sea-level pressure map. When the mountains were raised into place, this high appeared in full strength. In their absence, development of baroclinic disturbances occurred between air heated over the Indian Ocean and radiatively cooled air over Siberia, advecting heat into the latter region.

In the Southern Hemisphere, the limited west-east extent of mountain massifs and the smallness of the land areas in middle latitudes make for a much more symmetrical mean circulation. Although the Andes Cordillera is very high, it is narrow, and there is little evidence of a lee trough in the mean flow patterns of either winter (Fig. 3.3) or summer (Fig. 3.4). This is in line with the finding by Bolin (1950) that the width of a mountain chain has a very important effect on its influence in setting up planetary waves.

In both hemispheres, Figs. 3.1 through 3.4 show that the westerly wind belt extends to considerably lower latitudes in winter than in summer. It should be noted that the zonality of the *mean* flow in the Southern Hemisphere does not connote that the synoptic disturbances are weak. At the 500-mb level, van Loon (1965) has shown that in the Southern, compared with the Northern Hemisphere, the geostrophic meridional flow associated with the daily wave patterns averages about 10% weaker in winter and 30% stronger in summer. The relative zonality of the southern mean flow is due to mobility of the disturbances, which tend to a lesser degree to amplify and remain fixed in favored longitudes. The regularity of the contours in Figs. 3.3 and 3.4 is also deceptive with regard to the strength of the mean flow in different parts of the hemisphere, as discussed later.

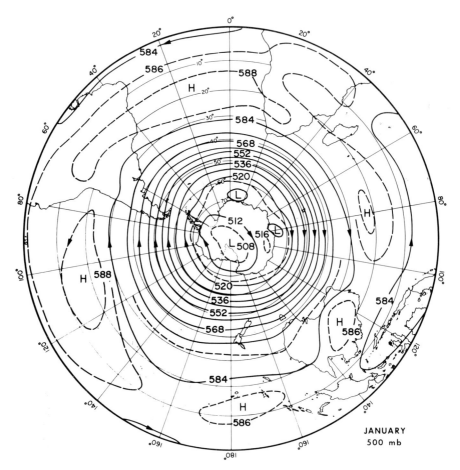

FIG. 3.3   Mean 500-mb contours (80-m interval) in January (summer), Southern Hemisphere. (After Taljaard *et al.*, 1969.)

In an individual January or July, there may, of course, be a substantial deviation from the long-term mean patterns in Figs. 3.1 and 3.2. Abnormal heights in one locality tend to be associated with abnormalities in other favored longitudes (Martin, 1953), reflecting a linkage between successive troughs and ridges of the long waves in the westerlies (Section 6.1). Evident-

ly a markedly abnormal heat source or sink in one locality—evidenced in
the sea-surface temperature, moisture characteristics of land surfaces, or
the presence of extensive snow cover in normally snow-free areas (Namias,
1962)—can influence the flow patterns in distant regions. The several ways

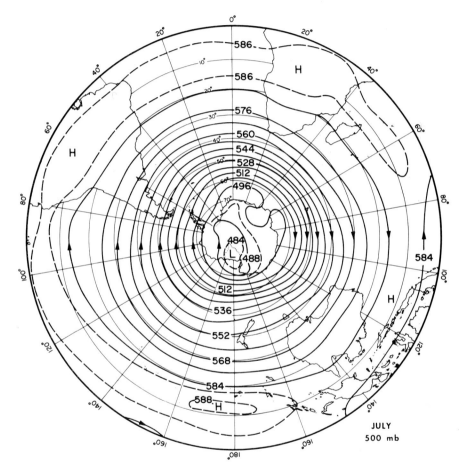

Fig. 3.4   Mean 500-mb contours in July (winter), Southern Hemisphere. (After
Taljaard *et al.*, 1969.)

in which an anomalous atmospheric flow pattern (such as excessive cyclonic
activity in a given region) can influence the surface characteristics, resulting
in many cases in a "feedback" that tends to perpetuate the atmospheric
flow once a favored pattern has been established, are discussed at length
by Namias and by Bjerknes (1964).

3.2 Seasonal Mean Structures in Meridional Sections

The details of the mean structure are best illustrated by sections (Fig. 3.5) along 80°W, where abundant observations are available. Kochanski (1955) also presents the temperature fields as well as sections for the intermediate seasons. In accord with the northward shift and weakening of the the middle-latitude baroclinic zone of the Northern Hemisphere between January and July (Fig. 1.1), corresponding changes are observed in the wind fields shown in Fig. 3.5. Aside from this, the principal seasonal change is a complete reversal of the meridional temperature gradient, and a corresponding reversal of the wind direction, in the extratropical stratosphere between the winter and summer seasons.

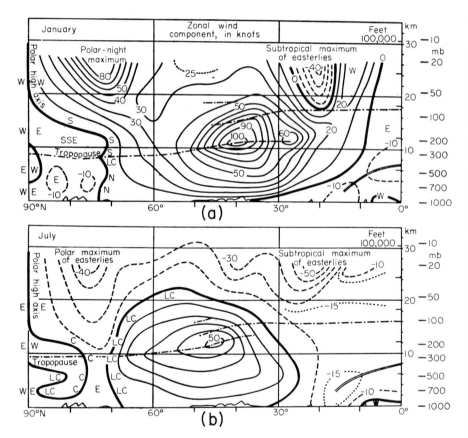

FIG. 3.5 Zonal component of mean geostrophic wind at 80°W in (a) January and (b) July. Isotachs in knots; dashed for east winds. Dash-dotted lines show mean tropopause heights. (After Kochanski, 1955.)

Figure 3.6 shows that the seats of the principal interseasonal temperature changes lie in the lower levels, reflecting largely the continental response to differences of insolation, and in the stratosphere where there is a strong radiative loss when subpolar latitudes are in darkness (the influence of ozone absorption increases upward). The more modest changes in the upper troposphere and lower stratosphere can also be ascribed partly to moderation of the winter cooling by meridional and vertical transfer of heat (Section 2.4). Over an oceanic region, the seasonal change in low levels would, of course, be smaller.

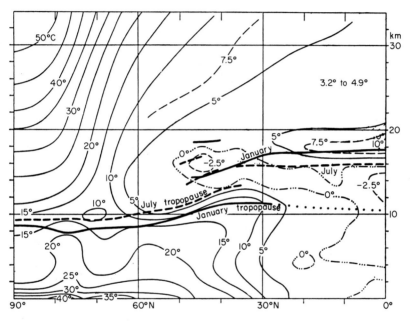

FIG. 3.6   July minus January differences of mean temperatures (°C) at 80°W. (After Kochanski, 1955.)

McClain (1960) has drawn attention to the ambiguous significance of "mean" soundings in the upper stratosphere of high latitudes. He demonstrated that the far-northern stations tend to display bimodal frequency distributions of temperature of high levels, even though a single mode is prominent in the lower stratosphere (Fig. 3.7). At the higher levels, the "mean temperature" may be seldom observed. An individual station may, during a particular season, remain for a long time under the influence of one or the other temperature regime. This, combined with the circumstance that only limited periods of record are available at high-latitude stations,

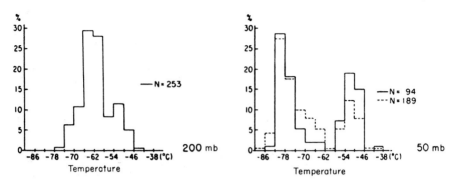

FIG. 3.7   Frequency distributions of January temperature at the 50- and 200-mb levels for Alert, in northern Canada. (After McClain, 1960.)

was pointed out by McClain as the source of the large differences in high latitude between mean sections, as constructed by different authors.

It is obvious from Figs. 3.1 and 3.2 that the structure shown by a meridional cross section must be considerably influenced by its longitude. To illustrate this, we may compare the winter mean zonal wind fields in Fig. 3.8a, along the east Asiatic trough, and in Fig. 3.8b near the crest west of Europe. In Fig. 3.8a, a single upper-wind maximum of 120 knots can be observed near latitude 35°N, whereas in Fig. 3.8b there are two separate speed maxima, a southern one of 75 knots at 25°N and a northern one of 35 knots near 60°N.

Although the contrast between these examples is extreme, qualitatively this difference between the wind structures near troughs and ridges is characteristic. This feature is to a large extent obscured on mean charts,[2] and is better illustrated by the schematic Fig. 3.9. Here the shaded area represents the region where the polar-front jet stream is most often located in winter, while the line farther south is the mean position of the subtropical jet.

These two wind systems most nearly approach each other at the longitudes of the semipermanent troughs of the middle-latitude westerlies (Fig. 3.1), where they are close enough together to appear as one maximum when averaged over a large number of situations. Where this occurs, downstream from the main regions of "confluence" (Namias and Clapp, 1949), the warm air from low latitudes is brought into juxtaposition with cold air

[2] For example, a single wind maximum appears at latitude 38°N on Crutcher's mean section at 80°W in winter. Essenwanger's daily statistics (1953) indicate that at this longitude the strongest winds at 500 mb are found most frequently near latitudes 35° and 48°, with distinctly lower frequency of occurrence at other latitudes.

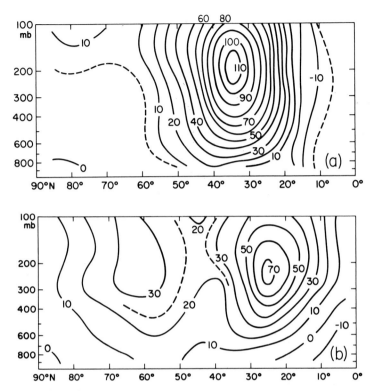

FIG. 3.8 Winter (December–February) zonal mean wind components (knots), Northern Hemisphere, at (a) 140°E and (b) 0° longitude. (Redrawn from Crutcher, 1961.)

masses from higher latitudes. Consequently, the meridional temperature gradient is much enhanced, and since the westerly flow in the upper troposphere is roughly proportional to the baroclinity of the atmosphere, the longitudes of the mean polar troughs are also the locations of the strongest mean west winds (Section 8.5). Over southern Japan in winter, three-month average wind speeds of 75 m/sec or more have been documented by Mohri (1953) and by Matsumoto *et al.* (1953).

It is obvious that the previous discussions of the mean meridional flux of angular momentum and energy (Chapters 1 and 2) would be strongly modified if applied in the localities of mean troughs or ridges. Also, it is evident that the existence of semipermanent troughs and ridges is of considerable importance for the meridional exchange of these properties on a hemispheric basis. A large part of the poleward eddy flux of angular momentum and energy is accomplished by these very large-scale disturbances in the mean circulation (Section 17.4).

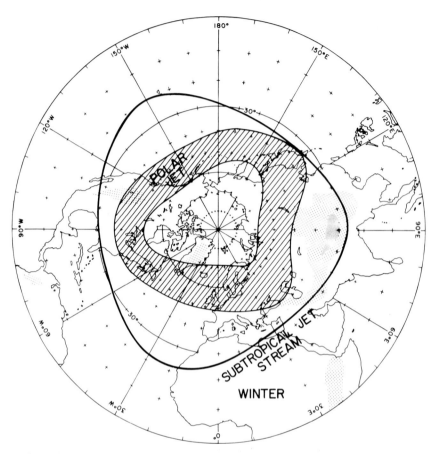

FIG. 3.9   Mean axis of subtropical jet stream during winter, and area (shaded) of principal activity of polar-front jet stream. (After Riehl, 1962.)

3.3   WIND AND TEMPERATURE IN HIGH ATMOSPHERE

Information on the structure of the atmosphere at levels above those reached by sounding balloons was first systematically integrated by Kellogg and Schilling (1951), who described the temperature distribution and principal wind systems in summer and winter. Murgatroyd (1957) constructed sections based on a very thorough analysis of all data of various types, with an attempt at consistency (geostrophic) between wind and temperature fields. Batten (1961) modified his analysis of the wind field, incorporating new observations and rejecting certain types of observations considered to be of a doubtful nature. Batten's analysis is shown in Fig. 3.10a.

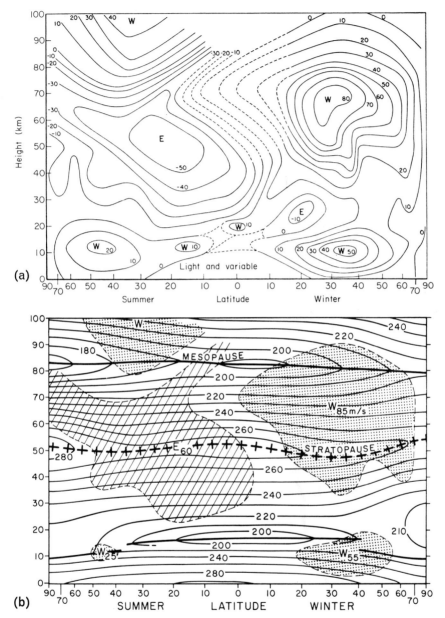

FIG. 3.10 (a) Mean zonal wind (m/sec) in summer and winter, after Batten (1961), an emendation of the analysis by Murgatroyd (1957). (b) Mean thermal structure, a compromise between analyses by Kochanski (1963) and Cole and Kantor (1963) as described in text. Hatching indicates mean easterlies and stippling mean westerlies, both in excess of 20 m/sec.

Kochanski (1963) has also presented analyses of the winter and summer temperature fields in the 40 to 100 km levels. An analysis of observations by rockets and rocket-released instruments is given by Cole and Kantor (1963). There are large differences in high levels, the rocket temperatures being systematically lower at 100 km, by as much as 30°K. Above 40 km, Fig. 3.10b represents an equal-weight compromise between these two analyses, with slight adjustments to secure partial consistency with the vertical shears.[3] Below 40 km, Cole's and Kantor's analysis is accepted. West or east winds greater than 20 m/sec are indicated by shading, and the cores of the mean currents are marked, according to Fig. 3.10a. It should be emphasized that the values shown are provisional for several reasons, including the obvious one of sparse data. However, the major features are well established.

The more-or-less complete reversal of the high-level wind systems between seasons is connected with an analogous reversal of the temperature gradients. At the stratopause, and in a layer of considerable depth above and below it, the air in high latitudes is warmer in summer and colder in winter than at low latitudes. The opposite is true at the mesopause and in the contiguous layers of the upper mesosphere and lower thermosphere. Correspondingly, in summer the tropospheric westerlies give over to easterlies already in the lower stratosphere (cf. Fig. 3.5b). These increase up to about the strato-pause level. Through the mesosphere and lower thermosphere, the pole-to-equator temperature difference is opposite to that below. Correspondingly, the easterlies decrease with height and change over to westerlies in the thermosphere. In winter the vertical shears are, in harmony with the changed temperature gradients, in a broad sense reversed from those in summer.

On the basis of analyses in the lower stratosphere, Julian et al. (1959) state that the "polar regime" (stratospheric westerlies increasing with height) usually begins in September, reaches peak intensity near the end of December and disappears at a variable time in January through March. The break-down from stratospheric westerlies to summer easterlies may be quite rapid, and has been found to be associated with the "explosive stratospheric warmings" first described by Scherhag (1952).

Kellogg and Schilling (1951) point out that in the layer comprising the

---

[3] Kochanski suggests that the meridional temperature gradient at latitudes 45° to 65° above the mesopause is reversed from that shown here. This feature is based on radio observations of meteor trails at Jodrell Bank (53°N). These indicate an increase of west wind upward from 82 to 102 km, which does not appear in Fig. 3.10a. The overall Pole-to-Equator temperature contrasts are, however, consistent in all analyses.

upper stratosphere and mesosphere, there is a level of maximum wind in both winter and summer hemispheres. Considering frictional losses, the winds should be subgeostrophic, with a component of flow toward lower pressure. On these grounds, and also due to intensification of the thermal and pressure gradients up to the times of the solstices, they suggested that in the corresponding levels there should be a systematic meridional flow from the summer pole toward the winter pole. The principle is the same as in the Hadley circulation (Section 17.1); flow in this sense acts to generate easterlies in the summer and westerlies in the winter hemisphere, thus counteracting the frictional losses.

Within the stratosphere and mesosphere, the Kellogg-Schilling scheme implies general ascending motions over high latitudes in summer and sinking motions in winter. The attendant adiabatic cooling or warming counteracts the rapid local temperature changes that would otherwise occur as a result of the strong net radiative warming in these latitudes in summer and the net radiative cooling in winter. The meridional and vertical circulations have been calculated by Murgatroyd and Singleton (1961) on the basis of heat transfer by the mean motions and mass continuity. Near the times of the solstices, flows in the sense postulated by Kellogg and Schilling are found, but with maximum meridional motions at somewhat higher levels near 70 to 80 km.

Near the solstices, the most intense net radiative heating ($7°K/day$) or cooling ($13°K/day$) is at the 60- to 70-km level at the summer and winter poles. According to Murgatroyd and Singleton, the most intense rising and sinking motions (about 1.5 cm/sec) are found at corresponding levels in high latitudes. This is evidently necessary, as suggested by several writers, to account for the pole-to-pole temperature gradient near the mesopause, which is the reverse of what would be expected from radiative processes alone. It may be noted that the heat sources and sinks involved (mainly absorption of solar radiation by O_3 and O_2, and long-wave eradiation by CO_2 and O_3) are distinct from those in the troposphere, indicating that the circulations of the high atmosphere are separately driven. Also, the summer pole (where heat is added) is warmer than the winter pole only below about 70 km. This suggests, according to the principles in Section 16.1, that kinetic energy is generated essentially in the upper stratosphere and lower mesosphere and that the circulation at higher levels is driven rather than being energy-producing.

Although Murgatroyd and Singleton stress that the deficiencies in their calculations are due to omission of the unknown eddy influences, the overall scheme is in agreement with observations. In the solstitial months, these computations indicate meridional velocities up to 5 to 6 m/sec near 75 km. At 80 to 100 km (Kochanski, 1963), the available observations indicate a strong meridional flow (up to 10 m/sec or more) from the summer toward the winter pole. It may also be noted that the heat-budget computations support an early deduction by A. W. Brewer that there are gentle rising motions through the level of the tropical tropopause, with poleward movement of air in the low stratosphere toward both poles.

Large long-term variations in some regions are not reflected in Fig. 3.10. These are in part regular seasonal changes and in part not tied to the seasons. In the latter category is the "26-month oscillation" of stratospheric winds, which was recognized independently by Veryard and Ebdon (1961) and by Reed et al. (1961). This oscillation, whose period varies, is between zonal easterlies and westerlies in the monthly mean flow. Near the Equator, it has an amplitude of 2 m/sec at 100 mb, increasing to about 20 m/sec between 40 and 10 mb; the amplitude at 10 mb decreases to 3 to 5 m/sec at latitude 25°. A transition from easterlies to westerlies (or vice versa) appears first at high levels and successively at lower levels, about a year being required for a wind reversal to work down from 10 mb (31 km) to 60 mb (17 km). For a comprehensive review of this phenomenon, see Reed (1965). Belmont and Dartt (1966) present monthly mean charts showing that the oscillation extends all around the equatorial belt, and also space-time analyses illustrating the transitions between this quasi-biennial oscillation and the annual changes that dominate the subtropical latitudes.

From an analysis of rocket wind data at heights of 28 to 64 km, Reed (1966) found evidence of a strong *semiannual* oscillation in equatorial latitudes.[4] In both midsummer and midwinter, he estimated that winds in this layer over the Equator are easterly, with maximum mean speed of about 35 m/sec near 47 km in the long-term average. This is in close agreement with Fig. 3.10b. However, in the intervening equinoctial months, westerly winds predominated, increasing upward to about 35 m/sec at 64 km. Kochanski (1963) cites evidence of a semiannual oscillation at high levels over Jodrell Bank (53°N), where the winds at 80 to 100 km are westerly near the solstices and easterly near the equinoxes. At Adelaide (35°S), only the annual variation is evident.

In addition to the longer-period variations summarized above, there are other large and rapid changes. These are partly diurnal variations, and

---

[4] This was strong at Ascension Island (8°S) but not evident at Barking Sands, Hawaii (22°N). In an individual season the mean winds would, of course, also be influenced by the quasi-biennial oscillation.

partly due to synoptic changes. The latter were first systematically investigated at the 50-mb level by Austin and Krawitz (1956), who discuss the
principal similarities and differences between stratospheric and tropospheric
circulations. Statistical analyses by Julian *et al.* (1959) indicate that the
short-wavelength disturbances characteristic of the middle latitude troposphere are essentially damped out at levels as low as 100 mb. This damping
is attributed to thermal compensation, owing to the association of warm
air with troughs and of cold air with ridges in the lower stratosphere. At
the same time, "the basic long-wave structure of the upper troposphere is
preserved in the lower stratosphere to at least 100 mb." Furthermore,
above the lower stratosphere, ridges are usually warm and troughs cold,
so that these large-scale systems tend to preserve their identities through
a deep layer of the middle and high stratosphere and to increase in intensity upward. Their movements are associated with very large changes
of temperature (reflected in the frequency distribution of 50-mb temperature in Fig. 3.7) and wind (which at high levels achieves speeds in the
"polar night jet" that are comparable with the strongest tropospheric jet
streams).

Austin and Krawitz suggested that these upper-stratospheric waves are
baroclinic disturbances of a basic nature similar to those of the troposphere,
but developing independently. This view has been substantiated in studies
by Boville *et al.* (1961) and by Julian and Labitzke (1965), in which it
was found that the disturbances convert available potential energy to kinetic
energy at stratospheric levels. As discussed elsewhere, this is also true in
the troposphere. However, in the intervening region of the lower stratosphere, computations by Miyakoda (1963) and by Oort (1964) indicate a
conversion of the opposite sort, suggesting that the circulations in this layer
are "driven" by those above and below.

Since in this volume we are not mainly concerned with high-atmosphere
systems, we refer for more general information to the reviews by Panofsky
(1961) and Kellogg (1964), and to the book by Craig (1965). Webb (1966)
presents the results of rocket measurements. Those interested especially in
the relations between the atmospheric circulation and the distribution of
chemical constituents may consult the comprehensive review by Sheppard
(1963) and an article by Newell (1963), which treats the transfer processes.
Pogosian (1965) presents a very complete analysis of the seasonal mean
temperature and wind fields up to the 90-km level. Thorough discussions
of the literature and data available on the high atmosphere, including
physical interpretations, are given in a collection of articles by Murgatroyd
et al. (1965).

3.4 Seasonal Characteristics of Upper-Tropospheric Wind Field

The annual course of the zonal flow averaged around the Northern Hemisphere is shown for the 700-mb level in Fig. 3.11 (Klein, 1958). This illustrates the well-known fact that the northern belt of maximum westerlies is farthest south and reaches greatest intensity in winter, and is farthest north and much weaker in summer. The latitudes of maximum frequency of both cyclogenesis and anticyclogenesis follow the same trend. This is in accord with the general preference for disturbances to form where the baroclinity and upper winds are strong (Chapters 6 and 11).

The seasonal variation in the Southern Hemisphere is very different from that in Fig. 3.11. As shown by Fig. 3.12, in middle latitudes the southern westerlies at 500 mb are significantly stronger in summer than in winter. As van Loon (1964) observes, this reflects the presence of a greater meridional gradient of the surface temperature in summer than in winter, owing to a larger annual temperature range in subtropical than in higher middle latitudes. In the Northern Hemisphere the percent coverage by land increases poleward in middle latitudes, and correspondingly (considering the thermal properties of the surface) the annual range of temperature

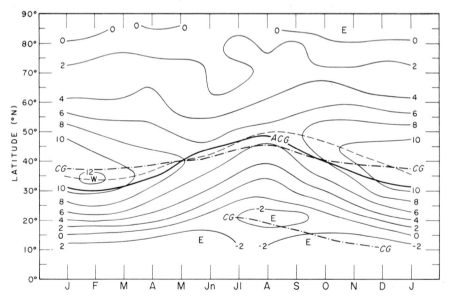

FIG. 3.11   Seasonal course of zonally averaged west wind speed (m/sec) at 700 mb in Northern Hemisphere. Dashed line indicates latitude of greatest wind speed; heavy lines are latitudes of greatest frequency of cyclogenesis (CG) and anticyclogenesis (ACG), respectively. (After Klein, 1958.)

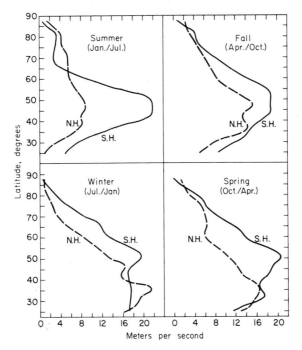

FIG. 3.12   Profiles of mid-season average zonal wind at 500 mb, in Northern (dashed lines) and Southern (solid) Hemispheres. In each case, the first month given refers to S.H. and the second to N.H. for the corresponding season. (After van Loon, 1964.)

increases with latitude. In the Southern Hemisphere, by contrast, land (aside from Antarctica) is virtually confined to the low latitudes. This contributes in part to a greater temperature range in subtropical than in high middle latitudes. In addition, however, it is observed that the *sea-surface* temperature in this predominantly oceanic hemisphere has a greater annual range in latitude 35°S than in 50°S.

This behavior has been explained by van Loon (1966) in terms of the heat budget of the southern oceans. Although at the higher latitude there is a greater seasonal variation of insolation at the top of the atmosphere, greater cloudiness near 50°S results in a smaller variation of the insolation received at the earth's surface. While the effect of this is somewhat offset by seasonal differences in the sensible and latent heat transfer at the two latitudes, there is an appreciably greater interseasonal heat storage and loss by the sea at 35°S than at 50°S. The effect is enhanced by stirring of heated or cooled surface waters to greater depths by the strong winds at the higher latitude.

The overall Equator-to-Pole temperature contrast is, of course, greater in winter than in summer. So, correspondingly, is the total hemispheric westerly momentum, the westerlies outside the belt discussed above being stronger in winter than in summer. Near latitude 30°S the westerlies in winter are about twice as strong as in summer at 500 mb (Fig. 3.12) and also at the level of maximum wind (Fig. 1.2).

In this discussion it has been tacitly assumed that there is a connection (through the thermal wind) between the sea-surface temperature gradient and the strength of the mid-tropospheric westerlies. The lower-tropospheric mean temperatures are indeed quite responsive to the sea-surface temperature variations, as is demonstrated by the fact that the 500- to 1000-mb thickness field (van Loon and Taljaard, 1958) displays the same variations of gradient as were discussed above for the surface temperature.

Reuter (1936), Schwerdtfeger and Prohaska (1956), and others have described *semiannual* oscillations of sea-level pressure that dominate the annual oscillation in some southern latitudes. These achieve greatest amplitude in latitudes 40°S to 50°S, where mean pressures are *highest* at the times of the equinoxes, and in latitudes 65°S to 70°S, where pressures are *lowest* at the equinoxes. Correspondingly, the meridional pressure gradient and the westerly winds at sea level in subantarctic latitudes are strongest near the spring and autumn equinoxes. Schwerdtfeger (1960) discovered also that there is a dominant second harmonic, of the same phase, in the 700 to 300 mb thickness gradient between 50°S and the South Pole.[5] From an examination of 500-mb temperatures, van Loon (1967) concluded that this is mostly due to variations within the belt 50°S to 65°S, where the second harmonic of the meridional temperature gradient has an amplitude 2.5 times that of the first harmonic.[6] Consonant with the semiannual variations mentioned above, the belt of cyclones around Antarctica is displaced farther poleward and somewhat more intense in the equinoctial months than in summer and winter (there being no consistent relation between their latitudes and the seaward extension of the pack ice).

A pronounced semiannual variation of zonal wind is also present in the

---

[5] Schwerdtfeger notes an analogous variation in the Northern Hemisphere at latitudes 60° to 80°.

[6] Owing to factors in the oceanic heat budget discussed by van Loon, the mid-tropospheric temperature at 50°S falls more rapidly in autumn and rises more slowly in spring than at 65°S. Highest temperature is reached later in summer and lowest temperature earlier in winter at 50°S than at 65°S. The annual temperature range at 500 mb is almost the same at the two latitudes, so that the temperature contrast across the belt is greatest near the equinoxes and smallest near the solstices.

Southern Hemisphere tropics (van Loon and Jenne, 1969). This reaches maximum amplitude near the 200-mb level in latitudes 10°S to 20°S, and is most pronounced in the eastern half of the hemisphere where the semi-annual variation exceeds the annual. Maximum westerlies (or minimum easterlies) occur in May and November. The amplitude diminishes above the 200-mb level; an oscillation of reversed phase is present between about 15 and 21 km. May and November westerly maxima are again observed in higher levels, increasing in amplitude upward in the stratosphere, as noted in Section 3.3. This semiannual oscillation is consistent with changes of the meridional temperature gradient; at tropical and subtropical stations there is a well-marked semiannual variation of temperature in upper levels, of opposite phase near 10°S and 20°S.

The world-wide distribution of upper winds is best illustrated at the 200-mb level, which lies somewhat below the mean level of strongest wind in subtropical latitudes and somewhat above it in higher latitudes. Except in the lowest levels, the broad features of the mean wind fields in the tropo-sphere and lower stratosphere are similar to the configurations at 200 mb. Figure 3.13, taken from Heastie and Stephenson (1960), shows the January and July isotachs and streamlines.

In the Southern Hemisphere the mean wind pattern is comparatively simple. A belt of nearly zonal westerlies, between 45°S and 55°S, is in both seasons considerably stronger south of Africa and in the Indian Ocean than in other longitudes.[7] A complete circumhemispherical belt of strong winds (subtropical jet stream) is found near 30°S in winter. In summer, this is indicated only east of Australia. However, later observations such as those during the IGY (U. S. Weather Bureau, 1962) show that the subtropical jet stream is present (and generally distinct from the polar-front jet) also over South America. It seems likely that it extends around the hemisphere in summer as well as winter, although in a weakened state (as indicated in the longitudes of New Zealand by Gabites, 1953). As noted in Chapter 1, the Hadley circulation, which is the main generator of kinetic energy for the subtropical jet stream, is well developed all year round in the Southern

---

[7] At 500 mb in January, according to van Loon (1964), the maximum wind averaged over an 80°-longitude sector of the Indian Ocean is 28 m/sec, while over the opposite sector of the South Pacific it is 16 m/sec. This reflects a corresponding difference in the meridional gradients of sea-surface temperatures in these regions (van Loon, 1967). In the Northern Hemisphere, the pronounced eastward decrease of the mean wind speeds over the Atlantic and Pacific oceans must also be strongly influenced by the marked weakening of sea-surface temperature gradient from the west to the east sides of the oceans.

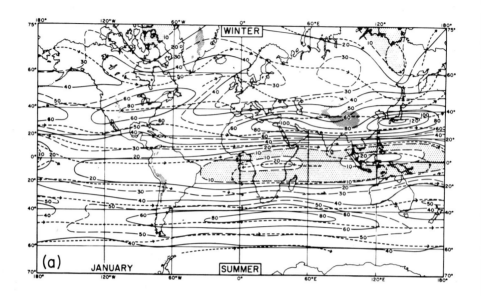

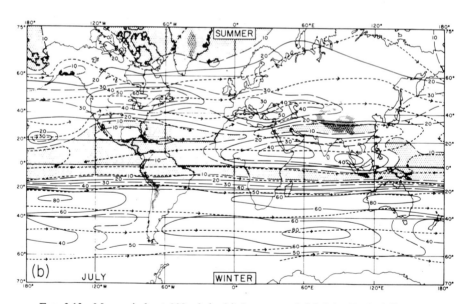

FIG. 3.13   Mean winds at 200 mb in (a) January and (b) July. Dashed lines, mean streamlines; solid lines, mean isotachs (knots). (From Heastie and Stephenson, 1960, by permission of the Controller, Her Britannic Majesty's Stationery Office.)

Hemisphere, whereas in the Northern Hemisphere this circulation is interrupted (or distorted) by the Asiatic monsoon in summer.

In the Northern Hemisphere, the distribution of winds is more irregular. In winter (Fig. 3.13a) the principal belts of strongest wind extend across the United States and into northern Europe, and across northern Africa and southern Asia into the western Pacific. Distinct speed maxima are seen near the continental east coasts and over the Near East, all being considerably weaker in summer than in winter.

A weak band of easterlies, observed near the Equator in January around half the circumference of the earth, extends all around the equatorial belt in July. In summer, the mean easterlies achieve maximum speeds in excess of 70 knots somewhat above 100 mb, near latitude 15°N, around longitude 80°E. Like the extratropical west-wind belts, this belt of upper easterlies is present throughout the year. In contrast, however, it is strongest when farthest from the Equator, owing to the arrangement of heat sources (Section 8.7).

Based on the winter and summer mean cross sections by Crutcher (1961), the latitudes and strengths of the maximum mean westerly wind in the Northern Hemisphere are summarized in Fig. 3.14. At all longitudes, the strongest mean winds are observed in the winter season and the weakest in summer. Also, the west-wind belt is farthest north in summer and farthest south in winter.[8]

Lahey *et al.* (1960) provide geostrophic-wind analyses at 300 mb for individual months; from these it is possible to examine the seasonal changes in more detail. Latitudes and strengths of the 300-mb wind are plotted in the lower part of Fig. 3.14 for three longitude regions where the greatest mean speeds are attained. In the west Pacific-Japan area wind maximum (inset VI), the seasonal trends of both speed and latitude are very regular, the speed variation being much greater than is observed elsewhere in the hemisphere (essentially the same behavior is observed at the *fixed* longitude 140°E). By contrast, over eastern North America (IV) the latitudinal shifts are fairly rapid both in spring and autumn. A marked eastward displacement of the wind maxima (by about 30° longitude) takes place, along with the northward movement of the wind belt between winter and summer, in regions IV and VI where winds are strongest. Near the west coasts of both North America and Europe (insets I and II), two wind maxima are evident at well-separated latitudes, as in Fig. 3.8b, except in midsummer.

---

[8] An exception is the longitudes of North America, where the strongest mean winds are farther south in spring than in winter.

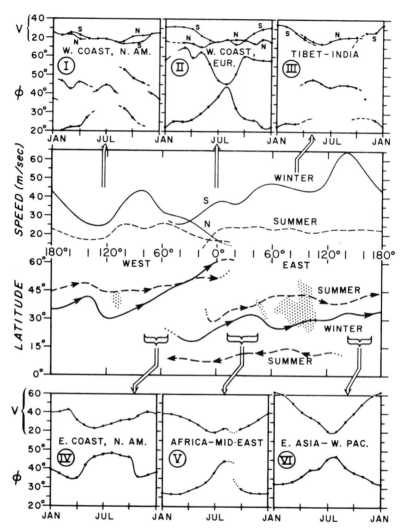

FIG. 3.14 Middle charts show total speed at level of maximum mean wind, and latitude of strongest winds, in winter (December–February) and summer (June–August) of Northern Hemisphere. Based on Crutcher's data (1961). Upper and lower diagrams show month-to-month variations of 300-mb geostrophic wind speed and latitude of maximum wind belt(s) at certain longitudes. Where indicated by brackets, the longitude of the strongest mean wind varies between winter and summer. Where wind maxima exist at two latitudes, profiles for the northern or southern one are marked N or S. (Based on mean maps by Lahey et al., 1960.)

Both over Asia and North America (central part of Fig. 3.14), there is a distinct preference for the jet stream to pass either north or south of the highest terrain. In the longitudes of Tibet and India (inset III), this preference for a favored low latitude during the cooler months and a high latitude in the warmer months is strikingly evident.[9] At longitude 120°W, over the Rocky Mountains, a similar large shift takes place northward from latitude 30° to 47° in May to July and southward from 50° to 29° in October and November.

At individual stations upstream from the Hindu Kush-Himalaya plateau region, Sutcliffe and Bannon (1954) found that in the spring transition season the winds change from a regime of steady westerlies to one of steady easterlies within a period of a few days. The indicated sudden northward movement of both the tropical easterly jet and the subtropical westerly jet, which occurred regularly within a three-week period of the six years they studied, was found to be closely allied with the onset of the Indian monsoon on the Malabar coast in late May to early June. An example of a similarly rapid change of the wind regime over northern India during the retreat of the monsoon (October) is given by Reiter and Heuberger (1960). The connection between the "burst of the monsoon" and the collapse of the westerly jet stream in subtropical latitudes was first noted by Yin (1949). Yin ascribed a westward shift of the trough over the Bay of Bengal (which occurred at the same time, and made conditions more favorable for the invasion of Burma and eastern India by low-tropospheric moist air) to a change in the way the westerlies are deflected by the Himalayas when the current shifts from the south to the north side.

Yeh *et al.* (1959), in a study of the circulation over and east of the plateau, show meridional sections that strikingly illustrate the abrupt northward shift of the wind systems (which they relate to the onset of the India-Burma monsoon and the spring rains in China and Japan) and the southward shift in October. They indicate that these changes are followed by similar rapid latitudinal migrations over North America (inset IV, Fig. 3.14). Radok and Grant (1957) present evidence suggesting that a decrease of upper-tropospheric winds over Australia at low latitudes, accompanied by an increase in middle latitudes, occurs rapidly in spring; it has not been established whether this is a hemisphere-wide feature. They indicate that this generally occurs close to the time of a rapid transition of the opposite sort in the Northern Hemisphere.

[9] It should be noted that although as early as April the average westerlies are stronger north of Tibet, mean westerlies persist over north India into June. These have an important effect on the character of the weather systems (Section 14.10).

3.5 DISTRIBUTION OF CYCLONES AND ANTICYCLONES

In a broad fashion, the occurrence of cyclones and anticyclones can be related to the large-scale circulations described above. The physical relationships as they apply to individual synoptic systems will be discussed in Chapters 6 and 11.

Petterssen (1950) has mapped the geographical frequencies of cyclones and anticyclones, and of the genesis of both, in the Northern Hemisphere as determined from 40 years of maps. The meridional distribution, averaged around latitude circles, is shown in Fig. 3.15. The greatest frequencies of both cyclogenesis and anticyclogenesis are centered around 38°N in winter

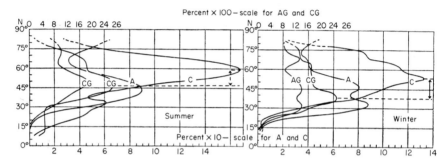

FIG. 3.15 Mean meridional distribution of percentage frequencies of anticyclogenesis (AG), cyclogenesis (CG), cyclones (C), and anticyclones (A) for Northern Hemisphere summer and winter. (After Petterssen, 1950.)

and 48°N in summer, roughly the latitudes of the strongest mid-tropospheric westerlies (Fig. 3.12). Comparing the shapes of curves *CG* and *C* and of curves *AG* and *A*, it is seen that in both seasons there is a strong tendency for cyclones to move poleward after formation, and a lesser tendency for anticyclones to move equatorward.[10] A similar result for cyclones is found in the Southern Hemisphere (van Loon, 1965). In both hemispheres and in different seasons the greatest frequency of cyclones is about 15° poleward of the latitude of most frequent cyclogenesis.

In his comparison of the circulations of the hemispheres, Gibbs (1953) found significant differences in the distributions of cyclones and anticyclones. These differences are borne out by the recent results of Taljaard (1967), who

---

[10] This comparison is not strictly conclusive, since the same effect on the statistics could result from longer lives of cyclones in high latitudes than in low latitudes, and from longer lives of anticyclones in low than in high latitudes. However, synoptic experience indicates that these tendencies are true ones.

gives a very detailed account of the behavior of Southern Hemisphere synoptic systems based on carefully analyzed maps during the IGY. Southern anticyclones are more concentrated in subtropical latitudes, with much smaller frequencies of closed anticyclones in higher latitudes than those shown in Fig. 3.15. The anticyclone frequency is near zero over much of the ocean region between 45°S and the Antarctic coast, although Antarctica itself has a predominantly anticyclonic circulation. Cyclogenesis frequencies are fairly uniform between seasons in most latitudes. Taljaard's maps indicate clearly that "the Antarctic Ocean is a region with many mature and occluded cyclones rather than the cradle of new systems." These cyclones originate in subtropical or lower middle latitudes and move poleward, with marked preference for two spiral paths across the Pacific and the Atlantic and Indian oceans.

When longitudinal as well as latitudinal variations are considered, the distributions of the circulation centers in the Northern Hemisphere become exceedingly complex, although as shown by Petterssen (1950), these complications can be related physically to topographic features (mountains, land-sea distribution, and large water bodies that act as heat and cold sources in different seasons). In the Southern Hemisphere, while there is an overall organization in the patterns of cyclone and anticyclone distribution, the embedded irregularities are much more marked than might be expected from the comparatively uniform circumpolar flow in Figs. 3.3 and 3.4. Taljaard describes the connection of these with the topographic features.

In addition to these more local influences, pronounced variations on a hemispheric scale are connected with variations in latitude of the maximum-wind belt. Barrett (1961) has described a planetary-scale asymmetry of the mean wind system, which he connects theoretically with the very large-scale orographic structure of the Northern Hemisphere. In winter, the axis of the belt of strongest mean wind (Fig. 3.16) spirals poleward in a cyclonic fashion, starting from around 19°N at 30°W longitude, and increasing in latitude until it is found at 52°N at this same longitude after passing around the hemisphere. This spiral is not quite regular; rather, there is a southward jog over the eastern Pacific, and the northward trend is much more pronounced over the Atlantic than elsewhere. As indicated by Fig. 3.14, this spiral is present also in summer, although less pronounced. A similar spiral is evident in the Southern Hemisphere in winter (van Loon, 1964). The regions where the ends of the mean "jet stream" overlap, over the eastern North Atlantic and southeast of Australia, are identified with the longitudes of greatest frequency of blocking highs in the respective hemispheres.

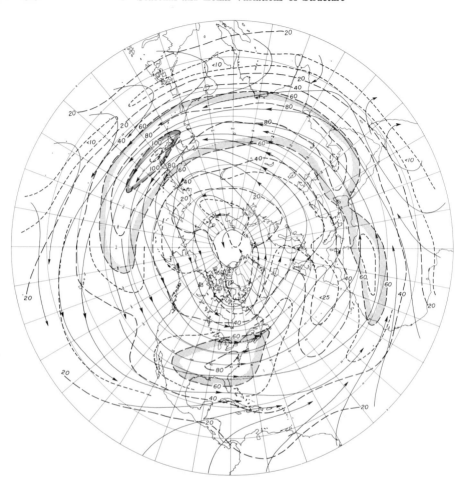

Fɪɢ. 3.16   Streamlines at surface of mean maximum wind, and isotachs (dashed, knots; stippled where vector mean speed exceeds 60 knots) in winter of Northern Hemisphere. Analyzed by Ralph L. Coleman from data in Crutcher (1961).

For reasons discussed in Chapter 6 and elsewhere, extratropical disturbances form most readily in the strongly barocline region near the jet stream. Hence it would be expected that the giant spirals mentioned above would be reflected in the frequency distribution of cyclones and anticyclones, at least in a broad fashion. This feature is apparent in Fig. 3.17, on which is drawn the axis of the mean maximum wind and the principal tracks of cyclones and anticyclones in winter. Two major cyclone tracks exist, from the southwestern to the northeastern parts of the Atlantic and Pacific oceans. On the average, these cyclones originate and move in the

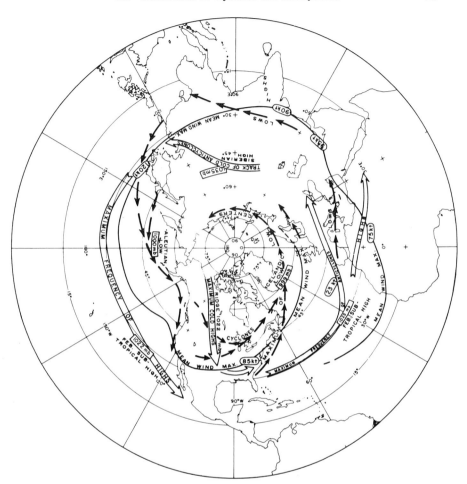

FIG. 3.17 Axis of mean maximum wind (heavy line; taken from Fig. 3.16) and simplified axes of maximum frequency of occurrence of cyclones (short arrows) and anticyclones (double-shafted arrows) based on the maps by Petterssen (1950) for the winter season. Rectangles show mean sea-level pressures in principal semipermanent low and high centers, and ellipses show regions of maximum upper winds according to Crutcher's sections (1961).

general southwesterly currents east of the mean troughs near the continental east coasts (Fig. 3.1). Note in both cases that the cyclone tracks start on the equatorward side of the axis of maximum mean upper wind and deviate toward the poleward side farther downstream. This behavior can be partly interpreted in light of the life histories of individual cyclones. Young wave cyclones (generally those farthest southwest) are typically located on the warm-air side of the jet stream. As a cyclone matures, and especially as it becomes occluded, the sea-level circulation center deviates increasingly to left of the jet stream. As discussed in Section 12.1, this change is due to the evolving thermal structure as the cyclone matures.

Other principal cyclone tracks exist over continental regions, each being affected by regional topography. Cyclones over North America in winter weaken as they pass inland from the Pacific and regenerate east of the Rockies, particularly in the Alberta and Colorado vicinities (see Section 11.8), thence moving on the average over the Great Lakes and Labrador into the Atlantic Ocean. The great frequency of winter cyclones over the Mediterranean is influenced by the heat source in this body of water (which also accounts for the preference of anticyclones to move either south or north of, but not over, this sea). These cyclones sometimes move north of the Eurasian mountain complex, in which event they mostly perish east of the Caspian Sea. On other occasions, they move eastward south of the mountains. Since it is blocked by the east-west mountain chain, cold air from the continental interior cannot enter these disturbances in lower levels. Real tropical air does not enter them either because of the dominance of the northeast monsoon circulation. Consequently, these westerly disturbances develop only feeble air-mass contrasts and generally remain weak in terms of the strengths of their circulations as well as in the amount of precipitation produced.

Cold highs migrate southeastward from the higher-latitude continental interiors, where their formation is influenced by intense low-level cooling.[11] Note that the two mean maximum-wind regions near the continental east coasts (Fig. 3.16) are located east of the regions of frequent cold-air outbreaks, reflecting the role of these in continually reinforcing the temperature contrast. This is an essential part of the mechanism of confluence, discussed in Section 8.5 in relation to the broad-scale upper currents.

Around latitude 30°N, anticyclogenesis is most frequent near the west sides of both oceans, hand in hand with the genesis of cyclones in the same region. Such anticyclones appear as offshoots of the continental highs; they typically migrate eastward and regenerate the semipermanent subtropical highs of the eastern oceans.

Petterssen (1950) discusses the role of atmospheric disturbances in the production and destruction of vorticity, and in its transport, in relation to

---

[11] In corresponding latitudes of the Southern Hemisphere, Taljaard (1967) observes that polar anticyclones rarely cross the ocean to join the subtropical anticyclones. However, the latter are commonly rejuvenated by closed systems, which "form in intensifying ridges consisting of strong surges of cold air in the rear of the last members of cyclone families."

the land-sea distribution and heat sources and sinks. Viewed in a very broad sense, it is evident that a poleward transfer of vorticity across the middle latitudes is necessary to maintain the mean state of the atmosphere because surface friction acts to destroy vorticity in the cyclone belts of the higher middle latitudes. At the same time, in the latitudes of the subtropical anticyclones, vorticity is generated by surface friction. Eady (1950), Kuo (1951), and Kao (1955) have given theoretical discussions of the vorticity transfer, which, as they point out, is related to the angular-momentum transfer discussed in Chapter 1. The production of absolute angular momentum in the surface easterlies south of the subtropical highs and its destruction in the westerlies to the north is tantamount to a destruction of anticyclonic relative vorticity in the subtropical belt or to the production of vorticity. Likewise, the destruction of angular momentum south of the subpolar cyclone belt and its production in the zone of easterlies farther north is equivalent to the destruction of vorticity.

In Fig. 3.17 it is seen that the tendency for cyclones to move to the left of the axis of mean maximum wind is coupled with a tendency for anticyclones to move toward the right of that axis. The systematic shedding of vortices in this manner expresses an eddy transfer of vorticity from source to sink regions (Petterssen, 1950) or, considered in light of the preferred equatorward movement of anticyclones and poleward movement of cyclones, a transfer up the gradient from latitudes where the vorticity is low to those where it is high (Kuo, 1951). This behavior of the circulation systems is, of course, partly connected with the fact that cyclones are most frequently under the influence of southwesterly currents downstream from troughs aloft, while anticyclones tend to be "steered" by the northwesterly currents in their rear. Thus, the planetary wave patterns play an essential role in the required vorticity transfer across the mean current.

REFERENCES

Austin, J. M., and Krawitz, L. (1956). 50-millibar patterns and their relationship to tropospheric changes. *J. Meteorol.* **13**, 152–159.

Barrett, E. W. (1961). Some applications of harmonic analysis to the study of the general circulation, I. *Beitr. Physik Atmosphäre* **33**, 280–332.

Batten, E. S. (1961). Wind systems in the mesosphere and lower ionosphere. *J. Meteorol.* **18**, 283–291.

Belmont, A. D., and Dartt, D. G. (1966). The non-repeating variations of the observed wind in the equatorial stratosphere. *Tellus* **18**, 381–390.

Berkofsky, L., and Bertoni, E. A. (1955). Mean topographic charts for the entire earth. *Bull. Am. Meteorol. Soc.* **36**, 350–354.

Bjerknes, J. (1964). Atlantic air-sea interaction. *Advan. Geophys.* **10**, 1–82.

Bolin, B. (1950). On the influence of the earth's orography on the westerlies. *Tellus* **2**, 184–195.

Boville, B. W., Wilson, C. V., and Hare, F. K. (1961). Baroclinic waves of the polar-night vortex. *J. Meteorol.* **18**, 567–580.

Budyko, M. I. (1963). "Atlas Teplovogo Balansa Zemnogo Shara," (Atlas of Heat Balance of the Earth's Surface), 69 pp. U.S.S.R. Glavnaia Geofizicheskaia Observatoriia, Moscow.

Charney, J. G., and Eliassen, A. (1949). A numerical method for predicting the perturbations of the middle latitude westerlies. *Tellus* **1**, No. 2, 38–55.

Cole, A. E., and Kantor, A. J. (1963). Air Force interim supplemental atmosphere to 90 km. *Air Force Surveys in Geophysics*, No. 153, 29 pp. AFCRL-63-936. Air Force Cambridge Res. Lab., Bedford, Massachusetts.

Craig, R. A. (1965). "The Upper Atmosphere—Meteorology and Physics," 509 pp. Academic Press, New York.

Crutcher, H. L. (1961). Meridional cross sections. Upper winds over the Northern Hemisphere. Tech. Paper No. 41, 307 pp. U. S. Weather Bur., Washington, D. C.

Eady, E. T. (1950). The cause of the general circulation of the atmosphere. *Cent. Proc. Roy. Meteorol. Soc.* pp. 156–172.

Essenwanger, O. (1953). Statistische Untersuchungen über die Zirkulation der Westdrift in 55° Breite. *Ber. Deut. Wetterdienstes* **7**, 1–22.

Flohn, H. (1960). Recent investigations on the mechanism of the "Summer Monsoon" of southern and eastern Asia. *In* "Symposium on Monsoons of the World" (S. Basu, P. R. Pisharoty, K. R. Ramanathan, U. K. Bose, eds.), pp. 75–88. India Meteorol. Dept., Delhi.

Frenzen, P. (1955). Westerly flow past an obstacle in a rotating hemispherical shell. *Bull. Am. Meteorol. Soc.* **36**, 204–210.

Fultz, D., and Long, R. R. (1951). Two-dimensional flow around a circular barrier in a rotating spherical shell. *Tellus* **3**, 61–68.

Gabites, J. F. (1953). Mean westerly wind flow in the upper levels over the New Zealand region. *New Zealand J. Sci. Technol.* **B34**, 384–390.

Gibbs, W. J. (1953). A comparison of hemispheric circulations with particular reference to the western Pacific. *Quart. J. Roy. Meteorol. Soc.* **79**, 121–136.

Goldie, N., Moore, J. G., and Austin, E. E. (1958). Upper air temperature over the world. *Geophys. Mem.* **13**, No. 101, 1–228.

Gutman, G. J., and Schwerdtfeger, W. (1965). The role of latent heat and sensible heat for the development of a high pressure system over the subtropical Andes in the summer. *Meteorol. Rundschau* **18**, 1–7.

Heastie, H., and Stephenson, P. M. (1960). Upper winds over the world, Parts I and II. *Geophys. Mem.* **13**, No. 103, 1–217.

Jacobs, I. (1958). 5-bzw. 40 jährige Monatsmittel der absoluten Topographien der 1000 mb-, 850 mb-, 500 mb- and 300 mb-Flächen sowie der relativen Topographien 500/1000 mb und 300/500 mb über der Nordhemisphäre und ihre monatlichen Änderungen. Folge 2. *Meteorol. Abhandl., Inst. Meteorol. Geophys. Freien Univ. Berlin* **4**, No. 2, 47 pp. (Part I); 121 pp. (Part II).

Jacobs, W. C. (1951). Large-scale aspects of energy transformation over the oceans. *In* "Compendium of Meteorology" (T. F. Malone, ed.), pp. 1057–1070. Am. Meteorol. Soc., Boston, Massachusetts.

Julian, P. R., and Labitzke, K. B. (1965). A study of atmospheric energetics during the January-February 1963 stratospheric warming. *J. Atmospheric Sci.* **22**, 597–610.

Julian, P. R., Krawitz, L., and Panofsky, H. A. (1959). The relation between height patterns at 500 mb and 100 mb. *Monthly Weather Rev.* **87**, 251–260.

Kao, S.-K. (1955). On total momentum vorticity with application to the study of the general circulation of the atmosphere (Article XII). Final Rept. Gen. Circ. Proj., Contr. AF 19(122)-48. Dept. Meteorol., Univ. California, Los Angeles.

Kasahara, A. (1966). The dynamical influence of orography on the large-scale motion of the atmosphere. *J. Atmospheric Sci.* **23**, 259–271.

Kellogg, W. W. (1964). Meteorological soundings in the upper atmosphere. *World Meteorol. Organ., Tech. Note* No. 60, 1–46.

Kellogg, W. W., and Schilling, G. F. (1951). A proposed model of the circulation in the upper stratosphere. *J. Meteorol.* **8**, 222–230.

Klein, W. H. (1958). The frequency of cyclones and anticyclones in relation to the mean circulation. *J. Meteorol.* **15**, 98–102.

Kochanski, A. (1955). Cross sections of the mean zonal flow and temperature along 80°W. *J. Meteorol.* **12**, 95–106.

Kochanski, A. (1963). Circulation and temperatures at 70- to 100-kilometer height. *J. Geophys. Res.* **68**, 213–226.

Kuo, H.-L. (1951). Vorticity transfer as related to the development of the general circulation. *J. Meteorol.* **8**, 307–315.

Lahey, J. F., Bryson, R. A., Wahl, E. W., Horn, L. H., and Henderson, V. D. (1958). "Atlas of 500 mb Wind Characteristics for the Northern Hemisphere." Univ. of Wisconsin Press, Madison, Wisconsin.

Lahey, J. F., Bryson, R. A., Corzine, H. A., and Hutchins, C. W. (1960). "Atlas of 300 mb Wind Characteristics for the Northern Hemisphere." Univ. of Wisconsin Press, Madison, Wisconsin.

McClain, E. P. (1960). Thermal conditions in the Arctic stratosphere in January. *J. Meteorol.* **17**, 383–389.

Martin, D. E. (1953). Anomalies in the Northern Hemisphere 700-mb 5-day mean circulation patterns. AWSTR 105-100. Air Weather Service, U. S. Air Force, Washington, D. C.

Matsumoto, S., Itoo, H., and Arakawa, A. (1953). On the monthly mean distribution of temperature, wind and relative humidity of the atmosphere over Japan from March 1951 to February 1952. *J. Meteorol. Soc. Japan* **31**, 248–258.

Mintz, Y. (1965). Very long-term global integration of the primitive equations of atmospheric motion. *World Meteorol. Organ., Tech. Note* No. 66, 141–167.

Miyakoda, K. (1963). Some characteristic features of winter circulation in the troposphere and lower stratosphere. Tech. Rept. No. 14, Grant NSF-GP-471, 93 pp. plus figs. Dept. Meteorol., Univ. Chicago.

Mohri, K. (1953). On the fields of wind and temperature over Japan and adjacent waters during the winter of 1950-1951. *Tellus* **5**, 340–358.

Murgatroyd, R. J. (1957). Winds and temperatures between 20 km and 100 km. A review. *Quart. J. Roy. Meteorol. Soc.* **83**, 417–458.

Murgatroyd, R. J., and Singleton, F. (1961). Possible meridional circulations in the stratosphere and mesosphere. *Quart. J. Roy. Meteorol. Soc.* **87**, 125–135.

Murgatroyd, R. J., Hare, F. K., Boville, B. W., Teweles, S., and Kochanski, A. (1965). The circulation in the stratosphere, mesosphere and lower thermosphere. *World Meteorol. Organ., Tech. Note* No. 70, 1–206.

Namias, J. (1962). Influences of abnormal surface heat sources and sinks on atmospheric behavior. *In* "Proceedings of the International Symposium on Numerical Weather Prediction in Tokyo" (S. Syono, ed.), pp. 615–627. Meteorol. Soc. Japan, Tokyo.

Namias, J., and Clapp, P. F. (1949). Confluence theory of the high-tropospheric jet stream. *J. Meteorol.* **6**, 330–336.

Newell, R. E. (1963). Transfer through the tropopause and within the stratosphere. *Quart. J. Roy. Meteorol. Soc.* **89**, 167–204.

Oort, A. H. (1964). On the energetics of the mean and eddy circulations in the lower stratosphere. *Tellus* **16**, 309–327.

Panofsky, H. A. (1961). Temperature and wind in the lower stratosphere. *Advan. Geophys.* **7**, 215–247.

Petterssen, S. (1950). Some aspects of the general circulation of the atmosphere. *Cent. Proc. Roy. Meteor. Soc.* pp. 120–155.

Pogosian, Kh. P. (1965). "Seasonal and Intraseasonal Variations of Temperature Geopotential and Atmospheric Circulation in the Stratosphere," 109 pp. Publishing House "Nauka," Moscow (in Russian).

Radok, U., and Grant, A. M. (1957). Variations in the high tropospheric mean flow over Australia and New Zealand. *J. Meteorol.* **14**, 141–149.

Reed, R. J. (1965). The present status of the 26-month oscillation. *Bull. Am. Meteorol. Soc.* **46**, 374–387.

Reed, R. J. (1966). Zonal wind behavior in the equatorial stratosphere and lower mesosphere. *J. Geophys. Res.* **71**, 4223–4233.

Reed, R. J., Campbell, W. J., Rasmussen, L. A., and Rogers, D. G. (1961). Evidence of a downward-propagating annual wind reversal in the equatorial stratosphere. *J. Geophys. Res.* **66**, 813–818.

Reiter, E. R., and Heuberger, H. (1960). Jet stream and retreat of the Indian summer monsoon and their effect upon the Austrian Cho-Oyu expedition 1954. *Geograph. Ann.* **42**, 17–35.

Reuter, F. (1936). Die synoptische Darstellung der halbjährigen Druckwelle. *Veroeffentl. Geophysik. Inst. Univ. Leipzig* **7**, 257–295.

Riehl, H. (1962). Jet streams of the atmosphere. Tech. Rept. No. 32, 117 pp. Dept. Atmospheric Sci., Colorado State Univ., Fort Collins, Colorado.

Scherhag, R. (1952). Die Explosionsartigen Stratosphärenerwärmungen des Spätwinters 1951/1952. *Ber. Deut. Wetterdienstes U. S. Zone* **38**, 51–63.

Schwerdtfeger, W. (1960). The seasonal variation of the strength of the southern circumpolar vortex. *Monthly Weather Rev.* **88**, 203–208.

Schwerdtfeger, W., and Prohaska, F. (1956). Der Jahresgang des Luftdrucks auf der Erde und seine halbjährige Komponente. *Meteorol. Rundschau* **9**, 33–43.

Sheppard, P. A. (1963). Atmospheric tracers and the study of the general circulation of the atmosphere. *Rept. Progr. Phys.* **26**, 213–267.

Smagorinsky, J. (1953). The dynamical influence of large-scale heat sources and sinks on the quasi-stationary mean motions of the atmosphere. *Quart. J. Roy. Meteorol. Soc.* **79**, 342–366.

Sutcliffe, R. C. (1951). Mean upper contour patterns of the Northern Hemisphere—the thermal-synoptic viewpoint. *Quart. J. Roy. Meteorol. Soc.* **77**, 435–440.

Sutcliffe, R. C., and Bannon, J. K. (1954). Seasonal changes in upper-air conditions in the Mediterranean-Middle East area. *In* "Scientific Proceedings, International Association of Meteorology," pp. 322–334. U.G.G.I., Rome. Butterworth, London and Washington, D.C.

Taljaard, J. J. (1967). Development, distribution, and movement of cyclones and anti-cyclones in the Southern Hemisphere during the IGY. *J. Appl. Meteorol.* **6**, 973–987.

Taljaard, J. J., van Loon, H., Crutcher, H. L., and Jenne, R. L. (1969). Climate of the upper air. Vol. I. Charts of monthly mean pressure, temperature, dew point, and geopotential in the Southern Hemisphere. NAVAIR 50-IC-55. Off. Chief of Naval Ops., Washington, D.C. In press.

Tucker, G. B. (1960). Upper winds over the world, Part III. *Geophys. Mem.* **13**, No. 105, 1–101.

U. S. Navy (1959). Upper wind statistics of the Northern Hemisphere, Vols. I and II. NAVAER 50-1C-535. Off. Chief of Naval Ops., Washington, D.C.

U. S. Weather Bureau. (1952). Normal weather charts for the Northern Hemisphere. Tech. Paper No. 21. U. S. Weather Bur., Washington, D.C.

U. S. Weather Bureau. (1962). Daily aerological cross sections pole to pole along meridian 75°W for the IGY period. U. S. Weather Bur., Washington, D.C.

van Loon, H. (1964). Mid-season average zonal winds at sea level and at 500 mb south of 25 degrees south, and a brief comparison with the Northern Hemisphere. *J. Appl. Meteorol.* **3**, 554–563.

van Loon, H. (1965). A climatological study of the atmospheric circulation in the Southern Hemisphere during the IGY, Part I: 1 July 1957-31 March 1958. *J. Appl. Meteorol.* **4**, 479–491.

van Loon, H. (1966). On the annual temperature range over the southern oceans. *Geograph. Rev.* **56**, 497–515.

van Loon, H. (1967). The half-yearly oscillations in middle and high southern latitudes and the coreless winter. *J. Atmospheric Sci.* **24**, 472–486.

van Loon, H., and Jenne, R. L. (1969). The half-yearly oscillations in the tropics of the Southern Hemisphere. *J. Atmospheric Sci.* **26**. In press.

van Loon, H., and Taljaard, J. J. (1958). A study of the 1000-500 mb thickness distribution in the southern hemisphere. *Notos* **7**, 123–158.

Veryard, R. G., and Ebdon, R. A. (1961). Fluctuations in tropical stratospheric winds. *Meteorol. Mag.* **90**, 125–143.

Webb, W. L. (1966). "Structure of the Stratosphere and Mesosphere," 382 pp. Academic Press, New York.

Wege, K., Leese, H., Groening, H. U., and Hoffmann, G. (1958). Mean seasonal conditions of the atmosphere at altitudes of 20 to 30 km and cross sections along selected meridians in the Northern Hemisphere. *Meteor. Abhandl., Inst. Meteorol. Geophys. Freien Univ. Berlin* **6**, No. 4, 28 pp. (Part I); 101 pp. (Part II).

Yeh, T.-C., and Chu, P.-C. (1958). "Some Fundamental Problems of the General Circulation of the Atmosphere," 159 pp. Inst. Geophys. Meteorol., Academia Sinica, Peking (in Chinese, extended English abstract, pp. 147–156).

Yeh, T.-C., Dao, S.-Y., and Li, M.-T. (1959). The abrupt change of circulation over the Northern Hemisphere during June and October. *In* "The Atmosphere and the Sea in Motion" (B. Bolin, ed.), pp. 249–267. Rockefeller Inst. Press, New York.

Yin, M. T. (1949). A synoptic-aerologic study of the onset of the summer monsoon over India and Burma. *J. Meteorol.* **6**, 393–400.

4

PRINCIPAL AIR MASSES AND FRONTS,
JET STREAMS, AND TROPOPAUSES

The thermal structure and field of motion of the atmosphere at a given synoptic time are generally very different from the mean conditions discussed in the previous chapters. This is especially true for the extratropical regions with their strong transient disturbances. But also in lower latitudes, averaging over long time periods is likely to suppress many characteristic and significant features of the thermal structures and flow patterns that appear at given synoptic times.

Two principal aims of synoptic meteorology are to recognize among the practically unlimited individual synoptic situations the most important and typical features, and to distill from these a systematic description of atmospheric processes. Some of the earlier steps in this direction, in which the Austrian-German school and British meteorologists played particularly important roles, are recounted briefly in Section 5.1. Further decisive contributions to progress resulted from introduction of the concepts of *air masses* and *fronts* by the "Norwegian school" (V. Bjerknes *et al.*, 1933) and of *jet streams* by the "Chicago school" (University of Chicago, 1947).

By using air masses, fronts, and jet streams as characteristic features of the atmospheric structure and flow patterns, some order was brought into the chaos of the manifold processes with which synoptic meteorology has to be concerned. The development of the polar-front theory and the concept of jet streams will be elaborated in subsequent chapters. Here, however, some of the essential features of these atmospheric phenomena will be anticipated and described in the simplest possible framework.

4.1 PRINCIPAL AIR MASSES

The properties of the air depend upon radiation and upon the exchange of heat and moisture with the underlying earth's surface, to which the air has been subjected. In this respect one can distinguish between air masses with typical properties defined by geographical locations, or "*source regions*," where the properties have been acquired, provided the air has stayed sufficiently long in these regions to take on the characteristic properties.

Air masses have been divided into many classes, depending on the source region and subsequent life history of the air in motion. Reference may be made here to the Northern Hemisphere air-mass classification given, e.g., by Bergeron (1928), Schinze (1932), and Petterssen (1940), and to the detailed studies of North American air masses by Willett (1933) and by Showalter (1939). By considering the large-scale advective processes in the atmosphere and the modifications of the different air properties due to radiation, exchange of heat and moisture between the earth's surface and the atmosphere, condensation processes, etc., the subdivision of air masses can be carried very far. Since, however, the purpose of this chapter is to deal only with a few of the most characteristic features of the atmospheric structure and flow patterns on a global or at least a hemispheric scale, we shall restrict the air-mass classification accordingly.

Most of the original air-mass classifications were deduced from analyses of synoptic surface maps and were strongly influenced by the pressure and flow patterns in very low levels. However, considering the strong vertical wind shear in many parts of the globe, and the height variation of the wind direction, the air at different levels in a given column may have come from widely different regions in the recent past. Owing to this, and to the great variations of the height to which the influence of the earth's surface can penetrate directly, it is often very difficult to classify different layers of an air column according to their original source regions and subsequent life histories. Examples illustrating this difficulty will be presented later, e.g., in the synoptic discussion of the different air masses of extratropical cyclones.

For our purpose it therefore seems convenient to divide the hemispheric air masses into only three principal classes: *tropical air (TA), middle-latitude air (MLA), and polar air (PA)*. The source regions of TA are the trade-wind zones of the Northern and Southern hemispheres, bounded poleward by the subtropical high-pressure belts of both hemispheres. The principal source regions of PA are the arctic and subarctic regions in the Northern Hemisphere and the antarctic and subantarctic regions of the

Southern Hemisphere. The equatorward extensions of these source regions are not very well marked and undergo large seasonal variations. The source regions of MLA are therefore also not well defined. This air mass must necessarily be the least homogeneous, since its source regions are characterized by large meridional variations in insolation and in the thermal influences from the earth's surface, as well as large longitudinal variations of the latter influences due to the continents and oceans.

In the original polar-front theory, the polar front was considered as the boundary between polar air and tropical (or subtropical) air. The actual position of this front on synoptic maps then showed the poleward extension of the tropical or subtropical air masses. The recognition of the relatively closed tropical circulation of the Hadley type and the reversed extratropical "Ferrel circulation," and of the existence of two principal jet streams in both hemispheres, have made it physically significant to distinguish between the real tropical air and the middle-latitude air, which has properties intermediate between those of TA and PA. This distinction is not so clear in the lowest atmospheric layers, but appears natural in the middle and upper troposphere with its strong "meandering" westerly current.

Some valuable information about the principal air masses can be deduced from the scheme of the mean circulation discussed in Chapter 1. In the poleward upper boundary region of the Hadley cell, the circulation pattern is characterized by a mean meridional convergence that contributes to concentration of thermal properties. A similar but less pronounced convergence zone appears near the poleward boundary of the Ferrel cell. There is, however, a striking difference in the vertical positions of these zones of convergence.

To illustrate this, we reproduce (Figs. 4.1 and 4.2) the three-cell meridional circulation model[1] after Rossby (1941) and the modified model by Palmén (1951). According to these models, the air-mass classification outlined above seems the most natural one. Owing to the low-level divergence in the boundary region between the Hadley and Ferrel cells, no well-marked air-mass boundary should be expected in the lower troposphere. A contrast

---

[1] In its general aspects, this resembles an earlier model by Bergeron (1928, 1930), who also indicated the frontogenetical aspects connected with the circulation cells. The three-cell meridional circulation scheme was originated by Ferrel (1856), who discussed it in terms of certain "forces" identifiable with the conservation of absolute angular momentum in the various branches. Ferrel also related the scheme to the belts of subtropical highs, the trade convergence zone, and "another meeting of the air at the surface near the polar circles similar to the one at the equator," where the air rises in the cyclone belts.

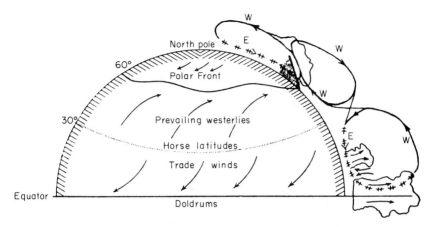

Fig. 4.1 The three-cell meridional circulation scheme, after Rossby (1941), with convection in the equatorial zone, subsidence in the horse latitudes, ascent in the polar-front zone (shown without disturbances), and subsidence near the pole. W and E denote branches with general westerly and easterly winds.

between TA and MLA is, however, favored in the upper troposphere, where there is meridional convergence between these circulation cells. On the same grounds, the contrast between MLA and PA on the poleward edge of the Ferrel cell could be expected to be relatively distinct in the low and middle troposphere and less distinct higher up. According to the model in Fig. 4.2, the aforementioned quasi-permanent air-mass boundaries should also be characterized by belts of strong zonal circulation and more-or-less pronounced discontinuities in the tropopause. This led to the assumption of the regular existence of two hemispheric jet streams, which have been termed the "polar-front jet stream" (or, for brevity, "polar jet stream") and the "subtropical jet stream," situated close to the tropopause breaks.

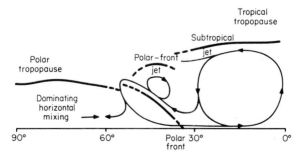

Fig. 4.2 Scheme for the mean meridional circulation in winter according to Palmén (1951), showing the principal tropopauses and jet streams.

The air in the middle and upper troposphere, in middle latitudes, circulates as a meandering stream around the hemisphere. Most of this current has not recently been under the influence of tropical conditions, and hence cannot properly be considered as tropical air. As indicated by Fig. 4.3, however, regular injections of TA occur on the western sides of the subtropical high-pressure cells, where low-level tropical air moves poleward as an ascending current until it reaches thermal equilibrium with the middle- and upper-tropospheric currents meandering in the temperate zone. This represents a regular transformation of TA into MLA. Because of this reasoning, we feel it is more correct to consider the air mass equatorward of the polar front as a middle-latitude mass rather than a tropical mass, despite the fact that in some preferred regions it is strongly intermixed with low-level TA.[2]

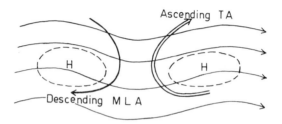

FIG. 4.3 Schematic upper-level streamlines (solid lines) and general locations of sea-level anticyclones (dashed lines). Heavy arrow shows general trajectory of descending middle-latitude air; double-shafted arrow is trajectory of ascending tropical air. The westward slope of the pressure systems is not taken into account.

Within the middle-latitude zone, the lower layers of the atmosphere are subject to frequent poleward incursions of TA and equatorward incursions of PA which, despite modifications due to energy exchange with the underlying surface and to vertical motions, retain to a large extent the properties of their source regions. Because of the influence of the meridional temperature gradient on the variation of wind with height, the trajectories emanating more or less directly from the tropical and polar source regions in low levels give over to the meandering westerly current of the upper levels. Thus, *a sounding in the middle-latitude belt may show TA or PA in the low troposphere, with overlying MLA that has resided in middle latitudes for a long time.* This characteristic (see Chapter 13) has an important influence on the stability properties of the composite air mass, which vary greatly with time and location.

---

[2] This intermixing of TA with MLA will be discussed in detail in Chapter 10, where the role of migratory disturbances is emphasized.

It should be emphasized that any air-mass classification necessarily must be somewhat arbitrary. On the other hand, synoptic experience shows that the regular appearance of air-mass boundaries is very characteristic of the behavior of the atmosphere. The most consequential attempt to classify the principal air masses of the Northern Hemisphere according to the principles outlined above has been made by Defant and Taba (1957, 1958a,b). In developing their classification, they considered all soundings at selected synoptic times in relation to careful analyses of fronts, jet streams, and tropopauses, arriving at the scheme in Fig. 4.4. In accord with the preceding

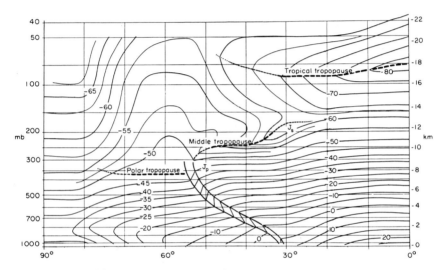

FIG. 4.4 Meridional cross section showing mean conditions around the Northern Hemisphere on Jan. 1, 1956, averaged relative to the principal zones of baroclinity. J_p and J_s denote locations of polar and subtropical jet streams. Isotherms in degrees Celsius. (After Defant and Taba, 1957.)

principles, the polar front is regularly associated with a jet stream in the upper troposphere and with a break in the tropopause or at least a narrow region of indistinct tropopause (marked also in Fig. 4.2). *The polar-front jet stream and the distinctive tropopause structure in its vicinity thus characterize the air-mass boundary in these upper levels.* In the middle troposphere, the corresponding air-mass boundary is typified by pronounced baroclinity, much larger than that observed if the temperature field is averaged with respect to space and time. In the surface layers, the polar front is well marked by characteristics known to all synopticians. Here, however, the thermal influence of the underlying earth's surface often so modifies the polar air that its temperature can no longer be considered characteristic of the source region. We shall return to this question later in the discussion of air-mass modifications and transformations, and also in the chapters dealing with fronts and jet stream structures.

According to Defant and Taba, *the subtropical tropopause break and jet stream determine the boundary between TA and MLA.* This break appears at relatively constant latitudes and is hence discernible in mean cross sections (Fig. 1.1). Below the jet-stream level, a baroclinic zone (the subtropical frontal zone) is taken as the air-mass boundary. This boundary is better marked in high altitudes, but tends to become less distinct in the lower troposphere. As pointed out earlier, in the lower troposphere one should not expect any clear air-mass boundary because of the low-level frontolytic effect of divergence prevailing in these latitudes.

The classification of the principal air masses outlined here necessarily represents a very simplified scheme. The scheme does not, e.g., include the "Arctic Front," which separates the coldest air of arctic origin from the more moderately cold PA. The subclassification of PA into "Arctic" (A) air and "Polar" (P) air was originally proposed by Bergeron (1928). Such an arctic front is often very distinct in certain regions of the Northern Hemisphere, and probably also in the Southern Hemisphere, especially during the colder season. However, it appears mostly as an air-mass boundary in low altitudes and is strongly influenced by the distribution of snow cover on continents and the extent of the ice cover of the Arctic and Antarctic oceans. Hence the distinction between air with true arctic origin and the often equally cold or even colder "continental polar" air originating at lower latitudes (e.g., over the northern Eurasian continent in winter) is at times hard to justify. In the upper parts of the troposphere, it is most difficult to maintain clear separation between arctic and polar air, since here the latter air mass is often just as cold as the arctic air. Fronts between continental and maritime air masses can also in general be readily distinguished in the lower troposphere, but they tend to diffuse in the upper troposphere.

In a classification founded on the fundamental features of the general atmospheric circulation, it is therefore convenient to restrict the air-mass classification to the simplest scheme that describes the overall hemispherical distribution. At the same time, we do not intend to minimize the merits of a more detailed air mass and frontal analysis in dealing with the properties of the atmosphere over more restricted regions. In Canada, e.g., it has been found advantageous to adopt a "three-front" model (see, e.g., Anderson *et al.*, 1955; McIntyre, 1958). This takes account of the four air masses (continental and maritime arctic, and maritime polar and tropical) that most commonly invade, or are formed within, the region.

The existence of such different air masses is corollary to the exchange processes discussed on a global scale in Chapters 1 and 2, which necessarily entail *modification* or *conversion* of air masses in a way that depends largely on the character of the surface over which they move. In this modification, there must always exist intermediate stages wherein an air mass cannot be categorized with assurance. Accordingly, it should be borne in mind that air-mass analysis is just a tool for a simple description of the atmospheric structure, which cannot be applied objectively in all synoptic situations, and care should be taken to avoid "inflation" in the analysis of air masses and fronts on synoptic charts.

4.2 POLAR FRONT AND POLAR-FRONT JET STREAM

The structure of the polar front and its jet stream will be treated in detail in later chapters; here only a few general features will be stressed. In their original paper on the polar-front theory, J. Bjerknes and Solberg (1922) pointed out that *the polar front should not be considered a global phenomenon in the respect that it appears as an air-mass boundary around the whole hemisphere because then a meridional air-mass exchange and the necessary air-mass conversions would be prohibited.* Analyses of surface maps also showed at that time that there exist regions of breaks in the front.

With the availability of upper-air data later on, it became clear that the polar front in the free atmosphere (e.g., at 500 mb) has the character of a hyperbaroclinic sloping layer within which a considerable part of the overall meridional temperature gradient is concentrated. This character of the frontal zone visible on upper air charts was confirmed by frequency analyses of temperature at fixed levels, performed by McIntyre (1950) and Berggren (1953).[3] It could also be shown that the crowding of isotherms generally appears around preferred absolute values of temperature depending on season. For instance, the polar front at the 500-mb surface in winter is found mostly in the isotherm band $-24°C$ to $-34°C$, while in summer it is characterized by $10°C$ to $15°C$ higher temperatures.

An example of the polar front at 500 mb is reproduced in Fig. 4.5, after

---

[3] At individual stations where fronts are frequently present, this is evidenced by a tendency for skewness in the frequency distribution, toward lower temperatures in lower levels and toward higher temperatures in higher levels, with a double peak in middle levels.

Bradbury and Palmén (1953). The thermal patterns north and south of the band of crowded isotherms identified as the polar front seem to have quite different characters. On the south side, the isotherms have the shape of a wave pattern with considerable meridional baroclinity of the atmosphere, whereas on the north side, the isotherms form closed regions of cold air

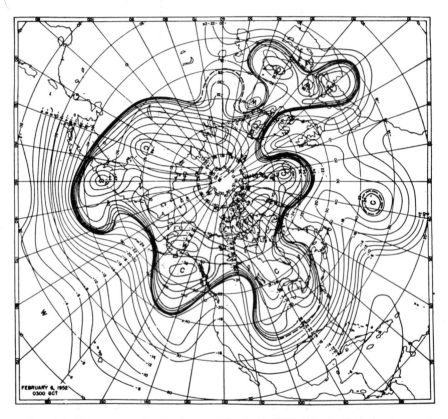

FIG. 4.5   Temperature distribution at 500 mb, 03 GCT Feb. 6, 1952; isotherms at 2°C intervals. The heavy line marks the approximate southern limit of the polar air, including the frontal zone. (From Bradbury and Palmén, 1953.)

separated by regions of somewhat warmer air. This difference in isotherm patterns should be expected if the front separates a zone of the atmosphere characterized by a strong westerly current with superposed disturbances, from a zone farther north with a weak zonal circulation in which, however, are embedded cyclonic and anticyclonic vortices.

Figure 4.5 also shows that the hyperbaroclinic zone at 500 mb extends essentially around the whole hemisphere, with a few interruptions where it is relatively diffuse. In this example, the mean position of the front is around latitude 48°N, with a maximum southward displacement to 30°N and maximum northward displacement to 72°N. Similar analyses have shown that the polar front can on rare occasions be displaced very far north to the vicinity of the North Pole, but that at the 500-mb level it can hardly ever be seen very much south of 30°N (see Chapter 10). This has been confirmed by the aforementioned analyses of temperature frequencies by McIntyre and by Berggren.

The thermal structure of Fig. 4.5 can be interpreted by a schematic cross section of the type in Fig. 4.6. Here the polar front is marked by a sloping hyperbaroclinic layer embedded in a less baroclinic atmosphere to

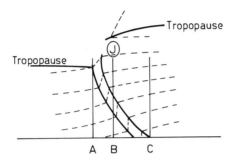

FIG. 4.6   Schematic vertical section showing the zone of strong integrated baroclinity through the troposphere (between *A* and *C*), the sloping frontal layer, and the jet-stream core (*J*) situated over the vertical *B*, where the integrated baroclinity has its greatest value. The gap in the tropopause is also indicated.

the south and north (the baroclinity in the air masses decreasing with distance from the front). The vertical shear of the geostrophic wind $\mathbf{V}_g$ is given by

$$\frac{\partial \mathbf{V}_g}{\partial z} = -\frac{g}{fT} \nabla_p T \times \mathbf{k} \qquad (4.1)$$

and the maximum geostrophic wind $\mathbf{V}_{g(\text{max})}$ at the level H is determined by

$$\mathbf{V}_{g(\text{max})} = \mathbf{V}_{g0} - \frac{g}{f} \int_0^H \frac{1}{T} \nabla_p T \times \mathbf{k} \, dz \qquad (4.2)$$

where $\mathbf{V}_{g0}$ represents the geostrophic wind at sea level.

The position of the jet core is mainly determined by the region where the integral in Eq. (4.2) achieves maximum value, since ordinarily the horizontal variation of $\mathbf{V}_{g0}$ is weak by comparison. It therefore follows that where there is a distinct polar front, there should also be a jet stream.[4] This connection between the polar front, the polar-jet stream, and the break between the middle-latitude tropopause and the polar tropopause are regular characteristics of the atmospheric structure.

When defining these as characteristic features, we also want to stress that strong deviations from any single "model" that illustrates these features will occur in the atmosphere. These deviations will be considered in the discussion of the polar-front disturbances in their different stages of development.

4.3 SUBTROPICAL JET STREAM AND "SUBTROPICAL FRONT"

In the higher-tropospheric boundary region between the Hadley cell and the Ferrel cell of the mean meridional circulation, a quasi-permanent "frontal zone" between air masses of different origin and life history develops in upper levels because, as noted earlier and discussed further in Section 9.7, these circulation cells must here act frontogenetically. This front is geographically much more permanent than is the polar front. In a synoptic cross section along a meridian, on the other hand, it is usually less distinct. Only in special regions (e.g., near the east coasts of Asia and North America) does the subtropical front frequently appear as a distinct front in the temperature and wind structure.

The existence of a semipermanent jet stream in subtropical latitudes was first shown by Namias and Clapp (1949). The distinction between the much less fixed polar-front jet stream and the subtropical jet stream was demonstrated by Palmén (1951), who also stressed the connection between the Hadley circulation of the tropics and the formation and maintenance of the subtropical jet stream (Section 1.4).

The subtropical jet stream, which is typically strongest near the 200-mb surface, is characterized by strong vertical wind shear in the upper troposphere. At the 500-mb level it is, by comparison, only weakly recognizable; above that level the baroclinic zone corresponding to strong vertical wind shear is on the average most pronounced in regions of maximum jet velocity, and more diffuse in regions of weaker velocity (Section 8.5). This characteristic upper-tropospheric baroclinic zone was first investigated by Mohri (1953) over Japan, where the subtropical jet stream in winter is especially

---

[4] The converse is not necessarily true. Examples in Chapter 9 illustrate that in some localities there may be no distinct front in the upper troposphere, while there may be a prominent jet stream; however, there is a "polar-front zone" where the baroclinity is strong.

strong. Extensive studies of the subtropical jet stream and the atmospheric structure in its vicinity have been made by Krishnamurti (1961) and by Defant and Taba (1957, 1958a, b), as already mentioned.

Especially during the colder season, the subtropical jet stream is a dominating phenomenon in both hemispheres. Loewe and Radok (1950), Hutchings (1950), and Gibbs (1953) have shown that the west-wind maximum of the upper troposphere, at least in the longitudes of Australia and New Zealand, is found at a latitude similar to that of the subtropical jet stream of the Northern Hemisphere, and that south of this wind maximum another belt of strong wind, corresponding to the polar-front jet of the Northern Hemisphere, can be discerned.

The relationship between the mean positions of the subtropical jet and of the polar-front jet was illustrated in Fig. 3.9. In the Northern Hemisphere the mean axis of the subtropical jet generally reaches its northernmost position at longitudes where the activity of the polar-front jet penetrates far southward, and vice versa. This indicates that there exists some kind of interaction between the polar and subtropical jets. The subtropical jet stream and the subtropical front will be discussed in more detail in Section 8.6.

4.4   SCHEME OF THE CHARACTERISTIC ATMOSPHERIC STRUCTURE

In Fig. 4.7 an attempt is made to present the typical structure of the Northern Hemisphere in a singular model. The model is grossly simplified and contains only the most characteristic features. The scheme is most applicable to the atmospheric structure during the cold season when all features are better pronounced than during the warm season.[5]

In the figure the tropospheric air is divided into three different air masses: TA, MLA and PA, according to the principles outlined earlier. However, in addition to this, the typical position of a low-tropospheric arctic front is considered. The mean positions of the polar front and the subtropical front are marked with their corresponding jet-stream cores. The three characteristic tropopauses are indicated, viz., the tropical tropopause, the middle-latitude tropopause, and the polar tropopause.[6] In the tropical re-

---

[5] Similar schemes have been presented by Mohri (1958, 1959), based on conditions around longitude 140°E, and by Newton and Persson (1962), characterizing the structure in winter near 80°W.

[6] In this scheme, the concept of the tropopause has been simplified. Very often the tropopause has a quite complicated multiple structure and is difficult to fix objectively. This will be apparent in synoptic examples presented later.

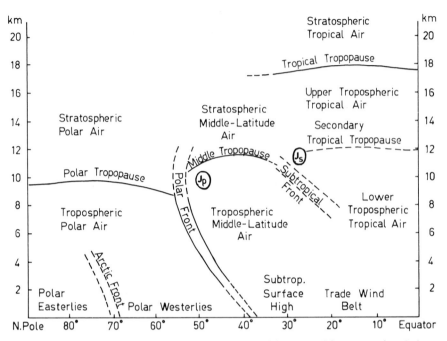

FIG. 4.7   The principal air masses, tropopauses and fronts, and jet streams in relation to the features of the low-level wind systems. Depending on the location and time, the fronts may be either well developed or relatively weak.

gion a secondary lower tropopause at about 12 km is marked as a dashed line. This corresponds to the level where a pronounced increase of vertical stability can generally be seen and to the level of maximum meridional velocity in the Hadley cell (Chapter 1).

The subtropical front is indicated as a sloping upper-tropospheric layer between the tropical air and the middle-latitude air. It is well marked in some regions, but ill-defined or absent in others. Because it overlies the low-level divergence region near or somewhat equatorward of the high-pressure belt in the lowest troposphere, the subtropical front mostly has no weather characteristics corresponding to those of the polar front.[7] Although the zonal mean temperature contrast may be attributed to the

---

[7] Monthly mean cross sections by Matsumoto *et al.* (1953) over the longitude of Japan demonstrate clearly that the tropospheric relative humidity is on the average low beneath the subtropical jet stream, with higher values both north and south. This is consistent with the generation of this jet stream by the Hadley circulation, with average descending motions in its poleward parts.

mean meridional field of motion, this contrast may be locally sharpened or weakened as a result of the confluence and diffluence pattern of the middle- and upper-tropospheric flow described in Sections 3.2 and 8.5. The subtropical jet and the connected baroclinic zone on the average have their maximum strength at longitudes where they meander farthest poleward. In this respect they behave differently from the polar jet and the polar front, where no corresponding rule is valid.

In the schematic figure (Fig. 4.7), the polar front in low and high levels is marked by dashed lines to indicate that those are the levels where some kind of air-mass exchange occurs between the different air-mass source regions. Parallel with this exchange by the synoptic-scale eddies, which largely accomplish the required meridional transfer of energy (Chapter 2), conversions between available potential and kinetic energy take place as discussed in Section 16.3.

It should be emphasized that the indicated positions of the fronts and the jet streams represent their approximate mean latitudes during the colder season. The latitude of the polar front, in particular, varies strongly with longitude and time; hence, it cannot be recognized on meridional cross sections constructed by averaging along entire latitude circles.

A meridional scheme of the type in Fig. 4.7 is primarily valid in synoptic situations where the temperature field does not deviate too strongly from mean conditions or where the air motion is predominantly zonal. In cases of "breakdown" of the zonal motion and formation of deep cold cyclones in lower latitudes and warm anticyclones in higher latitudes, the thermal and flow patterns fundamentally deviate from the scheme shown in Fig. 4.7. In extreme cases of such a "low-index" type of circulation, a meridional cross section sometimes shows the existence of two westerly polar jets at widely separated latitudes, with an easterly jet between them. Illustrations of such a meridional structure of the atmosphere have been presented, for example, by Berggren *et al.* (1949) and by Defant and Taba (1958b). Because of the great variety of the possible atmospheric disturbances, discussion of such more-or-less extreme synoptic situations is deferred to later chapters.

REFERENCES

Anderson, R., Boville, B. W., and McClellan, D. E. (1955). An operational frontal contour-analysis model. *Quart. J. Roy. Meteorol. Soc.* **81**, 588–599.

Bergeron, T. (1928). Über die dreidimensional verknüpfende Wetteranalyse. *Geofys. Publikasjoner, Norske Videnskaps-Akad. Oslo* **5**, No. 6, 1–111.

Bergeron, T. (1930). Richtlinien einer dynamischen Klimatologie. *Meteorol. Z.* **7**, 246–262.

Berggren, R. (1953). On temperature frequency distribution in the free atmosphere and a proposed model for frontal analysis. *Tellus* **5**, 95–100.

Berggren, R., Bolin, B., and Rossby, C.-G. (1949). An aerological study of zonal motion, its perturbations and breakdown. *Tellus* **1**, No. 2, 14–37.

Bjerknes, J., and Solberg, H. (1922). Life cycle of cyclones and the polar front theory of atmospheric circulation. *Geofys. Publikasjoner, Norske Videnskaps-Akad. Oslo* **3**, No. 1, 1–18.

Bjerknes, V., Bjerknes, J., Solberg, H., and Bergeron, T. (1933). "Physikalische Hydrodynamik," 797 pp. Springer, Berlin.

Bradbury, D. L., and Palmén, E. (1953). On the existence of a polar-front zone at the 500-mb level. *Bull. Am. Meteorol. Soc.* **34**, 56–62.

Defant, F., and Taba, H. (1957). The threefold structure of the atmosphere and the characteristics of the tropopause. *Tellus* **9**, 259–274.

Defant, F., and Taba, H. (1958a). The breakdown of the zonal circulation during the period January 8 to 13, 1956, the characteristics of temperature field and tropopause and its relation to the atmospheric field of motion. *Tellus* **10**, 430–450.

Defant, F., and Taba, H. (1958b). The details of wind and temperature field and the generation of the blocking situation over Europe (January 1 to 4, 1956). *Beitr. Physik Atmosphäre* **31**, 69–88.

Ferrel, W. (1856). An essay on the winds and the currents of the oceans. *Nashville J. Med. Surg.* **11**, 277-301, and 375-389.

Gibbs, W. J. (1953). A comparison of hemispheric circulations with particular reference to the western Pacific. *Quart. J. Roy. Meteorol. Soc.* **79**, 121–136.

Hutchings, J. W. (1950). A meridional cross section for an oceanic region. *J. Meteorol.* **7**, 94–100.

Krishnamurti, T. N. (1961). The subtropical jet stream of winter. *J. Meteorol.* **18**, 172–191.

Loewe, F., and Radok, U. (1950). A meridional aerological cross section in the southwest Pacific. *J. Meteorol.* **7**, 58–65.

McIntyre, D. P. (1950). On the air-mass temperature distribution in the middle and high troposphere in winter. *J. Meteorol.* **7**, 101–107.

McIntyre, D. P. (1958). The Canadian 3-front, 3-jet stream, model. *Geophysica (Helsinki)* **6**, 309–324.

Matsumoto, S., Itoo, H., and Arakawa, A. (1953). On the monthly mean distribution of temperature, wind and relative humidity of the atmosphere over Japan from March 1951 to February 1952. *J. Meteorol. Soc. Japan* **31**, 248–258.

Mohri, K. (1953). On the fields of wind and temperature over Japan and adjacent waters during the winter of 1950–1951. *Tellus* **5**, 340–358.

Mohri, K. (1958, 1959). Jet streams and upper fronts in the general circulation and their characteristics over the Far East. *Geophys. Mag. (Tokyo)* **29**, 45–126; 333–412.

Namias, J., and Clapp, P. F. (1949). Confluence theory of the jet stream. *J. Meteorol.* **6**, 330–336.

Newton, C. W., and Persson, A. V. (1962). Structural characteristics of the subtropical jet stream and certain lower-stratospheric wind systems. *Tellus* **14**, 221–241.

Palmén, E. (1951). The rôle of atmospheric disturbances in the general circulation. *Quart. J. Roy. Meteorol. Soc.* **77**, 337–354.

Petterssen, S. (1940). "Weather Analysis and Forecasting," 1st ed., Chapter 3. McGraw-Hill, New York.

Rossby, C.-G. (1941). The scientific basis of modern meteorology. *In* "Yearbook of Agriculture, Climate and Man" (G. Hambidge, ed.), pp. 599–655. U. S. Gov't. Printing Office, Washington, D. C.

Schinze, G. (1932). Troposphärische Luftmassen und vertikaler Temperaturgradient. *Beitr. Physik Freien Atmosphäre* **19**, 79–90.

Showalter, A. K. (1939). Further studies of American air mass properties. *Monthly Weather Rev.* **67**, 204–218.

University of Chicago, Staff Members, Department of Meteorology. (1947). On the general circulation of the atmosphere in middle latitudes. *Bull. Am. Meteorol. Soc.* **28**, 255–280.

Willett, H. C. (1933). American air mass properties. *Papers Phys. Oceanog. Meteorol., Mass. Inst. Technol. Woods Hole Oceanog. Inst.* **2**, No. 2, 1–116.

5

THE POLAR-FRONT THEORY
AND THE BEGINNINGS OF SYNOPTIC AEROLOGY

Analyses of the structure and behavior of extratropical disturbances were initiated more than a century ago with the introduction of daily synoptic weather maps. Considering the extensive literature accumulated since that time, we cannot attempt to review it thoroughly. Those interested in a broader account of the contribution of scientists in the past may refer to von Hann (1915), Shaw (1926–1931), and Exner (1925) for the time up to World War I, and to Chromow (1942) and Scherhag (1948) for the development between World Wars I and II. Bergeron's fascinating historical critique (1959) embodies a thorough but concise documentation, showing how the study of weather systems has proceeded alternately in spurts of progress or periods of lull or even of decline. The rate of advance has been conditioned partly by the gradual increase of data available, but also to a very significant extent by the kind of emphasis placed on those observational data, in terms of use or disuse of existing concepts.

In this chapter only a few essential steps in the unfolding of our present knowledge of atmospheric disturbances will be stressed. Especially strong influences upon progress evolved during and after World Wars I and II, owing largely to the immediacy of demands for weather forecasts imposed by military operations. Hence, it seems natural to consider separately the events prior to World War I and the period between World Wars I and II.

During the latter period, emphasis will be placed on the discovery of the polar-front theory. Development of this concept must be considered as an event of first magnitude, not because it provided the final answer on the detailed structures of weather systems, but because it furnished for the first time a logical framework into which the observations could be systematically fitted. The polar-front theory thus served as an essential starting point for later aerological studies. Further, this theory clearly demonstrated the role of disturbances in the global atmospheric circulation. For these reasons, we devote special attention to the polar-front ideas.

5.1   A BRIEF HISTORICAL REVIEW

As early as a century ago, FitzRoy (1863) presented a strikingly realistic picture of the structure of the surface air currents in extratropical cyclones and anticyclones (for illustrations, see Petterssen, 1956, or Bergeron, 1959). The essential feature of his model was a recognition that *cyclones form at the zone of interaction of air masses having different properties, which originate in subtropical and polar regions.* Dove (1837) had earlier discovered that temperate-latitude weather systems were associated with the interaction of warm and cold currents.

FitzRoy's contribution had relatively little influence on practicing meteorologists in succeeding decades. The reason for this was, no doubt, partly rooted in the limited amount of observational material available for day-to-day analysis. Nevertheless, a few investigators paid attention to the linkage between cyclonic disturbances and the baroclinic regions between different air masses, which eventually became the central theme underlying modern concepts of circulation systems in middle latitudes. Notable contributions came from Blasius (1875), von Helmholtz (1888, 1889), and Bigelow (1904), all of whom in one way or another recognized the importance of shearing currents and temperature contrasts for the development of cyclones and the associated weather.

Early cyclone theories dealt with the problem from two distinctly different standpoints. One concept, due largely to Espy (1841), may be called the *"convectional" hypothesis.* From the association of depressions with precipitation, it was inferred that low-pressure centers comprised regions of ascending motion. Cyclones were regarded as heat engines driven by the supposedly warm, rising columns in their centers, the heat being derived from release of latent heat of condensation. Heating and vertical expansion led to a raising of the upper isobaric surfaces in the central portion, causing a lateral outflow aloft. Owing to the removal of air from the central column, a pressure fall took place in low levels, resulting in an inrush of air in response to the low-level pressure gradient. Espy attached no significance to the rotation of cyclones, although this was known. The idea that this gyration is due to the deflection of currents approaching the low-pressure center, induced by the earth's rotation, was introduced by Ferrel (1856). C. H. D. Buys Ballot was the first to formulate (in 1857) the rule: "Winds always blow, in the Northern Hemisphere, with high barometer to the right and low barometer to the left of the direction in which they blow." He stated also that strong pressure gradients are connected with strong winds, a rule later formalized into the geostrophic wind equation relating the pressure gradient to the deflecting force.[1]

The convectional theory was criticized by Hann as early as 1876 (see von Hann, 1915). His most substantial objection was based on mountain-top temperature observations, which indicated temperatures higher over anti-cyclones than over cyclones. Hann proposed an alternative that may be classed among various *"dynamical" hypotheses*. According to this concept, cyclones and anticyclones originate in some way as eddies in the general strong westerly circulation of middle latitudes, and feed off the energy derived from the basic current. As will be seen later, there is substance to both the convectional and dynamical theories, but neither is likely to ac-count exclusively for the development of disturbances. The convectional theory is most closely realized in tropical storms (Chapter 15).

The hypotheses mentioned above were based principally on observations made at the earth's surface, and there was at the time little knowledge of the vertical structure of the atmosphere or of the currents aloft. In the last years of the nineteenth and the early decades of the twentieth centuries, investigations of the vertical structure were undertaken through the use of balloons and kites. These demonstrated *the general lapse of temperature with height, the existence of inversions, and the presence of the "isothermal layer" of the stratosphere* (Assmann, 1900; Teisserenc de Bort, 1902).

The most extensive series of upper-air observations, and the most thorough analyses were provided by Dines (1912, 1919, 1925) and by Schedler (1921). A remarkable result of these analyses was the discovery that *low-pressure centers were, on the whole, considerably colder aloft, and high-pressure centers were warmer aloft than the average temperature of all soundings*. This came generally as a surprise (despite Hann's earlier observations), since it had been concluded—from the surface temperature distribution (especially dur-

---

[1] Coriolis, in his earlier general treatment of this deflection, did not mention its ap-plication to the atmosphere. Ferrel (1889) also observed that the westerlies flow around a vast polar cyclone, to which the same principle applies. Teisserenc de Bort's global map of the 4-km pressure distribution, published in 1889 (reproduced by Bergeron, 1959), shows this feature very clearly in both hemispheres.

ing the colder season) and from the supposition that high pressure must be due to the weight of overlying colder air—that the opposite was true.

Dines considered that his observations dealt a death blow to the convectional theory of cyclone formation, since according to that theory the central portion of a cyclone was supposed to be warm. It was not evident until much later that the thermal structures of cyclones change during their life cycles and that this variation and the thermal asymmetry of cyclones were both obscured by the statistical treatment of the data.[2]

Dines worked out extensive correlations between the pressure and the temperature at various levels, finding this correlation to be positive at levels below 9 km and negative above 9 km. The height of the tropopause was also positively correlated with surface pressure and with the mean temperature in the troposphere. Dines concluded that the low tropospheric temperature of the cyclone and the higher temperature of the anticyclone were due to forced ascent and descent in the respective systems. In the stratosphere and at the tropopause level, vertical motions of an opposite sort were deduced as having taken place.

A valuable aspect of Dines' work was the demonstration that, *on the average, the intensity of the horizontal circulations of migratory cyclones and anticyclones increases upward to about the tropopause level, and decreases higher up.* This feature is in agreement with the limited wind observations available at the time. Concerning the physical explanation of the temperature distribution in cyclones and anticyclones, no simple interpretation can be given, as Exner (1925) and many others have pointed out. For a satisfactory explanation of Dines' correlations it was, according to Exner, necessary to consider both vertical motions and horizontal temperature advection, as well as diabatic influences such as radiation or condensation of water vapor.

The statistical studies of the pressure and temperature variations connected with migratory disturbances drew attention to the importance of processes in the upper troposphere and lower stratosphere, leading to the emphasis of the German-Austrian school upon lower-stratospheric advection as an important cause of the pressure changes in lower levels. The negative correlation between temperature and pressure in the lower stratosphere was interpreted as the result of stratospheric advection. The upper-level pressure changes were therefore considered as "primary", and the lower-

---

[2] Until Hanzlik (1909) demonstrated the distinction between migratory cold anticyclones and the slow-moving warm highs that were more common over western Europe, it was apparently not recognized that there are different kinds of anticyclones.

level changes as "secondary" and caused by the former (Ficker, 1920; Defant, 1926). As will be seen later, the complex problem of the cause of the pressure changes in different altitudes cannot be solved by purely statistical methods.

Valuable contributions to the discussion of the "seat" of the pressure variations observed at fixed levels were at that time given by several other investigators (e.g., Haurwitz, 1927; Rossby, 1927). In these papers the combined effect of temperature advection and the vertical movement induced by this was subjected to systematic investigations.

An important study by Lempfert and Shaw (1906) demonstrated the way in which the air at the surface spirals inward toward cyclone centers and outward from anticyclone centers. Trajectories of air particles converging toward various parts of the cyclone were found to originate in widely different regions. Shaw's simple model of cyclone structure (1911), drawing on the results of these studies, embodies some of the essential features of the later Norwegian cyclone models, although in less definitive form.

Margules (1903) was the first to provide a thorough theoretical analysis showing that *the kinetic energy of disturbances is derived from the realizable potential energy associated with horizontal temperature contrasts.* Margules' work on energy transformations and the precise formulation of the circulation theorem by V. Bjerknes (1898, 1902) provided much of the theoretical background and inspiration for later hypotheses concerning atmospheric disturbances.

5.2 THE POLAR-FRONT THEORY

As Bergeron (1959) has emphasized, progress in weather analysis and prognosis has stemmed from improvements in three major ingredients: *observations*, *tools*, and *models*. The remarkable success of the Norwegian (Bergen) school in the years following World War I resulted from conscious and successful efforts at simultaneous development and blending together of these essential ingredients.

The fundamental philosophy expounded by V. Bjerknes, and also by Richardson (1922), was that *if the initial state of the atmosphere were completely described, it would then be possible to make a prognosis of the future states by use of the fundamental laws of hydrodynamics and thermodynamics.*

Application of these theoretical *tools* required the development of a description of the atmospheric circulation more accurate than that existing at the time. Since it was evident that observations would never be widespread enough or sufficiently dense to describe the state of the atmosphere in all details, it was necessary to develop *models*. An objective of a model is to provide a generalized framework in which, by use of observations available at a relatively small number of locations, the overall structure of the circulation system may be described with a considerable degree of accuracy. In order to develop models and to utilize them on a day-to-day basis, improvements in *observations* had to take place.

Painstaking analysis of the data from a special dense network of stations resulted in the first clear-cut description of the structure of cyclones and of

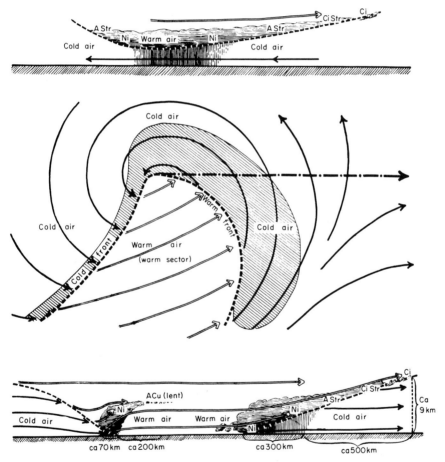

Fig. 5.1   Idealized cyclone, from J. Bjerknes and Solberg (1921). In middle diagram, dash-dotted arrow shows direction of motion of cyclone; other arrows are streamlines of air flow at earth's surface. Top and bottom diagrams show cloud systems and air motions in vertical sections along direction of cyclone movement north of its center and across the warm sector south of its center.

their role in the general circulation. Whereas earlier the weather map had been viewed as a more-or-less amorphous collection of observations, now the new models provided a framework that connected the different observations and which was amenable to theoretical treatment.

The basic cyclone model developed by J. Bjerknes (1919), and slightly modified by J. Bjerknes and Solberg (1921, 1922), is illustrated in Fig. 5.1. The salient feature is that *the extratropical cyclone forms on a frontal surface, wherein a major part of the temperature contrast between neighboring air masses is concentrated into a narrow transition layer* which, on the scale of the weather map, amounts essentially to a surface of discontinuity in temperature or density.

Margules (1906) had earlier demonstrated that in a rotating framework, such a discontinuity between air masses of different temperature can exist in equilibrium, with a slope departing from the horizontal, as long as there is a cyclonic wind shear across the discontinuity. Helmholtz' theory had suggested that perturbations on a shear zone might, under certain circumstances, become unstable and grow, eventually developing into vortices. The Bergen group found by detailed analyses of the streamline field that wind discontinuities were in fact associated with the zones of temperature contrast and that cyclones did originate on the shear zone called a *front*.

As noted earlier, the existence of fronts had to some degree been recognized by earlier investigators. This was true of the *cold front*, where wind shifts and temperature contrasts are most obvious. A particularly thorough description of this phenomenon was given by Lempfert and Corless (1910), who called it a "line squall" or "linear front" whose regular progression they demonstrated by use of isochrone charts, discussing also the cyclonic shift of the geostrophic wind and the character of the vertical circulation. However, the *warm front* on the advancing side of the cyclone was discovered only after painstaking analysis by the Bergen group.[3]

In the cyclone model illustrated in Fig. 5.1, the cyclone is shown in the wave shape that was commonly observed and most easily recognized on

---

[3] Von Ficker (1911), in his important studies of outbreaks of warm and cold air masses in northern Europe and Asia, developed a scheme quite close to a frontal concept of air masses. It is not surprising that the warm front and the combination of warm and cold fronts in the central part of a cyclone were first recognized in northwestern Europe, where especially warm fronts are well developed during the colder season. The principles of analysis used by the Norwegian school were laid out by V. Bjerknes *et al.* (1911). Some of their synoptic examples, in which are analyzed the surface streamlines and isotachs and their derivatives, rather strongly suggest the frontal structure of a mature cyclone.

its passage over Scandinavia. A *"warm sector"* is bounded on the advancing side by a warm front and on the rear side by a cold front. From observations of clouds and precipitation, Bjerknes and Solberg concluded that the warm-frontal cloud system is intimately connected with the sloping frontal surface in the manner indicated in the vertical cross sections of Fig. 5.1. Warm moist air, overtaking the warm front, was visualized as ascending the sloping surface and forming an extensive cloud sheet. At the cold front, the air in upper levels, also moving faster than the front, was pictured as blowing with a component down the frontal surface, leading to clearing soon after the frontal passage. On or just ahead of the cold front, however, a narrow band of precipitation resulted from lifting by the front, which overtook the moist air in lower levels.

The model in Fig. 5.1 depicts the structure of an extratropical cyclone only at a particular time midway in its development. Cyclone development was visualized as starting as a small perturbation (Figs. 5.2a, b) on a quasi-

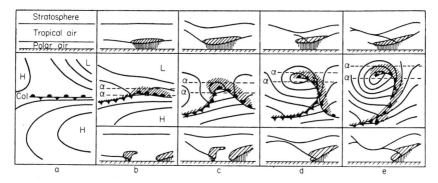

FIG. 5.2   Life cycle of cyclone from wave to vortex, after J. Bjerknes, from Godske *et al.* (1957). In middle figures, thin lines are sea-level isobars. Top and bottom figures shown schematic clouds, frontal surfaces, and tropopauses along lines α, a little north and south of cyclone center. The times between stages (a) and (c), and between stages (c) and (e), correspond approximately to one day in each case.

stationary front, characterized by cyclonic shear. Once a small initial wave had started, with the warm air encroaching slightly on the cold air, the pressure fell near the crest of the wave. Following the initial perturbation, the pressure distribution favored development of a cyclonic circulation about the wave crest. An important feature of this circulation (Fig. 5.2c) is that the wind has a component blowing from cold toward warm air behind the wave crest and from warm toward cold air ahead of it. The advance of the cold front and the retreat of the warm front cause the frontal wave as a whole to progress more or less along the direction of the flow in the warm sector above the friction layer.[4] This rule was in agreement with previous finding of Hesselberg (1913) who had shown that cyclones move in the direction of the general upper-air flow (at the cirrus level).

Observations of the structures of cyclones at different times in their development showed that the initial small perturbation gradually increases in amplitude, with further reduction of pressure at the cyclone center and an increase of the circulation about it. Furthermore, it was observed that cold fronts typically advance at a rate faster than warm fronts. Eventually the cold front overtakes the warm front, and the warm-sector air is entirely lifted away from the earth's surface. This process (discovered by Bergeron in 1919) was given the name *"occlusion,"* and the resulting front, bounded on both sides by cold air masses of slightly differing properties, was called an *"occluded front"* (Fig. 5.2d). At a still more advanced stage (Fig. 5.2e) the cold front overtakes the warm front at increasing distances from the cyclone center, the occlusion grows in extent, and the cyclone size increases and it is transformed into a large cold vortex in the lower troposphere, with the warm air still existing higher up. The whole process comes to an end when the cyclone ultimately takes on the character of a more-or-less barotropic whirl, which loses its frontal character and gradually dissipates as a result of friction.

An essential feature of this conception is that it demonstrates an important aspect of the energy conversion involved in cyclogenesis. *During the occlusion process, the warm air, which is at first extensive, gradually diminishes in extent, and is replaced by encroaching cold air masses.* In the neighborhood of the cyclone center, this means that in effect *the center of gravity of the overall atmosphere has been lowered, with a decrease of potential energy simultaneous with an increase of the kinetic energy of the cyclonic system.* This is in broad accord with the principle demonstrated by Margules (1903). J. Bjerknes and Solberg recognized this conversion of energy as being appropriate to the process of cyclogenesis; they stated that the kinetic energy of a cyclone could increase only as long as an appreciable air-mass (frontal) contrast was available. In the final stage, when the cyclone has become thoroughly occluded, its failure to develop further was considered as being due to the absence of available potential energy in the weakened air-mass contrast near the cyclone center. At this stage, essentially all warm air has

---

[4] Kasahara *et al.* (1965) have succeeded in simulating this process in a very realistic way by the numerical computation of air trajectories and frontal motions, based on a two-layer model of an incompressible fluid on a rotating earth.

been lifted, and the cold air has descended and spread out in lower levels in the entire region occupied by the cyclone. Because a cyclone cannot be considered a dynamically and thermodynamically closed system, the energy processes of a developing cyclone are in reality more complicated than the corresponding processes in Margules' simple models. With regard to energy development in real cyclones, the reader is referred to Chapter 16.

5.3 CYCLONE FAMILIES

In the course of the analysis of cyclones, it was realized that these entered the west coast of Europe in series, in fairly quick succession, and that the structure of an individual cyclone varied according to its place in the series. From these circumstances it was deduced that cyclones are generally arranged in *"families,"* as in Fig. 5.3. Individual cyclones were viewed as forming on a *"polar front,"* which separated air masses of polar origin from warmer

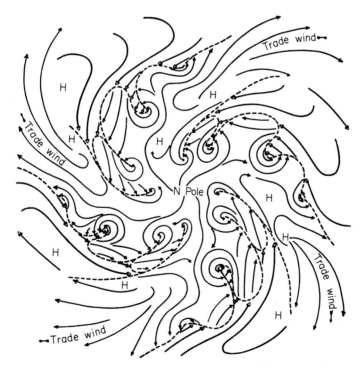

FIG. 5.3 General extratropical circulation of the atmosphere, from J. Bjerknes and Solberg (1922). Dashed lines show frontal systems of cyclone families; arrows are surface streamlines.

masses of subtropical or tropical origin, and moving generally northeast-ward while developing according to the scheme in Fig. 5.2. Successive cyclones then formed on the trailing cold fronts behind the earlier ones. Generally, the cyclone farthest northeast in a given family was the most mature, being more-or-less fully occluded, while the cyclones farthest south-west, being youngest, were open-wave cyclones.

After the passage of a cyclone family, it was observed that there was a prolonged period of general northerly winds with relatively little cyclone activity, followed eventually by the passage of another cyclone family. The cyclone families and their individual members were regarded as a mechanism for the meridional exchange of warm- and cold-air masses.

5.4 Relation of Cyclone Families to General Circulation

Different processes might be imagined which could effect the heat exchange between tropical and polar regions, necessary to balance the net radiative gain of heat in low latitudes and the net radiative loss in high latitudes. One of these might be a simple meridional circulation, symmetrical about a hemisphere, with cold air sinking near the pole and flowing equatorward, and warm air rising in low latitudes and flowing poleward in upper levels.

J. Bjerknes and Solberg (1922) pointed out that this scheme would be impossible in higher latitudes because, owing to the earth's rotation, extremely large zonal velocities would be attained if the air moved all the way from very low to very high latitudes, or vice versa. Only in low latitudes, where the deflection due to the earth's rotation is moderate, would the simple trade-wind (low level) and antitrade (upper level) circulation be realistic. An alternative for the higher-latitude circulation, suggested by Bjerknes and Solberg, is shown in simplified form in Fig. 5.4.

In this scheme the meridional movement takes place in broad tongues of warm air moving poleward and of cold air moving equatorward. Although each current would tend to be deflected toward the right, this deflection would lead to a piling-up of air and development of high pressure in the zone between the right-hand flanks of the currents, and to a depletion of air and formation of a low-pressure zone between the left-hand flanks (in the Northern Hemisphere). The resulting zonal pressure gradients would oppose the generation of extreme eastward and westward velocities.

Thus, *in higher latitudes, equivalents of the "trades" and "antitrades" takes the form of deep southerly and northerly currents, alternately spiraling poleward and equatorward*, as shown in Fig. 5.4. Cyclones form in the low-pressure zone, and "the cyclone family is thus a boundary phenomenon between the left flank of a polar current and the adjacent tropical current."

When the cyclonic disturbances are added as perturbations on the left flank of the broad-scale tropical current, the simplified scheme of Fig. 5.4 takes on the aspect shown in Fig. 5.3. Polar air sweeps southward on the west side of each cyclone, but in general has only limited penetration south-

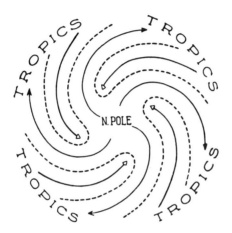

FIG. 5.4 Schematic picture of the general atmospheric circulation in extratropical regions, simplified by omitting individual perturbations of Fig. 5.3, to show broad-scale tongues of cold air moving equatorward and tongues of warm air moving poleward. (After J. Bjerknes and Solberg, 1922).

ward between the cyclones of a family. Only in the region between the last member of one cyclone family and the high separating it from the next family farther west is there a massive and uninterrupted equatorward penetration of polar air into the trade winds of subtropical latitudes.

The tropical air that moves northward on the west sides of the subtropical highs (Fig. 5.3) starts out as a broad current, but gradually diminishes in lateral extent as the tropical air in the warm sector of each cyclone is forced upward by the encroaching cold air in low levels during the process of occlusion. *Tropical air that has moved to upper levels in this process, and which at the same time has been cooled adiabatically or pseudoadiabatically, gradually undergoes an additional cooling by radiation in high latitudes, sinks, and returns southward as polar air.*

It had been observed at various places in middle latitudes that wet spells tend to occur with a certain periodicity, being interspersed with spells of little or no precipitation. Defant (1912) determined that in the year 1909 the wet spells occurred with an average period of about 5.7 days in the Northern Hemisphere and 7.2 days in the Southern. Bjerknes and Solberg found that during one year, 66 cyclone families entered western Europe (each family comprising an average of four cyclones that crossed the coast). Thus, the mean periodicity of the cyclone families was about 5.5 days. It was clear that the passage of the cyclone families and of the extensive highs separating them caused the alternate wet and dry spells observed at a given location.

Prior to the discovery of cyclone families, Defant (1912) had deduced (mainly on the basis of the precipitation periodicities mentioned above) that *the upper-air circulation is dominated by large-scale, eastward-moving waves*, the length of which he concluded on the average to be 90° longitude. The existence of such long waves was verified much later when extensive upper-air data became available. These waves and their relation to sea-level disturbances are discussed in Chapter 6.

The polar-front theory was essentially constructed after a careful analysis of surface weather maps, without satisfactory aerological observations. Despite this and the somewhat oversimplified character of the theory, it has remained for a half-century a cornerstone of synoptic meteorology, a tribute to the imagination displayed by the Bergen school in interpreting observational data and in constructing rational atmospheric models. Of special importance in this matter was the use of the concept of "*indirect aerology*," described in detail by Bergeron and Swoboda (1924) and by Bergeron (1928). Through the method of indirect aerology, it became possible to some extent to substitute for the lack of aerological measurements by conclusions drawn from surface observations. It was, for instance, possible to deduce certain characteristics of air masses (e.g., stability) from the types of clouds and hydrometeors observed. Combining these with the thermodynamic changes that air masses were assumed to have undergone during their prehistory, valuable conclusions were obtained concerning the three-dimensional structure, without any direct upper-air observations.

Later on, when aerological data became available, some modifications of the original theory were necessary. However, the present-day reader of the original papers of the Bergen school may be surprised at the freshness of the original theory in so many points. This success can be explained only by the soundness the Bergen school showed in the combination of skillful analysis with an essentially correct interpretation of the large-scale dynamic and thermodynamic processes in the atmosphere.

5.5   EARLY STUDIES OF THE UPPER-AIR STRUCTURE OF CYCLONES

Before the introduction of the radiosonde technique, the data available for synoptic studies of the upper-air structure of cyclonic disturbances were inadequate. Nevertheless the increasing number of observations from balloon sondes and aircraft permitted some general conclusions concerning the upper structure. These, on the whole, supported the deductions made from surface observation data obtained by the method of indirect aerology.

A number of serial and "swarm" ascents, beginning in Europe in 1928, using small meteorographs of the type constructed by Jaumotte, gave valuable results. These are described by J. Bjerknes (1932, 1935), Palmén (1935), J. Bjerknes and Palmén (1937), Van Mieghem (1939), J. Bjerknes et al. (1939), Nyberg and Palmén (1942), and others. Through these synoptic studies a satisfactory idea of the frontal structure, the coupling between lower and upper disturbances, and the structure and variation of the tropopause was achieved.

Figure 5.5 shows an example, taken from Van Mieghem (1939), of a time cross section constructed with the careful use of serial ascents together with all available surface data. This illustrates the structure of a warm sector, followed by a shallow cold-air tongue, and conforms well with the original Bjerknes scheme (Fig. 5.1), although there was in this case extensive precipitation in the warm sector. Figure 5.6 (J. Bjerknes and Palmén, 1937) gives an illustration, deduced from swarm ascents around a cyclone, of the flow in the warm air at the frontal surfaces and above the surface friction layer in the warm sector. This study brought out clearly the cyclonic curvature aloft behind the cold front and showed the pronounced tendency toward anticyclonic curvature over the warm front[5] (in which the advancing cirrus edge was closely identified with the region ahead of the front where the streamlines became parallel to the frontal contours, indicating negligible vertical motion). Figure 5.5 also illustrates a common feature: In the warm air of the upper troposphere the isotherms are most elevated above the warm front and lowest some distance behind the cold front (the same being true of the tropopause).

---

[5] The association of this feature with horizontal divergence above the warm front is indicated in Section 10.8. J. Bjerknes and Palmén, and Hewson (1937) independently, used the wet-bulb potential temperature to show that air above warm fronts is identifiable with that in the warm sectors of depressions, and Hewson made estimates of the divergence undergone in the ascent. Hewson (1936) used the same method to study the vertical motions along trajectories in subsiding air masses.

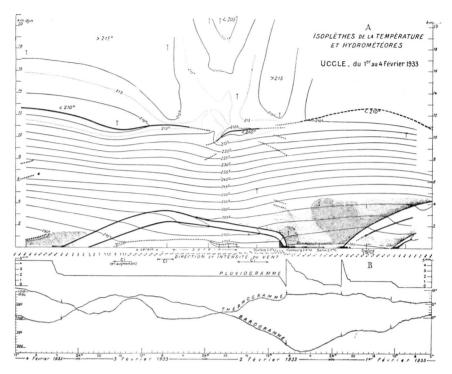

FIG. 5.5   Time cross section based on serial soundings at Uccle, Belgium, during the period Feb. 1–4, 1933. Thin lines are isotherms (°K); heavy lines are tropopause or boundaries of frontal layers. At bottom are surface winds and rainfall, and temperature and pressure traces. (After Van Mieghem, 1939.)

During the same period before World War II, synoptic-aerological studies of cyclonic disturbances and fronts were also undertaken in the United States, especially by Willett (1933, 1935). The possibilities for aerological analyses of cyclones developed here somewhat later than in Europe, but were spurred by the inauguration of a radiosonde network, largely through the influence of F. W. Reichelderfer. During the preradiosonde period and during the first years of radiosondes before World War II, synoptic meteorology was gradually developed into a synoptic aerology. Important contributions to this evolution came from several countries, but special mention should be given to the numerous papers by Scherhag. Scherhag's book, "*Wetteranalyse und Wetterprognose*" (1948) contains, in addition to its advanced upper-air analyses, a very complete list of publications in the field of synoptic meteorology, especially from the time covered by this brief and limited historical review.

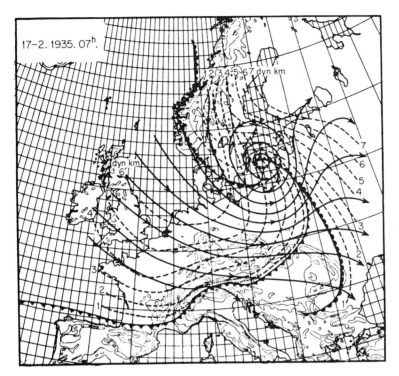

Fig. 5.6 Dashed lines show the frontal contours (kilometers above sea level) of a cyclone over western Europe at 02 GCT Feb. 17, 1935. Solid lines are streamlines inferred from the geostrophic wind field at various levels, corresponding to the flow in the warm air at the frontal surfaces or, in the warm sector, at the top of the surface friction layer. (After J. Bjerknes and Palmén, 1937.)

5.6 Linkage between Upper Waves and Surface Disturbances

The general nature of the problem of cyclogenesis has been elaborated upon by a large number of authors, e.g., by Brunt (1930): First, the pressure at any point on the earth's surface reflects the weight of the entire overlying atmosphere. In order to cause the pressure to fall, as is necessary to create a cyclone, a certain amount of the air must be removed horizontally (Margules, 1904). Second, removal of air horizontally must result in creation of an anticyclonic circulation due to the influence of the coriolis acceleration upon the outflowing air particles. Third, owing to the same influence, the air must be made to converge horizontally in order to produce a cyclonic circulation in low levels. It follows that *if the surface pressure is to fall, horizontal divergence must take place in upper levels, removing more air from the total atmospheric column than is accumulated through horizontal convergence in lower levels.*

Consequently, the physical cause for the deepening of cyclones must be ultimately sought in an examination of the higher levels.[6] This reasoning formed the basis for Ryd's (1923, 1927) and Scherhag's (1934) ideas on forecasting lower-level development from the upper-level flow patterns. They observed that sea-level cyclones frequently intensified where the upper-level isobars diverged laterally in a downstream direction. Especially with regard to this pattern, Scherhag expounded the proposition that "divergent upper winds must produce in general a fall of pressure if they are not compensated by strong convergence below." The work of Scherhag and others of the German school represented the first systematic application of upper-air charts to the forecasting of migratory weather systems, and constituted an important step forward despite the inadequacy of upper-air observational data.

The basic nature of the upper-level divergence field was elucidated by J. Bjerknes (1937) in a remarkably simple and far-reaching concept. By use of the gradient-wind relationship, he inferred that in general there is upper-level divergence downstream from and convergence upstream from troughs in the westerlies. With the aid of the "tendency equation," Bjerknes linked the upper-level divergence pattern to the lower-level pressure changes.

The tendency equation, obtained from the hydrostatic and continuity equations, is

$$\left(\frac{\partial p}{\partial t}\right)_h = - \int_h^\infty g\nabla_h\cdot(\varrho\mathbf{V})\,dz + (g\varrho w)_h \tag{5.1}$$

where h is the height of a fixed horizontal surface in the atmosphere and $\nabla_h\cdot(\varrho\mathbf{V})$ is the horizontal mass divergence. If applied on the ground level (over flat terrain), the last term in Eq. (5.1) vanishes, and the pressure tendency can be expressed by

$$\left(\frac{\partial p}{\partial t}\right)_0 = - g \int_0^\infty \varrho\nabla_h\cdot\mathbf{V}\,dz - g \int_0^\infty \mathbf{V}\cdot\nabla_h\varrho\,dz \tag{5.2}$$

Here the first r.h.s. term represents the influence of the horizontal wind divergence, and the second term is the contribution to the pressure tendency by the density advection.

It the case of a moving cyclone, the surface pressure generally falls ahead

---

[6] A partial exception to this general rule (namely, orographic cyclogenesis) is discussed in Section 11.8.

of the center and rises behind it. The negative pressure tendency is largely, but not wholly, determined by warm-air advection ahead of the moving warm front; the positive tendency, by the cold-air advection behind the cold front. In the inner part of the warm sector, where thermal advection is of minor influence, the sign of $(\partial p/\partial t)_0$ is essentially determined by the divergence term in Eq. (5.2). Falling pressure here indicates a deepening of the cyclone. This deepening, along with increasing cyclonic circulation in low levels, is possible only if the high-level divergence ahead of the upper trough exceeds the low-level convergence, thus providing a net mass divergence.

Bjerknes' scheme for the structure of the cyclone, based on the afore-mentioned aerological studies (e.g., Figs. 5.5 and 5.6), is shown in Fig. 5.7.

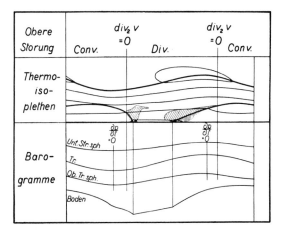

FIG. 5.7 Schematic vertical section (top) through the warm sector of a young wave cyclone. Thin lines are isotherms; thick lines, fronts and tropopauses. In lower part, pressure profiles in the lower stratosphere, near the mean level of the tropopause, in the upper troposphere, and at the ground. (After J. Bjerknes, 1937.)

The upper part shows a cross section through a cyclone, taken along a line south of its center and parallel to its direction of movement. In the lower part of the figure, pressure profiles are shown for level surfaces in the lower stratosphere, at the mean level of the tropopause (Tr.), in the upper tropo-sphere, and at the earth's surface. The vertical separations of the pressure profiles are indicative of the mean temperatures of the bounded layers. Thus, for example, the two uppermost profiles are relatively far apart near the upper trough where the lower-stratospheric air is warmest, and the thickness of the lowest layer reflects the presence of the cold-air mass in varying depths at different places.

In line with the earlier reasoning, Bjerknes sought the key to surface-pressure changes in the divergence fields associated with the upper waves (Section 6.1 and 6.2). Near the troughs and ridges of these upper waves, divergence is zero. Thus, if the lower-level cyclones and anticyclones were in phase with the upper-level pattern, it is clear that the upper-level divergence and convergence required to generate and maintain the surface pressure systems would be absent. If, however, there were a westward tilt, as in Fig. 5.7, *the surface cyclone would lie beneath the region of upper-level divergence and the anticyclone beneath the region of convergence aloft.* At the same time, if the upper- and lower-level systems were nearly in phase, surface-pressure falls ahead of and pressure rises behind cyclones (which are essential if the systems move) are provided for by the field of divergence in upper levels. Hence, deepening of a polar-front cyclone requires a phase lag between the surface depression and the upper trough, with the latter lagging. Hydrostatically, this is equivalent to a corresponding phase difference between the low-level pressure trough and the thermal trough, which in turn is consistent with the pattern of warm and cold advection around the cyclone.

Margules (1904) indicated that because only a very small net divergence or convergence is needed to account for the observed changes in surface pressure, the limited accuracy of observations aloft precludes the use of Eq. (5.1) in a quantitative sense. Nevertheless, Bjerknes was able to use this tool along with inferences of the upper divergence, based on the configuration of the pressure field, to link in a logical fashion the processes going on in the various levels of the atmosphere. Further significant contributions on this subject came especially from Sutcliffe (1938, 1939). He emphasized particularly *the requirement for ageostrophic motions, without which the divergence would be virtually absent, and the essential importance of vertical shear for the development of disturbances.* Also in 1939, Rossby *et al.* presented a theory of the upper-level waves, which was to play a very important role in the future development of meteorology. At this point in history, theory had outrun application, and full realization of the practical value of the accumulated theories had to await the vast improvement of the aerological network, which took place during and after World War II.

REFERENCES

Assmann, R. (1900). "Beiträge zur Erforschung der Atmosphäre mittels des Luftballons." Mayer & Müller, Berlin.
Bergeron, T. (1928). Über die dreidimensional verknüpfende Wetteranalyse. I. *Geofys. Publikasjoner, Norske Videnskaps-Akad. Oslo* **5**, No. 6, 1–111.

Bergeron, T. (1959). Methods in scientific weather analysis and forecasting. An outline in the history of ideas and hints at a program. *In* "The Atmosphere and the Sea in Motion" (B. Bolin, ed.), pp. 440-474. Rockefeller Inst. Press, New York.

Bergeron, T., and Swoboda, G. (1924). Wellen und Wirbel in einer quasistationären Grenzfläche über Europa. *Veroeffentl. Geophys. Inst. Univ. Leipzig* [2] **3**, No. 2, 63–172.

Bigelow, F. H. (1904). The mechanism of countercurrents of different temperatures in cyclones and anticyclones. *Monthly Weather Rev.* **31**, 72.

Bjerknes, J. (1919). On the structure of moving cyclones. *Geofys. Publikasjoner, Norske Videnskaps-Akad. Oslo* **1**, No. 1, 1–8.

Bjerknes, J. (1932). Exploration de quelques perturbations atmosphèriques à l'aide de sondages rapprochés dan le temps. *Geofys. Publikasjoner, Norske Videnskaps-Akad. Oslo* **9**, No. 9, 1–52.

Bjerknes, J. (1935). Investigations of selected European cyclones by means of serial ascents. *Geofys. Publikasjoner, Norske Videnskaps-Akad. Oslo* **11**, No. 4, 1–18.

Bjerknes, J. (1937). Theorie der aussertropischen Zyklonenbildung. *Meteorol. Z.* **54**, 462–466.

Bjerknes, J., and Palmén, E. (1937). Investigations of selected European cyclones by means of serial ascents. *Geofys. Publikasjoner, Norske Videnskaps-Akad. Oslo* **12**, No. 2, 1–62.

Bjerknes, J., and Solberg, H. (1921). Meteorological conditions for the formation of rain. *Geofys. Publikasjoner, Norske Videnskaps-Akad. Oslo* **2**, No. 3, 1–60.

Bjerknes, J., and Solberg, H. (1922). Life cycle of cyclones and the polar front theory of atmospheric circulation. *Geofys. Publikasjoner, Norske Videnskaps-Akad. Oslo* **3**, No. 1, 1–18.

Bjerknes, J., Mildner, P., Palmén, E., and Weickmann, L. (1939). Synoptisch-aerologische Untersuchung der Wetterlage während der Internationalen Tage vom 13 bis 18 Dezember 1937. *Veroeffentl. Geophysik. Inst. Univ. Leipzig* **12**, No. 1, 1–107.

Bjerknes, V. (1898). Über einen hydrodynamischen Fundamentalsatz und seine Anwendung besonders auf die Mechanik der Atmosphäre und des Weltmeeres. *Kgl. Svenska Vetenskapsakad. Handl.* **31**, No. 4.

Bjerknes, V. (1902). Zirkulation relativ zur Erde. *Meteorol. Z.* **19**, 97–108.

Bjerknes, V., Hesselberg, T., and Devik, O. (1911). "Dynamic Meteorology and Hydrography. Part II. Kinematics," Chapter 11. Carnegie Inst. Washington, D. C.

Blasius, W. (1875). "Storms, their Nature, Classification and Laws," 342 pp. Porter & Coates, Philadelphia, Pennsylvania.

Brunt, D. (1930). Some problems of modern meteorology: I. The present position of theories on the origin of cyclonic depressions. *Quart. J. Roy. Meteorol. Soc.* **56**, 345–350.

Chromow, S. P. (unter Mitwirkung von N. Konček) (1942). "Einführung in die synoptische Wetteranalyse" (German ed. by G. Swoboda), 532 pp. Springer, Vienna.

Defant, A. (1912). Die Veränderungen der allgemeinen Zirkulation der Atmosphäre in den gemässigten Breiten der Erde. *Wiener Sitzber.* **121**, 319.

Defant, A. (1926). Primäre und sekundäre, freie und erzwungene Druckwellen in der Atmosphäre. *Sitzber. Akad. Wiss. Wien., Math.-Naturn. Kl. Abt. IIa* **135**, 357-377.

Dines, W. H. (1912). The vertical temperature distribution in the atmosphere over England, with some remarks on the general and local circulation. *Geophys. Mem.* No. 1.

Dines, W. H. (1919). The characteristics of the free atmosphere. *Geophys. Mem.* No. 13.

Dines, W. H. (1925). The correlation between pressure and temperature in the upper air with a suggested explanation. *Quart. J. Roy. Meteorol. Soc.* **51**, 31–38.

Dove, H. W. (1837). "Meteorologische Untersuchungen." Berlin.

Espy, J. P. (1841). "Philosophy of Storms." Boston, Massachusetts.

Exner, F. M. (1925). "Dynamische Meteorologie," 2nd ed., 421 pp. Springer, Vienna.

Ferrel, W. (1856). An essay on the winds and the currents of the ocean. *Nashville J. Med. Surg.* **11**, 277–301 and 375–389.

Ferrel, W. (1889). "A Popular Treatise on the Winds." Wiley, New York.

Ficker, H. (1911). Das Fortschreiten der Erwärmungen (der "Wärmewellen") in Russland und Nordasien. *Sitzber. Akad. Wiss. Wien., Phys.-Math. Kl. II* **120**, 745.

Ficker, H. (1920). Beziehung zwischen Änderungen des Luftdruckes und der Temperatur in den unteren Schichten der Troposphäre (Zusammensetzung der Depressionen). *Sitzber. Akad. Wiss. Wien., Math. Naturn. Kl. Abt. IIa* **129**, 763–810.

FitzRoy, R. (1863). "The Weather Book, A Manual of Practical Meteorology," 2nd ed. London.

Godske, C. L., Bergeron, T., Bjerknes, J., and Bundgaard, R. C. (1957). "Dynamic Meteorology and Weather Forecasting," p. 536. Am. Meteorol. Soc., Boston, Massachusetts.

Hann, J. (under Mitwirkung von R. Süring) (1915). "Lehrbuch der Meteorologie." Tauschnitz Verlag, Leipzig.

Hanzlik, S. (1909). Die räumliche Verteilung der meteorologischen Elemente in den Antizyklonen. *Denkschr. Akad. Wiss. Wien* **84**, 163–256.

Haurwitz, B. (1927). Beziehungen zwischen Luftdruck und Temperaturänderungen. *Veroeffentl. Geophysik. Inst. Univ. Leipzig* **3**, No. 5.

Helmholtz, H. (1888). Über atmosphärische Bewegungen (I). *Sitzber. Preuss. Akad. Wiss.* pp. 647–663; (II) *Meteorol. Z.* **5**, 329–340.

Helmholtz, H. (1889). Über atmosphärische Bewegungen (III). *Sitzber. Preuss. Akad. Wiss.* pp. 761–780.

Hesselberg, T. (1913). Über die luftbewegung im Zirrusniveau und die Fortpflanzung der barometrischen Minima. *Beitr. Physik Freien Atmosphäre* **19**, 198–205.

Hewson, E. W. (1936, 1937). The application of wet-bulb potential temperature to air mass analysis. *Quart. J. Roy. Meteorol. Soc.* **62**, 387–420; **63**, 7–20.

Kasahara, A., Isaacson, E., and Stoker, J. J. (1965). Numerical studies of frontal motion in the atmosphere. I. *Tellus* **17**, 261–276.

Lempfert, R. G. K., and Corless, R. (1910). Line-squalls and associated phenomena. *Quart. J. Roy. Meteorol. Soc.* **36**, 135–170.

Lempfert, R. G. K., and Shaw, W. N. (1906). Life history of surface air currents. Publ. No. 174, 107 pp. Meteorol. Office, London.

Margules, M. (1903). Über die Energie der Stürme. *Jahrb. Zentralanst. Meteorol. Wien* 1–26; translation in C. Abbe (1910). *Smithsonian Inst. Meteorol. Coll.* **54**, 533–595.

Margules, M. (1904). Über die Beziehung zwischen Barometerschwankung und Kontinuitätsgleichung, Boltzman Festschrift, Leipzig, 585–589.

Margules, M. (1906). Über Temperaturschichtung in stationär bewegter und ruhender Luft. *Hann-Band. Meteorol. Z.* pp. 243–254.

Nyberg, A., and Palmén, E. (1942). Synoptische-aerologische Bearbeitung der internationalen Registrierballonaufstiege in Europa in der Zeit 17–19 Oktober 1935. *Statens Meteorol.-Hydrol. Anstalt, Medd., Ser. Uppsater* No. 40, 1–43.

Palmén, E. (1935). Registrierballonaufstiege in einer tiefen Zyklone. *Soc. Sci. Fennica, Commentationes Phys.-Math.* **8**, No. 3.

Petterssen, S. (1956). "Weather Analysis and Forecasting," 2nd ed., Vol. I, Chapter 12. McGraw-Hill, New York.

Richardson, L. F. (1922). "Weather Prediction by Numerical Process," 236 pp. Cambridge Univ. Press, London and New York.

Rossby, C.-G. (1927). Zustandsänderungen in atmosphärischen Luftsäulen. *Beitr. Physik. Freien Atmosphäre* **13**, 662–673.

Rossby, C.-G., and Collaborators (1939). Relation between the intensity of the zonal circulation of the atmosphere and displacements of the semipermanent centers of action. *J. Marine Res.* **2**, 38–55.

Ryd, V. H. (1923). Meteorological problems. II. Travelling cyclones. *Medd. Dansk Meteorol. Inst.* No. 5, 1–124.

Ryd, V. H. (1927). The energy of the winds. *Medd. Dansk Meteorol. Inst.* No. 7, 1–96.

Schedler, A. (1921). Die Beziehung zwischen Druck und Temperatur in der Freien Atmosphäre. *Beitr. Physik. Freien Atmosphäre* **9**, 181–201.

Scherhag, R. (1934). Die Bedeutung der Divergenz für die Entstehung der Vb-Depressionen. *Ann. Hydrograph. (Berlin)* **62**, 397.

Scherhag, R. (1948). "Neue Methoden der Wetteranalyse und Wetterprognose," 424 pp. Springer, Berlin.

Shaw, Sir Napier (1926-1931). "Manual of Meteorology," Vols. 1-4. Cambridge Univ. Press, London and New York.

Shaw, W. N. (1911). "Forecasting Weather," 3rd ed. Constable, London.

Sutcliffe, R. C. (1938). On the development in the field of barometric pressure. *Quart. J. Roy. Meteorol. Soc.* **64**, 495–509.

Sutcliffe, R. C. (1939). Cyclonic and anticyclonic development. *Quart. J. Roy. Meteorol. Soc.* **65**, 518–524.

Teisserenc de Bort, L. (1902). Variation de la température de l'air libre dans la zone comprise entre 8 km et 13 km d'altitude. *Compt. Rend.* **134**, 987.

Van Mieghem, J. (1939). Sur l'existence de l'air tropical froid et de l'effect de foehn dans l'atmosphère libre. *Mem. Inst. Met. Belg.* **12**, 1–32.

Willett, H. C. (1933). American air mass properties. *Papers Phys. Oceanog. Meteorol., Mass. Inst. Technol. Woods Hole Oceanog. Inst.* **2**, No. 2, 1–116.

Willett, H. C. (1935). Discussion and illustration of problems suggested by the analysis of atmospheric cross-sections. *Papers Phys. Oceanog. Meteorol., Mass. Inst. Technol. Woods Hole Oceanog. Inst.* **4**, No. 2, 1–41.

6

EXTRATROPICAL DISTURBANCES IN
RELATION TO THE UPPER WAVES

During World War II and the postwar years, there was a very considerable expansion of the global aerological network and of communications
facilities. This made it possible for the first time to carry out routine analysis
on a hemisphere-wide basis. Publication of the U.S. Weather Bureau *Northern Hemisphere Daily Series, Sea-Level and* 500-mb *Charts*, and of aerological data by some national services, greatly stimulated research on this
scale. It became increasingly evident, in line with earlier but less substantial
evidence, that major changes of the flow pattern in one part of the hemisphere often presage significant events in other parts. The increase in density
of aerological observations, together with their extension to higher elevations, led to renewed attacks upon the relationships between higher-level
flow patterns and the development and behavior of weather systems. In
this chapter we briefly review some of the findings.

6.1 Planetary Control of Upper Waves

Except in the "friction layer" near the earth's surface, the wind blows
predominantly parallel to the isobars on a level surface, or to the contours
on an isobaric surface. The gradient wind relationship is given as

$$KV^2 = -fV + g\left(\frac{\delta z}{\delta n}\right)_p \tag{6.1}$$

where K is *trajectory* curvature (positive when cyclonic), The distance δn between contours at an interval δz is thus prescribed by the wind speed, the latitude, and the trajectory curvature. The trajectory curvature can be related to the streamline curvature, and thus the wind and its variations can be connected with the geometry of the isobaric contours.

Based upon this relationship, Bjerknes (1937) deduced the general pattern of divergence connected with wave systems. His analysis was elaborated by Bjerknes and Holmboe (1944) in a paper that provides a particularly straightforward and lucid account of the essential features of upper-tropospheric waves and their connection with lower-tropospheric cyclones and anticyclones. In the ensuing discussion, we adopt the approach of Bjerknes and Holmboe to arrive at some generalizations about wave properties. For the sake of simplicity, we make the special assumption that the wind blows strictly parallel to the contours. Although this assumption is not everywhere valid, it is sufficiently close to reality so that the resulting generalizations are broadly correct.

Let $\bar{D}_A$ denote the mean horizontal divergence over the area A in Fig. 6.1, enclosed by two contours and extending ahead of the trough line to the downstream ridge line. The mean divergence may be evaluated from the fundamental definition of divergence, $\bar{D} = (1/A)\, dA/dt$. If both contours are sinusoidal,

$$A = \frac{(\delta n_R + \delta n_T)}{2} \cdot \frac{L}{2}$$

With the assumption of no cross-contour flow,

$$\frac{dA}{dt} = (V_R\, \delta n_R - V_T\, \delta n_T)$$

By the same assumption, no speed change occurs between trough and ridge, so $V_R = V_T = V$. Then, substituting the above,

$$\bar{D}_A = \frac{4V}{L} \frac{(\delta n_R - \delta n_T)}{(\delta n_R + \delta n_T)} \tag{6.2}$$

From Eq. (6.1),

$$\delta n = \frac{g\, \delta z}{V(f + KV)} \tag{6.3}$$

Thus the transport $V\, \delta n$ through trough or ridge line, prescribed by the width of the contour channel, is governed by two principal factors. Neglecting the influence of curvature, the contours would be closest together in

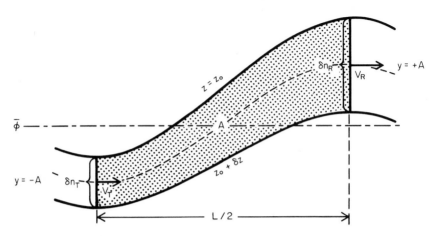

Fig. 6.1   A portion of an upper-level wave pattern, over which the mean divergence is computed (see text).

the ridge of a wave in the westerlies, owing to the northward increase of the coriolis parameter. Considering only the influence of curvature, the contours would be farthest apart in the ridge, since the centripetal acceleration (expressed by the curvature) is the resultant of the coriolis and pressure-gradient accelerations. This is so if the wind speed exceeds the speed of movement of the wave pattern, in which case the trajectory and streamline curvatures have the same sign. Accordingly, *the influences of the variation of curvature along the wave and of the variation of the coriolis parameter with latitude act in opposite senses, the former contributing toward divergence and the latter toward convergence east of troughs.*

Substitution from Eq. (6.3) into Eq. (6.2) gives

$$\bar{D}_A = \frac{4V}{L} \left\{ \frac{(f_T - f_R) + V(K_T - K_R)}{(f_T + f_R) + V(K_T + K_R)} \right\} \tag{6.4}$$

In the center of the contour channel, let a streamline be prescribed by

$$y = A \sin 2\pi \left(\frac{x - ct}{L} \right)$$

where A is streamline amplitude and c is the eastward speed of wave movement. At the wave crests and troughs, the *streamline* curvature $K_s = \partial^2 y / \partial x^2$ is, by differentiation of this expression,

$$K_s = \mp \frac{4\pi^2 A}{L^2} \qquad (y = \pm A)$$

From Blaton's formula, $K = K_s(V - c)/V$, the corresponding trajectory curvatures are

$$K_T = - K_R = \frac{V - c}{V} \cdot \frac{4\pi^2 A}{L^2} \tag{6.5}$$

Also, $(f_T + f_R) \approx 2\bar{f}$, where $\bar{f}$ denotes the coriolis parameter at the central latitude of the wave and $f_R - f_T \approx 2\beta A$ where $\beta \equiv df/dy$. Making these substitutions along with Eq. (6.5) into Eq. (6.4),

$$\bar{D}_A \approx \frac{4VA}{\bar{f}L} \left\{ (V - c) \frac{4\pi^2}{L^2} - \beta \right\} \tag{6.6}$$

At the level of nondivergence ($\bar{D} = 0$), denote the wind speed by V_L. Then, from Eq. (6.6),

$$V_L - c = \frac{\beta L^2}{4\pi^2} \tag{6.7}$$

This expression was originally derived by C.-G. Rossby (1939) on the basis of conservation of absolute vorticity, and played an important role in the evolution of modern meteorology. Demonstration of the practical applicability of Rossby's formula strengthened his conviction that the principal changes of the atmospheric circulation could be predicted by considering readjustments of the velocity field, expressed by advective changes of the vorticity field, at a level of nondivergence in the middle troposphere, without taking into account the complications arising from the changes of structure with height. This proposition, which must be considered bold in light of the strongly baroclinic structure of the atmosphere, made it practical to carry out the first numerical predictions by using the vorticity equation. These calculations formed the cornerstone for the elaborate models that have since been developed, eventually taking baroclinic influences into account.

Equation (6.7) corresponds to the condition that contour spacings are equal at trough and ridge, the transports of air into and out of area "A" in that case being the same. According to Eq. (6.3), this condition is satisfied if the difference of f between the trough and the ridge at a higher latitude is compensated by an equal difference in the curvature term KV. Since for a prescribed wavelength both the curvature difference and the difference of the coriolis parameters increase with the amplitude, at this level (only) the divergence is independent of wave amplitude. As pointed out in the prologue of Rossby's paper, when the lateral shear is neglected, these same characteristics of the wave, which determine the transport capacity between isobars and thus the divergence, also determine the variations of absolute vorticity along the current. Consequently, his derivation using the vorticity equation yields the same result.

Bjerknes and Holmboe called the quantity on the right of Eq. (6.7) the "critical velocity" V_c. Transposing terms,

$$c = V_L - \frac{\beta L^2}{4\pi^2} = V_L - V_c \qquad (6.7')$$

This formulation by Rossby, and by Bjerknes and Holmboe, introduced the sphericity of the earth (i.e., the latitude variation of f) as an influence that fixes the movement of a wave in relation to its dimensions. Equation (6.7) states that, *for a given wavelength and latitude, waves move faster if the wind speed is greater. Also, for a given wind speed, the wave speed is greater for short than for long wavelengths.* While very short waves tend to move nearly with the wind speed at the level of nondivergence, in the case of long waves (commonly 4 or 5 around the hemisphere), V_c approaches the value of V_L. Long waves thus move slowly and may even be stationary.

6.2   BAROCLINIC WAVES

By substitution from Eq. (6.7) into Eq. (6.6), the mean divergence at any level ahead of a sinusoidal wave trough is

$$\bar{D}_A \approx \frac{4VA\beta}{\bar{f}L} \left\{ \frac{V-c}{V_L-c} - 1 \right\} \qquad (6.8)$$

Figure 6.2a shows the typical divergence pattern in the baroclinic westerlies, with an increase of wind upward through the troposphere. At the level of nondivergence, the west wind exceeds the wave speed by the critical velocity specified by Eq. (6.7'). At higher or lower levels, $(V - c) \gtrless (V_L - c)$, since normally the wave speed varies little with elevation.

Accordingly, Eq. (6.8) shows that if the wind speed increases with height, there is upper-tropospheric divergence and lower-tropospheric convergence ahead of troughs; the converse behind them, as in Fig. 6.2a. From the principle of mass continuity, ascending motions must be present ahead of troughs (reaching greatest magnitude at the level of nondivergence), with descending motions behind troughs.

Equation (6.8) may, by substitution for $(V_L - c)$ from Eq. (6.7), be written in the equivalent form

$$\bar{D}_A \approx \frac{16\pi^2 A}{\bar{f}L^3} V(V - V_L) \qquad (6.9)$$

This shows that *the magnitude of the gradient-wind divergence is greatest if the wave amplitude is large, the wavelength small, and the wind speed is large and significantly different from that at the level of nondivergence.*

Since $(V - V_L)$ is greatest at the level of maximum wind, Eq. (6.9) indicates that the divergence reaches greatest magnitude at the level of maximum wind in a sinusoidal wave. Also, since $(V - V_L)$ is a measure

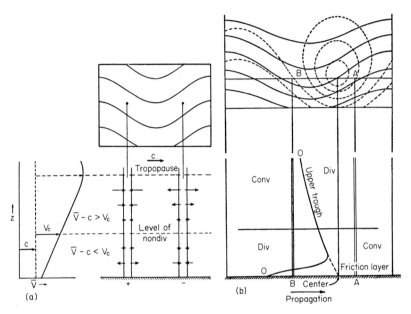

FIG. 6.2 (a) At top, a wave pattern on an upper-level chart; below, the divergence and convergence in vertical columns. At lower left, the vertical profile of west-wind speed $\bar{V}$ in the basic baroclinic current (see text). (b) At top, solid contours correspond to an upper-tropospheric wave pattern; dashed lines, simplified sea-level isobars. Lower diagram, the distribution of divergence in a west-east vertical section. (After Bjerknes and Holmboe, 1944.)

of the vertical shear, it follows that *the magnitude of the upper-tropospheric divergence, and correspondingly of the associated middle-tropospheric vertical motion, depend upon the baroclinity of the basic current.* Essentially for this reason, migratory cyclones and anticyclones, which are characterized by appreciable divergence and vertical motions, are observed in close association with the baroclinic region of the polar-front jet stream and do not achieve great intensities elsewhere.

The preceding analysis is based upon the assumption of a sinusoidal wave pattern and is thus applicable only to the middle and upper troposphere. For a discussion of systems with closed isobars, characteristic of the low troposphere, reference may be made to Bjerknes and Holmboe (1944). The general conclusion, based upon Eq. (6.3), is that in general the gradient wind distribution implies convergence on the east and divergence on the west sides of closed cyclones. The reverse may be true if a cyclone is strongly asymmetrical, with closely packed isobars on the equatorward side and widely spaced isobars on the poleward side. However, the required degree of asymmetry is not observed in sea-level cyclones of middle latitudes; in fact many cyclones have asymmetries of the opposite kind.

The general scheme arrived at by Bjerknes and Holmboe is shown in Fig. 6.2b. Owing to the presence of colder air to the west of a cyclone, the upper trough normally slopes toward west with increasing height. Thus, the cyclone at the surface is located beneath the divergent part of the upper wave. Considering a vertical column over point A, according to the tendency equation (Section 5.6), the pressure fall usually required for eastward movement of the surface cyclone depends upon a net mass divergence in the upper levels which exceeds the mass convergence in the lower troposphere. At B, an analogous conclusion holds to account for the surface-pressure rise.

Within the layer affected by surface friction, there is cross-isobaric flow toward the cyclone center. Consequently, low-level convergence is present at the cyclone center and for some distance behind it, the convergence and divergence regions being separated by the solid line rather than the dashed line corresponding to the trough in Fig. 6.2b in lower levels. In the earlier analysis, $\bar{D}_A$ corresponds to the mean divergence between trough and ridge. By symmetry, the divergence is zero at troughs and ridges and greatest at the inflections of the upper wave where the variation of $(f + KV)$ along the current is greatest. Because of the phase lag of the upper trough in Fig. 6.2b, the surface cyclone center lies beneath the region of appreciable divergence. This condition, which is also an expression of the baroclinity of the basic current, is evidently necessary either for the initial deepening of the cyclone (see Chapter 11), or for its maintenance when it has reached considerable intensity and would otherwise fill as a result of the lower-tropospheric convergence.

Equation (6.9) indicates a strong dependence of divergence upon wavelength. On the average, there is some proportionality between A and L, such that waves of great length tend also to have larger amplitudes than those of shorter waves. If we consider two waves of similar shape but of different lengths, Eq. (6.9) may be put into a form that more truly reflects the significance of wavelength in relation to divergence. For example, in the case of well-developed waves in which the amplitude is a quarter-wavelength, Eq. (6.9) becomes

$$\bar{D}_{(L=4A)} \approx \frac{4\pi^2}{\bar{f}L^2} V(V - V_L) \tag{6.10}$$

In general, since V and V_L are essentially the same for short waves and long waves (upon which they may be superposed), this expression shows that *for waves of similar shapes, the magnitude of the upper-tropospheric divergence (or mid-tropospheric vertical motion) is inversely proportional to the square of the wavelength.* Typically, a middle-latitude long wave has a length of around 6000 to 8000 km, and a short wave has a length of 2000 to 3000 km. Thus, the divergence in short waves is characteristically about an order of magnitude larger than the mean divergence in a long wave.

The preceding equations are based upon rather specialized assumptions of a simple wave system, and thus should not be expected to give numerical accuracy.[1] Nevertheless, they provide a basis of comparison for the properties of waves that depend upon their dimensions, together with the strength and baroclinity of the basic current. Reference may be made to Bjerknes and Holmboe (1944) or Holmboe, Forsythe, and Gustin (1945) for a more general discussion, and to Charney (1947) for a cognate theoretical analysis of the properties of baroclinic disturbances. Among the features not taken into account above is the velocity concentration in the jet stream, with its attendant lateral shears. For discussions of currents of restricted width, reference may be made to the original treatment of the problem by Haurwitz (1940) and to Petterssen (1952, 1956); Petterssen deals specifically with the propagation of waves in the jet stream.

6.3 Divergence and Vertical Velocity in Actual Disturbances

With the aid of the improved aerological soundings, efforts were made during the postwar years, particularly at New York University and Massachusetts Institute of Technology, to compute the three-dimensional fields of motion and to relate them to the occurrence of clouds and precipitation (Haurwitz *et al.*, 1945; Houghton and Austin, 1946; Panofsky, 1946; Miller, 1948; Fleagle, 1947, 1948). Figures 6.3 and 6.4, taken from Fleagle, are of particular interest in relation to the foregoing discussion.

---

[1] Partly because the "transversal divergence" connected with cross-contour flow has not been considered here. Its inclusion (see qualitative discussion by Bjerknes and Holmboe) does not fundamentally alter the generalizations made above.

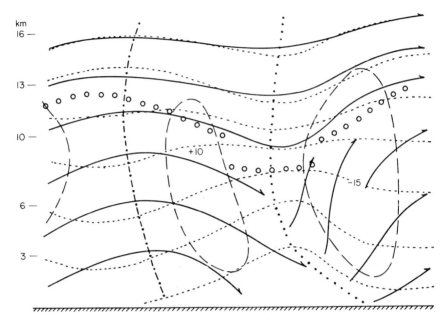

FIG. 6.3   Vertical cross section showing idealized composite field of potential temperature (short-dash lines), local pressure change (long-dash lines are isallobars for ±6 mb/
12 hr), and streamlines projected into the section. Generalized from several synoptic
cases by Fleagle (1948). Dotted curve represents trough line; dash-dotted curve, ridge
line; lines of circles, tropopauses.

Fleagle computed vertical motions from 12-hr trajectories on isentropic
surfaces. Owing to the length of this time interval between upper-air charts
and to the neglect of release of latent heat in precipitation areas, the ex-
tremes of vertical motion and divergences have been considerably underestimated. Nevertheless, the general patterns may be accepted as essentially
correct.

A generalization of the three-dimensional fields of motion, potential
temperature, and pressure change in a zonal section through traveling
cyclones and anticyclones is shown in Fig. 6.3, based on computations in
several synoptic cases. The overall characteristics are in accord with the
foregoing discussion. Vertical motions are largest in the middle troposphere
and decrease upward, but keep the same sign in the lower stratosphere as
in the troposphere. Ascent is generally associated with regions of pressure
fall at all levels; descent, with pressure rises. Referring to the tendency
equation, Eq. (5.1), this indicates that of the two r.h.s. terms, which tend
to compensate each other, the first term slightly dominates the second.

Consonant with the earlier conclusion from Eq. (6.9), Fig. 6.4a shows that the greatest upper-level divergence and convergence occur near tropopause level, approximately midway between troughs and ridges. In the lower troposphere, the strongest mass divergence is observed in the surface layers, convergence being greatest somewhat in advance of surface cyclones and divergence greatest some distance ahead of surface anticyclones. Allowing for the fact that pressure is represented on a logarithmic scale, Fig. 6.4 brings out *the near compensation of mass divergences of different signs in the lower troposphere and in upper layers.* The surfaces of nondivergence

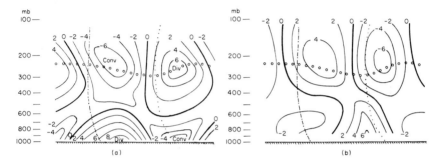

FIG. 6.4   (a) Vertical cross section showing average spatial distribution of horizontal divergence $\overline{\varrho^{-1}V_h \cdot \varrho V_h}$ relative to trough and wedge lines, in units of 10^{-6} sec^{-1}. (b) The proportionate local density change $\overline{\varrho^{-1}\,\partial\varrho/\partial t}$, in units of 10^{-7} sec^{-1}. (After Fleagle, 1948.)

are not horizontal, but divergence is on the whole near zero in the middle troposphere around 600 mb. This mean level is in good agreement with theoretical findings by Charney (1947) and empirical results of Cressman (1948) based on Rossby's wave formula (Eq. 6.7). Figure 6.4a is remarkably similar to an analogous section derived through a mathematical analysis by Charney, employing a model atmosphere with vertical but no horizontal shear.

The average spatial density-change distribution is shown for comparison in Fig. 6.4b. In general, there is close correspondence between horizontal divergence and density decrease, and vice versa, in upper levels. Since the quantity represented in Fig. 6.4b is, by the continuity equation, proportional to the three-dimensional mass divergence, this shows that in most regions the horizontal divergence overcompensates the vertical convergence. Alternatively, the local density changes show that, in most places, *the horizontal advection of warmer or colder air dominates the generally opposing influence of adiabatic cooling or warming associated with vertical motions* ahead of or behind the trough, respectively.[2] In the lower levels, local density increases reach a maximum slightly behind the trough line, where cold advection is pronounced; and, for a short distance behind the trough, adiabatic cooling due to ascending motions contributes further to a density increase. At the ridge line in lower levels, the local decrease of density is due almost entirely to subsiding motions with adiabatic warming.[3]

6.4   Long Waves and Short Waves, and Their Relation to Synoptic Disturbances

Investigations by Fultz (1945) and Cressman (1948) demonstrated that there are two characteristically different scales of waves in the upper westerlies. One of these is the "long" or "major" wave, with cold troughs and warm ridges, which is slow moving. The other "short" or "minor" waves display isotherm patterns that are typically not in phase with the streamlines. These waves are of comparatively small amplitude in the upper troposphere, and they move rapidly.

When the long waves are well defined, in winter there are most often four or five waves around the Northern Hemisphere.[4] An idealized sketch showing the most typical relationship of cyclone families to the upper waves is shown in Fig. 6.5. Most commonly, cyclone families exist between long-wave troughs and the downstream ridges, i.e., beneath the region of

---

[2] It should be noted that the horizontal temperature advection cannot be judged from Fig. 6.3, since the meridional temperature gradients and meridional wind components are not represented in that section.

[3] Vederman (1952) found that tropospheric temperatures over the surface high center increased markedly on successive days (particularly in lower levels) in winter anticyclones over the United States, whose central pressure was decreasing. For increasing anticyclones, only a slight tropospheric warming was observed on the average. The decreasing (strongly warmed) highs move generally with a southward component. This might suggest a systematically greater intensity of subsidence over their centers than over the centers of intensifying highs that moved generally with a northward component, in line with considerations in Section 10.4. As Wexler (1951) stresses, changes of anticyclone structure must be viewed in light of several interacting processes, which especially include dynamical influences in the upper levels.

[4] It will be recalled that the winter mean 500-mb chart (Fig. 3.1) shows only three waves. This is a statistical result of the fact that long-wave troughs are most often well developed in the particular longitudes concerned and show a lesser preference to become fixed in other longitudes. As a matter of incidental interest, the "stationary" wavelengths, computed by Eq. (6.7) from the mean zonal geostrophic wind profiles derived from Figs. 3.1 and 3.2 (after reduction to a level of nondivergence assumed at 600 mb), correspond to hemispheric wave numbers of 3.8 in winter and 4.8 in summer.

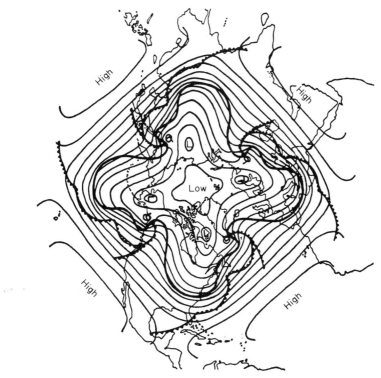

FIG. 6.5 Schematic circumpolar chart showing a simple four-wave pattern. Thin lines represent contours of 500-mb surface; heavy lines, the 500-mb polar-front intersection. Surface-front lines corresponding to cyclone families are superposed. (After Palmén, 1951.)

average upper-tropospheric divergence east of the major troughs. Super-posed on the long waves are the short-wave perturbations, which Bjerknes calls "cyclone waves" because of their intimate association with surface cyclones and migratory ridges. Because of their shorter wavelength, accord-ing to Eq. (6.7′), these waves progress more rapidly (about 20 to 40 knots) than the long waves (about 0 to 15 knots). Thus, the associated individual cyclones move forward relative to the long waves and are largely "steered" by the broad-scale flow patterns of the major waves.[5] Also (Section 6.2),

[5] The concept of "steering" is quite old in meteorology. As early as 1913, Hesselberg pointed out that low-level cyclones (or rapidly moving perturbations) move along the direction of the upper-tropospheric current. This concept was further developed by Schereschewsky and Wehrlé (1923), Mügge (1931), and Ficker (1929, 1938). The steering principle has been extensively used by Scherhag (1948), to whose work we especially refer the reader. A broad interpretation of the processes involved in steering is given in Chapter 11.

short waves in a baroclinic current are characterized by greater divergence and vertical motion than are the long waves. Although cyclones commonly form downstream from long long-wave ridges (particularly in western Canada and western Europe), these generally do not achieve the intensity of the cyclones forming downstream from long-wave troughs. An exception to this rule is the orographic cyclone, discussed in Chapter 11.

Because of the intimate connection between long waves and the behavior of cyclone waves, it is a matter of practical importance to predict movements and changes in pattern of the long waves. Although this is now done by numerical forecasting techniques, some of the earlier investigations based on Eq. (6.7) are still of interest in getting a simple picture of the linkage between the motions of waves, their dimensions, and the strength of the westerlies.

Practical application of Rossby's long-wave concept was first described by Namias and Clapp (1944), who demonstrated its usefulness in prognosticating the movements of waves on 5-day mean charts. Later, Cressman (1948, 1949) carried out extensive investigations in relation to shorter-period forecasting. He found that the latitude and strength of the maximum westerlies showed a considerable degree of conservatism from day to day. This fact established a firm basis for using Eq. (6.7), together with certain relationships derived from it, in predicting the movements and certain types of changes of the major waves.

By setting $c = 0$ in Eq. (6.7), a wavelength $L_s = 2\pi(V_L/\beta)^{\frac{1}{2}}$ can be defined as one for which the waves would be stationary. Then Eq. (6.7) can be written

$$c = \frac{\beta}{4\pi^2}\,(L_s^{\,2} - L^2) \qquad\qquad (6.11)$$

Waves are then eastward-progressive if the actual wavelength L is smaller than L_s, and Eq. (6.11) gives a useful measure of their speed.

Probably the most significant and interesting application of the theory of long waves is in appraising the likelihood and probable place of formation of new wave troughs. Cressman (1948) found that such new troughs did not develop when waves are, according to Eq. (6.11), progressive. New trough formations were found to occur only when the actual wavelength significantly exceeded the stationary wavelength. In that case, Eq. (6.11) suggests a westward movement of the troughs.

This occurs, however, not by the movement of existing troughs westward, but by a process of "discontinuous retrogression." Where the wavelength between one trough and the next trough downstream becomes excessively great (compared to L_s), the downstream trough is frequently replaced by a new one forming farther west in a location approximately L_s distant from the upstream trough.

This process is seldom localized, since if one trough retrogrades, the wavelength to the east of it becomes excessive, and in this event the next trough to the east is likely to retrograde in the same fashion. "The process repeats itself downstream until finally the wave number is increased by the addition of a new major trough, which is not compensated by the disappearance of an old trough." Such an event requires typically about a week for readjustment of the entire wave pattern around the hemisphere. When the wave number increases in this manner, the wavelength correspondingly diminishes and the waves become progressive.

Such readjustments of the upper-level waves have an important influence on the development of cyclones and anticyclones at sea level. For example, in the process of discontinuous retrogression, the former major trough accelerates eastward, and while it typically diminishes in intensity, cyclogenesis frequently takes place at sea level. A second cyclogenesis, often of major proportions, frequently occurs in connection with the deepening (farther west) of a minor trough to become the new major trough. These, along with the development of a new cold trough at the end of a retrogression period when the wave number increases, are stated by Cressman (1948) to be the developments most frequently associated with major cyclogenesis. The relationships between changes in the upper-level wave patterns and sea-level developments are elaborated by Riehl, LaSeur, *et al.* (1952), and interactions between short and long waves are discussed fully by Petterssen (1956).

The generalized structures of the waves of Fig. 6.5 are shown in somewhat more detail in Fig. 6.6. Figure 6.6a represents an idealized sea-level chart with its polar fronts and cyclone families, these being separated by an extensive anticyclone over central North America. Outflow of low-level polar air into low latitudes, partly into the trade-wind belt, occurs east and south of such anticyclones where the polar front at the surface is interrupted.

Figure 6.6b shows the same frontal system in relation to the 500-mb flow pattern. As mentioned in Section 6.2, the overall region between trough and downstream crest of the long wave is characterized by weak, average, upper divergence. Although the individual cyclones are connected with intense upper divergence and convergence regions, these (when averaged over the whole region between major trough and ridge) also constitute weak mean divergence. It is common to observe more than one surface

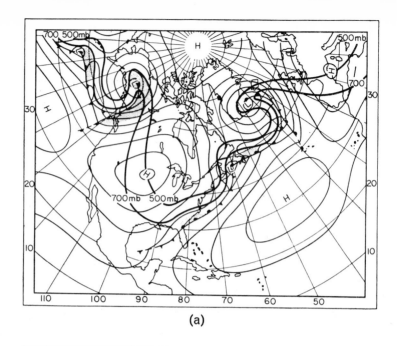

(a)

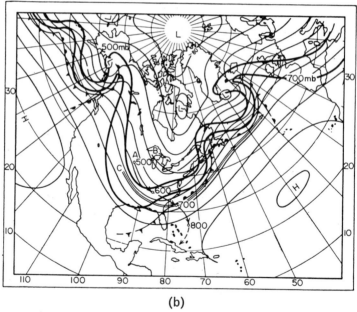

(b)

FIG. 6.6 (a) Schematic sea-level map showing cyclone families and topography of frontal surface. Cold air streams equatorward through the break between the polar-front systems in the lower layers (cf. Fig. 5.3). (b) The corresponding contours of the 500 mb surface, frontal contours, and sample three-dimensional trajectories (pressures on trajectory A at 12-hr intervals). Trajectory A does not overtake the surface front; rather, it remains in the cold air while the front moves eastward. (From Palmén, 1951.)

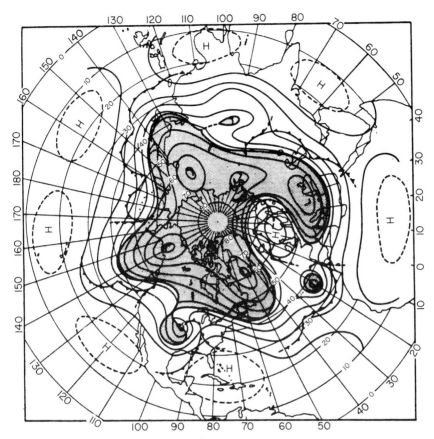

FIG. 6.7 Schematic circumpolar chart illustrating a highly perturbed wave pattern. System of lines as in Fig. 6.5. Stippling indicates areas occupied by polar air at the 500-mb level. (From Palmén, 1951.)

high to the west of a long-wave trough, but it is also not unusual to find this region (especially in winter) dominated by a single high, as in Fig. 6.6a. The mean low-level divergence in such anticyclones, even though they are of massive extent, is comparatively weak, in accord with the feeble upper-level convergence over such a region in the long-wave system.[6]

---

[6] Panofsky (1951) has stated the general rule that in the expression for divergence $(\Delta \bar{u} + \Delta \bar{v})/L$, where L is the horizontal dimension of an area occupied by a disturbance, the magnitudes of $\Delta \bar{u}$ and $\Delta \bar{v}$ tend to be almost independent of the scale of the disturbance. Hence, the mean divergence tends to be proportional to $1/L$. With an anticyclone of about 4000-km diameter, as in Fig. 6.6a, a mean outflow velocity of 5 m/sec would correspond to a lower-level divergence of 0.5×10^{-5} sec^{-1}. This is about the magnitude

Synoptic investigations have revealed many exceptions to the idealized patterns shown in Figs. 6.5 and 6.6. The atmosphere is frequently characterized by waves of irregular shape, as in Fig. 6.7, departing markedly from the form shown in Fig. 6.5 (which corresponds nearly to a sinusoidal shape of the upper contours when these are referred to a parallel of latitude). The different forms taken by disturbances are discussed further in Chapter 10.

6.5 Three-Dimensional Motions and Meridional Exchange of Air Masses

When three-dimensional motions are considered, the horizontal projections of the paths followed by air parcels may differ greatly from the paths suggested by isobaric charts. This is illustrated schematically by trajectory *A* in Fig. 6.6b, which originates west of the trough at 500 mb on the cold-air side of the polar front and jet stream. The curvature is at first slightly cyclonic, becoming anticyclonic at the end. The curvature change is consistent with the circumstance that in the cold air, the geostrophic wind backs with height so that as the air particle descends, it finds itself at levels where the wind blows more from a northerly or northeasterly direction. The implied mean vertical velocity during the approximately 36 hr required for the air to descend from 500 to 800 mb is about -3 cm/sec, a reasonable value.

Under conservation of absolute vorticity, a particle moving toward lower latitudes, where the earth's vorticity is smaller, would have to gain relative vorticity. The curvature change, suggestive rather of a decrease of relative vorticity, thus indicates marked horizontal divergence. The strong descent results in adiabatic (compressional) heating of the air column in the lower troposphere. At the same time, diabatic heating from the surface takes place as the air moves southward (especially if over the sea). From both considerations, it is apparent that the cold-air mass is rapidly modified as it moves southward. Motions of this character will be discussed in detail in Chapter 10, where real examples are shown.

On the warm-air side of the polar front at 500 mb, the schematic trajectory

---

of the upper convergence in the corresponding part of a long-wave system, whereas the upper divergence in cyclone-scale systems may locally be ten times as large (for an example, see Newton and Palmén, 1963). It is also often observed that a single, very large cyclone, rather than several smaller ones, are found in association with a long-wave trough (see Fig. 6.9). In this case also, the low-level convergence averaged over the area of the cyclone is likely to be small even though the cyclone has an intense circulation.

C follows the isobaric contours more closely. In this case, descent takes place as the air parcel approaches the trough from the west side, and ascent occurs after it moves to the east side. Since the wind speed is much greater on the warm-air than on the cold-air side of the front at this level, the air parcel moves through the wave pattern rapidly, and the total vertical displacement is relatively small. Within the warm air in the middle troposphere, the turning of wind with height is weak. Consequently, the influence of vertical movement on the direction followed by the air parcel is minor, in contrast to this influence in the cold air.

Parcel *B*, farther removed from the front, is in a region of weaker wind. According to the reasoning of Section 6.2, the divergence and vertical motions in this region distant from the jet stream are small, so that the departure of the wind from the trajectory suggested by the 500-mb contours is much less pronounced than for parcel *A*.

The overall picture is one in which the cold air to the west of disturbances undergoes marked vertical shrinking while being warmed through both adiabatic and diabatic influences. Although the warm-air trajectories in the middle and upper troposphere approximately follow the direction indicated by the contours, the subsiding cold-air masses partly cross beneath the jet-stream axis in their southward migration. Accompanying this is a general ascent of warm air from lower levels on the east side of the upper trough. The processes affecting the trajectories here will be described in Chapter 10, where we can go into more detail. In a situation such as illustrated by Fig. 6.6, ascent of warm air will take place primarily over the warm fronts of each individual cyclone, downstream from the connected short-wave troughs in the upper troposphere. While rising, the warm air takes on an anticyclonic curvature in these regions where the wind veers with height. Ascent is enhanced by the release of latent heat of condensation. This heat is mostly realized in the lower half of the troposphere and is carried upward as sensible heat, while some additional latent heat is liberated by further condensation and partial freezing of water substance in the upper troposphere.

This is part of the eddy transfer that, as discussed in Chapter 2, is necessary to counteract the radiative loss of heat in the upper troposphere. *Warm air rises and spreads out in the upper troposphere as it goes poleward, and at the same time cold air sinks and spreads out in the lower troposphere as it moves equatorward.* The meridional transfer of heat in upper levels is particularly effective when warm "blocking highs" of the kind over western Europe in Fig. 6.7 become cut off in higher latitudes and steer cyclones around their poleward boundaries. Counterpart to these are the cold "cutoff lows" in lower latitudes.

A model summarizing the principal aspects of the general circulation is shown in Fig. 6.8, after Defant and Defant (1958). This embodies the dominance of synoptic disturbances in middle latitudes and of the meridional Hadley circulation in lower latitudes. Although the latter is "closed" insofar as the mean meridional circulation is concerned, the surface streamlines in this picture show how polar air in the eddies enters the tropical region where it is modified in the trades, from which tropical-air branches also enter the extratropical latitudes.

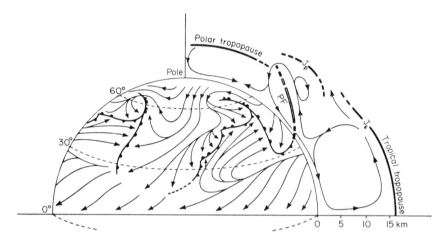

Fig. 6.8   Schematic model of the general circulation in a meridional section, as in Fig. 4.2, and schematic fronts and streamlines at the earth's surface (By permission from A. Defant and F. Defant, *Physikalische Dynamik der Atmosphäre*, Akademische Verlagsgesselschaft mbH, 1958.)

6.6   EXAMPLES OF WAVES OF DIFFERENT SCALES

Figures 6.9 and 6.10 illustrate the great range of scales of disturbances encountered in the atmosphere and serve to indicate some of the corresponding differences in the behaviors of the synoptic systems.

In the case of Fig. 6.9a, four long waves are evident in the form of large closed vortices separated by well-developed ridges. For the most part, the isotherms at 500 mb are very similar to the contours, air essentially as cold as the at that pole being found in the middle-latitude lows at this level. The "waves" were, as is characteristic with patterns of this type, very slow-moving and persistent. Corresponding to the upper-level vortices, there are at sea level (Fig. 6.9b) four significant regions occupied by cyclones, each very large and well defined except for the one off the east coast of Asia.

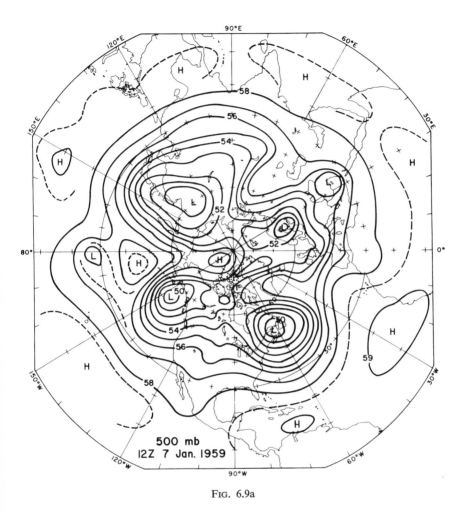

FIG. 6.9a

A strongly developed sea-level anticyclone is seen over the pole, with extensions into middle latitudes between the cyclones. The subtropical highs are only weakly evident.

The example in Fig. 6.10 offers marked contrasts in almost every respect. In Fig. 6.10a, nine almost equally spaced waves are present in the main belt of westerlies, all in varying stages of development but with generally small amplitudes. These moved rapidly eastward, with average speeds (about 20° longitude per day) in good accord with Eq. (6.7). At sea level (Fig. 6.10b), cyclones can be connected with each of the upper waves, with the exception of the one over Siberia. The subtropical highs are well

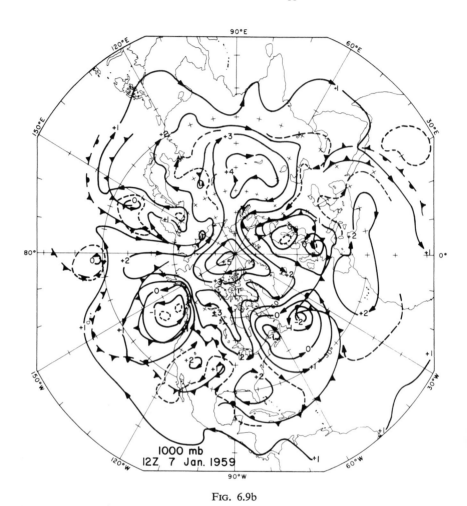

FIG. 6.9b

FIG. 6.9   500-mb and 1000-mb charts (contours at 100-m intervals) for 12 GCT
Jan. 7, 1959, illustrating a four-wave pattern with large closed cyclones. (From U.S.
Weather Bureau, 1959.)

developed. In harmony with the movement of the upper waves, the middle-
latitude cyclones move rapidly.

These examples bring out some of the striking differences characterizing
"low index" and "high index" situations. With low index (Fig. 6.9), there
is pronounced meridional flow; the zonally averaged west wind speed is
weak, and this combined with large wavelengths results in sluggish movement of the waves. With high index (Fig. 6.10), the meridional development

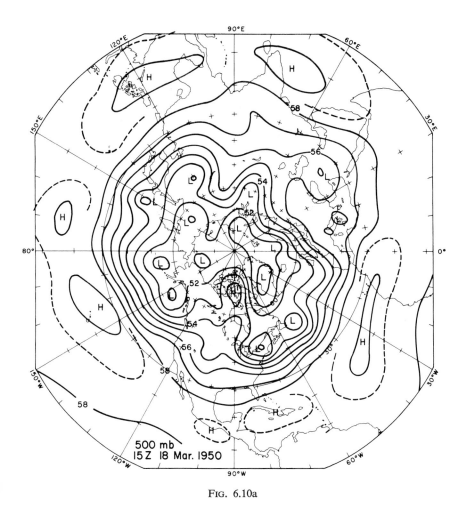

FIG. 6.10a

is by comparison weak; the zonally averaged west wind is large,[7] and this along with short wavelengths means rapid progression.

[7] Note in Fig. 6.10a that the belt of strong westerlies is centered about a point significantly displaced from the pole toward Alaska. As brought out by LaSeur (1954), such a displacement is characteristic and is especially pronounced in high-index situations. Computations mentioned in Section 17.4 indicate that hemispheric wave number 1, which expresses this eccentricity, contributes a substantial amount to the meridional transfer of heat. In Fig. 6.10 this eccentricity is pronounced, while the disturbances are of relatively small meridional extent; in Fig. 6.9, on the other hand, it seems likely that a greater proportion of the heat transfer is accomplished by the disturbances with their long north-south trajectories of warm and cold air.

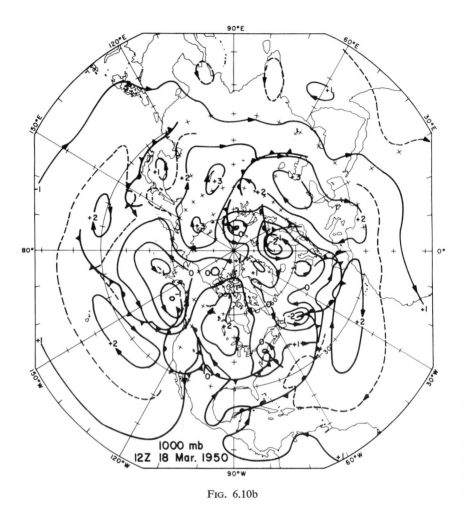

FIG. 6.10b

FIG. 6.10   The same, for 12-15 GCT Mar. 18, 1950, illustrating predominantly zonal upper-level flow with many waves. (From U.S. Weather Bureau, 1950.)

In the situation of Fig. 6.9, the four-wave pattern was clearly defined between Jan. 6 and 14, 1959, during which time the large vortices remained at practically fixed longitudes. However, the shapes of these vortices changed appreciably from day to day as short-wave perturbations, accompanied by successively developing sea-level cyclones, passed eastward through them. Between Jan. 12 and 13, the major trough over the eastern Pacific was replaced by a trough 20° farther west, where an intense surface cyclone developed. This event was followed by successive discontinuous retrogressions downstream and an increase of the hemispheric wave number from 4 to 5. An account of the cyclogenesis over North America, which resulted from this readjustment, is given in Section 11.5.

The significance of the varying wave patterns in terms of weather events is evident. While, in a situation like that shown in Fig. 6.10, cyclones progress rapidly all around the hemisphere, in the case of Fig. 6.9 certain regions are dominated for long periods by cyclonic activity while extensive areas remain relatively cyclone-free. Massive statistical investigations by Willett *et al.* (1947 and later) revealed significant contemporaneous correlations between certain features of the circulation at different latitudes. However, aside from a small degree of persistence, lag correlations revealed no reliable predictors of the general circulation in the indices of the zonal or meridional flow. An "index cycle," or alternation between periods of mainly zonal flow and highly disturbed flow, is a regularly observed feature (Rossby and Willett, 1948; Namias and Clapp, 1951). Such a fluctuation typically takes place over a period of several weeks. Julian (1966) carried out a statistical analysis of the sequential variations of the zonal index and found that there was no evidence for a particular periodicity; rather, the significant changes were related to a broad range of periods.

6.7 THE CONCEPT OF LATERAL MIXING

A concept that has played a major role in earlier studies of the general circulation is that of lateral mixing of atmospheric properties. In a broad sense, this is the process described by the Reynolds stresses and by the eddy transfers of properties other than momentum. In computations of these transfers, discussed in Chapters 1, 2, and 16, no assumption is made concerning the form of the transfer. Rather, appeal is made to the observations within the framework of physical equations.

By analogy with G. I. Taylor's formulation of vertical mixing, Defant (1921) introduced the idea that the meridional exchange of properties could be considered as an effect of large-scale horizontal turbulence in which the turbulence elements are cyclones and anticyclones. Thus, e.g., the poleward transfer of sensible heat could be expressed in the form $S_\varphi = - c_p A \, \partial \bar{T}/\partial y$, where $\bar{T}$ is the mean temperature at a latitude and A is an "Austausch" coefficient. This coefficient (evaluated empirically) can be expressed as the product of density, a characteristic meridional velocity, and a "mixing length" corresponding to the distance traveled in the meridional direction, prescribed by the typical dimensions of disturbances. Thus, A is many orders of magnitude greater for horizontal than for vertical transfer, while the horizontal gradients of the properties are also correspondingly smaller (see Haurwitz, 1941, for a full discussion). Evaluations by Lettau (1936) indicated that, on the average, $A \approx 8 \times 10^7$ gm cm^{-1} sec^{-1} in latitudes 50°N to 55°N where it is greatest, with about half this magnitude near the latitude of the subtropical highs. A corresponding value for middle latitudes was computed by Elliott and Smith (1949) from a long period of data (at sea level). For a comprehensive discussion of the principle of horizontal exchange processes, see Defant and Defant (1958).

The importance of lateral mixing in atmospheric disturbances was further emphasized by the studies of Namias *et al.* (1940), utilizing isentropic charts (the principle of which was introduced by Shaw, 1933). These analyses demonstrated the combined effects of vertical and horizontal motions in the synoptic-scale eddies, which appeared in the form of large branches of rising moist air and descending dry air, mainly turning in an anticyclonic sense on the isentropic charts (see discussion of three-dimensional trajectories in Chapter 10).

The concept of an exchange coefficient has also been employed by Adem (1964) in numerical experiments relating to the heat balance. While his assumption of a constant exchange coefficient over the hemisphere is not entirely justified, his experiments give reasonable results and offer valuable comparisons of the relative influences of heat sources and lateral transfer processes upon the hemispheric distribution of temperature.

The difficulty of applying the concept of an exchange coefficient to characterize a given season may be appreciated by a comparison of Figs. 6.9 and 6.10. Since either of these extreme types of patterns may occur within a particular season and may dominate for long periods in a given year, it is evident that the exchange coefficient may vary greatly from one occasion to another. Elliott and Smith (1949) demonstrated that the winter mean values of A (at the surface) vary by about $\pm 25\%$ around the long-period means and that these variations and those of the poleward heat transport are significantly correlated with the extent of "blocking action" of the kind illustrated over the Atlantic-Europe sector in Fig. 6.7.

Principles based on lateral mixing processes were emphasized over a period of many years by C.-G. Rossby in their application to both atmospheric and oceanic problems. In particular regard to the distribution of the westerlies with latitude, Rossby (1947) suggested that the broad-scale features poleward of the latitude of maximum wind could be accounted for on the basis of a north-south mixing of absolute vorticity. Starting with a polar cap in which the atmosphere is in solid rotation with the earth, such a mixing eventuates in an increase of relative vorticity (or meridional shear) in the lower-latitude parts and a decrease of relative vorticity in the poleward parts of the cap in which mixing takes place, due to transfer of high (earth) vorticity equatorward and low (earth) vorticity poleward. The resulting velocity profile, characterized by constant absolute vorticity with latitude, is in broad agreement with the observed wind profile within this polar cap.

While Rossby's theory reasonably describes the statistical effect of eddy transfers by synoptic systems within the polar cap, the zonal-velocity distribution in lower latitudes is accounted for by the different mechanism discussed in Chapter 1. Platzman (1949) indicates that when semisynoptic rather than mean wind data are considered, there is essentially a discontinuity of *potential* absolute vorticity across the jet stream, with relatively uniform high values over a range of latitudes poleward and uniform low values on the equatorward side. The absolute vorticity itself has greater values near the polar-front jet stream than at the poles.

References

Adem, J. (1964). On the normal state of the troposphere-ocean-continent system in the Northern Hemisphere. *Geofis. Intern.* **4**, 3–32.

Bjerknes, J. (1937). Theorie der aussertropischen Zyklonenbildung. *Meteorol. Z.* **54**, 462–466.

Bjerknes, J., and Holmboe, J. (1944). On the theory of cyclones. *J. Meteorol.* **1**, 1–22.

Charney, J. G. (1947). The dynamics of long waves in a baroclinic westerly current. *J. Meteorol.* **4**, 135–162.

Cressman, G. P. (1948). On the forecasting of long waves in the upper westerlies. *J. Meteorol.* **5**, 44–57.

Cressman, G. P. (1949). Some effects of wave-length variations of the long waves in the upper westerlies. *J. Meteorol.* **6**, 56–60.

Defant, A. (1921). Die Zirkulation der Atmosphäre in den gemässigten Breiten der Erde. *Geograf. Ann.* (*Stockholm*) **3**, 209–266.

Defant, A., and Defant, F. (1958). "Physikalische Dynamik der Atmosphäre," Chap. 11 and p. 348. Akad. Verlagsges., Frankfurt.

Elliott, R. D., and Smith, T. B. (1949). A study of the effects of large blocking highs on the general circulation in the Northern-Hemisphere westerlies. *J. Meteorol.* **6**, 67–85.

Ficker, H. (1929). Der Sturm in Norddeutschland 4 Juli 1928. *Sitzber. Preuss. Akad. Wiss., Physik.-Math. Kl.* **22**, 290–326.

Ficker, H. (1938). Zur Frage der "Steuerung" in der Atmosphäre. *Meteorol. Z.* **55**, 8–12.

Fleagle, R. G. (1947). The fields of temperature, pressure and three-dimensional motion in selected weather situations. *J. Meteorol.* **4**, 165–185.

Fleagle, R. G. (1948). Quantitative analysis of factors influencing pressure change. *J. Meteorol.* **5**, 281–292.

Fultz, D. (1945). Upper-air trajectories and weather forecasting. *Dept. Meteorol., Univ. Chicago, Misc. Rept.* No. 19, 1–123.

Haurwitz, B. (1940). The motion of atmospheric disturbances on the spherical earth. *J. Marine Res.* **3**, 254–267.

Haurwitz, B. (1941). "Dynamic Meteorology," Chapters 11 and 13. McGraw-Hill, New York.

Haurwitz, B. *et al.* (1945). Advection of air and the forecasting of pressure changes. *J. Meteorol.* **2**, 83–93.

Hesselberg, T. (1913). Über die Luftbewegungen im Zirrusniveau und die Fortpflanzung der barometrischen Minima. *Beitr. Physik. Freien Atmosphäre* **5**, 198–205.

Holmboe, J., Forsythe, G. E., and Gustin, W. (1945). "Dynamic Meteorology," Chapter 10. Wiley, New York.

Houghton, H. G., and Austin, J. M. (1946). A study of nongeostrophic flow with applications to the mechanism of pressure changes. *J. Meteorol.* **3**, 57–77.

Julian, P. R. (1966). The index cycle: A cross-spectral analysis of zonal index data. *Monthly Weather Rev.* **94**, 283–293.

LaSeur, N. E. (1954). On the asymmetry of the middle-latitude circumpolar current. *J. Meteorol.* **11**, 43–57.

Lettau, H. (1936). Luftmassen und Energieaustausch zwischen niederen und hohen Breiten der Nordhalbkugel während des Polarjahres 1932/33. *Beitr. Physik. Atmosphäre* **23**, 45–84.

Miller, J. E. (1948). Studies of large scale vertical motions of the atmosphere. *Meteorol. Papers, New York Univ.* No. 1, 1–48.

Mügge, R. (1931). Synoptische Betrachtungen. *Meteorol. Z.* **48**, 1–11.

Namias, J., and Clapp, P. F. (1944). Studies of the motion and development of long waves in the westerlies. *J. Meteorol.* **1**, 57–77.

Namias, J., and Clapp, P. F. (1951). Observational studies of general circulation patterns. *In* "Compendium of Meteorology" (T. F. Malone, ed.), pp. 551–567. Am. Meteorol. Soc., Boston, Massachusetts.

Namias, J., Haurwitz, B., Bergeron, T., Stone, R. G., Willett, H. C., and Showalter, A. K. (1940). "Air mass and Isentropic Analysis" (R. G. Stone, ed.), 232 pp. Am. Meteorol. Soc., Milton, Massachusetts.

Newton, C. W., and Palmén, E. (1963). Kinematic and thermal properties of a large-amplitude wave in the westerlies. *Tellus* **15**, 99–119.

Palmén, E. (1951). The aerology of extratropical disturbances. *In* "Compendium of Meteorology" (T. F. Malone, ed.), pp. 599–620. Am. Meteorol. Soc., Boston, Massachusetts.

Panofsky, H. A. (1946). Methods of computing vertical motion in the atmosphere. *J. Meteorol.* **3**, 45–49.

Panofsky, H. A. (1951). Large-scale vertical velocity and divergence. *In* "Compendium of Meteorology" (T. F. Malone, ed.), pp. 639–646. Am. Meteorol. Soc., Boston, Massachusetts.

Petterssen, S. (1952). On the propagation and growth of jet-stream waves. *Quart. J. Roy. Meteorol. Soc.* **78**, 337–353.

Petterssen, S. (1956). "Weather Analysis and Forecasting," 2nd ed., Vol. 1, Chapters 8–10. McGraw-Hill, New York.

Platzman, G. W. (1949). The motion of barotropic disturbances in the upper troposphere. *Tellus* **1**, No. 3, 53–64.

Riehl, H., LaSeur, N. E. *et al.* (1952). Forecasting in middle latitudes. *Meteorol. Monographs* **1**, No. 5, 1–80.

Rossby, C.-G. (1947). On the distribution of angular velocity in gaseous envelopes under the influence of large-scale horizontal mixing processes. *Bull. Am. Meteorol. Soc.* **28**, 53–68.

Rossby, C.-G., and Willett, H. C. (1948). The circulation of the upper troposphere and lower stratosphere. *Science* **108**, 643–652.

Rossby, C.-G. *et al.* (1939). Relation between variations in the intensity of the zonal circulation of the atmosphere and the displacements of the semi-permanent centers of action. *J. Marine Res.* **2**, 38–55.

Schereschewsky, P., and Wehrlé, P. (1923). Les systèmes nuageux. *Mem. Office Natl. Meteorol. (Paris)* No. 1.

Scherhag, R. (1948). "Neue Methoden der Wetteranalyse und Wetterprognose," 424 pp. Springer, Berlin.

Shaw, N. (1933). "Manual of Meteorology," Vol. 3, p. 259. Cambridge Univ. Press, London and New York.

U. S. Weather Bureau. (1950, 1959). "Daily Series Synoptic Weather Maps, Northern Hemisphere Sea Level and 500 Millibar Charts." U. S. Weather Bur., Washington, D.C.

Vederman, J. (1952). The growth and decline of anticyclones. *Bull. Am. Meteorol. Soc.* **33**, 315–321.

Wexler, H. (1951). Anticyclones. *In* "Compendium of Meteorology" (T. F. Malone, ed.), pp. 621–629. Am. Meteorol. Soc., Boston, Massachusetts.

Willett, H. C. *et al.* (1947). Final report of the Weather Bureau—M.I.T. extended forecasting project for the fiscal year July 1, 1946-July 1, 1947. M.I.T., Cambridge, Massachusetts.

7

THERMAL STRUCTURE OF FRONTS AND
CORRESPONDING WIND FIELD

In this chapter, we outline some of the basic characteristics of front and tropopause structure, and the relationships connecting the temperature and wind fields, especially in the neighborhood of the polar-front zone. In addition to the examples given here, numerous others will be given in Chapters 8 through 11. These illustrate the considerable variations observed in the atmosphere; however, they generally have in common certain basic features that are stressed here. Some aspects of the relations between fronts and weather phenomena are treated in Chapter 12.

7.1 FRONTS AS SURFACES OF DISCONTINUITY OR LAYERS OF TRANSITION

In Chapter 4 the principal fronts were described as layers separating air masses of different "origin," or "prehistory," characterized by certain physical properties. Those ordinarily considered are density, temperature, water vapor, and the wind implied by the density field that prescribes the pressure distribution. General discussions of the interrelations between these properties have been given by Palmén (1948) and Godson (1951).

If the transition layer between the air masses is sufficiently thin, a front approximates a true surface of discontinuity in density and may be defined as a discontinuity surface of *zero order* (Fig. 7.1a). The *dynamic boundary condition* specifies that the pressure is equal just adjacent to the two sides of the front; hence, the gas law implies that a discontinuity in density is also a

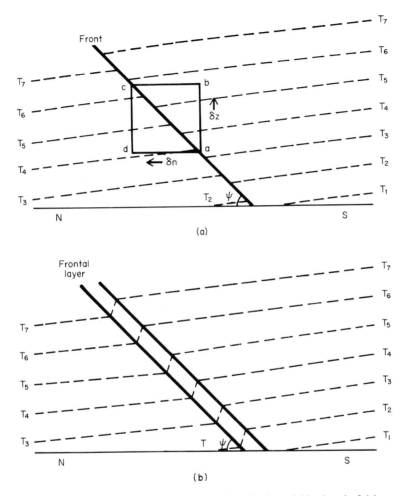

Fig. 7.1   Isotherm distribution in vertical section, in the neighborhood of (a) a zero-order frontal discontinuity; (b) a frontal layer with discontinuities in the temperature gradient at its upper and lower surfaces.

discontinuity of virtual temperature,[1] as indicated by the schematic isotherms.

In the following discussion it is convenient to make use of a general expression for the frontal slope. Referring to rectangle *abcd* in Fig. 7.1a,

[1] In the following, the temperature T should, to be strictly correct, be replaced by the *virtual* temperature. Ordinarily, the difference of specific humidity across a front in lower levels does not exceed the order of 6 gm/kg, which corresponds to a virtual temperature correction of 1°C, while the sensible temperature difference is of the order of 10°C. In the upper troposphere, this refinement is even less significant.

and letting Q be *any property that is continuous across the surface* (but whose *gradient* is discontinuous), we write the identity

$$(Q_c - Q_b) + (Q_b - Q_a) = (Q_c - Q_d) + (Q_d - Q_a)$$

For small δz and δn, the differences along the boundaries may be written $(Q_c - Q_b) = (\partial Q/\partial n)\, \delta n$, etc., from which it follows that

$$\tan \psi \equiv \frac{\delta z}{\delta n} = \frac{-[(\partial Q'/\partial n) - (\partial Q/\partial n)]}{(\partial Q'/\partial z) - (\partial Q/\partial z)} \tag{7.1}$$

where the primed and unprimed terms distinguish properties on opposite sides of the surface immediately adjacent to it, respectively. Here ψ is the angle between the front and a *level* surface, direction n being normal to the intersection between the frontal surface and the level surface, and positive toward colder air, as in Fig. 7.1a.

Considering a surface of *zero-order* discontinuity of density, since the pressure is continuous, p may be inserted in place of Q in Eq. (7.1) to give

$$\tan \psi = \frac{-[(\partial p'/\partial n) - (\partial p/\partial n)]}{(\partial p'/\partial z) - (\partial p/\partial z)} \tag{7.2}$$

where primes refer to the warm-air side and unprimed quantities are on the cold-air side of the front.

In the real atmosphere, the boundary separating air masses of different properties is never a real surface of discontinuity in density even though—especially near the earth's surface where the properties are usually more concentrated, owing to the influence of surface friction on the wind field (Section 9.1)—it can for some purposes often be treated as a true discontinuity. The temperature distribution has rather the character of Fig. 7.1b, where two surfaces now separate a frontal layer of finite width from the more homogeneous air masses outside the frontal layer. Since at these two surfaces the density and temperature are continuous but their *gradient* is discontinuous, they may be considered surfaces of *first-order* discontinuity. This assumption is, of course, also somewhat artificial, but is better applicable to reality.

Considering that the pressure gradient must be equal on both sides just adjacent to such a surface,[2] either $\partial p/\partial n$ or $\partial p/\partial z$ may be inserted for Q

---

[2] This follows from the equation of motion, $\mathbf{g} + f\mathbf{V} \times \mathbf{k} - \varrho^{-1}\nabla p = \dot{\mathbf{V}}$. Since the velocity field is continuous, there is no discontinuity in the first two terms. If, in addition, ϱ is continuous, a discontinuity in ∇p would imply that immediately adjacent particles would have different accelerations and the velocity field would become discontinuous.

in Eq. (7.1) to give the following expressions for its slope:

$$\tan \psi = \frac{-[(\partial^2 p'/\partial n^2) - (\partial^2 p/\partial n^2)]}{(\partial^2 p'/\partial n\, \partial z) - (\partial^2 p/\partial n\, \partial z)} \tag{7.3a}$$

$$\tan \psi = \frac{-[(\partial^2 p'/\partial n\, \partial z) - (\partial^2 p/\partial n\, \partial z)]}{(\partial^2 p'/\partial z^2) - (\partial^2 p/\partial z^2)} \tag{7.3b}$$

In these equations, p' denotes the pressure outside the frontal layer and p is the pressure inside it, the terms being evaluated immediately adjacent to one of the bounding surfaces.

7.2 THE SLOPE OF A ZERO-ORDER DISCONTINUITY SURFACE

If we denote the horizontal wind component parallel to the front at a fixed level by V, the acceleration in the direction normal to the front is expressed by

$$\frac{dv_n}{dt} = f(V_g - V) = -\left(\frac{1}{\varrho} \frac{\partial p}{\partial n} + fV\right) \tag{7.4}$$

wherein dv_n/dt may be construed to include the influence of frictional acceleration normal to the front; or, for example, in the case of frictionless flow entirely parallel to a curved front, it may account for the acceleration associated with curvature of the flow. By substitution from Eq. (7.4) for $\partial p/\partial n$ and from the hydrostatic equation for $\partial p/\partial z$, Eq. (7.2) can be replaced by

$$\tan \psi = \frac{f(\varrho' V' - \varrho V) + [(\varrho'\, dv_n'/dt) - (\varrho\, dv_n/dt)]}{-g(\varrho' - \varrho)} \tag{7.5}$$

which is given by Bjerknes (1924).

The well-known formula for frontal slope, derived by Margules (1906), can be obtained by substitution of the middle expression in Eq. (7.4) for the normal acceleration terms in Eq. (7.5). Then, substituting p/RT for ϱ,

$$\tan \psi = \frac{f(TV_g' - T'V_g)}{g(T' - T)}$$

or, by further manipulation,

$$\tan \psi = \frac{f\bar{T}}{g} \frac{V_g' - V_g}{T' - T} - \frac{f\bar{V}_g}{g} \tag{7.6}$$

where $\bar{T}$ denotes the mean temperature of the two air masses and $\bar{V}_g$ is the mean geostrophic wind parallel to the front. The last term corresponds to the slope of an isobaric surface so that the slope ψ_i relative to such a surface is

$$\tan \psi_i = \frac{f\bar{T}}{g} \frac{V_g' - V_g}{T' - T} \tag{7.6'}$$

The difference is small compared with the ordinary slope of a front.

In the special case of frictionless straight-line flow with balance between the coriolis and horizontal pressure-gradient forces, the slope of a front conforms to Eqs. (7.6) and (7.6') if the real wind replaces V_g.

The expressions above are simply a form of the thermal wind equation and may be derived directly from it. Letting δT denote the *isobaric* temperature differential between two air masses separated by a zone of transition of width δn, and with δV_g as the difference in V_g *vertically* through the transition layer of depth δz, the geostrophic thermal wind equation states that

$$\frac{\delta V_g}{\delta z} = -\frac{g}{f\bar{T}} \left(\frac{\delta T}{\delta n} \right)_p \tag{7.7}$$

Upon rearrangement to give the slope $(\delta z/\delta n)$, the result is identical to Eq. (7.6'), being valid either for a front of finite thickness or for a zero-order discontinuity if δz and δn are imagined to shrink to zero.

In cases of strong curvature of the air flow, Eq. (7.6') should be substituted by a formula expressing the influence of the normal acceleration. In its simplest form, this influence is expressed by the relationship

$$V_g = V + \frac{kV^2}{f} \tag{7.8}$$

where V is the actual wind and k is the horizontal curvature of an air trajectory. If k is taken to be the same on both sides of the front (which with a zero-order discontinuity would be strictly true only if the system were not moving), substitution into Eq. (7.6') gives

$$\tan \psi_i = (f + 2k\bar{V}) \frac{\bar{T}}{g} \frac{V' - V}{T' - T} \tag{7.9}$$

which is similar to an expression given by Exner (1925).

In the case of geostrophic balance for straight flow or gradient wind balance for curved flow, Eqs. (7.6') and (7.9) state that *a frontal surface has an equilibrium slope (relative to an isobaric surface) different from zero, provided the wind increases upward through the front. For a given ratio be-*

tween the wind and temperature differentials, the equilibrium slope is larger for cyclonically curved flow and smaller for anticyclonically curved flow. In the case of a zero-order discontinuity, this statement implies a cyclonic wind shift across the front. If the accelerations in a vertical plane normal to the frontal surface are considered, the tendency inherent in the density distribution for a direct solenoidal circulation to take place (with consequent decrease in frontal slope) is, in this equilibrium condition, balanced by an opposite tendency for an indirect circulation, owing to the combined coriolis and centrifugal accelerations acting on the different winds above and below the front.

7.3   The Slope of a First-Order Discontinuity Surface

Figure 7.2 shows a first-order discontinuity surface, the isotherm configuration being characteristic of either a sloping tropopause or the lower boundary of a frontal layer, as in Fig. 7.1b. As already mentioned, both pressure and pressure gradient are continuous across such a surface, and its slope may therefore be determined from Eq. (7.3a) or Eq. (7.3b). This implies that both temperature and geostrophic wind are continuous, although the temperature *gradient* and geostrophic wind *shear* are discontinuous.

Rather than utilize Eqs. (7.3), the simplest way to derive a useful expression for the slope is to consider the quadrilateral *ABCD* in Fig. 7.2, in which the wind difference between points *B* and *D* can be expressed as the sum of the horizontal shear times δn plus the vertical shear times δz, this sum being equal on the two sides of the discontinuity. If the quasi-horizontal sides are taken along isobaric surfaces, the slope takes the form

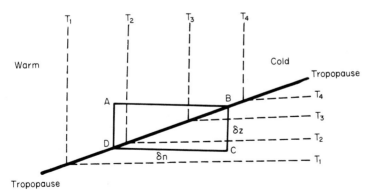

Fig. 7.2   Isotherms in the vicinity of a sloping tropopause surface.

of Eq. (7.1), in which Q is replaced by V_g and ψ by ψ_i. Substitution from the thermal wind equation into the denominator of the resulting expression then gives

$$\tan \psi_i = \frac{fT}{g} \frac{(\partial V_g'/\partial n) - (\partial V_g/\partial n)}{(\partial T'/\partial n) - (\partial T/\partial n)} \tag{7.10}$$

Here the primed quantities may refer to either side of the first-order discontinuity surface, the expression being valid for either the upper or lower surface of a frontal layer (Fig. 7.1b) or for a tropopause.

Alternatively, the temperature T may be inserted for Q in Eq. (7.1) to give the corollary expression

$$\tan \psi_i = \frac{fT}{g} \frac{(\partial V_g'/\partial z) - (\partial V_g/\partial z)}{(\partial T'/\partial z) - (\partial T/\partial z)} \tag{7.11}$$

which, for some purposes, is more descriptive of a tropopause, considering this as a discontinuity of the vertical lapse rate of temperature.

Equation (7.10) states that *a discontinuity of horizontal shear across a first order discontinuity surface is such that the shear is more strongly cyclonic (or less anticyclonic) within a frontal layer than it is in the bounding air masses.* Equation (7.11) states that, *accompanying the discontinuity of vertical lapse rate at a tropopause or frontal surface, there is a corresponding discontinuity of vertical shear.* The sense of this discontinuity is obvious from the horizontal temperature gradients above or below the discontinuity surfaces in Fig. 7.1b or Fig. 7.2.

If the flow is curved and variations of trajectory curvature with height are neglected, the following expressions for the inclination of the first-order discontinuity surface replace Eqs. (7.10) and (7.11), from which they are derived by substitution from Eq. (7.8):

$$\tan \psi_i = \frac{(f + 2kV)T}{g} \frac{(\partial V'/\partial n) - (\partial V/\partial n)}{(\partial T'/\partial n) - (\partial T/\partial n)} \tag{7.12}$$

$$\tan \psi_i = \frac{(f + 2kV)T}{g} \frac{(\partial V'/\partial z) - (\partial V/\partial z)}{(\partial T'/\partial z) - (\partial T/\partial z)} \tag{7.13}$$

Here the slopes are expressed relative to an *isobaric* surface. This is generally the most convenient representation, since the isobaric coordinate system is commonly used. If it is desired to express the slope ψ relative to a *level* surface, this may be done by the conversion

$$\tan \psi - \tan \psi_i = \alpha = -\frac{fV_g}{g} = -\frac{V}{g}(f + kV) \tag{7.14}$$

This additional term is mostly small compared with the slope of a front. Considering the equilibrium slope of a tropopause, however, it should be taken into account because of the generally weaker inclination of a tropopause and the commonly very strong wind at that level. If V_g is 75 m/sec and $f = 10^{-4}$ sec^{-1}, e.g., α amounts to a slope of 750 m in 1000 km horizontal distance.

Equations (7.12) and (7.13) indicate that, other things being equal, *for a given slope of the discontinuity surface the discontinuity of the vertical or horizontal shear is accentuated when the flow is anticyclonic.* Equation (7.13) was first derived by Ertel (1938) and was applied by him especially to the structure of "tropopause funnels," where the influence of curvature is considerable.

In extratropical latitudes the absolute value of $2kV$ is in most regions much smaller than f. There are, however, cases with strong winds and large cyclonic or anticyclonic curvature, where $|2kV|$ approaches f; hence, the curvature has a very strong influence on the equilibrium slope of a front or a tropopause. In the tropics, where f is small and cyclonic perturbations may become very concentrated, $2kV$ often dominates. An instance is the tropical hurricane, in which the boundary between the "eye" and the outer "cloud wall" may be considered as a sloping transition layer (Section 15.4).

Equations (7.10) and (7.11), or (7.12) and (7.13), can, of course, be applied independently of the orientation of the current. At the tropopause, the denominator of Eq. (7.11) is always positive if the primes stand for stratospheric conditions. Likewise, if V_g (or V) is taken as intrinsically positive and the wind decreases with height above the tropopause, the numerator is negative. In that case the tropopause rises, as in Fig. 7.2, toward the right of the current. This represents the most common condition in a vertical section across the west-wind belt. If, however, V_g increases

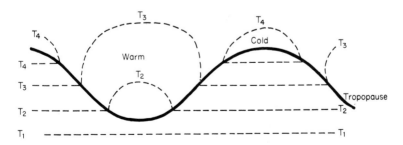

FIG. 7.3   Schematic wave form of tropopause surface characteristic of upper-level troughs (low tropopause) and ridges (high tropopause), and the corresponding isotherm configuration. The structure is simplified, assuming a single continuous tropopause.

more rapidly with height in the lower stratosphere than in the troposphere, the tropopause rises toward the left of the current. This is often typical of the structure just to the left of the jet stream, as illustrated by later synoptic examples (see, e.g., Fig. 7.5).

Let Fig. 7.3 be a west-east vertical section in which the left half transects a trough and the right half a ridge in the westerlies, with the strongest wind near tropopause level. Equation (7.13) then prescribes an undulated form as shown, with a warm, low tropopause over the trough and a high, cold tropopause over the ridge.

7.4   GENERALIZED ATMOSPHERIC STRUCTURE NEAR POLAR FRONT

The results of the preceding discussion can now be utilized to describe the general aspects of the wind field as related to the temperature field in the neighborhood of a front. In Fig. 7.4 is presented a schematic vertical section normal to the polar front and the upper westerlies. This shows, in idealized form, the most commonly observed features when the polar front

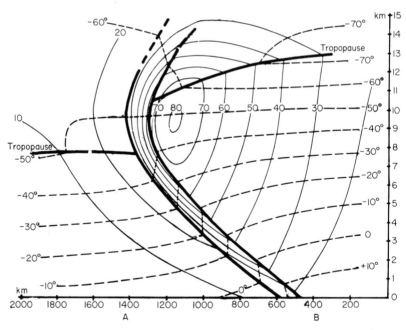

Fig. 7.4   Schematic isotherms (dashed lines, °C) and isotachs (thin solid lines, meters per second) in the polar front zone. Heavy lines are tropopauses and boundaries of frontal layer. (Adapted from analysis model by Berggren, 1952.)

is well developed throughout the whole depth of the troposphere and lower stratosphere. No essential features would be altered if the orientation of the front and of the general air flow were not directed west-east.

Although there are different opinions as to how fronts and tropopauses should be analyzed in upper levels, we have chosen the special style in Fig. 7.4 because it is pedagogically most useful. The front is drawn as a continuous layer that extends through the tropopause, becomes vertical at the level of maximum wind, and then continues with a reversed slope into the lower stratosphere, where it gradually becomes diffuse. This style of analysis, proposed by Nyberg and Palmén (1942), has been elaborated particularly by Berggren (1952, 1953), upon whose analyses the model in Fig. 7.4 is based.

This interpretation has the advantage that the front separates both warm tropospheric air and cold stratospheric air of the right-hand air mass from the cold troposphere and warm stratosphere of the left-hand air mass. It thus embodies an extension of the air-mass concept to the lower stratosphere.

Both in the troposphere and in the lower stratosphere, the slope of the front is such that warmer air overlies colder. Also, as indicated by Eq. (7.10), the slope becomes vertical (tan $\psi_i = \infty$) where the horizontal temperature gradient vanishes. At this level, where the usual definition of a front as a discontinuity in density is not applicable, *the front is defined as the region where the horizontal cyclonic wind shear is greatest* (after Berggren, 1953). Although earlier it was convenient to begin the discussion with the more usual concept of a front as a discontinuity or layer of transition of density, the preceding definition is more general and, from Eq. (7.10), is applicable at all levels.

As prescribed by Eq. (7.11), there is a discontinuity of vertical shear at the warm-air tropopause, which is characteristically observed to slope upward toward the right of the current in this region near the jet stream. Discontinuities of both the horizontal and vertical shear are also observed at the upper and lower frontal boundaries; this, too, is in accord with Eqs. (7.10) and (7.11).

It is worth noting that, contrary to the dictates of Margules' formula as applied to a *zero-order* discontinuity, it is possible for the lateral shear in some portions of a front of finite width to be anticyclonic. If primed quantities represent conditions in the warm air mass above and adjacent to the frontal layer, the shear within the front may, by rearrangement of Eq. (7.10), be written as

$$\frac{\partial V_g}{\partial n} = \frac{\partial V_g'}{\partial n} + \frac{g \tan \psi_i}{fT} \left(\frac{\partial T}{\partial n} - \frac{\partial T'}{\partial n} \right) \tag{7.15}$$

With n positive toward left in Fig. 7.4, since the temperature gradient is strongest within the frontal layer, the last term is negative and contributes toward cyclonic shear inside the frontal layer. To the right of the jet core in Fig. 7.4, however, $\partial V_g'/\partial n \gg 0$ in the warm air. If that term is larger in magnitude than the second term of Eq. (7.15), $\partial V_g/\partial n > 0$. Thus, as pointed out by Palmén (1948), *the combined conditions of strong anticyclonic shear in the warm-air mass, and a relatively weak horizontal temperature gradient within the front, may result in an anticyclonic shear across the front.* Weak cyclonic or sometimes anticyclonic shear is most likely to be present near the 700-mb level, since the portion of the front at that level generally lies beneath the zone of strongest anticyclonic shear to the right of the jet core. Higher up in the front (for instance, at 500 mb and especially at 300 mb), strong cyclonic shear is the rule.

Although the baroclinity is strongest in the frontal layer, it is usually observed that both the warm and the cold air are more barocline, both above and below the frontal layer, than are the air masses far removed from the front. Thus, approximately between A and B in Fig. 7.4, baroclinity is significant through most of the troposphere. A belt of this sort, extending more or less vertically within the bounds prescribed by the polar front, has been called a *Frontalzone* by Scherhag (1948). Godson (1951) distinguishes between the "frontal zone," as a broad baroclinic zone without sharp boundaries, and the "hyperbaroclinic zone," constituting a well-defined frontal layer.

From the discussion in Section 4.2 it is evident that the jet stream should be located somewhere near the middle of the frontal zone as defined by Scherhag. Here, in addition to the pronounced baroclinity of the front itself, there is the added effect of the moderate baroclinity within the adjacent air masses that, when integrated through the deep layers above and below the front, give an overall large thermal wind. *On the average, the jet-stream axis is located nearly above the intersection of the frontal layer with the* 500-mb *surface* (Palmén, 1948). Since the polar front in this level is characterized by a band of crowded isotherms (e.g., Fig. 4.5), the approximate configuration of the polar-front jet can easily be found from 500-mb charts. However, there are exceptions to this general rule, and therefore it should be used with restraint.

It may also be noted that the baroclinity ordinarily reverses sign within the upper warm-air troposphere so that (Austin and Bannon, 1952; Coudron, 1952) the strongest wind is most often found about 1 km below the tropopause, as in Fig. 7.4. However, the really strong decrease of wind occurs mostly just above the tropopause, where the strongest reversed baroclinity is observed.

A striking example of front and jet-stream structure, analyzed by L. Oredsson, is shown in Fig. 7.5. This illustrates several of the features discussed above, such as: the strong shear through the frontal layer, and also the large contribution to the overall vertical wind change within the neighboring air masses, especially in the warm air; the intense cyclonic shear between the air masses in the upper troposphere and lower stratosphere; the changes of vertical shear at the tropopauses in accord with their different slopes; and the corresponding changes of temperature gradient between troposphere and stratosphere.

According to the most common interpretation, fronts are characterized by the amount of baroclinity, or vertical stability. It is true that most tropospheric fronts are typified by a vertical stability considerably exceeding that of tropospheric-air masses in general. However, as we have seen, this rule can no longer be applied to fronts near the level of maximum wind where the horizontal temperature gradient vanishes and the vertical lapse rate is disturbed by the transition to the stratosphere. Partly because of this, it has been customary not to extend frontal analyses above the tropopause. However, there is no cause to assume that frontogenetic processes acting in the atmosphere should cease just at the level of no horizontal temperature gradient. The existence of jet streams just at that level, with their pronounced vorticity field, must be related to frontogenetic processes acting upon this part of the atmosphere. In the zone of strong horizontal wind shear near the jet-stream level, air masses of approximately the same temperature, but with quite different origins, have obviously been brought close together, a feature characteristic of frontogenetic processes. As brought out in Chapter 9, these processes involve rearrangements of both the temperature and momentum fields, the relative influence on the two fields being different at different levels in the atmosphere.

From the structure in Figs. 7.4 and 7.5 it is evident that a tropospheric warm front would, higher up in the atmosphere, change over into a stratospheric cold front, and vice versa. It is, however, questionable how far up one should extend the frontal analysis. In most cases, even in the lowest stratosphere, fronts are diffuse compared with tropospheric fronts, and they should therefore not be extended very far up above the tropopause.

7.5   DIFFERING INTERPRETATIONS OF FRONTS AND TROPOPAUSES

From the preceding discussion it follows that there must be a close connection between upper-tropospheric fronts and the overall structure of the tropopause. This was stressed in Chapter 4 as a distinguishing feature

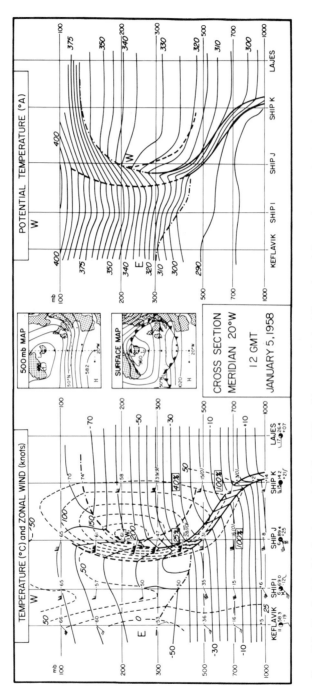

FIG. 7.5   On the left side, isotherms (solid lines, °C) and isotachs (dashed lines, knots) in relation to polar-front layer (heavy solid lines) and tropopauses (dash-dot lines), in a meridional section across the westerly current along line indicated on inset maps. On right side, potential temperature (°K) in same cross section. (Reproduced through courtesy of L. Oredsson, University of Stockholm.)

of the transition zones between the principal air masses, and is well illustrated by Fig. 7.5, where a very large change in tropopause height across the polar-front zone is evident.

The tropopause is distinguished as a surface of discontinuity in the vertical lapse-rate of temperature.[3] In synoptic analyses, however, several difficulties arise in fixing the tropopause. First, the transition between troposphere and stratosphere is not always abrupt, but may extend over a layer of transition in which two or several surfaces with discontinuities of lapse rate can be noted. Secondly, no real discontinuities in lapse rate can be found in certain parts of the atmosphere. Because of these and other difficulties in delineating a distinct bounding surface between troposphere and stratosphere, the whole layer with a pronounced change of temperature lapse rate may often better be characterized as a "substratosphere." In such a substratospheric layer, several "tropopause leaves" may be found, and among these individual leaves or surfaces one or the other can on occasion become the most prominent and therefore be selected as the "real" tropopause. This difficulty has to be kept in mind when treating the tropopause as a true boundary between troposphere and stratosphere.[4]

On the average, the tropopause or substratospheric layer slopes downward from the Equator toward the Poles. If frontogenetic processes now act for a sufficiently long time on the upper air masses, resulting in a concentration of the tropospheric baroclinity into a relatively narrow zone, the same processes must also result in an increased slope of the tropopause around the same latitude. In one interpretation (Fig. 7.6a) the slope was

[3] We shall not discuss the processes involved in formation of the tropopause. That it is basically a radiation phenomenon was demonstrated many years ago by Humphreys, Gold, and Emden. Recent calculations of an atmosphere in radiative equilibrium, by Manabe and Möller (1961), resulted in a sharply defined tropopause. In these computations, the surface temperature and vertical distributions of water vapor and ozone were prescribed. Since only radiation was considered, tropopause temperatures computed were much lower than those observed, especially in the cooler months in high latitudes. When their radiation model was incorporated into numerical forecasting schemes, in which vertical and meridional transfers of heat by air motions were included, realistic temperatures were obtained. As is well known, tropopauses may also be formed or dissolved as a result of the vertical motions and divergence fields in synoptic disturbances.

[4] The WMO criterion is that the tropopause is located at the first point on a sounding, above which (for at least 2 km) the lapse rate is 2°C/km or less. While such an objective definition serves a useful purpose, there are obviously many cases wherein confusion may arise. For example, a clearly defined tropopause may be evident above which the air is slightly less stable, or where there are shallower layers of alternating greater or smaller stability, or, with a literal application, the base of a deep frontal layer could be classified as a tropopause.

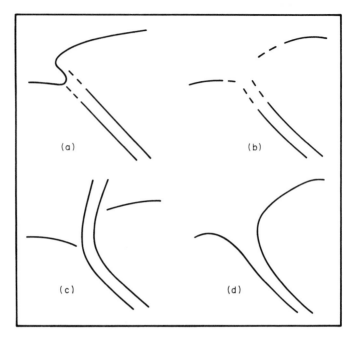

FIG. 7.6   Schematic diagrams of four methods of analyzing fronts and tropopauses. (After Reed and Danielsen, 1959.)

visualized as being concentrated in this way, even to the extent of a "folding" of the tropopause, which was regarded as a continuous surface (Bjerknes and Palmén, 1937). But it was difficult to find evidence for such an interpretation, or even to locate a distinct tropopause in the region of transition, and for a while it was fashionable to analyze the region as in Fig. 7.6b.

As discussed earlier, the interpretation in Fig. 7.6c (as in Fig. 7.4) has the advantage that the middle-latitude air mass as a whole, including its tropopause and lower stratospheric properties, is separated from the polar air mass as a whole. Analyses of serial soundings by Van Mieghem (1937) suggested that the lower boundary of the polar front, with the connected stability above, could be regarded as continuous with the polar-air tropopause. Mohri (1953) indicated a similar continuity between the middle-latitude tropopause and the lower boundary of the subtropical front (Fig. 4.7). This in part supports the interpretation in Fig. 7.6d, which has been utilized by Newton (1954), Reed and Danielsen (1959), and others. The connection between the upper surface of the polar front and the warm-air tropopause is more difficult to motivate.

Generally, the available observations permit analyses in either of the fashions of Fig. 7.6c or 7.6d, both of which will be seen in later examples. Usually, the choice of one or the other of these models, although perhaps meaningful with regard to physical processes, affects only minor details of the temperature and wind analysis. As illustrated by later examples, the tropopause structure is often much more complex than that indicated in the simple schematic pictures of Figs. 7.4 and 7.6, frequently with two or more main tropopause surfaces in both air masses. How complicated the analysis of the tropopause really can be was clearly shown by Danielsen (1959) in his study of the microstructure of layers with stronger and weaker vertical stability in a selected synoptic example.

7.6 SIMPLIFIED THERMAL STRUCTURE OF A DISTURBANCE ON THE POLAR FRONT

In view of the complex structures of real disturbances and the large variations from one case to another, it is considered desirable at this point to present a schematic description of the typical main features of a disturbance, shorn of these complications. Figure 7.7a represents the thermal structure of a wave disturbance on the polar front in the middle troposphere (for instance, in the 500-mb level). In order to simplify the discussion, a thermal field symmetrical about a N-S oriented trough line is considered. Real examples of both quasi-symmetrical and asymmetrical structures will be given later.

Figure 7.7a shows the typical crowding of the isotherms in the frontal zone and also the characteristic baroclinic field outside the front. The temperature distribution along the trough line N-S and along the three zonal cross sections *A-A*, *B-B*, and *C-C* are presented in Figs. 7.7b through e. In sections N-S, *A-A*, and *B-B* also, the positions of the jet-stream cores are marked. This is, however, absent in the zonal cross section *C-C* because this is somewhat south of the jet-stream core in Fig. 7.7a.

The frontal layer in Figs. 7.7b and 7.7c, in the vicinity of the tropopause, is constructed according to the principle of Fig. 7.4. In both sections are indicated the characteristic break in the tropopause observed in cases of very deep cold air. In these cross sections we can therefore distinguish between two different tropopauses: the polar-air tropopause and the tropopause of the warmer air. The cross sections farther south, Figs. 7.7d and 7.7e, are in the upper troposphere, entirely outside the influences of the really cold air. Hence, the tropopause here appears as a continuous line with a depression over the trough line and with highest elevation over the warm ridges of both sides of the cold trough, as in Fig. 7.3.

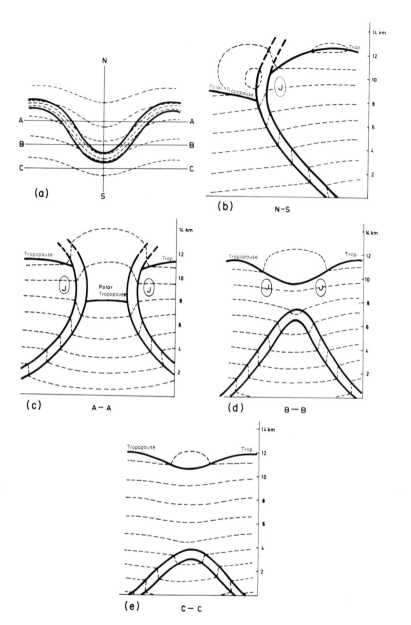

FIG. 7.7   (a) Frontal boundaries and isotherms in a middle-tropospheric wave pattern;
(b) fronts, tropopauses, isotherms, and jet-stream core in vertical section along N-S;
(c) section along *A-A*; (d) section along *B-B*; and (e) section along *C-C*. A simplified
symmetrical structure is assumed. The isotherms are drawn at approximately 4°C intervals in (a) and 10°C intervals in the vertical sections.

In many cases the real structure of cold upper troughs and warm upper ridges differs considerably from that shown here. For instance, the temperature field is often much more diffuse and the frontal structure not at all so clear as in the scheme above. However, this scheme serves the purpose of outlining the leading principles for a consistent analysis of upper air charts and cross sections when fronts are well developed, and should not be considered a "straitjacket" for all analyses.

7.7 SYNOPTIC EXAMPLE OF FRONTAL STRUCTURE IN A LARGE-AMPLITUDE WAVE

The various physical features discussed above are well illustrated by Figs. 7.8 through 7.10, showing the structure in a large-amplitude wave. At this time (Fig. 7.9), the front was clearly evident both east and west of the trough. However, as indicated in Chapters 9 and 11, the front may be

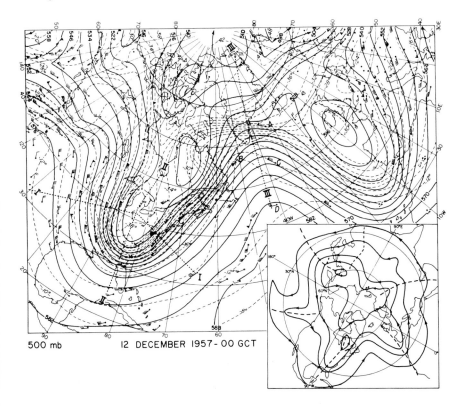

500 mb   12 DECEMBER 1957- 00 GCT

FIG. 7.8a

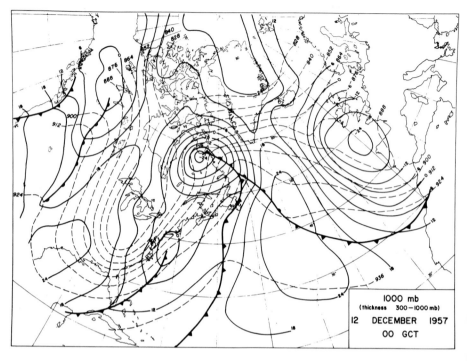

FIG. 7.8b

Fɪɢ. 7.8   (a) 500-mb and (b) 1000-mb charts at 00 GCT Dec. 12, 1957. In (a), contours are at 60-m intervals; isotherms (dashed) at 2°C intervals. Inset shows hemispheric wave pattern, contours at 240-m interval; outer contour is 5820 m. In (b), contours at 60-m intervals; dashed lines represent 300- to 1000-mb thickness, at intervals of 120 m corresponding to layer–mean-temperature interval of about 3°C. (From Newton and Palmén, 1963.)

diffuse on one side or the other at different times in the lifetime of such a trough.

Even in this case, the horizontal temperature gradient within the front was much stronger on the east than on the west side, although the overall temperature contrasts were comparable. Correspondingly, although strong cyclonic shear was present inside the front on the east side, on the west the shear was weakly anticyclonic near 700 mb. This is in accord with the earlier discussion of Eq. (7.15), wherein the weak slope of the front also contributes to make the influence of the second (cyclonic shear) term smaller than the first term, which in this case expresses the anticyclonic shear above the front. As discussed in Sections 10.4 through 10.5, such an asymmetry with a shallow cold-air mass in the western part is characteristic of outbreaks of polar air.

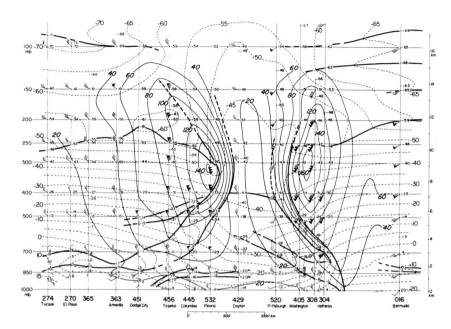

FIG. 7.9 Cross section along line *I-I* in Fig. 7.8a, approximately normal to the upper-tropospheric current on the west and east sides of the trough. Dashed isotherms in °C; isotachs represent total wind speed (knots). On the east side, the significant front reaching the surface is identified with the one just east of the coast in Fig. 7.8b. (From Newton and Palmén, 1963.)

Figure 7.9 also illustrates the break across the polar front between the polar-air tropopause at about 400 mb and the middle-latitude tropopause near 250 mb; the tropical tropopause is also evident near 100 mb. Most of the wind-speed contrast between the 160-knot jet stream and the weak winds of the polar-air stratosphere is concentrated within the front, which is vertical at 300 to 350 mb where the horizontal temperature gradient vanishes.

In the meridional section of Fig. 7.10a, approximately along the trough line, the frontal boundaries are not sharply defined except in the lower part, but the front has a considerable depth and embraces a large overall temperature contrast in both troposphere and stratosphere. This section shows an extreme example wherein the real tropical tropopause as well as the middle-latitude tropopause and the polar tropopause have been brought to almost the same latitude.

Along the ridge line to the east (Fig. 7.10b), the polar front is very clearly defined both in temperature and wind fields. In this case, most of the over-

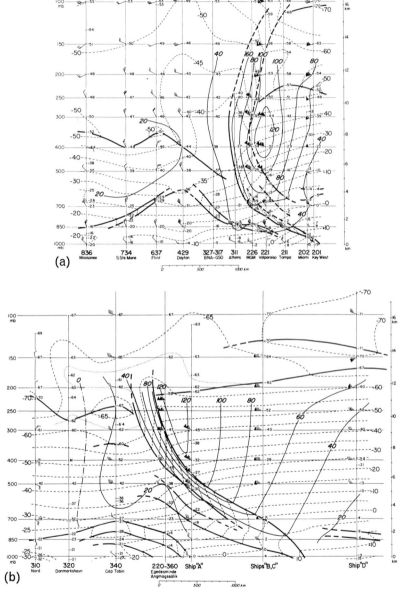

FIG. 7.10 Cross sections along (a) line II in trough, and (b) line III in ridge, of Fig. 7.8a. In (b), isotachs along the line of the section were taken from isobaric charts and do not necessarily agree with plotted winds, which in some cases correspond to stations at a distance east or west of section. (From Newton and Palmén, 1963.)

all temperature contrast between the cold and warm air is contained within the frontal layer, the individual air masses being relatively barotropic in the upper troposphere. This structure may be contrasted with that upstream, in the right side of Fig. 7.9, where the front, although quite distinct, includes a smaller temperature differential and the temperature gradient and vertical shear are pronounced within the air masses, particularly in the warm air above the front.

The most striking difference between Figs. 7.10a and 7.10b is in the slopes of the frontal layer. From the expression equivalent to Eq. (7.9),

$$\tan \psi_i = \frac{(f + 2kV)T}{g} \frac{\delta V}{\delta T}$$

this difference in slopes is seen to be mainly due to the opposite curvatures of the current at the two locations. Considering conditions near the 500-mb level, the temperature contrast δT is about 15°C in both sections, and the vertical wind difference δV is also nearly the same, about 60 knots.

In the front, f is about 0.6×10^{-4} sec^{-1} greater at the ridge than at the trough, so that if curvature were neglected, the equilibrium slope should be greater at the ridge. However, the variation of curvature overcompensates this variation of the coriolis parameter. Corresponding to the radius of trajectory curvature, which is about 1500 km in both trough and ridge, and a wind speed of 40 m/sec, $|2kV| = 0.53 \times 10^{-4}$ sec^{-1}. The curvature term is thus 1.06×10^{-4} sec^{-1} larger in trough than in ridge, and correspondingly $(f + 2kV)$ is almost twice as large in the trough than in the ridge. From the cross sections, the frontal slope at 500 mb is estimated to be 1.1×10^{-2} in the trough, and 0.5×10^{-2} in the ridge, in good agreement with the theoretical values of the difference in slope. At the intermediate latitude of the section on the right side of Fig. 7.9, where the current was nearly straight, the calculated slope of 0.7×10^{-2} is likewise in close agreement with the actual slope.

Considering the various uncertainties in observations and analysis, the example in Figs. 7.8 through 7.10 demonstrates that the relationships presented earlier, linking the various features of the wind and temperature fields in the frontal zones between the air masses, serve to provide a consistent picture of the overall atmospheric structure. The use of these relationships, especially in regions where (e.g., by use of a sounding from an isolated ship) some features (such as the overall wind and temperature contrast through a front) are known with some confidence, but direct analysis of the total structure is impossible, can considerably amplify the value of the limited observational material.

7.8   ADDITIONAL CONSIDERATIONS CONCERNING FRONTS

In the foregoing we have discussed in some detail only those aspects of fronts that are most relevant to later chapters. In the interest of brevity, we shall not deal with such matters as frontal classifications (different kinds of warm, cold, and occluded fronts). Extensive descriptions of these and their associated weather characteristics, pressure and flow patterns, etc., may be found, for example in the books by Petterssen (1940), Chromow (1942), Saucier (1955), and Godske *et al.* (1957). Here we mention only an additional few pertinent features.

The emphasis in discussing the formulas above has been placed upon equilibrium conditions between the wind and density (or temperature) distributions. Departures from equilibrium are, of course, allowed by the general Eq. (7.5). One interpretation of this equation was given in Section 7.2, where the equilibrium slope of a front in curved flow was considered.

Another interpretation was advanced by Bjerknes (1924) and is summarized by Brunt (1944). As the simplest example, we may consider the case of a plane front with a certain slope, but not in equilibrium according to Margules' formula, the temperature difference being too large in comparison with the wind shear across the front. In that case, the direct solenoidal acceleration due to the density distribution would exceed the indirect acceleration associated with the coriolis forces. According to Eq. (7.5), after some time $v_n' > v_n$ and according to the kinematic boundary condition ($w/v_n = \tan \psi$), upslope motion would take place in the warm air, relative to the cold air. Although the required imbalance between wind and density field is generally too delicate to establish by observation, the principle is qualitatively useful.

In the earlier discussion we considered only the wind components parallel to, and the temperature gradients normal to, fronts. More generally, there is a component of motion normal to the front except at special locations. This is a general characteristic of moving synoptic systems in which (discounting the influence of vertical motions) a front, being essentially an impermeable surface, moves approximately with the component of wind normal to it.

Simple geometry shows that in case of a zero-order discontinuity, the normal geostrophic wind components on both sides of such a front must be the same, whereas the front-parallel geostrophic wind components are different in the sense described previously. Any difference in the real wind components normal to the front is therefore an expression for the deviation from geostrophic flow and results in vertical motion, mostly different on the warm and the cold sides. From geometrical considerations it also follows that the isobars must show a cyclonic kink at any front as long as this is considered a zero-order discontinuity.

Real fronts, especially above the disturbed friction layer, are transition layers with horizontal widths varying between tens and hundreds of kilometers and vertical depths varying between, e.g., 100 to 2000 m. Because of the gradual change of the geostrophic wind inside a frontal zone, which is characterized by stronger cyclonic wind shear than the regions outside it, the isobaric kink of a zero-order discontinuity is replaced by a finite zone of *stronger cyclonic curvature* of the isobars inside the zone than outside it. This, however, does not necessarily imply that the isobar curvature always must be cyclonic in the frontal zone itself. Just as a stationary front under special conditions can be characterized by anticyclonic shear, depending on the shear in the surrounding air masses and the temperature distribution (Section 7.4), a moving front can also have anticyclonically curved isobars inside the frontal zone. This type of isobar curvature is occasionally noted if the frontal zone is relatively broad and the warm current above it shows sufficiently strong anticyclonic curvature. The type is not uncommon for weak warm fronts far away from a cyclone center, but it hardly ever occurs in connection with cold fronts because these are ordinarily associated with an upper warm current with cyclonic curvature.

Although warm fronts are quite generally characterized by upglide motion,[5] Bergeron (1937) distinguishes between cold fronts as "anafronts," above which the relative motion is upslope, and "katafronts," above which it is downslope; the latter is characteristic where the wind component normal to the front (except in the lowest levels) is greater than the rate of advance of the cold front.

Sansom (1951), in a study of 50 cold fronts passing over the British Isles, found this distinction to be valid. In the case of anafronts, the thermal wind was on the average nearly parallel to the surface front, the wind backing strongly with height and its component normal to the front being weaker than the frontal movement at every level up to 400 mb. With katafronts, the average thermal wind was inclined 30° to the front, and the

---

[5] As discussed in Section 12.4, frontal cloud masses do not usually have the same slopes as the fronts with which they are associated; thus, the air trajectories in the upgliding warm air are not directly identified with the frontal surface in the upper troposphere.

wind backed weakly with height and had a normal component surpassing the movement of the front except within the friction layer (by an average of 40 knots at 400 mb). As would be expected from these structures, ana- fronts were characterized by prolonged postfrontal rain, and katafronts by slight or no rain and rapid clearing. There were indications of a general transition from anafront to katafront as a cyclone developed.

A rather remarkable distinction brought out by Sansom's study concerns the frontal slopes. Examining the changes across the front at the 800- to 900-mb levels (above the layer of strong surface frictional influence), Sansom found the temperature differential comparable for the two types of fronts; however, the average horizontal differential of the front-parallel wind component was very much greater for anafronts than for katafronts. Correspondingly, from Margules' formula, he concluded that katafronts typically have shallow slopes in comparison with anafronts.

This feature also appeared as a result in an analytical model investigation by Ball (1960). In this, the front was considered as a zero-order discontinuity between air masses of uniform density. The geostrophic wind was prescribed in the warm air, that in the cold air being determined hydrostatically by this and the slope of the front in combination with the density differential. Thus, the horizontal pressure distribution within the cold air depended upon the configuration of the front in a vertical section. A surface fric- tional drag was imposed proportional to the wind velocity in the cold air, and mass continuity was prescribed in the vertical plane normal to the front.

Among the solutions obtained, the most relevant comparison is between those in which a southwesterly wind was imposed in the warm air, the normal component of this being set greater than the movement of the cold front (for a katafront), or smaller (for an anafront). In both cases the frontal inclination was very steep near the ground and decreased higher up, but in agreement with Sansom's analyses, the anafront had a large slope in higher levels whereas the katafront slope declined rapidly toward the horizontal, with a shallow cold-air mass.

The very steep slope of cold fronts near the ground, indicated by Ball's theory, is supported by observations cited in Section 9.6. This is in accord with what would be expected from surface friction (Bergeron, 1937). Since this imposes a drag opposite to the wind direction, its effect (considering a vertical plane normal to a cold front) is in the sense of an indirect cir- culation, adverse to the acceleration due to the solenoids. Thus, the slope of a front in the friction layer must be steeper than is indicated by Mar- gules' formula.

REFERENCES

Austin, E. E., and Bannon, J. K. (1952). Relation of the height of the maximum wind to the level of the tropopause on occasions of strong wind. *Meteorol. Mag.* **81**, 321–325.

Ball, F. K. (1960). A theory of fronts in relation to surface stress. *Quart. J. Roy. Meteorol. Soc.* **86**, 51–66.

Bergeron, T. (1937). On the physics of fronts. *Bull. Am. Meteorol. Soc.* **18**, 265–275.

Berggren, R. (1952). The distribution of temperature and wind connected with active tropical air in the higher troposphere, and some remarks concerning clear air turbulence at high altitude. *Tellus* **4**, 43–53.

Berggren, R. (1953). On frontal analysis in the higher troposphere and the lower stratosphere. *Arkiv Geofysik* **2**, 13–58.

Bjerknes, J. (1924). Diagnostic and prognostic applications of mountain observations. *Geofys. Publikasjoner, Norske Videnskaps-Akad. Oslo* **3**, No. 6.

Bjerknes, J., and Palmén, E. (1937). Investigations of selected European cyclones by means of serial ascents. *Geofys. Publikasjoner, Norske Videnskaps—Akad. Oslo* **12**, No. 2, 1–62.

Brunt, D. (1944). "Physical and Dynamical Meteorology," 2nd ed., Chapter 10. Cambridge Univ. Press, London and New York.

Chromow, S. P. (in collaboration with N. Konček, German transl. by G. Swoboda) (1942). "Einführung in die Synoptische Wetteranalyse," Chapter 5. Springer Verlag, Vienna.

Coudron, J. (1952). Le jet et les courbes de variation de la vitesse moyenne du vent avec l'altitude. *Jr. Sci. Meteorol.* **4**, 143–148.

Danielsen, E. F. (1959). The laminar structure of the atmosphere and its relation to the concept of a tropopause. *Arch. Meteorol., Geophys. Bioklimatol.* **A3**, 293–332.

Ertel, H. (1938). Methoden und Probleme der dynamischen Meteorologie. *Ergeb. Math.* **3**, 1–122.

Exner, F. M. (1925). "Dynamische Meteorologie," 2nd ed., 421 pp. Springer, Vienna.

Godske, C. L., Bergeron, T., Bjerknes, J., and Bundgaard, R. C. (1957). "Dynamic Meteorology and Weather Forecasting," Chapter 14. Am. Meteorol. Soc., Boston, Massachusetts.

Godson, W. L. (1951). Synoptic properties of frontal surfaces. *Quart. J. Roy. Meteorol. Soc.* **77**, 633–653.

Manabe, S., and Möller, F. (1961). On the radiative equilibrium and heat balance of the atmosphere. *Monthly Weather Rev.* **89**, 503–532.

Margules, M. (1906). Über Temperaturschichtung in stationär bewegter und in ruhender Luft. *Hann-Band. Meteorol. Z.* pp. 243–254.

Mohri, K. (1953). On the fields of wind and temperature over Japan and adjacent waters during winter of 1950–1951. *Tellus* **5**, 340–358.

Newton, C. W. (1954). Frontogenesis and frontolysis as a three-dimensional process. *J. Meteorol.* **11**, 449–461.

Newton, C. W., and Palmén, E. (1963). Kinematic and thermal properties of a large-amplitude wave in the westerlies. *Tellus* **15**, 99-119.

Nyberg, A., and Palmén, E. (1942). Synoptisch-aerologische Bearbeitung der internationalen Registrierballonaufstiege in Europa in der Zeit 17–19 Oktober 1935. *Statens Meteorol.-Hydrol. Anstalt, Medd., Ser. Uppsater* No. 40, 1–43.

Palmén, E. (1948). On the distribution of temperature and wind in the upper westerlies. *J. Meteorol.* **5**, 20–27.

Petterssen, S. (1940). "Weather Analysis and Forecasting," 1st ed., Chapter 6. McGraw-Hill, New York.

Reed, R. J., and Danielsen, E. F. (1959). Fronts in the vicinity of the tropopause. *Arch. Meteorol., Geophys. Bioklimatol.* **A11**, 1–17.

Sansom, H. W. (1951). A study of cold fronts over the British Isles. *Quart. J. Roy. Meteorol. Soc.* **77**, 96–120.

Saucier, W. J. (1955). "Principles of Meteorological Analysis," Chapters 6 and 9. Univ. of Chicago Press, Chicago, Illinois.

Scherhag, R. (1948). "Neue Methoden der Wetteranalyse und Wetterprognose," 424 pp. Springer, Berlin.

Van Mieghem, J. (1937). Analyse aérologique d'un front froid remarquable. *Mem. Inst. Meteorol. Belg.* No. 7.

8

PRINCIPAL TROPOSPHERIC JET STREAMS

In the preceding chapter, the jet stream was related in a general way to the barocline fields associated with frontal zones. In this chapter we describe the structures in greater detail. The general processes that maintain the large-scale wind systems are treated in Chapters 1 and 16, and the circulations connecting some of the more local variations of jet-stream structure with the process of frontogenesis will be taken up in Chapter 9.

The discussion in this chapter is confined to the jet streams of the troposphere and lowest stratosphere. Comparably strong jet streams exist in the upper stratosphere, and their lower parts are often evident down to tropopause levels. As indicated in Section 3.3, these systems are driven by radiative heat sources and sinks in the high atmosphere, which are quite separate from the processes in the troposphere. These high-level systems and other aspects of jet streams not covered in this chapter are described in reviews by Pogosian (1960) and Riehl (1962), and in the books by Reiter (1961, 1963) which deal very thoroughly with jet-stream related phenomena.

8.1 General Features of Jet Streams

In Chapter 6 it was indicated that the upper-level divergence and convergence associated with waves in the atmosphere should in general be most pronounced in regions of strong winds. Since medium-scale disturbances (cyclones and migratory anticyclones) are characterized by appreciable divergence and vertical motions, it is natural to expect these systems to show an affinity for jet streams. An association of this kind is illustrated

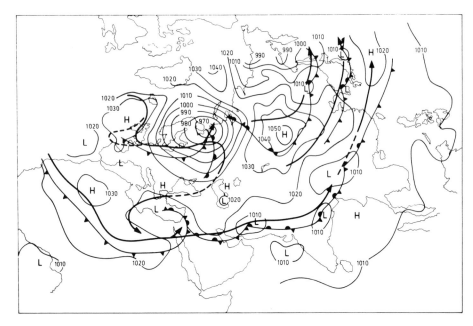

FIG. 8.1   Sea-level isobars, surface fronts, and axes of jet streams, Jan. 10, 1957.
(After Pogosian, 1960.)

by the example in Fig. 8.1, taken from Pogosian (1960). Each of the frontal
systems on the surface map has an accompanying jet stream. A significant
feature is the existence in different latitudes of multiple jet-stream systems,
some of which extend over long distances, often in a more-or-less connected
stream all the way around the hemisphere. Although trains of synoptic disturbances are found particularly along the polar-front jet stream, the fronts
associated with these disturbances are in some places distinct and in others
diffuse (see Sections 9.4 and 11.6).

In Chapter 4, emphasis was placed on the proposition that there are
two main jet-stream systems that have some characteristics in common,
but which are dissimilar as regards their relation to the atmospheric general
circulation. One of these is the *subtropical jet stream*, related to the poleward boundary of the Hadley circulation where low-level fronts tend to
be obscure or absent; the other is the *polar-front jet stream* (or polar jet
stream) connected with frontal zones in extratropical latitudes.

This generic classification will be adopted in this chapter. At the same
time, it is recognized that in a given region, more than one jet stream (of
either of these basic types) may be present (as in the eastern portion of
Fig. 8.1). McIntyre and Lee (1954) give the characteristic elevations of,

and 500-mb temperatures most frequently observed beneath, four different jet streams commonly found in connection with the complex of air-mass boundaries over Canada in winter.[1]

Examples of the subtropical jet stream are given later in this chapter, where certain distinctions between its structure and that of the polar jet stream are pointed out. A clear-cut specimen of a polar-front jet stream is shown in Fig. 8.2. This example illustrates the sharp increase of wind with height, which is found in frontal layers where the horizontal temperature gradient is strong, and shows in addition the appreciable vertical shear that normally characterizes the air masses outside the frontal layer, particularly in the warm air.[2] The pronounced decrease of wind speed with height above the tropopause, where the horizontal temperature gradient is reversed from that in the troposphere, is also demonstrated. Within the frontal layer, the cyclonic wind shear is very large, in accord with Eq. (7.15).

Horizontal wind profiles at various levels, typical of the polar jet stream in winter, are shown in Fig. 8.3. On the anticyclonic flank, the lateral shear is limited by the criterion for dynamic instability (Solberg, 1936; Klein-schmidt, 1941), which for a *straight* current prescribes that (n positive toward left of wind direction)

$$\frac{\partial V_g}{\partial n} \leq f \tag{8.1}$$

Palmén (1948), and Palmén and Nagler (1948), gave the first synoptic demonstration that this value of the shear is closely approached in a zone up to about 300 km wide on the warm-air side of the jet stream.[3] On the cyclonic flank the maximum shear is, as in Fig. 8.3, typically greater than f.

---

[1] McIntyre and Lee illustrate a strong front between maritime and continental arctic-air masses, in which the temperature on the warm-air side at 500 mb was $-34°C$. The structure is in all major respects similar to that of the polar front, adjacent to which the warm-air temperature at a lower latitude is commonly about 20°C higher.

[2] Observe that in the central part of the section, the wind in the upper troposphere increases with height, although the temperature decreases toward the right. This indicates a deviation from the thermal wind, discussed in Section 8.6, which in this case results from the presence of cyclonic curvature at 500 mb and anticyclonic curvature at 300 mb.

[3] The more general criterion, based on the principle of conservation of absolute angular momentum, is that a particle given an initial lateral displacement will undergo an acceleration in the direction of this displacement if the geostrophic absolute vorticity is negative (see Van Mieghem, 1951). In a cyclonically curved current, the anticyclonic shear may exceed f; in an anticyclonically curved current, the shear must be smaller than f. The criterion is properly applicable to conditions in an isentropic surface (or, in saturated air, to a surface of constant θ_e). Under geostrophic conditions, the anticyclonic shear along a tilted isentropic surface exceeds that along an isobaric surface.

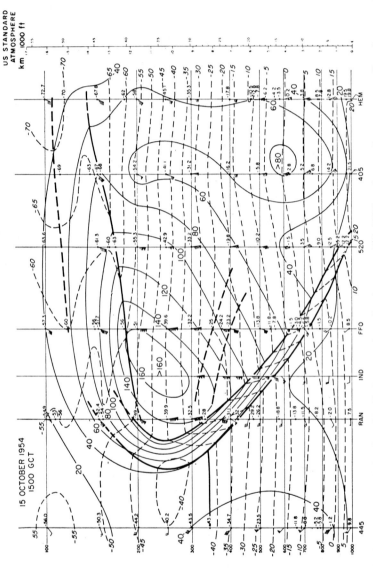

FIG. 8.2 Vertical section, 15 GCT Oct. 15, 1954, approximately along parallel 40°N and normal to strongest flow in upper levels. In this and following figures, the convention is to plot the winds as if north were at top of figure. Dashed lines, isotherms (°C); heavy lines, frontal boundaries or tropopauses; thin solid lines, isotachs at intervals of 20 knots of observed wind, without regard for direction. (From Palmén, 1958.)

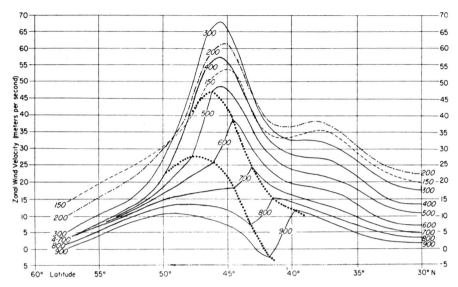

FIG. 8.3   Geostrophic wind profiles at isobaric surfaces, averaged from 12 cross sections at 80°W in December, 1946, across relatively straight flow. Heavy dotted line identifies intersections of frontal surface with wind profiles at different levels. (From Palmén and Newton, 1948.)

Shears up to $5f$ have been observed in cases described by Vuorela (1953) and Berggren (1953).

Techniques for analyzing the fields of wind and vertical shear, useful especially for aviation purposes, were introduced by Gustafson (1949), Johannessen (1956a), Bundgaard *et al.* (1956), Harmantas and Simplicio (1958), and Reiter (1958). An example illustrating the continuity of certain features, taken from Reiter, is shown in Fig. 8.4. Here the speeds and heights refer to centroid values in the "layer of maximum wind," which is bounded above and below by surfaces where the wind speed is 80% of the peak speed. The elevation of this layer (Fig. 8.4b) shows a close correspondence to the isotach pattern. On the average, Reiter found this elevation to be lowest about 100 km to the left of the jet-stream axis and also lowest near the speed maximum, rising about 1 km in a distance of 1000 km upstream or downstream from this.

Isotach maxima in the polar jet stream, as in Figs. 8.4a and c, are generally observed to progress in a regular manner and to a considerable extent to conserve their speeds (Fig. 8.4d). The alternation of streaks of high wind speed with regions of lower speeds, along the axis of the jet stream, is very characteristic. For convenience, isotach maxima of the kind shown by this example will be referred to as "*jet streaks.*"

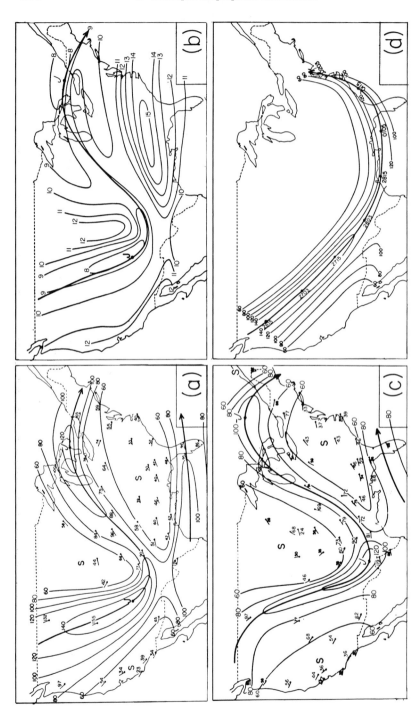

FIG. 8.4 Isotachs (a) and mean height (b) of layer of maximum wind, 15 GCT Feb. 27, 1954, and isotachs (c) 12 hr later. Part (d) is a space-time section, for the period 15 GCT Feb. 26, to 03 GCT Mar. 2, of the isotachs of the wind maximum west of the trough in (a) and (c). This analysis, slightly smoothed, was obtained by plotting winds along a line drawn through the jet-streak center normal to the jet.

8.2 STRUCTURE IN RELATION TO UPPER WAVES

In relating the wind to the geopotential field, it is important to realize that *near the jet-stream core, the ageostrophic wind may be a large fraction of the actual wind speed* when both the wind speed and the curvature of the current are large. This has been demonstrated by Väisänen (1954). The profiles of geostrophic and observed wind speeds along the trough and ridge lines of the large-amplitude wave system of Fig. 8.5 are compared in Fig. 8.6. The differences are in accord with the relationship

$$f(V_g - V) = kV^2 \tag{8.2}$$

where k is the trajectory curvature.

In this case, the maximum *geostrophic* wind speed at 300 mb in the trough is twice as great as the *actual* wind speed. In the ridge, the maximum wind speed is about 1.7 times the geostrophic speed.[4] Although the peak speeds are nearly equal in trough and ridge, the maximum geostrophic wind in the trough is nearly three times as great as that in the ridge. This requires essentially that the horizontal temperature gradients, integrated through the troposphere, differ in a corresponding way. As discussed in Section 7.7 with reference to the vertical sections in Fig. 7.10 for this same case, this condition is partly expressed by a slope of the polar-front layer in the trough (where the vertical depth of the frontal layer is very large) that is steeper than that in the ridge. In addition, there is more baroclinity within the air masses, especially in the warm air, at the trough line.

It is well known that the axis of the jet stream, although following the meanders of the contour pattern in a general fashion, does not exactly parallel the isobaric contours. The most familiar instance of such a deviation is strikingly illustrated by Fig. 8.7. On the upwind side of the isotach maximum, the wind speed increases strongly along the jet-stream axis, and there is a very pronounced deviation of that axis toward lower contours. The nature of such a deviation is well known. Omitting influences of movement of the velocity field, and assuming that an air particle moves along the jet axis in an *isobaric* surface, its trajectory should obey the simplified relationship

$$\Delta \left(\frac{V^2}{2} \right) = -g\,\Delta z \tag{8.3}$$

---

[4] It can be shown (see J. Bjerknes, 1951) that the wind speed in ridges cannot exceed twice the geostrophic wind speed if the wind follows the contours in a stationary system.

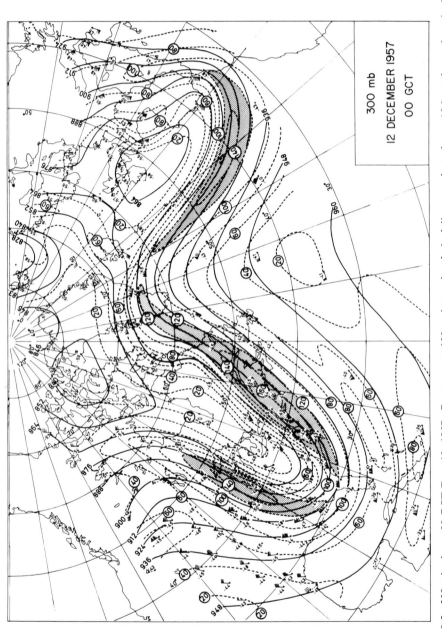

300 mb
12 DECEMBER 1957
00 GCT

FIG. 8.5   300-mb chart, 00 GCT Dec. 12, 1957. Contours at 120-m intervals; dashed lines are isotachs at 20-knot intervals; stippling indicates wind speed in excess of 120 knots. Temperatures plotted to left of stations. (From Newton and Palmén, 1963.)

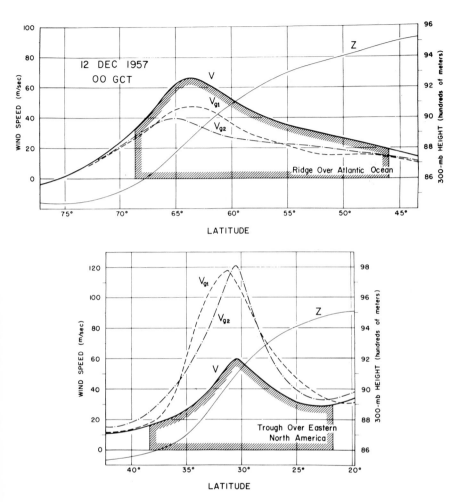

FIG. 8.6   Along ridge and trough lines of Fig. 8.5, profiles of 300-mb height (Z) and wind speed (V). V_{g1} and V_{g2} are geostrophic wind profiles, determined (1) from the isobaric contours and (2) from the actual winds, making use of Eq. (8.2). Hachures outline portion of profile between 8560-m and 9480-m contours in each case; the included areas measure the flow through this isobaric channel at ridge and at trough. (From Newton and Palmén, 1963.)

Considering the portion of the jet axis between the 100- and 160-knot isotachs, Δz computed from Eq. (8.3) is 210 m, in close agreement with the change of height along the jet axis in Fig. 8.7b.

With a large-amplitude wave system, another type of deviation may be present. This is illustrated by Fig. 8.5, where the amplitude of the jet-

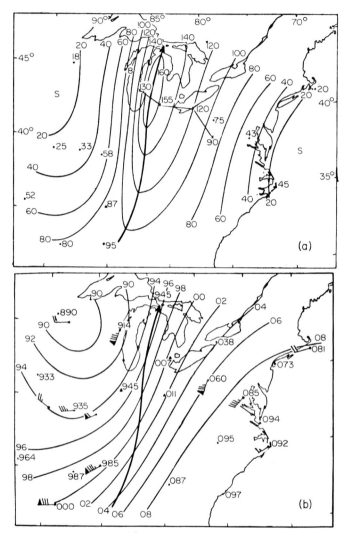

Fig. 8.7 (a) isotachs (knots), and (b) jet axis and contours (interval 200 ft) at 300 mb on Mar. 23, 1953. (After Riehl, 1954.)

stream axis is seen to be appreciably greater than that of the contours. Considering the region west of the trough over North America, on the upstream side of the jet streak the jet axis deviates toward *higher* contours despite an increase of wind speed downstream. Downwind from the same jet streak, the deviation toward higher contours is in the right sense but much more pronounced than would be judged from Eq. (8.3). These apparent inconsistencies are understood if the movement of the wave pattern is taken into account.[5]

On the left side of the jet axis, the absolute vorticity and vertical stability are both large, near the level of maximum wind. On the anticyclonic flank, the absolute vorticity is small and the stability weak. For an air parcel to cross from one side to the other, a very large change of potential vorticity would be required. On this basis, it is reasonable to assume that *near the level of maximum wind, the jet axis moves essentially as a material curve,* when times of the order of a day are considered. This viewpoint is supported by comparisons with air trajectories in a synoptic study by Newton and Omoto (1965) and by a result of Houghton's numerical computations (1965) relating to a theoretical jet-stream wave.

It is then evident from Fig. 8.8 that, in a wave system progressing along the direction of the basic current, the jet-stream axis must have a greater amplitude than the streamlines. Since an air particle originating on a crest of the jet-stream wave would follow a trajectory arriving at the downstream trough on the jet-stream wave at a later time, we may employ the familiar relationship for the trajectory-streamline amplitude ratio (Rossby, 1942) and write

$$\frac{A_J}{A_s} = \frac{V}{V - c} \tag{8.4}$$

where A_J and A_s are the amplitudes of the waves on the jet-stream axis and on a streamline at a corresponding mean latitude, c is the celerity of the wave system, and V is the wind speed in the jet core.

The configuration in Fig. 8.8 is also compatible with energy requirements, which state (V. Bjerknes, 1917; Haurwitz, 1941; Danielsen, 1961) that for trajectories on an isentropic surface,

$$\frac{d}{dt}\left(\frac{V^2}{2} + \psi\right) = \frac{\partial \psi}{\partial t} \tag{8.5}$$

where ψ is the Montgomery stream function ($c_p T + gz$). If the thin line in Fig. 8.8 is considered a stream-function isopleth moving with the wave system, then in the region west of the trough, $\partial \psi/\partial t > 0$. Then, according to Eq. (8.5), an air particle moving from ridge to trough must experience an increase of specific energy ($V^2/2 + \psi$). In the special case of gradient flow (wind parallel to stream-function isopleths), V would be constant and the air particle would simply undergo an increase of ψ.

---

[5] The same deviation is evident in *isentropic* surfaces near the level of maximum wind, so that this configuration is not due to the choice of the surface of representation.

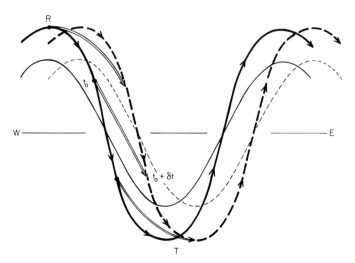

FIG. 8.8   Showing relationship of jet axis (heavy arrow) to a streamline (thin line) in a wave system progressing downstream. Solid lines refer to initial time and dashed lines to a time δt later. Air particle trajectories between these times are indicated by double-shafted arrows. (After Newton and Omoto, 1965.)

This interpretation of Eq. (8.5), together with Eq. (8.4), proves that, *to satisfy both energy and kinematic requirements in a progressive wave system with uniform wind speed along the jet axis, the stream function at the jet axis must be greater in the troughs than in the ridges.* The same is approximately true of the height in an isobaric surface near jet-stream level. In this case the amplitude of the jet-stream axis exceeds that of the contours, as in Fig. 8.5 and in numerous examples given by Raethjen and Höflich (1961). If the wind speed varies along the jet axis, this statement may be appropriately modified. Newton and Omoto (1965) have demonstrated that, within the limits of observational uncertainties, Eq. (8.5) is satisfied for air particles in the jet-stream core. In summary, the discussions above show that systematic deviations of the jet-stream axis with respect to the contour pattern may be associated either with movement of the wave pattern or variations of speed along the current, or both.

8.3   Nature and Dimensions of Jet Streaks

As mentioned earlier, jet streams cannot be considered as uniform currents around the globe; rather they are typified by concentrations of stronger wind in the jet streaks, alternating with weaker winds. As illustrated by Fig. 8.4, the individual streaks tend to progress along the current; how-

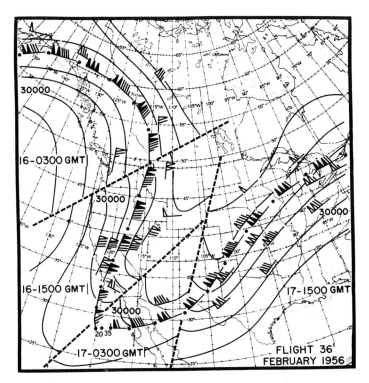

FIG. 8.9 Portion of trajectory of a transosonde balloon released from Japan and floating approximately at the 300-mb surface. Individual wind symbols correspond to smoothed positions of balloon at 2-hr intervals. Contours are taken from segments of 300-mb charts centered on 12-hr intervals of the trajectory. Near the cusp west of Lower California, speed (knots) is given by number. (After Angell, 1962.)

ever, they move very much slower than the wind. Consequently, air parcels move through them and gain speed as they pass through their upwind portions, losing speed in their downwind parts. This behavior has been demonstrated directly by the tracking of constant-pressure balloons. Figure 8.9 shows a portion of a balloon trajectory in a case wherein the velocity maxima were in the wave crests and the balloon underwent a remarkable deceleration and acceleration in passing through a sharp trough.[6]

The analysis in Fig. 8.10 further illustrates these velocity variations. It is

---

[6] For a summary of the findings from constant-pressure balloon (CPB) data, see the review by Angell (1961). In addition to providing wind information over data-sparse areas, CPB's afford the possibility of calculating directly such quantities as the geostrophic departure from the observed accelerations, without independent knowledge of the contour field.

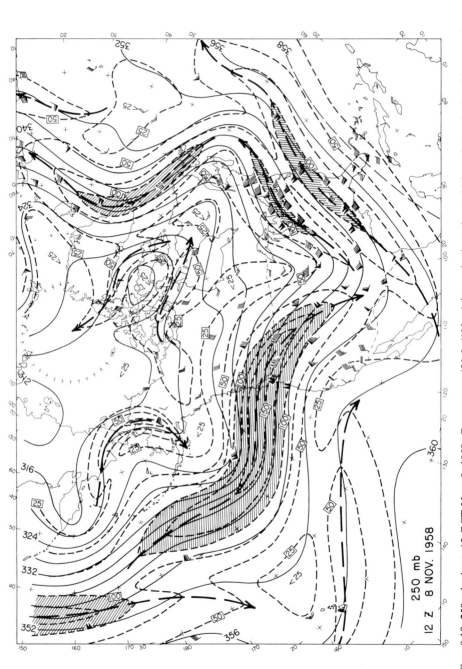

FIG. 8.10   250-mb chart, 12 GCT Nov. 8, 1958. Contours at 400-ft (120-m) intervals; isotachs at 25-knot intervals; hatching, speed over 100 knots; heavy arrows, jet axes. (From Newton, 1959)

natural to inquire whether the varying dimensions of the jet streaks can be connected with any general physical rule. One feature suggested by inspection of this figure is that the streaks of greatest length and breadth appear to be those with the highest maximum speeds. Since the streaks are slow-moving, a proportionality between their longitudinal dimensions and wind speeds would imply a preferred period of time required for air particles in the jet-stream core to move through them.

The character of the velocity oscillation undergone by such particles, required to produce the observed accelerations and decelerations during their passage through jet streaks, is similar in nature to an inertial oscillation. From the expression

$$\frac{d\mathbf{V}}{dt} = f\mathbf{V}_a \times \mathbf{k} \tag{8.6}$$

for frictionless flow (where $\mathbf{V}_a = \mathbf{V} - \mathbf{V}_g$), Petterssen (1956) showed that an air particle moving through a field of *constant* geostrophic wind would undergo an oscillation of velocity with a pure inertial period of a half pendulum day, the ageostrophic wind vector $\mathbf{V}_a$ remaining constant in magnitude and rotating in an anticyclonic sense with time. The trajectory described in such a case (Johannessen, 1956b) is a cycloid, with a wavelength corresponding to the inertial period times the geostrophic wind speed. The theory of trajectories in geostrophic fields having cyclonic or anticyclonic shear (in which case the oscillation periods, wavelengths, and amplitudes are different) has been treated extensively by Kao and Wurtele (1959), Raethjen (1958), and Raethjen and Höflich (1961). The results of these theories are suggestive, but not immediately applicable to the atmosphere, largely because they do not take into account any variation of the geostrophic wind along the direction of the basic current.

A jet streak superimposed on a straight basic current is sketched in Fig. 8.11. An air particle moving downstream from point A of minimum wind speed through the maximum-speed region to B undergoes a cyclical fluctuation about a mean velocity. This is indicated at the bottom of Fig. 8.11. In actual cases, if the basic current upon which the jet streak is superimposed does not have an excessively large curvature, it is usually observed that the maximum actual wind speed corresponds to the region of maximum geostrophic wind speed. In that case, variations of the actual and geostrophic wind velocities have the same sign. Considering the curvatures of the jet axis in Fig. 8.11, consonant with the velocity variations, it may be inferred from Eq. (8.2) that the strongest winds are supergeostrophic and the weakest winds are subgeostrophic. Consequently, the real wind velocity varies

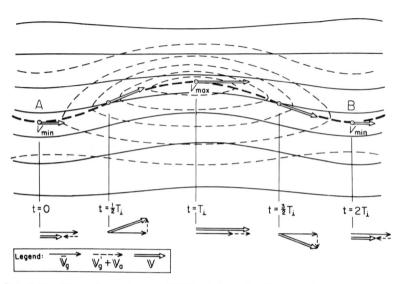

FIG. 8.11   Schematic contours (solid lines), isotachs (thin dashed lines), and vector winds at various locations along jet axis. An air particle in the jet core undergoes a fluctuation about a mean velocity in the manner shown at bottom of figure. (From Newton, 1959a.)

at a faster rate than the geostrophic wind velocity as the air particle moves along the jet stream.

More generally, these statements are valid with respect to the relationship between actual and *gradient* winds. The "planetary" waves discussed in Section 6.1 are of a basically different character, the winds in them being predominantly gradient, in which case the geostrophic wind is weaker in the crests than in the troughs. The "inertial type" waves discussed here, which in extratropical latitudes are superimposed on the planetary waves, basically express an oscillation about the gradient-wind condition. The same principles apply, with some modification.

As a basis for testing, it was hypothesized by Newton (1959a) that *in the core of the jet stream at the level of maximum wind*, the velocity fluctuations are described by the relationship

$$\frac{d\mathbf{V}}{dt} = 2\frac{d\mathbf{V}_g}{dt} \tag{8.7}$$

There is no theoretical basis for choosing the factor 2. The velocity oscillation prescribed by Eqs. (8.7) and (8.6) is illustrated in Fig. 8.12. The variable part of the velocity, $2\mathbf{V}_a$, rotates in an anticyclonic sense, as in a pure inertial oscillation, but with a different period. From Fig. 8.12, the rate of rotation is $fV_a/2V_a$, and the period defined by the time to rotate through 2π radians is

$$T_a = \frac{4\pi}{f} = 2T_i \tag{8.8}$$

or a full pendulum day (T_i being the period of a pure inertial oscillation). If the length of the jet streak between V_{min} points is denoted by L_J, Eq. (8.8) prescribes that, for a stationary jet streak,

$$L_J = \frac{4\pi}{f}\bar{V} = \frac{1 \text{ day}}{\sin \varphi}\bar{V} \tag{8.9}$$

where $\bar{V}$ is the mean wind speed in the core of the jet.

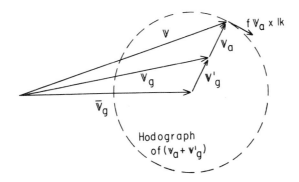

FIG. 8.12   Variation in velocity of a particle in jet-stream core when Eq. (8.7) is satisfied (see text). (From Newton, 1959a.)

Tests against three months of 250-mb charts showed that although there was a considerable scatter (only part of which could reasonably be attributed to uncertainties in analysis), the dimensions of two-thirds of the jet streaks corresponded to periods of between 1.5 and 2.5 T_i for the velocity oscillation in the jet core. Appropriate grouping of the data revealed that *the lengths of jet streaks at a given latitude are on the average proportional to the wind speed, and for a given wind speed are greater in low than in high latitudes,* in accord with Eq. (8.9), which is based on a semi-inertial period.

The lateral dimensions of jet streaks may be prescribed in terms of the "width" W_J, defined by Petterssen (1952) as the distance between points on either side of the jet-stream core where the wind speed is half the maximum speed V_{max} in the core. The maximum shear on the flanks of the jet stream is of the order of the coriolis parameter, being somewhat smaller on the anticyclonic-shear side and appreciably greater on the cyclonic-shear side, the shear tending on both sides to diminish with distance from the jet axis. We write

$$W_J \approx \frac{V_{max}}{f} \qquad (8.10)$$

as a crude approximation. Thus, with a maximum wind speed of 60 m/sec at latitude 42°, the width specified in this manner is about 600 km.[7]

From Eqs. (8.9) and (8.10), the length-to-width ratio of jet streaks is

$$\frac{L_J}{W_J} \approx \frac{4\pi \bar{V}}{V_{max}} \qquad (8.11)$$

For 183 jet streaks in October to December, 1958, over and near North America, the average values of V_{max} and V_{min} (defined in Fig. 8.11) were 128 and 64 knots, with $\bar{V} = 96$ knots. Equation (8.11) then indicates that a jet streak is about ten times as long as it is wide, on the average.

Studies of a large sample of rawin data by Endlich *et al.* (1955), supported by Reiter (1958), indicate that in the mean, the wind speed decays in a nearly linear fashion with height above and below the level of maximum speed, the percentile change with height being essentially independent of the maximum wind speed. Decay to half the speed at jet-stream level takes place in about 5 km on the average, being slightly more rapid above than below the level of strongest wind.

8.4 BROAD-SCALE ASPECTS OF SUBTROPICAL JET STREAM

In addition to being the most powerful wind system of the globe, in which wind speeds up to 260 knots have been observed over southern Japan (Arakawa, 1956), *the subtropical jet stream is also characterized by great steadiness, both in wind direction and in geographical location.* Figure 8.13 indicates that over southern Asia during winter and spring, 1949–1950, the subtropical jet stream was located more than 80% of the time between 25°N and 30°N, while the polar-front jet stream fluctuated over a wide range of latitudes. Ramage (1952) attributed the great steadiness in this locality to the influence of the Tibetan plateau as a cold region, the coldness resulting both from mechanical lifting with adiabatic cooling of the currents

---

[7] Checks against various examples in the literature indicate that Eq. (8.10) generally gives an underestimate, by up to 20% in the case of Fig. 8.3. For "strong" jets measured by Project Jet Stream (Saucier, 1958), which probably correspond to traverses near the cores of jet streaks, Eq. (8.10) agrees closely with the actual width.

impinging on it from the west and from a strong radiative deficit due to its extensive snow cover. The influence of the long west-east mountain complex of southern Eurasia as a barrier to the northward excursion of warm air was mentioned in Section 3.1.

Utilizing a dynamical model based on vorticity considerations, Mohri (1959) achieved a considerable degree of fidelity in reproducing the characteristic winter upper-level flow patterns over southeast Asia. He concluded that the Himalaya-Tibet plateau "establishes two branches of the jet stream

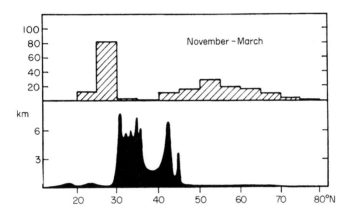

FIG. 8.13   Percentage frequency of jet-stream axis in 5° latitude belts at 500-mb level in the cool season 1949–1950, and the profile of topography, along longitude 80°E. (After Ramage, 1952.)

[one to the north and one on the south side], and forms a strong jet stream on the lee side of the mountains by the confluence of the northern branch and the southern one." While this analysis and Fig. 8.13 indicate a general aversion to the plateau, on occasion the subtropical jet stream may pass over it, especially during the transitions between the monsoon seasons. A striking example, given by Pogosian (1960), is shown in Fig. 8.14.

The mean winter configuration of the subtropical jet stream is illustrated by Fig. 8.15. Krishnamurti (1961) found only slight deviations of the hemispherically averaged latitude of the jet core, on a day-to-day basis, from the seasonal mean latitude. These excursions were somewhat more prominent over the America-Atlantic sector than in the Africa-Asia region. The basic character is a quasi-standing three-wave pattern. Temporary dislocations occur particularly when strong middle-latitude troughs extend into subtropical latitudes (the jet streaks of the subtropical system commonly moving eastward in association with these troughs), but there is a marked predilection for the subtropical features to be quickly reestablished in the mean positions shown.

It was stressed in Section 1.4 that the subtropical jet stream is generated as a result of the systematic poleward drift of air in the upper branch of the Hadley cell of the general atmospheric circulation, with partially conserved absolute angular momentum (Palmén, 1954). In accord with this concept,

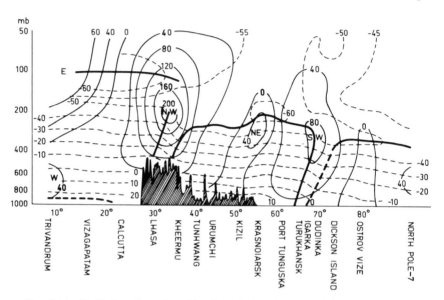

FIG. 8.14 Vertical section along line from Trivandrum (India) to North Pole—7, 15 GCT May 14, 1958 (isotachs in kilometers per hour). (After Pogosian, 1960.)

the subtropical jet stream is located near the poleward boundary of this cell. The near constancy in latitude of the circumhemispherical jet is thus tied to the circumstance that the Hadley cell (as a mechanism for maintaining the energy balance in a broad belt within which the heat loss or gain at different latitudes is relatively constant over long periods of time) is itself relatively steady in location and intensity.[8]

[8] An estimate of the energy generation is given in Section 16.2, based on meridional circulations averaged zonally. Krishnamurti calculated this circulation in a coordinate system relative to the meandering axis of the subtropical jet stream and obtained a much larger conversion from potential to kinetic energy than is suggested by Table 16.1. This difference is partly, but not wholly, accounted for by the fact that this method of averaging includes a generation that, in the other system, would correspond to an eddy generation by the standing waves. For three separate winter months, the mass circulations determined by Krishnamurti were slightly greater than the Hadley circulation in Fig. 1.5.

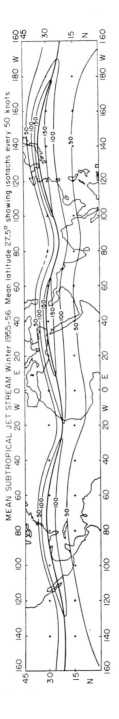

FIG. 8.15 Mean subtropical jet stream for winter 1955–1956. Isotachs (50-knot interval) at 200-mb surface. The mean latitude of the jet axis is 27.5°N. (After Krishnamurti, 1961.)

From the same viewpoint, there is a large variation in strength of the subtropical jet stream between seasons. Especially in the Northern Hemisphere, where the trades are reversed over a large region in summer (Fig. 14.1), the hemispherically averaged Hadley circulation is very much weaker in summer than in winter (cf. Figs. 1.5 and 1.6). In the Southern Hemisphere, although there is an appreciable seasonal variation in strength of the subtropical jet stream (Radok and Clarke, 1958), this is evidently much smaller than in northern latitudes. This is obviously connected with the smaller seasonal variation in intensity of the Hadley circulation, due basically to the different continentality influences.

As noted in Section 4.1, the subtropical jet stream is identified with a "break" between the middle tropopause (about 250 mb, $\theta \approx 330°K$) and the tropical tropopause (about 100 mb). Defant and Taba (1958a) describe a jet stream having these general characteristics and which exhibited a strong meandering into higher latitudes while in places closely paralleling the polar jet stream. Analyses by Krishnamurti (1961) for the same dates also show that the subtropical jet stream was present at a lower latitude in a configuration similar to Fig. 8.15. In a later paper, Defant and Taba (1958b) denoted a similar feature as the "subtropical jet branch," a separate current between the main polar and subtropical jet streams identified with a secondary break in the middle tropopause. Strong northward meandering of this intermediate jet stream occurs in "blocking" situations in which air masses having nearly tropical characteristics migrate far poleward on the east sides of the major troughs in the middle-latitude westerlies.

8.5   Confluence and Diffluence in the Major Wind Systems

Namias and Clapp (1949) related the regions of strongest winds appearing on mean winter wind charts to the process of confluence. This brings the cold- and warm-air masses together; as a result of the intensified solenoid field, a cross-stream circulation in a direct sense takes place, and the wind speed in the upper troposphere increases. In regions of diffluence, the opposite process occurs. The strongest mean winds (see Fig. 3.13a) are observed at quite low latitudes, whereas the strongest baroclinity, in the region of the polar front, is at a given time to be found at a considerably higher latitude around most of the hemisphere. This apparent inconsistency is resolved, as pointed out by Palmén (1954), if one considers the existence of two separate wind systems, a steady subtropical jet stream and a polar-front system that fluctuates in latitude and consequently does not so strongly influence the average winds at a given location (Fig. 3.9).

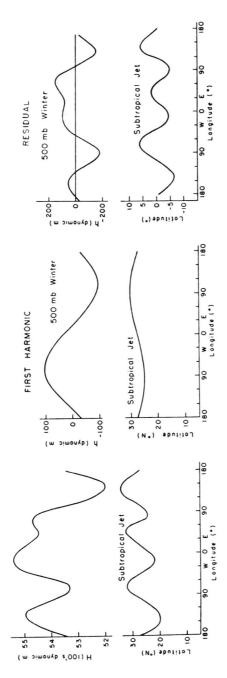

FIG. 8.16 Left: The 500-mb height (mean for winters 1945–1953) at latitude 47.5°, and the mean latitude of the subtropical jet axis from Fig. 8.15. Middle: The first harmonic of the curves on the left. Right: Residual, obtained from subtracting the first harmonic from the curves on the left. (After Krishnamurti, 1961.)

Wind data show that the mean wave patterns of middle and tropical latitudes are out of phase and that *the strongest winds are found in the crests of the subtropical jet stream at longitudes where the polar jet is closest to the Equator.* These features stand out very clearly in Krishnamurti's analyses, Figs. 8.15 and 8.16. Furthermore, the first harmonic of the 500-mb height in middle latitudes, which expresses the off-center circumpolar vortex, is also out of phase with the first harmonic of the subtropical jet latitude. Correspondingly, in Fig. 8.15, the crest over southeast Asia is located farther north and contains stronger mean winds than do the other two crests of the subtropical jet stream.

The relation between confluence and the variation of zonal wind is further illustrated by Fig. 8.17. The profiles in Fig. 8.17a illustrate an out-of-phase relationship between the meridional wind component at 15°N (near the middle of the Hadley cell) and at 50°N (near the middle of the Ferrel westerlies). The difference between these two curves, giving a measure of

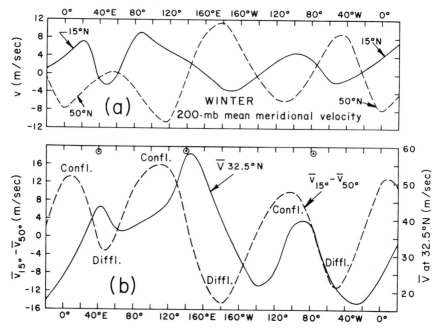

FIG. 8.17   Characteristics of the winter mean 200-mb flow, constructed from Crutcher's data (1961), around the Northern Hemisphere. (a) Meridional wind velocities at 15°N and at 50°N. (b) Dashed curve shows the difference of the two curves of (a) as a measure of the confluence or diffluence between the two latitudes. The solid curve shows the total wind speed at the intermediate latitude 32.5°N. At top, circles show longitudes of subtropical jet-stream crests from the independent analysis in Fig. 8.15.

the large-scale confluence and diffluence at 200 mb, is shown in Fig. 8.17b along with the profile of west-wind speeds at the intermediate latitude 32.5°N. *The major wind maxima are in each case located downstream from the strongest confluence,* and correspondingly there is a general decrease of the zonal wind downstream in regions of diffluence.

Krishnamurti averaged the temperature field for the three mean jet streaks in Fig. 8.15, with the result shown in Fig. 8.18a. This shows very clearly

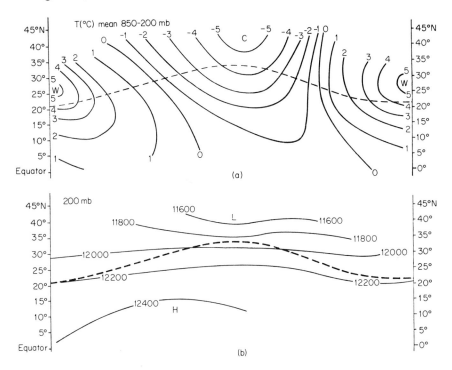

FIG. 8.18 Mean structure of subtropical jet waves, December, 1955: (a) For the average temperature in the 850- to 200-mb layer, the departure from the mean along the direction of the jet axis (in a curvilinear coordinate system); and (b) mean contours (meters) in relation to axis of subtropical jet stream. The latter is the dashed line in both figures. (After Krishnamurti, 1961.)

that the baroclinity is most pronounced in the crests and weakest in the troughs of the subtropical jet stream, as would be expected from the process of confluence and diffluence. This process is discussed in more detail in Section 9.2.

The variations in temperature gradient in Fig. 8.18a are reflected in corresponding variations of the contour gradient at 200 mb, Fig. 8.18b.

This pattern is broadly similar to that in the schematic Fig. 8.11. Also, a good agreement is found (Newton, 1959a) between the dimensions deduced in Section 8.3 and the mean dimensions of the meanders in the subtropical jet stream axis.[9] This suggests that these waves have the essential character of unalloyed semi-inertial oscillations. On daily maps also, the speed maxima seem always to be identified with the wave crests. In the polar jet stream, on the other hand, oscillations of this same nature are present, but these are superimposed on planetary waves of the Bjerknes-Rossby type. Jet streaks are commonly found near the crests of extratropical waves, but they may be located in any portion of the long-wave system (see, e.g., Fig. 8.5), including the troughs.

8.6   DETAILED STRUCTURE OF SUBTROPICAL JET STREAM

Saucier (1958), in discussing the results of extensive investigations of jet streams by use of aircraft and regular aerological observations, has emphasized that the structure varies greatly from one case to another. With this in mind, Figs. 8.19 and 8.20 are presented as an informative, but not necessarily "typical," example. One striking aspect evident in Fig. 8.20, noted by Saucier, is that the ratio of the width to the thickness is appreciably greater in the subtropical jet than in the polar jet. This is partly to be expected, according to Eq. (8.10), from the difference in the latitudes of the jet streams. In addition, however, the vertical depth of the subtropical jet tends to be comparatively small, most of the tropospheric shear sometimes being concentrated in a layer less than 100 mb deep beneath the jet stream, as in this example.

Such a concentration results from a combination of two influences. In the first place, the baroclinity in the neighborhood of the subtropical jet is generally greatest in the upper troposphere (for reasons discussed in Sections 4.1 and 9.7), in contrast to the polar-front region where the baroclinity is generally strong in the middle and lower troposphere. Thus, the geostrophic vertical shear is greatest in the corresponding layer. Secondly, the influence of curvature is very significant. Following Bellamy (1945) and Forsythe (1945), an equation for the vertical shear may be obtained by taking the derivative with height of Eq. (8.2). If, for the moment, the

---

[9] Based on data given by Krishnamurti, the mean wavelength agrees closely with Eq. (8.9). Equation (8.7) prescribes a cycloid whose amplitude is $4V_a/f$, wherein $4V_a$ is, according to Fig. 8.12, given by the difference of wind speed between crests and troughs. This value corresponds approximately with the observed amplitude of about $5°$ latitude.

variation of curvature with height is ignored, this gives

$$\frac{\partial V}{\partial z} = \left(\frac{f}{f + 2kV}\right)\frac{\partial V_g}{\partial z} = \left(\frac{1}{1 - (2V_a/V)}\right)\frac{\partial V_g}{\partial z} \qquad (8.12)$$

in the regions where $\mathbf{V}_a$ is parallel to $\mathbf{V}$.

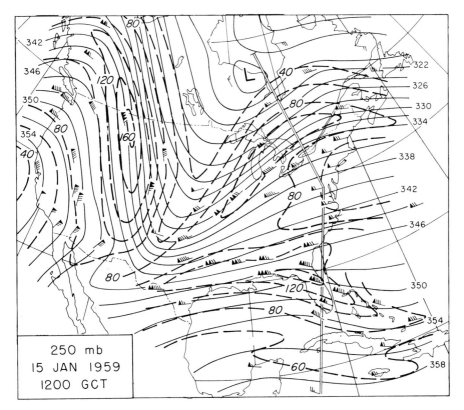

FIG. 8.19   Contours (hundreds of feet) and isotachs (knots) at 250 mb, 12 GCT Jan. 15, 1959. (From Newton and Persson, 1962.)

The middle expression indicates that the actual vertical shear is large in comparison with the geostrophic shear if the curvature is anticyclonic and the wind speed is large. Thus, the circumstance noted above, in which the baroclinity tends to be large in the higher troposphere where the wind speed is also large, accounts for the tendency for the vertical shear near the crests to be very much greater in the upper than in the lower troposphere. According to the last expression in Eq. (8.12), when the real wind speed

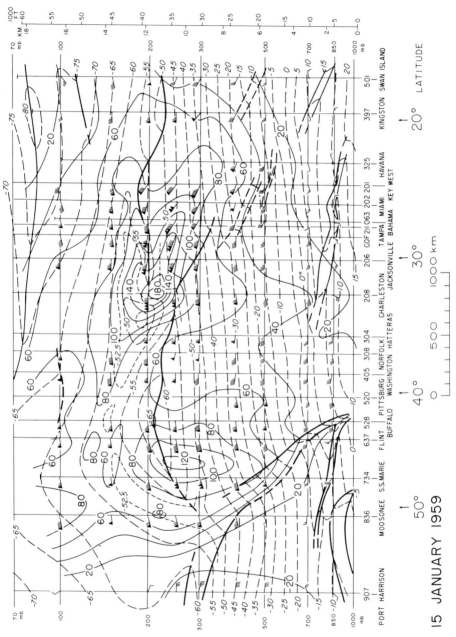

FIG. 8.20   Vertical section along line indicated in Fig. 8.19. Isotachs (knots) are for total wind speeds. (From Newton and Persson, 1962.)

exceeds the geostrophic wind by only one-third, the real vertical shear is double the geostrophic thermal shear. Saucier (1958) found the latter condition to be common.

If the variation of curvature is not neglected, the height derivative of Eq. (8.2) gives

$$\frac{\partial V}{\partial z} = \frac{1}{(f + 2kV)} \left(f \frac{\partial V_g}{\partial z} - V^2 \frac{\partial k}{\partial z} \right) \tag{8.13}$$

This expresses the difference between the variations of the actual and the geostrophic wind, arising from the variation with height of the centripetal acceleration. According to the last term, the increase of wind with height is augmented (relative to the thermal shear) if the anticyclonic curvature increases with height, while the decrease of wind speed above jet-stream level is enhanced if the anticyclonic curvature decreases with height. The effect of this can be as large as the influence mentioned earlier, so that *vertical shears differing by a factor of 3 to 4 from the "thermal wind" are sometimes observed.*

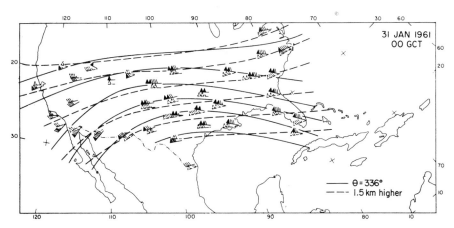

FIG. 8.21   Solid streamlines and wind symbols depict conditions at 336°K isentropic surface, which approximately corresponded to a surface of maximum wind speed, on the south flank of the subtropical jet, 00 GCT Jan. 31, 1961. Dashed symbols show wind field 1500 m higher. (From Newton and Persson, 1962.)

An example showing the large variation of curvature with height (in this case through a shallow layer above a surface of maximum wind) is shown in Fig. 8.21. In the cases studied by Newton and Persson (1962), the strongest anticyclonic curvature was located nearly at the surface of maximum wind. This is what would be expected if, as indicated earlier, the subtropical jet stream waves consist of ageostrophic meanders of the type of Fig. 8.11. These meanders are connected essentially with the variations of wind speed between troughs and ridges. Independent of the term in $\partial k/\partial z$, both the curvature influence and the changes of $\partial V_g/\partial z$ along the current (Fig. 8.18a) imply that this variation of wind speed is greatest at the level of maximum wind.[10] The inertial influence represented by the last term of Eq. (8.13) appears to contribute systematically toward augmenting the vertical shear, both above and below the level of maximum wind, in crests of the sub-tropical jet stream.

In the cases analyzed by Newton and Persson, there were often two levels of maximum wind south of the jet core. The upper layer, lying nearly horizontally, was on some occasions dominant, whereas the lower layer, which appeared to lie nearly on an isentropic surface, dominated on other occasions, as shown in Fig. 8.20. A peculiarity of this arrangement, which appears to be common, is that south of the subtropical jet core, the lower surface of maximum wind slopes downward toward south.[11] In a layer 1 to 2 km deep above this surface, the actual wind decreases with height, although the geostrophic wind increases slightly with height. This condition is possible, according to Eq. (8.13), if

$$\frac{\partial k}{\partial z} > \frac{f}{V^2}\frac{\partial V_g}{\partial z} \tag{8.14}$$

It is common to find that the surface of maximum wind rises north of the subtropical jet stream, being located 2 to 3 km above the middle tropo-pause, as in Fig. 8.20. This is consistent with the thermal field, the lower-stratospheric air being coldest over Buffalo, where the middle tropopause was highest. In this example, this sheet of maximum wind extends as a continuous layer northward to polar-front latitudes; more commonly it is broken up into separate cores. The presence of two discrete layers of maximum wind between, and their connection with, the subtropical and polar-front jet streams was pointed out by Bundgaard *et al.* (1956).

Endlich and McLean (1957) have drawn attention to a phenomenon that they term the "jet front," namely, a stable layer in the higher tropo-

---

[10] In higher latitudes also, the analyses by Murray and Daniels (1953) and by Briggs and Roach (1963) indicate that the transverse (ageostrophic) flow in the "entrance" and "exit" regions of jet streaks is most pronounced near the level of maximum wind.

[11] However, Singh (1964), in analyzing the subtropical jet stream over India, stated that he could not find evidence in that locality for two jet cores in the vertical, for any association of the layer of maximum wind with an isentropic surface, or for the sub-tropical tropopause (shown south of the jet core near 200 mb in Fig. 8.20).

sphere beneath and on the anticyclonic flank of the jet stream. This is found in association with the subtropical as well as the polar jet stream. Although in many cases the upper-tropospheric frontal layer becomes joined to a lower-tropospheric polar front, this is not so in the case of the "subtropical front" (Mohri, 1953), which is seldom evident below the 400- to 500-mb level.

Like the polar front, the subtropical front is not continuous around the hemisphere. An example of the local variation in its strength during a two-day period in a ridge over eastern North America is shown in Fig. 8.22. In Fig. 8.22a, the front is barely discernible at one station, while in Fig. 8.22b an extensive frontal layer is evident, with an overall temperature contrast comparable to that observed in the polar front.[12]

Other differences in structure are apparent. In Fig. 8.22a, wherein the front was absent, the lower subtropical tropopause at 200 to 250 mb was well developed, with quite stable air between it and the primary tropical tropopause above 100 mb. In Fig. 8.22b, however, the lower tropopause was absent, and lapse rates characteristic of the troposphere were present above the subtropical front. Where the subtropical front was strong, temperatures at its upper surface were higher than those farther south. Although in Fig. 8.22a the layer of maximum wind was nearly horizontal, in Fig. 8.22b a pronounced downward slope toward the south was observed, in accord with the difference in thermal structure. Multiple cores of maximum wind, as illustrated by these figures, are common in the subtropical jet-stream region.

Particularly over southeast Asia in winter, the polar-front and subtropical systems are often in close proximity. Matsumoto *et al.* (1953), in their extensive detailed analyses over the region of Japan, have demonstrated the extreme complexity of frontal structures, observing that stable layers may be up to 300 mb in depth. In discussing the example of Fig. 8.23, Mohri (1953) indicated that such an extreme stable layer (up to 6 km deep in this case) results when the polar front moves far south and combines with the subtropical front. Note that, as in Fig. 8.22b, the level of reversal of baroclinity was quite low on the south side of the jet stream. The mean level of the subtropical jet stream is near 200 mb, but in cases where the

---

[12] At the time of Fig. 8.22a, the subtropical front was strong on the west side of a trough over central North America and absent on its east side. At the time of Fig. 8.22b, the front was strong on the east side of the trough and weak in the west. The frontal layer moved eastward through the wave pattern of the subtropical jet stream in a manner similar to the way frontal layers in the upper troposphere migrate through polar-front troughs (Chapter 9).

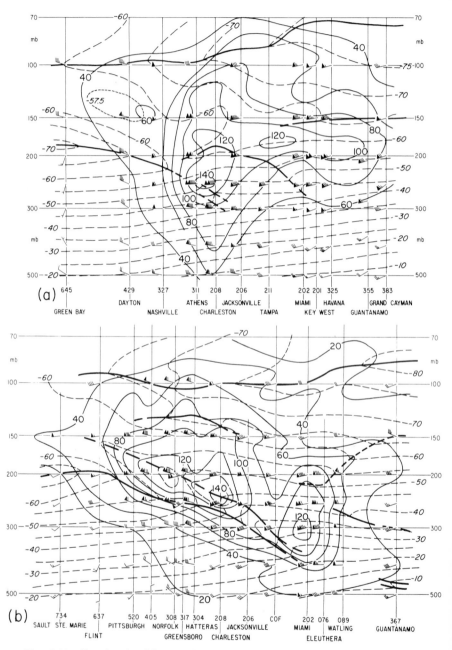

FIG. 8.22   Showing the differences in structure of the subtropical current system over eastern North America at 12 GCT Feb. 2, 1960 and at two days later when the subtropical front was well developed in this location. (From Newton, 1965.)

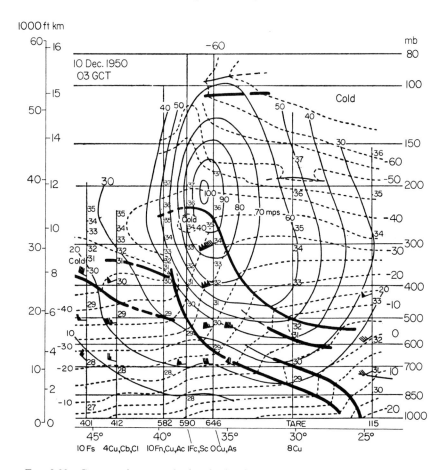

FIG. 8.23   Cross section over the longitude of Japan at 03 GCT Dec. 10, 1950 when the subtropical and polar fronts were combined. Isotachs in meters per second. (After Mohri, 1953.)

subtropical front is strong, the level of maximum wind in the southern portion of the jet stream is sometimes observed to be as low as 400 mb.

8.7   THE TROPICAL EASTERLY JET STREAM

With the northward shift of the westerlies in Northern Hemisphere summer, the easterly winds of the equatorial belt (Fig. 3.13) also shift northward. The easterly jet stream (Fig. 8.24), with strongest winds at 100 to 150 mb, is well developed, especially south of Asia. Here (Koteswaram, 1958) "the equatorial region is covered entirely by ocean, the region north

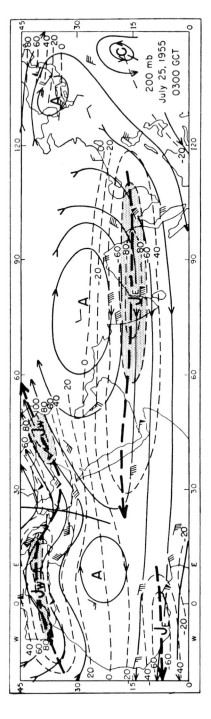

FIG. 8.24 Streamlines and isotachs (knots, negative for easterly component) at 200-mb surface, 03 GCT July 25, 1955. Heavy dashed lines are axes of easterly and westerly jets. (After Koteswaram, 1958.)

of 20°N by land. This arrangement prescribes maximum surface heating far to the north of the equator in the northern summer." Both Koteswaram, and earlier Flohn (1950), attributed importance to the presence of an elevated heat source over the Tibet-Himalaya massif, in accounting for the semipermanent summer anticyclone extending to high levels over this region (Fig. 8.24). This viewpoint has been disputed by Rangarajan (1963), who finds that mid-tropospheric temperatures in July are highest in a longitudinally extensive belt just south of latitude 30° rather than directly over the plateau to the north. This is compatible with the large release of latent heat in the ascending branch of the monsoon circulation (Section 14.5), although the additional contribution of high-level sensible heating should not be discounted.

The thermal field associated with this belt of warm air, which extends through the whole troposphere, favors a strong easterly current in upper levels. In midsummer the strongest mean easterlies, about 70 knots, are located near latitude 10° to 15° over the Arabian Sea, where their steadiness (defined by the ratio of the resultant wind speed to the mean scalar speed) reaches 96 to 100% (Flohn, 1964). While the directional steadiness is great, "wind speeds exceeding 100 kt occur usually in groups of several days simultaneously at more than one station; such strong wind periods are usually separated by other periods with generally weak winds." The influence of traveling perturbations in the upper easterlies upon the development of low-tropospheric disturbances is discussed in Section 14.8. For a thorough summary of the literature on the tropical easterly jet and a discussion of its three-dimensional circulation and influence on regional climates, the reader is referred to Flohn's monograph.

Migratory easterly jets over the longitudes of North America have been investigated by Alaka (1958), who found that their properties, such as vertical and horizontal shear, are comparable to those of middle-latitude westerly jets. As a steady phenomenon, the tropical easterly jet (in the Northern Hemisphere) is strikingly evident only over the region south of Asia. The case studies by Koteswaram and the mean charts by Flohn indicate that the easterly current is much weaker upstream (east of Asia) and downstream (west of Africa). The pronounced longitudinal variation of the easterly jet and its absence over the Atlantic and Pacific oceans are a result of the influence of the continents and oceans on the heat budget of the subtropical regions. Over the continents the heating of the earth's surface is immediately transferred to the atmosphere, largely in the form of sensible heat, whereas over the oceanic regions, a considerable part goes to heating of the upper ocean layers and to latent heat that is released partly over the continents. As a result, the subtropical continents comprise strong local heat sources in summer and temperature changes are suppressed over the oceans.[13]

8.8   THE LOW-LEVEL JET STREAM

A phenomenon of particular importance for the region east of the Rocky Mountains is the "low-level jet stream." As noted in Section 13.2, this southerly current plays a special role in the rapid advection of heat and moisture in convective situations. Properties of the low-level jet have been discussed extensively by Means (1954); a thorough statistical study of its geographical frequency of occurrence and physical structure is provided by Bonner (1965). An example is shown in Fig. 8.25. When the jet is well developed, as in Fig. 8.25f, the lateral shears are comparable with those in the high-tropospheric jet stream.

Wexler (1961) has drawn an analogy between the low-level jet stream and the Gulf Stream. He suggests that the Rocky Mountains play the role of a western boundary to the Bermuda High circulation, analogous to the role of the east coast of North America as a western boundary of the anticyclonic gyre of the oceanic circulation. Among other conditions, low-level jet formation is favored when the subtropical high is well developed and relatively far west. Then the westward transport of air over the Caribbean region is large, and the easterly current, being deflected northward on approaching the mountain barrier, furnishes an abundant supply of air for a strong southerly current east of the Rockies.

In spring, when currents of this type are best developed, it is characteristic that the air on the west side, heated over the western plateau, is warmer than the low-level air on the east side, which is marine air that has a relatively short trajectory over land. Thus the southerly component of the geostrophic wind is strongest at the surface and decreases upward through the lowest 1 to 2 km. This in part accounts for the rapid decay of wind speed above the jet in Fig. 8.25 (the large vertical shear beneath being due to surface friction).

In addition, however, it is significant to note that the marked diurnal variation of wind speed in Fig. 8.25 is characteristic. In an analysis of three

[13] Naturally the effect is most evident south of Eurasia, where land extends around more than one-third of the hemisphere. The very large dimensions of the easterly jet streak in Fig. 8.24, in comparison with the dimensions in higher latitudes (Figs. 8.4, 8.5, 8.10), are compatible with Eqs. (8.9) and (8.10), considering the difference in latitudes.

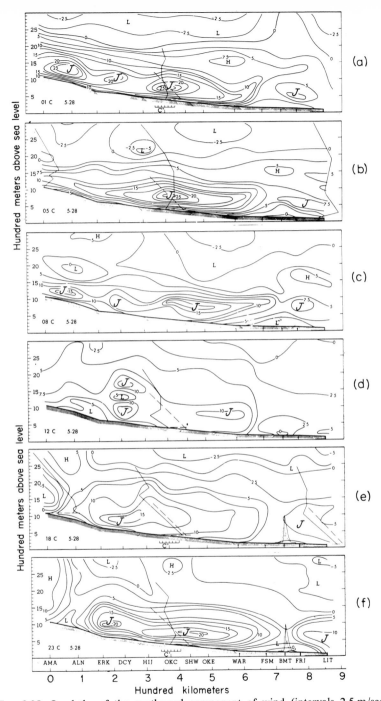

FIG. 8.25 Isopleths of the northward component of wind (intervals 2.5 m/sec) in vertical section along W-E line from Amarillo, Texas, to Little Rock, Arkansas, at selected times indicated in lower left of each figure, on May 28, 1961. Spacing of pilot-balloon stations shown at bottom. Temperature-height curves are shown for Amarillo (AMA), Oklahoma City (OKC) and Little Rock (LIT); dashed lines with 45° slope are dry adiabats. Cores of low-level jets indicated by J; other relative maxima or minima of southerly wind, H or L. (After Hoecker, 1963.)

situations using closely spaced serial wind observations, Hoecker (1963) found that maximum wind speeds were subgeostrophic during the day and strongly supergeostrophic (up to 1.75 times the surface geostrophic wind speed) in the early morning. The strong vertical shears in Figs. 8.25a, b, and f are thus largely ageostrophic above as well as below the jet.

The physical nature of the diurnal variation of wind speed has been indicated by Blackadar (1957). In the warmer part of the day when the lower layers have an unstable lapse rate, vigorous eddy exchange results in a momentum loss such that, at the top of the friction layer, the winds are subgeostrophic. In the evening when stability sets in, the strong frictional coupling with the ground is relieved and an inertial oscillation is initiated, owing to the imbalance between coriolis and pressure-gradient forces. At the latitude of the southern United States, the period of a pure inertial oscillation is approximately a calendar day; weakest winds are observed in midafternoon and strongest winds around 0300 local time.

8.9   GENERALITY OF JETS IN ROTATING FLUID SYSTEMS

Fast currents of restricted width have been produced in rotating fluid systems in the laboratory, which are very similar to those in the atmosphere (Riehl and Fultz, 1957) and in the oceans (von Arx, 1952).[14] Increasingly abundant observations in the two media have shown that there is a very marked similarity between the structures of the Gulf Stream, the Kuroshio current, and the atmospheric jet stream. Rossby (1951) remarked that the similarity between "apparently unrelated current systems... [suggests] that the factors controlling the shape and behavior of jets must be fairly independent of their driving mechanism, and derivable from quite general dynamic principles."

Some points of similarity brought out by comparisons of the Gulf Stream and jet stream (see, e.g., Iselin, 1950; Newton, 1959b) are the following: Direct velocity measurements indicate that the lateral shears on the flanks of the Gulf Stream are of the same magnitude as is characteristic of the jet stream. The slopes of fronts in the Gulf Stream, and the vertical and horizontal shears through the frontal layers, are essentially the same as are

[14] It should, however, be noted that—owing to the heat and cold sources used, the constancy of the coriolis parameter (in some types), and the absence of latent heat release —these laboratory phenomena cannot in all aspects be compared quantitatively with atmospheric jet streams. Despite these limitations, the overall physical processes are in most essentials similar, and much has been learned about these processes through the laboratory simulations.

observed in the polar front, despite the difference in dimensions of the systems. Successive cross sections of the upper layers of the Gulf Stream show that its detailed structure is quite variable, currents and fronts being at some times strongly developed and at other times diffuse, as is the case in the atmosphere (Section 9.4). Meanders in the Gulf Stream exhibit a process of "cutting off" and isolation of cold masses, which is like that in atmospheric developments (Chapter 10); superimposed upon the meandering current are velocity variations similar to atmospheric jet streaks, as in Fig. 8.5.

All these similarities suggest that, in line with Rossby's viewpoint, any theories proposed to account for the formation and behavior of jetlike currents and fronts should in their basic physical aspects be applicable to both the atmospheric and the oceanic systems.

References

Alaka, M. A. (1958). A case study of an easterly jet stream in the tropics. *Tellus* 10, 24–42.
Angell, J. K. (1961). Use of constant level balloons in meteorology. *Advan. Geophys.* 8, 137–219.
Angell, J. K. (1962). The influence of inertial instability upon transosonde trajectories and some forecast implications. *Monthly Weather Rev.* 90, 245–251.
Arakawa, H. (1956). Characteristics of the low-level jet stream. *J. Meteorol.* 13, 504–506.
Bellamy, J. C. (1945). The use of pressure altitude and altimeter corrections in meteorology. *J. Meteorol.* 2, 1–79.
Berggren, R. (1953). On frontal analysis in the higher troposphere and the lower stratosphere. *Arkiv Geofysik* 2, 13–58.
Bjerknes, J. (1951). Extratropical cyclones. *In* "Compendium of Meteorology" (T. F. Malone, ed.), pp. 577–598. Am. Meteorol. Soc., Boston, Massachusetts.
Bjerknes, V. (1917). Theoretisch-meteorologische Mitteilungen, 4. Die hydrodynamisch-thermodynamische Energiegleichung. *Meteorol. Z.* 34, 166–176.
Blackadar, A. K. (1957). Boundary layer wind maxima and their significance for the growth of nocturnal inversions. *Bull. Am. Meteorol. Soc.* 38, 283–290.
Bonner, W. D. (1965). Statistical and kinematical properties of the low-level jet stream. SMRP Res. Paper No. 38, 54 pp. Satellite and Mesometeorol. Res. Proj., Univ. Chicago.
Briggs, J., and Roach, W. T. (1963). Aircraft observations near jet streams. *Quart. J. Roy. Meteorol. Soc.* 89, 225–247.
Bundgaard, R. C. *et al.* (1956). The Black Sheep system of forecasting winds for long-range jet aircraft. *Air Weather Serv. Tech. Rept.* No. 105–139, 48 pp. U. S. Air Force, Washington, D. C.
Crutcher, H. L. (1961). Meridional cross sections. Upper winds over the Northern Hemisphere. Tech. Paper No. 41, 307 pp. U. S. Weather Bur., Washington, D. C.
Danielsen, E. F. (1961). Trajectories: Isobaric, isentropic and actual. *J. Meteorol.* 18, 479–486.
Defant, F., and Taba, H. (1958a). The strong index change period from January 1 to January 7, 1956. *Tellus* 10, 225–242.

Defant, F., and Taba, H. (1958b). The breakdown of zonal circulation during the period January 8 to 13, 1956, the characteristics of temperature field and tropopause and its relation to the atmospheric field of motion. *Tellus* **10**, 430–450.

Endlich, R. M., and McLean, G. S. (1957). The structure of the jet stream core. *J. Meteorol.* **14**, 543–552.

Endlich, R. M., Solot, S. B., and Thur, H. A. (1955). The mean vertical structure of the jet stream. *Tellus* **7**, 308–313.

Flohn, H. (1950). Tropische und aussertropische Monsun-Zirkulation. *Ber. Deut. Wetterdienstes U. S. Zone* **18**, 34–50.

Flohn, H. (1964). Investigations of the tropical easterly jet. *Bonner Meteorol. Abhandl.* **4**, 1–69.

Forsythe, G. E. (1945). A generalization of the thermal wind equation for arbitrary horizontal flow. *Bull. Am. Meteorol. Soc.* **26**, 371–375.

Gustafson, A. F. (1949). Final report on the upper wind project. Dept. Meteorol., Univ. Calif., Los Angeles.

Harmantas, L., and Simplicio, S. G. (1958). A suggested approach to the problem of providing high-altitude wind forecasts for jet transport operations. *Bull. Am. Meteorol. Soc.* **39**, 248–252.

Haurwitz, B. (1941). "Dynamic Meteorology," p. 240. McGraw-Hill, New York.

Hoecker, W. H., Jr. (1963). Three southerly low-level jet systems delineated by the Weather Bureau special pibal network of 1961. *Monthly Weather Rev.* **91**, 573–582.

Houghton, D. D. (1965). A quasi-Lagrangian study of the barotropic jet stream. *J. Atmospheric Sci.* **22**, 518–528.

Iselin, C. O'D. (1950). Some common characteristics of the Gulf Stream and the atmospheric jet stream. *Trans. N. Y. Acad. Sci.* [2] **13**, 84–86.

Johannessen, K. R. (1956a). Three-dimensional analysis of the jet stream through shear charts. *In* "Proceedings of the Workshop on Stratospheric Analysis and Forecasting, 1–3 February 1956," 168 pp. U. S. Weather Bur., Washington, D. C.

Johannessen, K. R. (1956b). Some theoretical aspects of constant-pressure trajectories. Air Weather Serv. Manual 105-47, Sect. II. U. S. Air Force, Washington, D. C.

Kao, S.-K., and Wurtele, M. G. (1959). The motion of a parcel in a constant geostrophic wind field of parabolic profile. *J. Geophys. Res.* **64**, 765–777.

Kleinschmidt, E., Jr. (1941). Stabilitätstheorie des geostrophischen Windfeldes. *Ann. Hydrograph. (Berlin)* **69**, 305–325.

Koteswaram, P. (1958). The easterly jet stream in the tropics. *Tellus* **10**, 43–57.

Krishnamurti, T. N. (1961). The subtropical jet stream of winter. *J. Meteorol.* **18**, 172–191.

McIntyre, D. P., and Lee, R. (1954). Jet streams in middle and high latitudes. *Proc. Toronto Meteorol. Conf.*, 1953 pp. 172–181. Roy. Meteorol. Soc., London.

Matsumoto, S., Itoo, H., and Arakawa, A. (1953). A study on westerly troughs near Japan (III). *Papers Meteorol. Geophys. (Tokyo)* **3**, 229–245.

Means, L. L. (1954). A study of the mean southerly wind maximum in low levels associated with a period of summer precipitation in the Middle West. *Bull. Am. Meteorol. Soc.* **35**, 166–170.

Mohri, K. (1953). On the fields of wind and temperature over Japan and adjacent waters during the winter of 1950–1951. *Tellus* **5**, 340–358.

Mohri, K. (1959). Jet streams and upper fronts in the general circulation and their characteristics over the Far East (Part II). *Geophys. Mag.* **29**, 333–412.

Murray, R., and Daniels, S. M. (1953). Transverse flow at entrance and exit to jet streams. *Quart. J. Roy. Meteorol. Soc.* **79**, 236–241.

Namias, J., and Clapp, P. F. (1949). Confluence theory of the high-tropospheric jet stream. *J. Meteorol.* **6**, 330–336.

Newton, C. W. (1959a). Axial velocity streaks in the jet stream: Ageostrophic "inertial" oscillations. *J. Meteorol.* **16**, 638–645.

Newton, C. W. (1959b). Synoptic comparisons of jet stream and Gulf Stream systems. *In* "The Atmosphere and the Sea in Motion" (B. Bolin, ed.), pp. 288–304. Rockefeller Inst. Press, New York.

Newton, C. W. (1965). Variations in structure of subtropical current system accompanying a deep polar outbreak. *Monthly Weather Rev.* **93**, 101–110.

Newton, C. W., and Omoto, Y. (1965). Energy distribution near jet stream, and associated wave-amplitude relationships. *Tellus* **17**, 449–462.

Newton, C. W., and Palmén, E. (1963). Kinematic and thermal properties of a large-amplitude wave in the westerlies. *Tellus* **15**, 99–119.

Newton, C. W., and Persson, A. V. (1962). Structural characteristics of the subtropical jet stream and certain lower-stratospheric wind systems. *Tellus* **14**, 221–241.

Palmén, E. (1948). On the distribution of temperature and wind in the upper westerlies. *J. Meteorol.* **5**, 20–27.

Palmén, E. (1954). Über die atmosphärischen Strahlströme. *Meteor. Abhandl., Inst. Meteorol. Geophys. Freien Univ. Berlin* **2**, 35–50.

Palmén, E. (1958). Vertical circulation and release of kinetic energy during the development of Hurricane Hazel into an extratropical storm. *Tellus* **10**, 1–23.

Palmén, E., and Nagler, K. M. (1948). An analysis of the wind and temperature distribution in the free atmosphere over North America in a case of approximately westerly flow. *J. Meteorol.* **5**, 58–64.

Palmén, E., and Newton, C. W. (1948). A study of the mean wind and temperature distribution in the vicinity of the polar front in winter. *J. Meteorol.* **5**, 220–226.

Petterssen, S. (1952). On the propagation and growth of jet-stream waves. *Quart. J. Roy. Meteorol. Soc.* **78**, 337–353.

Petterssen, S. (1956). "Weather Analysis and Forecasting," 2nd ed., Vol. I, pp. 60–61. McGraw-Hill, New York.

Pogosian, Kh. P. (1960). "Struinye techeniia v atmosfere," 178 pp. Gidrometeorologicheskoe Izdatel'stvo, Moscow. (Transl. by R. M. Holden, "Jet Streams in the Atmosphere," mimeograph, 187 pp. Am. Meteorol. Soc., Boston, Massachusetts).

Radok, U., and Clarke, R. H. (1958). Some features of the subtropical jet stream. *Beitr. Physik. Atmosphäre* **31**, 89–108.

Raethjen, P. (1958). Trägheitsellipse und Jet Stream. *Geophysica (Helsinki)* **6**, 439–453.

Raethjen, P., and Höflich, O. (1961). Zur Dynamik des Jet Streams. *Hamburger Geophys. Einzelschriften* **4**, 1–186.

Ramage, C. S. (1952). Relationship of general circulation to normal weather over southern Asia and the western Pacific during the cool season. *J. Meteorol.* **9**, 403–408.

Rangarajan, S. (1963). Thermal effects of the Tibetan Plateau during the Asian monsoon season. *Australian Meteorol. Mag.* No. 42, 24–34.

Reiter, E. R. (1958). The layer of maximum wind. *J. Meteorol.* **15**, 27–43.

Reiter, E. R. (1961). "Meteorologie der Strahlströme," 473 pp. Springer, Vienna.

Reiter, E. R. (1963). "Jet-Stream Meteorology," 515 pp. Univ. of Chicago Press, Chicago, Illinois.

Riehl, H. (1954). Jet stream flight, 23 March 1953. *Arch. Meteorol., Geophys. Bioklimatol.* A7, 56–66.

Riehl, H. (1962). Jet streams of the atmosphere. Tech. Rept. No. 32, 117 pp. Dept. Atmospheric Sci., Colorado State Univ., Fort Collins, Colorado.

Riehl, H., and Fultz, D. (1957). Jet stream and long waves in a steady rotating-dishpan experiment. *Quart. J. Roy. Meteorol. Soc.* 83, 215–231.

Rossby, C.-G. (1942). Kinematic and hydrostatic properties of certain long waves in the westerlies. *Misc. Rept.* No. 5, 37 pp. Dept. Meteorol., Univ. Chicago.

Rossby, C.-G. (1951). A comparison of current patterns in the atmosphere and in the ocean basins. *Intern. Union Geod. Geophys., Assoc. Meteorol., Brussels,* 1951 9th Assembly Mem., p. 9–31.

Saucier, W. J. (1958). A summary of wind distribution in the jet streams of the Southeast United States investigated by Project Jet Stream. Final Rept., pp. 6–57, Proj. AF 19-(604)-1565. A. and M. College of Texas.

Singh, M. S. (1964). Structural characteristics of the subtropical jet stream. *Indian J. Meteorol. Geophys.* 15, 417–424.

Solberg, H. (1936). Le mouvement d'inertie de l'atmosphère stable et son rôle dans la théorie des cyclones. *Procés Verbaux Météorol., Union Géod. Géophys. Intern. II* pp. 66–82. Edinburgh.

Väisänen, A. (1954). Comparison between the geostrophic and gradient wind in a case of a westerly jet. *Geophysica (Helsinki)* 4, 203–217.

Van Mieghem, J. (1951). Hydrodynamic instability. *In* "Compendium of Meteorology" (T. F. Malone, ed.), pp. 434–453. Am. Meteorol. Soc., Boston, Massachusetts.

von Arx, W. S. (1952). A laboratory study of the wind-driven ocean circulation. *Tellus* 4, 311–319.

Vuorela, L. A. (1953). On the air flow connected with the invasion of upper tropical air over northwestern Europe. *Geophysica (Helsinki)* 4, 105–130.

Wexler, H. (1961). A boundary layer interpretation of the low-level jet. *Tellus* 13, 369–378.

9

FRONTOGENESIS AND RELATED CIRCULATIONS

In earlier chapters, the thermal structure of the atmosphere was discussed in terms of zonally averaged properties, and also with regard to the large variations of the mean structure with geographical location. These variations were broadly related in Section 8.5 to the confluence and diffluence of the major current systems.

In the present chapter, we are concerned with the vertical and horizontal circulations that account for the even more pronounced variations of the thermal and wind fields of transient synoptic disturbances. As will be seen in Chapter 11, these structural variations have a bearing upon the development of cyclones and the related energy-transformation processes. Frontogenesis seems to be an essential accompaniment of the development of disturbances in a baroclinic fluid. Sharply defined fronts are observed as a characteristic feature of cyclones generated in "rotating dishpan" experiments (Fultz, 1952; Faller, 1956), which exhibit a remarkable similarity to atmospheric disturbances.

In Section 8.9 it was noted that oceanic fronts strikingly resemble those in the atmosphere. This similarity in thermal and kinematic structure[1] suggests, in line with Rossby's viewpoint, that the processes (other than heat sources) that lead to frontogenesis and frontolysis in the atmosphere should also apply to oceanic fronts. That such processes are active, super-

---

[1] Allowing for differences due to the near-incompressibility of water and effects of salinity upon density. Isotherms in an oceanic section correspond nearly to isentropes in the atmosphere. With these distinctions, the principles outlined in Chapter 7 apply equally to the oceanic structures.

posed upon those that govern the large-scale structures of ocean currents, is indicated by the analyses of von Arx *et al.* (1955). Their sections at frequent time intervals across the Gulf Stream show at times a clearly defined front and at other times a broad barocline zone, as in the example of Section 9.4.

Frontogenesis (or frontolysis) is a highly complex process, linked not only to deformation fields but also to the fields of translation and rotation. Correspondingly, since the structures of atmospheric eddies are so variable, no simple description can be given. Our objective must therefore be restricted to summarizing the various factors involved, and illustrating them with reference to some of the most common patterns, rather than presenting a final solution to the problem.

9.1   FRONTOGENESIS THROUGH HORIZONTAL MOTIONS

Since fronts mark discontinuities between air masses, they may be expected to show a preference for locations between the source regions (as, for example, during winter over the southeastern coasts of Asia and North America and the northern shore of the Mediterranean Sea) and the regions of maximum sea-surface temperature gradient. Such a preference is broadly evident in the frequency distributions of fronts.[2]

Fronts are, however, observed to form in and migrate over diverse locations, showing that their generation and maintenance depends in part upon factors other than the immediate influence of surface properties. Bergeron (1928, 1930) discovered that flow patterns with special characteristics (both in the mean flow and in individual situations) are typical of locations where frontogenesis takes place at the earth's surface. The special characteristic is a field of deformation, illustrated in Fig. 9.1.

It is evident that a pure deformation field of this kind not only concentrates properties, but also rotates the isopleths toward parallelism with the axis of dilatation.[3] In addition, if the zone of strongest concentration is initially displaced from the axis of dilatation, such a field would translate it bodily toward that axis. If S denotes a conservative property and v_n is

---

[2] For the Northern Hemisphere, see Schumann and van Rooy (1951) and Reed (1960); and for the Southern, van Loon (1965).

[3] This is true with respect to Fig. 9.1, but does not hold in arbitrary fields in which deformation is present. For example, in straight flow with horizontal shear (the sum of fields of rotation and deformation having equal strengths) the axis of dilatation is 45° from the streamlines (Saucier, 1955), but a material line initially parallel to that axis would in this case approach parallelism with the flow.

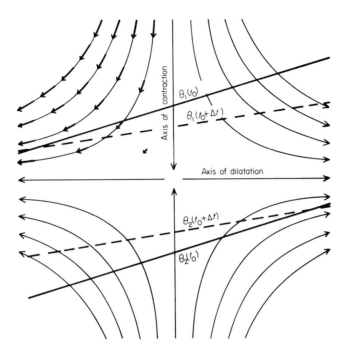

FIG. 9.1 Streamlines of a pure deformation field, with successive positions of two potential-temperature isopleths. (After Bergeron, 1928.)

the wind component normal to the isopleths, then (Petterssen, 1936, 1956) the rate of concentration may be expressed by

$$\frac{d}{dt}\left(\frac{\partial S}{\partial n}\right) = -\frac{\partial v_n}{\partial n}\frac{\partial S}{\partial n} \tag{9.1}$$

As Petterssen demonstrated, concentration takes place only if the S-isopleths are oriented within a certain angle from the axis of dilatation, about 45° in most situations. In general, since horizontal divergence is a small quantity, contraction of the wind components along one axis is nearly balanced by stretching along the other.

Commonly, especially near cold fronts with a meridional orientation, the presence of *shear* contributes to strengthening the frontal contrast in the manner illustrated in Fig. 9.2. In the coordinate system indicated, the effect (also representing deformation) is expressed by

$$-\frac{d}{dt}\left(\frac{\partial S}{\partial y}\right) = \frac{\partial u}{\partial y}\frac{\partial S}{\partial x} \tag{9.2}$$

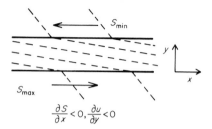

$$\frac{\partial S}{\partial x} < 0, \frac{\partial u}{\partial y} < 0$$

FIG. 9.2   Increase of contrast of a property S across a frontal zone, through differential advection by the winds parallel to the zone.

Figure 9.3 shows a common pattern, the col between two cyclones and two anticyclones, which is readily recognized as frontogenetic.[4] If the winds were *geostrophic*, the axis of dilatation would be approximately along line AA. However, the influence of surface friction, as indicated by the heavy arrows, is to cause the axis of outflow to be oriented somewhat counter-

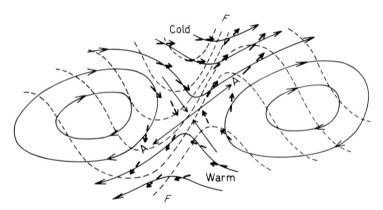

FIG. 9.3   Isobars and isotherms illustrating frontogenesis in the neighborhood of a col. Geostrophic streamlines are solid; deviations due to surface friction are indicated by heavy short arrows.

clockwise from this line. Concentration of the isotherms along zone FF is influenced partly by the systematic northward flow of warm air on the east side and southward flow of cold air on the west side, as in Fig. 9.2, and partly by contraction of the air motions normal to the frontal zone,

[4] Saucier suggests that the hyperbolic streamline patterns associated with such cols are overemphasized. More generally, the added fields of rotation, translation, and/or divergence tend to obscure the presence of deformation even where it may be pronounced. For examples of deformation relative to different types of flow, see Saucier (1955).

as in Fig. 9.1. It is evident from Fig. 9.3 that, considering the orientation of the isotherms, the cross-isobar flow due to surface friction enhances contraction across the frontal zone.

According to Section 7.2, once a front has formed, there is a trough or cyclonic shear across it, in the geostrophic wind field. This favors operation of the process characterized by Fig. 9.2. Also, especially considering the influence of surface friction, the effect described by Fig. 9.1 is enhanced in the immediate neighborhood of the frontal zone (both $\partial v_n/\partial n$ and $\partial T/\partial n$ being larger, the sharper the front). For these reasons, after a front has formed near the earth's surface it tends to be a self-maintaining phenomenon.

The processes discussed above concern only the kinematic influences. In general, migratory fronts are subjected to a tendency for weakening of thermal contrast in the surface layers, since except in special regions the transfer of properties from the earth's surface acts toward effacing the air-mass contrast.[5] Thus the air behind a cold front usually receives heat from the surface, and the lower-level air behind a warm front may be chilled in moving over ground recently overlain by a cold-air mass, or over a water surface with a strong temperature decrease along the direction of flow. Therefore, in order for a front to persist, it must generally be subjected continuously to kinematic frontogenesis. The tendency for transfer of heat to erase the thermal contrast across fronts is naturally much more pronounced over water than over land, owing to the small heat storage and slow conduction in solid earth (see Priestley, 1959). Also, fronts are more difficult to maintain in low than in high latitudes because (Section 10.4) cold air moving to low latitudes generally becomes shallow and thus requires less heat transfer to increase its temperature, in addition to being warmed by subsidence.

Since, according to Margules' relationship, surface fronts are zones of cyclonic shear, they are (especially cold fronts with strong thermal contrasts and steep slopes) characterized by high values of vorticity. In order to maintain the vorticity against dissipation by friction, actual horizontal convergence is required. Consequently, *convergence as well as a favorable deformation field is necessary for the formation and preservation of a surface front* (Petterssen and Austin, 1942).

[5] Although, as discussed later, the release of latent heat is important, for obvious reasons it has no direct influence on heating at the earth's surface, which would contribute to frontogenesis. However, precipitation falling from warm air and evaporating in the cold air may significantly sharpen a front and influence its movement (Oliver and Holzworth, 1953).

9.2   Circulations Associated with
Confluence and Release of Latent Heat

The discussion above was concerned only with frontogenesis near the earth's surface. It is appropriate to examine how this may be related to the three-dimensional circulation in the overlying layers. The necessity for vertical circulation is indicated by the fact noted above, i.e., that although nondivergent deformation could concentrate the isotherms, generation of *vorticity* in the surface front requires convergence with the attendant vertical motion field. In Sections 3.2 and 8.5 it was pointed out that in certain geographical regions, the horizontal temperature gradient is concentrated by the confluence of warm- and cold-air currents, with a corresponding influence on the strength of the upper-tropospheric winds. This effect is present in mobile disturbances as well as in the mean large-scale flow patterns.

The general principle of confluence of differing air masses was outlined by Sutcliffe (1940), and applied specifically to the intensification of jet streams as described by Namias and Clapp (1949). The sketches in Fig. 9.4 illustrate this principle. If strong translation along the direction of the

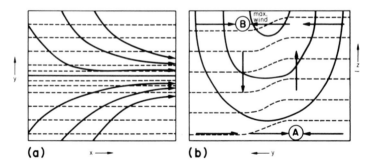

(a) $x \longrightarrow$ **(b)** $\longleftarrow y$

Fig. 9.4   (a) Horizontal streamlines and isotherms in a frontogenetic confluence; (b) vertical section across the confluence, showing isotachs (solid), isotherms (dashed), and vertical and transverse motions (arrows). (After Sawyer, 1956.)

axis of dilatation is superimposed on a deformation field such as Fig. 9.1, a confluent current system of the type of Fig. 9.4a results, which also contributes to a concentration of the isotherms. This leads to an increased thermal wind and to an adjustment of the vertical shear through the transverse circulation, indicated by the horizontal arrows in the central part of Fig. 9.4b. The circulation in the horizontal branches is then assumed to be compensated in the manner indicated by the vertical arrows.

The scheme in Fig. 9.4b (see Sawyer, 1956; Eliassen, 1959) results in an induced lateral contraction at A in lower levels and at B on the cold-air side of the circulation cell in upper levels. Thus, considering the induced circulations, frontogenesis is favored in the *sloping* layer between A and B rather than in a vertical barocline zone. At the same time, the horizontal convergence in lower levels at A and in upper levels at B, implied by the vertical circulation, results in generation of vorticity in a similarly sloping layer.

Sawyer (1956) carried out calculations of the transverse circulations in a zone of confluence, under the assumption that the flow along the direction of the isotherms in Fig. 9.4a is geostrophic, while an ageostrophic v-component (necessary for geostrophic adjustment of the u-component) is allowed. Assuming a confluence rate expressed by $\partial u/\partial x \approx 3$ m/sec per 100 km, and various density distributions (specifying the vertical stability and a horizontal temperature distribution similar to that in Fig. 9.4a), Sawyer presented several numerical solutions, two of which are shown in Fig. 9.5.

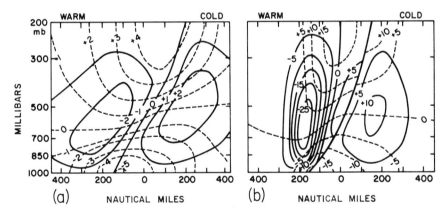

FIG. 9.5   Theoretical distribution of vertical motion (solid lines, mb/hr) and transverse horizontal velocity (dashed lines, kt), resulting from confluence. In (a) the air is dry; in (b) the rising air is assumed to be saturated. In both examples, the magnitude of the basic confluence was assumed to be the same; however, the stabilities and horizontal temperature gradients differed. (After Sawyer, 1956.)

In both cases a direct solenoidal circulation resulted, as anticipated from the qualitative discussion above. In Fig. 9.5a, no condensation was allowed, and both vertical motions (maximum about 1 μb/sec) and ageostrophic lateral motions are fairly small. Figure 9.5b illustrates a case wherein condensation was permitted to occur, with neutral stability in the rising warm air at a distance from the frontal zone. The horizontal density gradient assumed for this case was also appreciably greater than that for Fig. 9.5a. Correspondingly, both the vertical and the horizontal branches of the transverse circulation are much stronger. The ascending saturated branch is also more intense than, and confined to a smaller region than, the descending branch. Sawyer's calculations of the local density change indicate that frontogenesis was strongest in upper and lower levels, under the assumed conditions.

Eliassen's extension of the theory (1962) confirmed the basic results of Sawyer. Eliassen treated the more general case wherein the geostrophic wind turns with height, thereby allowing for the distinction between circulations around cold fronts and warm fronts. He found that the "source strength" of the circulation is

$$Q = 2 \left(\frac{\partial u_g}{\partial p} \frac{\partial v_g}{\partial y} - \frac{\partial u_g}{\partial y} \frac{\partial v_g}{\partial p} \right) \tag{9.3}$$

where u_g and v_g are the components of the geostrophic wind (u being normal to Fig. 9.6, out of the page). This expression is of particular interest in

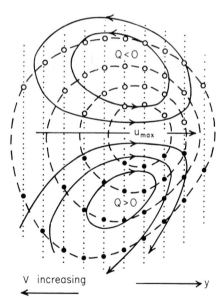

V increasing   ⟶ y
⟵

FIG. 9.6   Transverse circulation in the vicinity of an idealized jet core, with uniform confluence in the geostrophic wind field at all heights. Dotted lines are isotachs of v_g, the y-component of the geostrophic wind, and solid lines are streamlines of the transverse (nongeostrophic) circulation. (After Eliassen, 1962.)

describing the transverse circulation in the neighborhood of the jet stream. In the special case wherein there is confluence ($\partial v_g/\partial y < 0$) and the isotherms are parallel to the isobaric contours ($\partial v_g/\partial p = 0$, nullifying the last term), the transverse circulations form a doublet, as in Fig. 9.6. These toroidal circulations are centered about the regions where the products of $\partial u_g/\partial p$ and $\partial v_g/\partial y$ are largest. This is an expression of the fact that where the isotherms are already close together and the confluence is strongest, the further concentration of the isotherms proceeds at the greatest rate, as expressed by Eq. (9.1), and the circulation required to bring the real wind into balance with the thermal wind must therefore be most intense.

The influence of friction upon frontogenesis has been stressed by Eliassen (1959). As noted in connection with Fig. 9.3, friction has a pronounced effect in concentrating the frontal contrast at the ground, an effect that is obviously greatest when the front is already narrow and strong, and which therefore lends to the front a "self-sharpening" character. According to Welander (1963), there must be a frictional inflow in the Ekman layer toward a front from both the cold- and warm-air sides, and he suggests that air from the Ekman layer is physically incorporated into the lower part of the frontal layer. This viewpoint is supported by analyses of observations in the lower portion of fronts (Section 9.6).

Eliassen (1959) has also provided a theoretical treatment of the frontogenetic influence of condensation. Release of latent heat intensifies the upward motion in a cloud region in the warm air above a front, as in Fig. 9.5b. Connected with this is an induced deformation field in the vertical plane normal to the front (as in Fig. 9.4b), with lateral contraction of the air motions beneath the condensation region. This induced field, enhanced by frictional inflow, "...may cause a diffuse frontal zone to develop into a very sharp front within a relatively short time.... This mechanism does not depend upon a coincidental juxtaposition between the frontogenetic transverse field of motion and the front itself; the two are bound together, the frontal cloud representing the linkage between them. The process is therefore self-maintaining and irreversible: once started, it must go on at an accelerating speed until the frontogenesis is completed."

9.3 GENERAL PROCESSES IN THE REDISTRIBUTION OF PROPERTIES

Before proceeding to synoptic descriptions of three-dimensional frontogenesis, it is convenient at this point to summarize the basic processes involved. Following Petterssen (1956), we define frontogenesis as the process that concentrates the gradient of a property S, such as potential temperature. Then a frontogenetic function $\mathbf{F}$ may be defined, such that in three dimensions

$$\mathbf{F}_3 = \frac{d}{dt}(\nabla_3 S) \tag{9.4}$$

Obviously, the frontogenetic process is most effective if it operates, over a long time, on the same air particles that remain in the frontal zone. This is probably the general case in the "free atmosphere," where observations suggest that air does not readily move across frontal boundaries, but not near the ground (Section 9.6). Expanding d/dt and ∇ and regrouping terms, Eq. (9.4) becomes

$$\mathbf{F}_3 = \nabla_3 \frac{dS}{dt} - \left(\frac{\partial S}{\partial x} \nabla_3 u + \frac{\partial S}{\partial y} \nabla_3 v + \frac{\partial S}{\partial z} \nabla_3 w \right) \tag{9.5}$$

an expression given by Haltiner and Martin (1957). This corresponds to Miller's original formulation in scalar form (1948), which is the basis for the following discussion.

Because the vertical gradients of properties in fronts are ordinarily about 100 times as great as the horizontal gradients, the magnitude of $\mathbf{F}_3$ is dominated by changes in the vertical structure. For most purposes, however, we are interested in the horizontal (or isobaric) as distinct from the vertical variations of property. If the x-axis is chosen parallel to the front, the corresponding components of Eq. (9.5) are

$$F_z = \underbrace{\frac{d}{dt}\left(\frac{\partial S}{\partial z}\right) = \frac{\partial}{\partial z}\left(\frac{dS}{dt}\right)}_{\text{A}} - \left(\underbrace{\frac{\partial S}{\partial x}\frac{\partial u}{\partial z}}_{\text{B}} + \underbrace{\frac{\partial S}{\partial y}\frac{\partial v}{\partial z}}_{\text{C}} + \underbrace{\frac{\partial S}{\partial z}\frac{\partial w}{\partial z}}_{\text{D}} \right) \tag{9.6a}$$

$$F_y = \underbrace{\frac{d}{dt}\left(\frac{\partial S}{\partial y}\right) = \frac{\partial}{\partial y}\left(\frac{dS}{dt}\right)}_{\text{E}} - \left(\underbrace{\frac{\partial S}{\partial x}\frac{\partial u}{\partial y}}_{\text{F}} + \underbrace{\frac{\partial S}{\partial y}\frac{\partial v}{\partial y}}_{\text{G}} + \underbrace{\frac{\partial S}{\partial z}\frac{\partial w}{\partial y}}_{\text{H}} \right) \tag{9.6b}$$

the expression for F_x being similar to that for F_y. These equations would, of course, take the same form if written with p as vertical coordinate and ω in place of w.

The interpretation of terms D and G in Eqs. (9.6) is straightforward. These represent the concentration of gradient by vertical and lateral confluence fields, as in Fig. 9.1, the form of these terms being equivalent to the right side of Eq. (9.1). Terms B and F have the character of Fig. 9.2, the interpretation of which was given in Section 9.1 for the horizontal plane.

Terms C and H, respectively, are illustrated by Figs. 9.7b and 9.7a, which represent the distribution of S in a vertical section normal to the frontal zone at an initial time (solid lines) and at a later time (dashed lines). As drawn, the horizontal gradient of vertical velocity (Fig. 9.7a) contributes to an increase of the horizontal gradient of S; in Fig. 9.7b the vertical gradient of transverse velocity (the flow being predominantly into the page) increases the vertical gradient of S. The two motion fields, when added

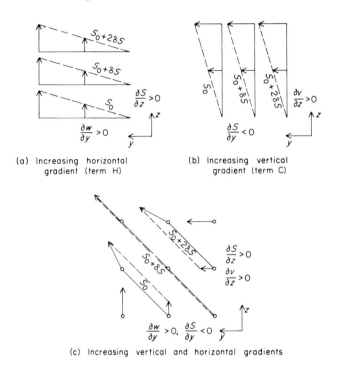

(a) Increasing horizontal gradient (term H)

(b) Increasing vertical gradient (term C)

(c) Increasing vertical and horizontal gradients

FIG. 9.7 (a) Increase of horizontal gradient due to vertical motion; (b) increase of vertical gradient due to transverse component of vertical shear; (c) the two influences combined. The S lines may, for example, correspond to isentropes in a cross section normal to the principal component of the wind velocity. (From Newton and Carson, 1953; essentially after Miller, 1948.)

together (Fig. 9.7c), contribute to a simultaneous vertical and horizontal concentration.[6] Terms A and E represent superimposed, nonconservative influences such as diabatic heating or cooling if θ is substituted for S, or particle velocity change (due to cross-contour flow or friction) if u is substituted for S.

[6] Note that although Fig. 9.7c portrays a deformation field in a vertical plane normal to the principal direction of flow, the vertical dimensions are smaller (by a factor of roughly 100) than the lateral dimensions; the same is true of the velocity components.

9.4   An Example of Three-Dimensional Frontogenesis and Frontolysis

As illustrated in Chapter 7, fronts often extend through the whole troposphere. However, it is not unusual for well-developed fronts to exist independently in the lower or upper troposphere, where the various processes involved in their origin differ in emphasis.

One of the significant features of pronounced fronts in the middle and upper troposphere is that the air in their vicinity tends to be very dry, as shown by Vuorela's (1953) analyses. As indicated in Section 12.4, even when there is extensive cloud overlying the lower portion of a front, the warm air next to the upper portion is characteristically dry. Computations by Reed and Sanders (1953) and by Newton and Carson (1953) indicated that upper-tropospheric frontogenesis is accompanied by a vertical-motion field in which the air on the warm side sinks relative to that on the cold side. The resulting different rates of adiabatic heating and of downward transport of momentum contribute to both an intensification of temperature gradient and an increase of vorticity.

To illustrate the processes of frontogenesis and frontolysis at different levels, we have chosen the example in Figs. 9.8 and 9.9. The structure shown here is commonly observed during the early stages of cyclone formation over North America. At this time a developing upper-tropospheric trough (Fig. 9.8a) was present over the western United States. A cyclone that had traversed from the Pacific was dissipating near the Great Lakes, and a new cyclone was forming on the cold front trailing from it, in the location of the heavy triangle in Fig. 9.8b. A pronounced front existed through the entire troposphere on the west side of the developing upper trough, but was absent in the upper troposphere farther east. The wind field at 850 mb (Fig. 9.8b), which has some of the characteristics of both Figs. 9.1 and 9.2, indicates that kinematic frontogenesis was active in the lower troposphere east of cross section C, but a well-developed front was present only below about 700 mb.

Later, the upper-tropospheric frontal layer progressed eastward and eventually surrounded the trough (Fig. 10.13), establishing a clear connection with the lower-level front, which also became much sharper. At the time shown here the frontal "exit" was moving eastward at about 30 knots, while at 500 mb, the winds in the exit region were about 40 knots on the cold side and 80 knots on the warm side of the front. This means that in the middle and upper troposhere, air particles were streaming eastward out of the frontal layer, undergoing "individual frontolysis" in the upper levels although frontogenesis was occurring in the lower troposphere.

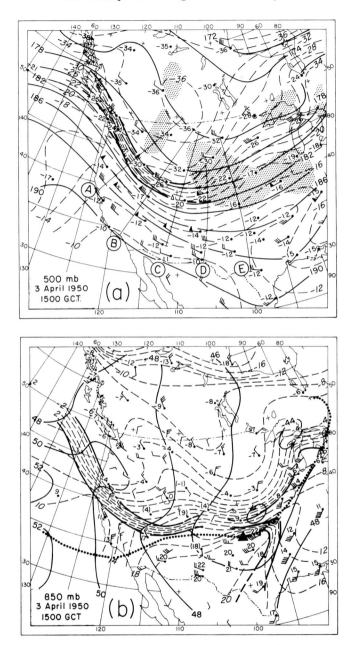

Fig. 9.8   (a) 500-mb and (b) 850-mb charts at 15 GCT April 3, 1950. Contours at intervals of 200 ft; isotherms 2°C intervals. Heavy lines are frontal boundaries. In (a), the precipitation area is stippled; in (b) dotted line marks surface front, and heavy triangle gives the location of surface cyclone. (From Newton, 1954).

The property changes experienced by a parcel leaving the frontal zone
in upper levels may be seen by comparing the structures at point *A* (representing the same air parcel moving from one section to the other) in
Figs. 9.9a and b. In Fig. 9.9a, the distinguishing characteristics of the
frontal layer are (1) great horizontal temperature gradient, (2) strong
cyclonic shear (high vorticity), (3) pronounced vertical stability, and (4)
strong vertical shear, all of which diminish downstream in the upper troposphere (Fig. 9.9b).

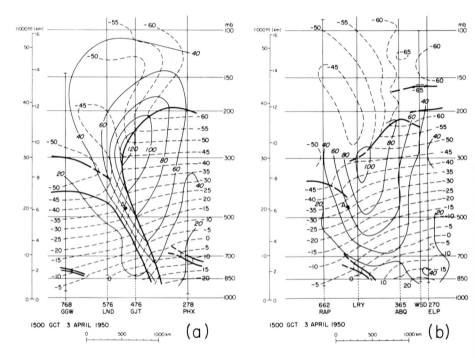

FIG. 9.9   Vertical sections along (a) line *C* and (b) line *D* in Fig. 9.8a. Wind speeds
in knots. (From Newton, 1954.)

As discussed in Section 7.3, the vertical and horizontal shears within the
frontal layer are related to the horizontal and vertical temperature gradients;
the changes of the various characteristics (1) to (4) between the cross section must be interrelated accordingly. The component equations expressing
these changes may be obtained by straightforward substitution, into Eqs.
(9.6), of θ and u for S. Observing that $du/dt = f(v - v_g)$ in Eq. (9.6b) and
$dv/dt = -f(u - u_g)$ in the corresponding equation for F_x, the resulting
expressions for F_x and F_y may be combined to give the vorticity equation.

The final equations, which are complete except for omission of frictional influences, are

$$\frac{d}{dt}\theta_z = \frac{\partial}{\partial z}\left(\frac{d\theta}{dt}\right) - \frac{\partial u}{\partial z}\theta_x - \frac{\partial v}{\partial z}\theta_y - \frac{\partial w}{\partial z}\theta_z \tag{9.7a}$$

$\qquad -1.67 \approx \qquad \ldots \qquad\qquad -0.52 \qquad -0.97 \qquad +0.20 \qquad (10^{-7} \text{ deg m}^{-1} \text{ sec}^{-1})$

$$\frac{d}{dt}\theta_y = \frac{\partial}{\partial y}\left(\frac{d\theta}{dt}\right) - \frac{\partial u}{\partial y}\theta_x - \frac{\partial v}{\partial y}\theta_y - \frac{\partial w}{\partial y}\theta_z \tag{9.7b}$$

$\qquad +1.39 \approx \qquad \ldots \qquad\qquad +0.41 \qquad +0.60 \qquad +0.54 \qquad (10^{-9} \text{ deg m}^{-1} \text{ sec}^{-1})$

$$\frac{d}{dt}\sigma = -\sigma D_{xz} + f\frac{\partial v}{\partial z} - f\frac{\partial v_g}{\partial z} - \frac{\partial v}{\partial z}\frac{\partial u}{\partial y} \tag{9.7c}$$

$\qquad -3.19 \approx \qquad +1.55 \qquad -1.89 \qquad -1.35 \qquad -1.97 \qquad (10^{-7} \text{ sec}^{-2})$

$$\frac{d}{dt}\zeta_a = -\zeta_a D_{xy} - \frac{\partial v}{\partial z}\frac{\partial w}{\partial x} + \frac{\partial w}{\partial y}\frac{\partial u}{\partial z} \tag{9.7d}$$

$\qquad -1.91 \approx \qquad -0.67 \qquad\quad \ldots \qquad -1.07 \qquad (10^{-9} \text{ sec}^{-2})$

Here $\sigma \equiv \partial u/\partial z$ is the principal component of the vertical shear (taken along the direction of the front), $D_{xz} \equiv (\partial u/\partial x + \partial w/\partial z)$ is the divergence in a vertical plane along x, $\zeta_a \equiv (f + \partial v/\partial x - \partial u/\partial y)$ is the component of absolute vorticity about the vertical, and $D_{xy} \equiv (\partial u/\partial x + \partial v/\partial y)$ is the horizontal divergence.

The numbers below the individual terms of Eqs. (9.7) correspond to an evaluation of their magnitudes based on the changes of shear and temperature gradients for particle A in Fig. 9.9 as it moves from the frontal layer into the region downstream where that layer vanishes. The inexact balance of the sums is due to uncertainties of analysis (for details, see Newton, 1954), but the individual magnitudes are approximately correct. The three terms for which no values are given are negligibly small.

All other terms are seen to be of some consequence, indicating the complicated nature of frontogenesis or frontolysis in the free atmosphere. One evident feature, in changing the vertical gradients, is the significance of ageostrophic motions normal to the front. Under geostrophic conditions, substitution from the thermal wind equation for θ_x and θ_y shows that the two middle r.h.s. terms of Eq. (9.7a) would cancel. Likewise, the corresponding terms in Eq. (9.7c) would cancel. In reality, these terms dominate. It may be noted also from Eq. (9.7c) that $\partial v/\partial z < 0$, indicating a veering of the actual wind with height, whereas $\partial v_g/\partial z > 0$, indicating a backing of the geostrophic wind with height (as shown by isotherms in this region, Fig. 9.8a).

According to the last term of Eq. (9.7a), vertical shrinking contributed to making the air more stable, whereas through the processes represented by the middle r.h.s. terms (whose nature is shown by Figs. 9.2., if viewed in the xz-plane, and 9.7b), a destabilization actually occurred. At the same time, the first term of Eq. (9.7d) indicates that the horizontal divergence accompanying the vertical shrinking was significant in decreasing the vorticity about a vertical axis. As shown by the last terms of Eqs. (9.7b) and (9.7d), lateral gradients of vertical motion helped to diminish both horizontal temperature gradient and vorticity through the process illustrated by Fig. 9.7a.

In Eq. (9.7c) it is seen that convergence in the xz-plane, which accompanies lateral diffluence in the frontal exit region, contributed to an increase rather than the observed decrease of vertical shear. The effect of this contribution was overcome by the influences of ageostrophic motions (two middle r.h.s. terms combined) and of the transverse differential advection of momentum (last term) illustrated by Fig. 9.7b.

While the discussion of Eqs. (9.7) is unwieldy because of the multiplicity of terms of different kinds, the essential features of the vertical and transverse circulations in the frontal exit zone fall into a coherent pattern. This is shown in Fig. 9.10, based on the preceding evaluations in the middle troposphere and similar deductions for other levels. While the motions shown are frontolytical in upper levels, they are frontogenetical in lower levels, as was pointed out earlier with regard to Fig. 9.8.

The motion field shown is naturally prescribed by the particular type of synoptic situation chosen; however, Fig. 9.10 serves to bring out the fact that *the relative importance of particular frontogenetical or frontolytical processes varies at different levels.* In lower levels, confluence and convergence dominate. In the middle- or upper-tropospheric part of the front (i.e., near 500 mb in Fig. 9.10), generation or destruction of vorticity and horizontal temperature gradient is largely accomplished by "tilting" or differential vertical motions across the front, although the effects of confluence (diffluence) and convergence (divergence) contribute significantly in the same sense. In the upper-tropospheric part, tilting assumes a decreasing role with increasing height, while divergence and diffluence increase in importance. As observed in Section 7.4, at the level of maximum wind where the horizontal temperature gradient is weak or absent, the property concentrated by confluence is momentum, and it is the wind shear (or vorticity) that identifies a front at this level.

At all levels, *vertical shear and stability are generated or dissipated mainly by a variation of the cross-stream circulation with height, essentially involving ageostrophic motions.* It is evident from the discussion above that in the immediate neighborhood of the front, upper-tropospheric frontogenesis

involves a circulation in which the horizontal branches are in a direct solenoidal sense, while the vertical branches are indirect; thus, both kinetic
and available potential energy are locally increased during the frontogenesis
process. This is not inconsistent with energy principles (Chapter 16), since
a closed system is not involved. If the components in Fig. 9.10 are composed, the result is similar to the right half of Fig. 9.1, or to a deformation
field in the vertical plane.[7]

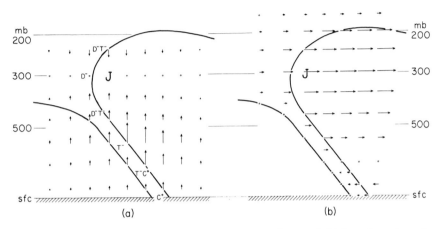

FIG. 9.10 Schematic distribution of (a) vertical motions and (b) lateral motions, in
cross section through upper frontal delta such as between lines C and D in Fig. 9.8a.
Regions are indicated where horizontal convergence (C), divergence (D), and "tilting"
by field of vertical motion (T) are important in changing vorticity, superscripts indicating
sign of contribution to vorticity change. The lateral motions have no absolute reference
and are intended only to show the vertical and horizontal gradients. (After Newton,
1954.)

Figure 9.10 is consistent with the circulations discussed in Section 9.2.
The lower part, where there is confluence, is comparable to the sense of
the circulation in Fig. 9.5. In the vicinity of the jet-stream core, the circulation is equivalent to that in Fig. 9.6 (considering that Fig. 9.10 represents diffluence rather than confluence in upper levels). Since Fig. 9.10
applies to a region where there was diffluence superposed over confluence,
the upper-tropospheric motions cannot be compared with those in Fig. 9.5,
in which Sawyer assumed confluence at all levels.

The general ascending motions in Fig. 9.10 are compatible with the

---

[7] Two-dimensional streamlines cannot be properly drawn because there is divergence
in this vertical plane associated with the confluent or diffluent velocity fields at different
levels.

presence of a developing cyclone beneath the diffluent zone. Also at this time and for the ensuing 36 hr, the precipitation shield of this cyclone was located downstream from the frontal exit where (although the surface front was distinct) no front was present in the middle troposphere. It may be noted that Staley (1960) derived an essentially identical pattern of vertical motions in a frontal exit region. Also, Hubert (1953) earlier computed the vertical motion field in a zone of upper diffluence that remained persistent over England and western Europe for a 36-hr period. His results were consistent with those in the preceding case; namely, ascent was most pronounced on the right side of the jet stream, as in Fig. 9.10. This example illustrates the essential connection between the frontogenetical processes and ageostrophic motions and vertical circulations, considering the troposphere as a whole.

9.5   THE POTENTIAL VORTICITY VIEWPOINT

Frontogenesis was discussed above within the framework of the xyz-(or xyp-) coordinate system most commonly used for other purposes. If, for example, θ is used as the vertical coordinate, the equations may be formulated in a different but consistent way. If potential temperature is conserved in frictionless motion, the potential absolute vorticity

$$P_\theta = - \frac{\partial \theta}{\partial p} \zeta_{\theta a} = \text{const.} \tag{9.8}$$

as shown by Rossby (1940). Here, $\zeta_{\theta a} \equiv (f + \zeta_\theta)$, where ζ_θ is the relative vorticity measured in a θ-surface. This is a component form of a more general expression derived by Ertel (1942).

According to Eq. (9.8), the absolute vorticity must increase if the stability decreases. Considering parcel A in Fig. 9.9, $- \partial\theta/\partial p$ decreases as the air streams out the frontal exit. Correspondingly, $\zeta_{\theta a}$ should increase. Since the variations of f and of curvature are small, the cyclonic shear measured along an isentropic surface should be weaker inside the front than in the nonfrontal region.

This condition is satisfied, as shown by Fig. 9.11. The distribution illustrated here is typical. Within a well-developed front, weak isentropic shear is present in the upper-tropospheric part, generally being nil in a region beneath or slightly on the polar-air side of the jet-stream core. Anticyclonic shear characterizes the frontal layer except for the uppermost part near stratospheric levels. Thus, strong stability is associated with weak absolute vorticity $(f + \zeta_\theta)$ in the tropospheric part of a front. Where a diffuse

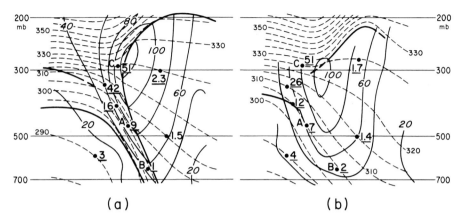

FIG. 9.11   Isotachs and isentropes in sections of Fig. 9.9. Potential absolute vorticity values are plotted at selected points, in 10^{-9} c.g.s. units (equivalent to units of 10^{-6} deg $mb^{-1} sec^{-1}$). Only at points A and C do these represent the same air parcel.

barocline zone rather than a front is present, cyclonic shear along the θ-surfaces (Fig. 9.11b) is typical of the general region to the left of the jet stream, at much lower tropospheric levels. In terms of Eq. (9.8), the differences of $\zeta_{\theta a}$ in this region conform with the differences of $\partial\theta/\partial p$.

In Section 9.4 it was concluded that changes of stability in the frontal zone are due to ageostrophic processes represented by the two middle r.h.s. terms of Eq. (9.7a). In the coordinate system used, an increase of stability during frontogenesis may be accompanied by *convergence in a horizontal plane.* As noted above, *divergence in an isentropic surface* is required. The consistency of these two conditions is illustrated by Fig. 9.12a, showing two isentropes in a vertical section normal to a frontal zone at an

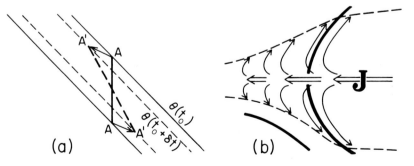

FIG. 9.12   (a) shows tilting and vertical stretching of an air column while the stability increases. (b) is a schematic illustration of cross-stream circulations near jet-stream level during frontogenesis (compare Fig. 9.6).

initial time (solid lines) and later (dashed lines) when the stability has increased. The initially vertical air column AA tilts ($\partial v/\partial z > 0$) and stretches vertically, while the bounding θ-surfaces come closer together. The vertical stretching indicates convergence in the xy-plane; however, it is evident that at the same time there can be horizontal divergence along the isentropic surfaces. Obviously, a decrease of stability cannot be uniquely connected with vertical stretching, or vice versa, if the isentropes have a large slope and the "stretching" is identified with an actual physical chain of air particles.

If ψ_θ is the slope of an isentrope from a horizontal surface and the x-axis is chosen along an isotherm in that surface, it can be shown that

$$\nabla_\theta \cdot \mathbf{V} = \nabla_h \cdot \mathbf{V} + \frac{\partial v}{\partial z} \tan \psi_\theta \qquad (9.9)$$

At point A in Fig. 9.9, $\nabla_h \cdot \mathbf{V}$ was evaluated as about 0.3×10^{-5} sec^{-1}, while $\partial v/\partial z \approx -2 \times 10^{-3}$ sec^{-1} and tan $\psi_\theta \approx 0.8 \times 10^{-2}$. Insertion into Eq. (9.9) gives $\nabla_\theta \cdot \mathbf{V} \approx -1.3 \times 10^{-5}$ sec^{-1}. This example illustrates in a real case that divergence in one system conformed with convergence in the other, as in Fig. 9.12.

On the basis of the high values of P_θ observed in upper-tropospheric fronts, Reed and Sanders (1953) suggested that the air within a front originates by subsidence from the lower stratosphere, "the tropopause assuming the character and appearance of a frontal surface." This viewpoint was elaborated by Reed (1955) and reiterated by Reed and Danielsen (1959). They demonstrated in a composite cross section that P_θ is much greater inside a frontal layer than in the neighboring air masses, with values intermediate between stratospheric and tropospheric in the upper part and diminishing downward through the front. Aircraft measurements have also shown that tongues of radioactive material (Danielsen, 1968) and of ozone and dry air (Briggs and Roach, 1963) extend from the lower stratosphere into the upper parts of fronts. In this region (Hering, 1966), considering seasonal mean conditions over the Northern Hemisphere, there is a remarkable similarity between the distributions of potential vorticity, ozone, and radioactivity and therefore these properties tend to be associated in synoptic systems.

Reed (1955) concluded, on the basis of potential vorticity analyses, that recently extruded stratospheric air was responsible for the formation of a front as a "folded tropopause" reaching levels as low as 700 to 800 mb. Although in some respects the evidence presented is appealing, we do not consider that the hypothesis has been conclusively demonstrated as a general process of frontogenesis.[8]

There is, however, substance to the viewpoint that a more limited extrusion of "stratospheric" air regularly takes place during upper-tropospheric frontogenesis. The process may be examined on the basis of the case described above. At point A, the isentropic shear was weakly anticyclonic in Fig. 9.11a and cyclonic in Fig. 9.11b. Correspondingly, $\zeta_{\theta a}$ was $0.82 \times 10^{-4} \sec^{-1}$ in the front and $1.24 \times 10^{-4} \sec^{-1}$ in the nonfrontal region, just above 500 mb. Considering the change of stability, P_θ was 9×10^{-9} (c.g.s. units) inside the front, and 7×10^{-9} units in the nonfrontal region. Along the same isentrope between 600 and 700 mb (point B), the corresponding values were 1×10^{-9} units in the front and 2×10^{-9} units in the diffuse barocline zone. This may be compared with values of 1.5 to 3×10^{-9} units in the neighboring warm- and cold-air masses. At C, in the region of strongest shear in the stratosphere, $P_\theta = 51 \times 10^{-9}$ units in both sections. The differences between the computed values at points A and B in the two sections are not meaningful because either value could be altered by a slight change in the analysis. Hence, in the middle troposphere, no significant change of potential vorticity is undergone by an air parcel in the process of frontolysis.

This example concerns the frontolysis experienced by air particles streaming out from the frontal layer in the upper troposphere. In the case studied by Reed and Sanders (1953), and in a similar analysis by Newton and Carson (1953), *genesis* of a front or a strong shear zone of analogous nature was accomplished by circulations substantially within the troposphere. As observed earlier, moderate cyclonic shear along the θ-surfaces (Fig. 9.11b) normally exists in the middle and upper troposphere, in the baroclinic zone when a front is absent. Thus, since moderately high potential vorticity is present in the polar-front zone even where this zone is diffuse, formation

---

[8] The argument is weakened by the fact, pointed out by Reed, that stable layers were present in the troposphere before (in his interpretation) the "folded tropopause" descended to corresponding levels. Soundings at several stations show that these stable layers were quite distinct in the range of potential temperatures corresponding to the strong front that later evolved. Our interpretation is that a front already existed in the troposphere prior to the time when, according to Reed, the extrusion of stratospheric air to low levels took place. There is no question that strong frontogenesis took place in Reed's example, but this appears to have involved a sharpening of fronts that were already present in the troposphere. The example appears to be broadly similar to that in Section 9.4, in which the upper-tropospheric frontal layer migrated from the west to the east side of the trough, where it combined with a strong low-level front between cold continental air and maritime tropical air (Newton, 1958).

of a tropospheric frontal layer does not require direct extrusion of air from the stratosphere except in the uppermost part, as noted below.

In Section 7.4 it was emphasized that a frontal zone should be considered as the whole region comprising both troposphere and stratosphere, marked by the zone of strongest cyclonic shear (in a horizontal surface). It was pointed out that the process of frontogenesis involves not only an intensification of the horizontal temperature gradient (and shear) in the levels above and below the jet stream, but also an increased shear at the level of the jet stream where the temperature gradient is nil. The gradient of momentum is concentrated by confluence, an effect augmented by ageostrophic motions, with greatest acceleration where the wind is strongest. The motions involved are indicated in the schematic Fig. 9.12b (the reverse of the frontolytic circulation in Fig. 9.10). Generation of vorticity near the level of maximum wind also requires vertical stretching on the cyclonic-shear side of the jet stream, which is connected with mid-tropospheric frontogenesis (of both temperature and momentum fields) in a manner evident from the figure, consistent with the discussion in Section 9.4.

Obviously, a circulation of this type must carry "stratospheric" air some distance downward into the tropospheric frontal layer, as an incidental part of the frontogenesis process. The distinction between stratospheric and tropospheric air in this region is vague; for example, the composite cross section by Reed and Danielsen (1959) shows a gradual variation of P_θ from high values in the upper part of a front to very much smaller values in the lower-tropospheric frontal layer, as in Fig. 9.11.

Considering the distance between the two sections in Fig. 9.11, an air particle moving at 60 knots would require 8 hr to move from the frontal to the nonfrontal section, in motion relative to the observed eastward movement of the frontal exit. In this time, with the rather substantial vertical motion of 5 cm/sec the vertical displacement would be 1.4 km. With strong frontogenesis, this may be somewhat greater. It may be concluded that *a limited extrusion of "stratospheric" air into the upper-tropospheric part of a frontal zone is a necessary accompaniment of the vertical circulations connecting the frontogenetic processes near the level of maximum wind and in the middle troposphere.* This conclusion is consistent with the observations cited earlier, concerning tongues of air with stratospheric properties extending downward into the upper parts of fronts. The concept of a "folded tropopause" cannot be involved, since observations show that the tropopause is discontinuous in this region.

Despite the preceding remarks, it should be acknowledged that Reed's contribution (1955) was important as one of the first indications of a significant mechanism of exchange between troposphere and stratosphere. Kleinschmidt (1955) independently concluded that masses of air with high P_θ, in high-level cyclones, are stratospheric in origin. Staley (1960) demonstrated that air could follow an isentropic trajectory from the lower stratosphere into the lower troposphere within a frontal layer, and suggested this as a means by which radioactive debris may be brought down from the stratospheric reservoir within a time as short as 24 hr. In the process, according to Staley's analyses, there are large individual changes of P_θ (within the front, a decrease), which he discusses theoretically in terms of friction and gradients of diabatic heating. Reiter (1963) demonstrated the movement of radioactive material, detected in the middle troposphere, to a region of intense surface deposition farther southeast. The material, probably of stratospheric origin, followed trajectories similar to those in Fig. 10.17, which are typical of the subsiding cold air behind well-developed troughs. Reiter observed that movements of this kind allow passage of air at lower levels from the poleward to the equatorward side of the jet-stream axis.

9.6 Frontogenetic Processes in Lower Levels

The contributions of the various processes to lower-tropospheric frontogenesis were investigated by Sanders (1955) for the situation shown in Fig. 9.13. The cross section in Fig. 9.14 illustrates a structure quite typical of the southern bounds of a low-level continental polar-air mass, with intense horizontal and vertical gradients near the ground gradually weakening aloft.

The surface winds (Fig. 9.13) indicate strong contraction toward the front. Figure 9.15b shows the effect of this contraction (which decreased upward) in concentrating the horizontal temperature gradient. In accord with the strong horizontal convergence in the lower part of the front (which also diminished upward), rising motions were large near the warm-air boundary and decreased markedly toward the cold air (Fig. 9.15a). Consequently, at modest elevations the influence of differential vertical advection (last term of Eq. 9.7b) countered the effect of horizontal contraction.

Figure 9.15c shows the overall effect (neglecting diabatic influences). The units are expressed as changes of horizontal temperature gradient in °C/100 km in 3 hr, which corresponds nearly to units of 10^{-9} deg m^{-1} sec^{-1}. Comparison with the magnitudes in Eq. (9.7b) shows that the rate of frontogenesis in a shallow layer near the ground is about 100 times as large

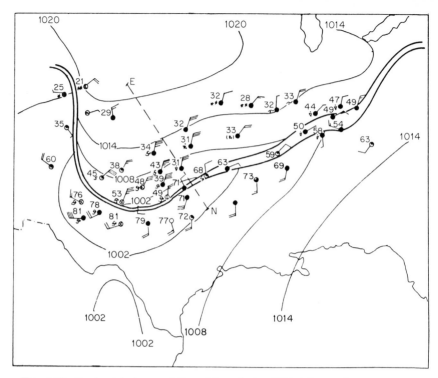

FIG. 9.13   Sea-level isobars and frontal boundaries, 0330 GCT April 18, 1953. (After Sanders, 1955.)

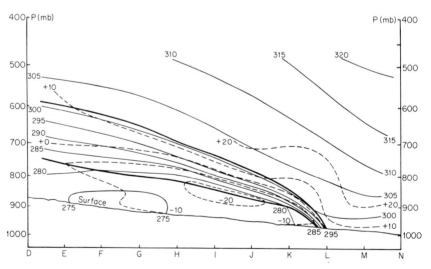

FIG. 9.14   Cross section along line *E-N* of Fig. 9.13, showing the distributions of potential temperature (solid thin lines, °K) and horizontal wind component normal to section (dashed lines, m/sec, positive for wind into page). (After Sanders, 1955.)

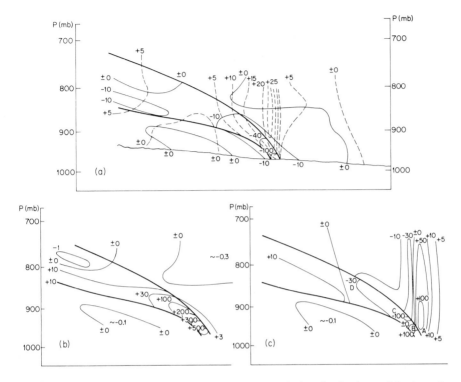

FIG. 9.15 For part of the section in Fig. 9.14: (a) the distributions of horizontal divergence (solid thin lines, 10^{-5} sec^{-1}) and vertical motion (dashed lines, cm/sec); (b) the frontogenetical effect of horizontal confluence; and (c) the frontogenetical effect combined with the influence of the horizontal gradient of vertical motion. Diabatic influences are not included. (After Sanders, 1955.)

as that characteristic of the free atmosphere.[9] In a corresponding analysis of the processes involving the vorticity about a vertical axis, Sanders found that production by the very strong convergence near the surface (Fig. 9.15a) was in a similar way counteracted by the influence of differential vertical transfer of momentum (third r.h.s. term of Eq. 9.7d), destruction of vorticity being indicated in the frontal layer above about 900 mb.

Sanders concluded that *in such a front, the characteristic process is an intense generation of frontal properties in the lowest part, an air parcel rising*

[9] The indicated intense rate of frontogenesis within the warm air (Fig. 9.13c) should apparently be discounted, since diabatic effects were not considered. The lapse rate in this region was near the moist adiabatic; with saturated air (suggested by the extensive stratocumulus overcast), vertical motions would have little influence in changing the temperature field in the warm-air mass.

within the front (along path *ABCD* in Fig. 9.15c) *losing these properties while passing through the higher portion of the front where frontolysis occurs.* The computations are thus consistent with the observation (Fig. 9.14) that this type of front typically becomes more diffuse with increasing elevation.

Bjerknes (1951) has indicated the importance of weak dynamic stability and a confluent velocity field for lower-tropospheric frontogenesis. As noted in Section 9.1, although deformation can concentrate the isotherms, low-level convergence is necessary to generate vorticity and maintain it against friction. Thus, any process that enhances removal of air from the lower part of the front, favors frontogenesis. This is a role played by the "isentropic upgliding" discussed by Bjerknes.

If the *x*-axis is taken along the direction of an isotherm and v_η indicates the component of lateral motion (toward the left and upward along a θ-surface in Fig. 9.14), Bjerknes shows that this is given by

$$v_\eta = \frac{u(\partial u_g/\partial x) + (\partial u_g/\partial t)}{f - (\partial u_g/\partial \eta)} \tag{9.10}$$

where η specifies the *y*-component along the θ-surface. Lateral upgliding movement is, according to the numerator, favored by confluence and is an ageostrophic effect resulting from air particles moving toward regions downstream where the geostrophic wind is stronger, with a consequent imbalance between the real and geostrophic wind. As Bjerknes demonstrates, frontogenesis associated with this process is especially favored in the regions behind fully developed cyclones (see Fig. 5.2a).

The process is enhanced when the denominator of Eq. (9.10) is small; that is, when the dynamic stability is weak. As Bjerknes shows, this condition is characteristic of the lower- and middle-tropospheric parts of fronts where (Fig. 9.11a) the isentropic shear is anticyclonic, and also of the anticyclonic-shear side of jet streams in the absence of a real front. Thus, both confluence and the low dynamic stability in this region favor isentropic upgliding and the removal of air from the lower layers that are necessary for frontogenesis.

Since in this book we are concerned mainly with the larger-scale circulation systems, we refrain from illustrating the detailed structures and circulations of fronts near the ground, while acknowledging the importance of studying these features. For instructive descriptions based on frequent observations, the reader is referred to Holland (1952), Berson (1958), Clarke (1961), and Brundidge (1965).

With regard to cold fronts, the analyses by Clarke and by Brundidge indicate that ascending motions in the range 50 to 100 cm/sec are attained within the lowest kilometer near the advancing edge. Observations also show that the boundaries of a front cannot be considered as substantial surfaces in the lower levels. Sanders (1955) suggested that warm air is incorporated into the advancing edge of a cold front; the analyses by Berson, Clarke, and Brundidge indicate a substantial entrainment from both cold- and warm-air sides (see also Welander, 1963). Clarke finds that within roughly the lower 500 m near a cold front, the solenoidal circulation tends to be balanced by frictional stresses rather than by coriolis accelerations. This is also indicated by Sanders' observation that, near the ground, the geostrophic vertical shear through the front is several times larger than the actual shear and is consistent with Ball's theoretical analysis (Section 7.8).

9.7 Large-Scale Processes

The studies reviewed above indicate that although other processes are also involved, confluent velocity fields typify the regions where strong fronts form, both in the upper and lower troposphere. The observation that distinct fronts are found in some regions, with other regions in between where there are broad barocline zones but no fronts, fits the consideration that strong confluence cannot be present all around the hemisphere, but rather that there are alternating regions of confluence and diffluence.

In order for hyperbaroclinic zones to form, a basic barocline field must obviously exist, the properties of which can be concentrated by the more localized fields of motion. One viewpoint on the broad-scale hemispheric distribution of atmospheric properties is that of Rossby (1949): "Lateral mixing within the polar cap north of the velocity maximum (jet stream) not only would equalize the vorticity but would otherwise tend to destroy the horizontal temperature gradients in the zone of mixing and to concentrate the temperature contrasts between high and low latitudes on the southern boundary of the mixing zone, i.e., under the jet stream. This zone of concentrated temperature contrast would take on the character of a frontal zone. Thus the fronts in the free atmosphere and the jet stream would simply be different manifestations of one and the same process."

In Chapter 4 it was indicated that there generally exist two principal fronts, with their accompanying jet streams, separating the three main hemispheric air masses. Although the intensity of the polar front varies from place to place, between about 700 and 400 mb there is a relatively unbroken zone of concentrated baroclinity around the hemisphere, associated with an upper jet stream of variable strength. Likewise, especially in Northern Hemisphere winter, there is a zone of relatively strong baroclinity in the upper troposphere (with a distinct front in some localities) associated with the subtropical jet stream.

The great steadiness of the subtropical jet stream suggests that the maintenance of its solenoid field can largely be considered as an effect of the mean atmospheric circulation, which in the Hadley cell of the tropics is itself very steady. In winter (Fig. 1.5), the strongest mean ascending motion of about 0.9 cm/sec occurs close to the thermal equator. The strongest descent, averaging about 0.7 cm/sec, is situated at about 20° N. Near the 200-mb level, a maximum mean northward velocity of about 2.5 m/sec is found around latitude 12° N (Fig. 1.3).

The meridional flow is thus characterized by persistent upper convergence between latitudes 15°N and 30°N, with divergence in lower levels in the same latitude belt. A simplified sketch of the vertical and horizontal branches of the Hadley circulation is shown in Fig. 9.16, in which are marked

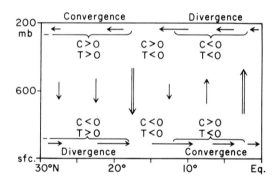

FIG. 9.16   Schematic representation of frontogenetical effects due to convergence (C) and to "tilting" (T), in mean Hadley-cell circulation.

the regions where the air motions contribute to frontogenesis due to hemispheric confluence (C, third r.h.s. term of Eq. 9.7b) or due to "tilting" or differential adiabatic heating through vertical motions (T, last term of Eq. 9.7b). In the northern upper part of the Hadley cell below 200 mb, the mean confluence tends to maintain a quasi-permanent frontal zone. Beneath the level of strongest meridional movement and north of the latitude of strongest descending motions, the tilting term is the most active one. These terms act in the same sense when referred to the upper northern part below the level of maximum wind. As seen by inspection of Fig. 9.16, in other portions of the Hadley cell these terms either act individually in an opposing manner or both contribute to frontolysis.

According to this very qualitative reasoning, the intensified barocline field due to the mean circulation should be confined essentially to the upper part of the troposphere and should gradually become more diffuse downward in the lower belt of divergence. Considering that the air acted upon frontogenetically is at the same time transported northward in the upper branch, the baroclinity should tend to be strongest somewhat north of the belt of strongest frontogenesis. The zone of strongest baroclinity (Fig. 4.4) must also be combined with the belt of strongest vertical shear beneath the subtropical jet stream core.[10]

Here, only the influence of the *mean* circulation has been considered. As indicated in Chapter 2, the northern upper branch of the Hadley cell is a belt of divergence in the poleward eddy flux of heat. Thus the horizontal *eddies* contribute to *frontolysis* of the mean temperature field. However, it should be considered that the strength of the eddy transport of heat depends in the first place partly on the strength of the meridional temperature gradient, which must originate from other processes in a region of eddy-heat flux divergence. The preceding discussion, while not quantitative, serves to identify the factors that contribute toward maintaining the necessary temperature gradient.

The frontogenetic and frontolytic processes acting to maintain the mean solenoid field in the polar-front zone are much more complicated than those in the subtropical frontal zone. Synoptic disturbances assume a dominant role. As elaborated in Chapter 10, synoptic evidence indicates that, on the average, the cold air sinks and the warm air rises in these disturbances, a necessary condition for upward transfer of heat and generation of kinetic energy (Chapters 2 and 16). Thus the mean circulation referred to the polar-front zone must be in a direct sense, which would contribute to frontolysis in the absence of heat sources and sinks. As mentioned in Section 9.1, the exchange of sensible heat between earth and atmosphere also in general acts frontolytically, especially over the oceans where strong heat transfer takes place into cold-air masses.

The influence of latent heat release, on the contrary, acts frontogenetically because (Fig. 9.17) the main condensation in the vicinity of the polar front occurs in the ascending warm air. To assess the significance of latent heat release, we make a very rough estimate: Suppose that the mean amount

---

[10] The relatively strong lower-tropospheric baroclinity around 20°N in that figure is probably in part due to the influence of sporadic, low-level, cold outbreaks.

of precipitation along a planetary polar front amounts to 2 mm/day and that this results from condensation in the warm air.[11] The heat released would then be 120 ly/day, or enough to raise the mean temperature of an air column between the 900- and 300-mb levels by 0.8°C per day.

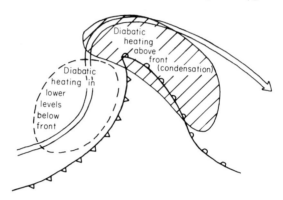

FIG. 9.17   Principal regions of strong diabatic heating in neighborhood of a cyclone, relative to surface front and jet stream.

Because this heat release occurs in a region where the air is at the same time cooled by lifting, its main influence is to counteract locally the cooling that would occur if the motions took place adiabatically. This was demonstrated for a synoptic case by Rao (1966). On evaluating the purely kinematical influences discussed in Section 9.6, he found them to have a net frontolytical effect at 800 and 600 mb, along most of a front associated with a wave cyclone. However, when the horizontal gradient of condensation heating was taken into account, the net effect was frontogenetical in the neighborhood of the front.

The significance of latent heat release in disturbances may be appreciated from Palmén's computations (1958) for the combined tropical and extra-tropical cyclone Hazel. From the observed precipitation, the release of latent heat was found to be 32×10^{18} cal/24 hr over an area of 8.5×10^5 km². This corresponds to an energy release of 2.6 ly/min, or about five times the maximum rate of transfer of sensible heat from the ocean surface into the cold air of an average cyclone (Section 12.2). This amount, applied to an air column between 900 and 200 mb, corresponds to an added heat

[11] Such condensation actually occurs, of course, only along restricted portions of the fronts. In those regions, however, it is much more intense. An average of the amount given above all along the frontal zone is not excessive, considering that this is equivalent to about 700 mm/year which is less than the typical middle-latitude precipitation (Fig. 2.3).

of 5.4 cal/gm per day, which in the absence of other influences (i.e, vertical motions) would increase the temperature by 23°C/day.

In the Hazel case, air rising from the surface layers in the central region of the cyclone would have a θ_w of about 21°C. The differences between this moist adiabat and the corresponding dry adiabat are 15° at 700 mb, 29° at 500 mb, and 42°C at 300 mb. From this extreme case we can see how much the condensation in the warm air helps to make possible the upward movement of the air, and how much it counteracts the frontolytic effect that otherwise would result from the last term in Eq. (9.7b).

The cross section in Fig. 8.2 was located about 700 km north of this cyclone, a branch of whose circulation is reflected in the lower troposphere east of the polar front in that section. While no three-dimensional trajectories are available for this case, these can be estimated from the published maps. It is evident that the air at the 300-mb level, within several hundred kilometers to the right of the front in Fig. 8.2, came from the vicinity of the tropical cyclone 12 to 18 hr earlier. At this level, the temperature of -32°C corresponds to the moist adiabat mentioned above, rendering it plausible that the air in the high troposphere originated in the outflow layer of this cyclone. On the other hand, the temperatures at 400 mb and below (e.g., at station FFO in Fig. 8.2) correspond to a much lower θ_w, and the winds indicate a trajectory originating within the main belt of westerlies. This example demonstrates clearly how the latent heat released in a disturbance can be utilized to warm the *upper* troposphere.[12] Since heating in this manner is an irreversible process, *the latent heat released by many such disturbances, and carried downstream from them in the upper westerlies, contributes to maintaining the strength of the hemispheric frontal zone against the destructive influences mentioned earlier.*

As observed in Section 9.2, an additional indirect influence of latent heat release should be considered. This has been outlined in broad terms by Bleeker (1950) and is illustrated schematically in Fig. 9.18. The most intense condensation and heating often occur in the vertical not too far from the surface front (see, e.g., Fig. 16.4). Consequent to the heating, the thickness between isobaric surfaces is increased within the cloud system. The heights of the isobaric surfaces in the upper part of the cloud are caused to rise, resulting in an ageostrophic outflow, as indicated in Fig. 9.18. Through an effect discussed by Durst and Sutcliffe (1938), such an outflow tends to

---

[12] Although in this particular example the heat released was transported to the high troposphere, as is characteristic of tropical storms (Chapter 15), much of the heat released in the stably stratified air of polar-front disturbances is retained at lower levels.

be asymmetrical in the manner shown. Air parcels ascending from the lower levels to a higher level, where the pressure-gradient force (geostrophic wind) is stronger, are no longer in gradient balance and are thus deflected toward the left of the isobaric contours. In confluent flow with low dynamic stability, this effect is enhanced (Section 9.6).

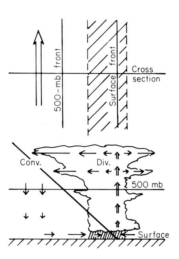

FIG. 9.18   General scheme of vertical and transverse motions associated with a precipitating cloud system, the double-shafted arrows indicating the upward motions resulting from release of latent heat; the other arrows, induced motions (see text).

As a result of these processes a lateral flow is set up from the cloud system toward the front, which contributes to confluence and strengthening of the front in upper levels. Although this type of ageostrophic motion can be seen on upper-air charts (as in Fig. 16.6, which relates to Fig. 8.2), it has not been systematically investigated, and therefore the preceding discussion is necessarily qualitative.[13] Since as noted above, ascent with precipitation is essentially an irreversible process, the circulation pattern in Fig. 9.18 contributes to an *average* frontogenetic process around the hemisphere.

A general scheme of the main processes acting toward frontogenesis and frontolysis in different parts of a wave system is sketched in Fig. 9.19. It

[13] It should not be overlooked that while the air is rising, it is also moving downstream horizontally. Thus, even with an average upward motion as large as 10 cm/sec, over a day would be required for an air parcel to rise from the surface to the 300-mb level. During this time, with a wind speed of 20 m/sec, the air parcel would move 2000 km horizontally. The difficulty of constructing three-dimensional trajectories and studying the frontogenesis process directly can be appreciated.

should be emphasized that this is intended to show "average" conditions, from which marked departures may appear at different times in the life cycles of synoptic disturbances.

On the eastward side of a cold trough, precipitation connected with a cyclone or a cyclone family results in addition of heat to the warm air. Thus, diabatic influences (*H*) contribute to frontogenesis, although "tilting" (*T*) or ascent of warm air and sinking of cold air acts in the opposite manner. The effect of the latent heat realized in this region is largely made evident

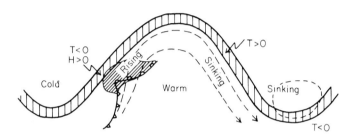

FIG. 9.19 Simplified scheme of a major wave in the polar front, showing where the influences of diabatic heating (*H*) and "tilting" (*T*) contribute to frontogenesis or frontolysis.

farther downstream (where there is typically more pronounced descent in the warm air than in the cold air) in the upper troposphere downwind from the ridge ($T > 0$). The reason for this has been discussed by Palmén and Nagler (1949). As noted in Section 10.7, in strong developments the most intense sinking of the cold-air mass occurs characteristically in the southern parts of cold-air tongues. This process contributes to frontolysis ($T < 0$) as indicated in this location.

Maintenance of the average strength of the front around the hemisphere is achieved principally as a balance between the frontolytic effects of adiabatic heating in sinking cold air and adiabatic cooling in rising warm air, and the regeneration of the temperature contrasts by release of latent heat of condensation in the warm air. To this is added the influence of meridional eddy transport of heat. Although the contributions of the various forms of heat transport have been established for zonal belts (Chapters 2 and 17), no investigation has yet been made of these processes referred to the meandering polar-front zone. This would be difficult to do, partly because of the ageostrophic motions that would have to be taken into account. Thus, no quantitative estimates can be given at this time. In the preceding discussion, the variability of radiative cooling as it affects the immediate neighborhood of the polar-front zone has also not been considered. This effect may be considerable over long periods, but evidently is a secondary influence in dealing with synoptic cases over short periods.

REFERENCES

Bergeron, T. (1928). Über die dreidimensional verknüpfende Wetteranalyse (I). *Geofys. Publikasjoner, Norske Videnskaps-Akad. Oslo* **5**, No. 6, 1–111.

Bergeron, T. (1930). Richtlinien einer dynamischen Klimatologie. *Meteorol. Z.* **7**, 246–262.

Berson, F. A. (1958). Some measurements in undercutting cold air. *Quart. J. Roy. Meteorol. Soc.* **84**, 1–16.

Bjerknes, J. (1951). Extratropical cyclones. *In* "Compendium of Meteorology" (T. F. Malone, ed.), pp. 577–598. Am. Meteorol. Soc., Boston, Massachusetts.

Bleeker, W. (1950). The structure of weather systems. *Cent. Proc. Roy. Meteorol. Soc.* pp. 66–80.

Briggs, J., and Roach, W. T. (1963). Aircraft observations near jet streams. *Quart. J. Roy. Meteorol. Soc.* **89**, 225–247.

Brundidge, K. C. (1965). The wind and temperature structure of nocturnal cold fronts in the first 1,420 feet. *Monthly Weather Rev.* **93**, 587–603.

Clarke, R. H. (1961). Mesostructure of dry cold fronts over featureless terrain. *J. Meteorol.* **18**, 715–735.

Danielsen, E. F. (1968). Stratospheric-tropospheric exchange based on radioactivity, ozone and potential vorticity. *J. Atmos. Sci.* **25**, 502–518.

Durst, C. S., and Sutcliffe, R. C. (1938). The importance of vertical motion in the development of tropical revolving storms. *Quart. J. Roy. Meteorol. Soc.* **64**, 75–84.

Eliassen, A. (1959). On the formation of fronts in the atmosphere. *In* "The Atmosphere and the Sea in Motion" (B. Bolin, ed.), pp. 277–287. Rockefeller Inst. Press, New York.

Eliassen, A. (1962). On the vertical circulation in frontal zones. *Geofys. Publikasjoner, Norske Videnskaps-Akad. Oslo* **24**, 147–160.

Ertel, H. (1942). Ein neuer hydrodynamischer Wirbelsatz. *Meteorol. Z.* **59**, 277–281.

Faller, A. J. (1956). A demonstration of fronts and frontal waves in atmospheric models. *J. Meteorol.* **13**, 1–4.

Fultz, D. (1952). On the possibility of experimental models of the polar-front wave. *J. Meteorol.* **9**, 379–384.

Haltiner, G. J., and Martin, F. L. (1957). "Dynamical and Physical Meteorology," pp. 287–296. McGraw-Hill, New York.

Hering, W. S. (1966). Ozone and atmospheric transport processes. *Tellus* **18**, 329–336.

Holland, J. Z. (1952). Time-sections of the lowest 5000 feet. *Bull. Am. Meteorol. Soc.* **33**, 1–6.

Hubert, W. E. (1953). A case study of variations in structure and circulation about westerly jet streams over Europe. *Tellus* **5**, 359–372.

Kleinschmidt, E. (1955). Die Entstehung einer Höhenzyklone über Nordamerika. *Tellus* **7**, 96–110.

Miller, J. E. (1948). On the concept of frontogenesis. *J. Meteorol.* **5**, 169–171.

Namias, J., and Clapp, P. F. (1949). Confluence theory of the high tropospheric jet stream. *J. Meteorol.* **6**, 330–336.

Newton, C. W. (1954). Frontogenesis and frontolysis as a three-dimensional process. *J. Meteorol.* **11**, 449–461.

Newton, C. W. (1958). Variations in structure of upper level troughs. *Geophysica (Helsinki)* **6**, 357–375.

Newton, C. W., and Carson, J. E. (1953). Structure of wind field and variations of vorticity in a summer situation. *Tellus* **5**, 321–339.

Oliver, V. J., and Holzworth, G. C. (1953). Some effects of the evaporation of widespread precipitation on the production of fronts and on changes in frontal slopes and motions. *Monthly Weather Rev.* **81**, 141–151.

Palmén, E. (1958). Vertical circulation and release of kinetic energy during the development of Hurricane Hazel into an extratropical storm. *Tellus* **10**, 1–23.

Palmén, E., and Nagler, K. M. (1949). The formation and structure of a large-scale disturbance in the westerlies. *J. Meteorol.* **6**, 227–242.

Petterssen, S. (1936). Contribution to the theory of frontogenesis. *Geofys. Publikasjoner, Norske Videnskaps-Akad. Oslo* **11**, No. 6, 1–27.

Petterssen, S. (1956). "Weather Analysis and Forecasting," 2nd ed., Vol. 1, Chapters 2 and 11. McGraw-Hill, New York.

Petterssen, S., and Austin, J. M. (1942). Fronts and frontogenesis in relation to vorticity. *Papers Phys. Oceanog. Meteorol., Mass. Inst. Technol. Woods Hole Oceanog. Inst.* **7**, No. 2, 1–37.

Priestley, C. H. B. (1959). "Turbulent Transfer in the Lower Atmosphere," Chapter 8. Univ. of Chicago Press, Chicago, Illinois.

Rao, G. V. (1966). On the influences of fields of motion, baroclinicity and latent heat source on frontogenesis. *J. Appl. Meteorol.* **5**, 377–387.

Reed, R. J. (1955). A study of a characteristic type of upper-level frontogenesis. *J. Meteorol.* **12**, 226–237.

Reed, R. J. (1960). Principal frontal zones of the Northern Hemisphere in winter and summer. *Bull. Am. Meteorol. Soc.* **41**, 591–598.

Reed, R. J., and Danielsen, E. F. (1959). Fronts in the vicinity of the tropopause. *Archiv. Meteorol., Geophys. Bioklimatol.* **A11**, 1–17.

Reed, R. J., and Sanders, F. (1953). An investigation of the development of a mid-tropospheric frontal zone and its associated vorticity field. *J. Meteorol.* **10**, 338–349.

Reiter, E. R. (1963). A case study of radioactive fallout. *J. Appl. Meteorol.* **2**, 691–705.

Rossby, C.-G. (1940). Planetary flow patterns in the atmosphere. *Quart. J. Roy. Meteorol. Soc.* **66**, Suppl., 68–87.

Rossby, C.-G. (1949). On the nature of the general circulation of the lower atmosphere. *In* "The Atmospheres of the Earth and Planets" (G. P. Kuiper, ed.), pp. 16–48. Univ. of Chicago Press, Chicago, Illinois.

Sanders, F. (1955). An investigation of the structure and dynamics of an intense surface frontal zone. *J. Meteorol.* **12**, 542–552.

Saucier, W. J. (1955). "Principles of Meteorological Analysis," Chapter 10. Univ. of Chicago Press, Chicago, Illinois.

Sawyer, J. S. (1956). The vertical circulation at meteorological fronts and its relation to frontogenesis. *Proc. Roy. Soc.* **A234**, 246–262.

Schumann, T. E. W., and van Rooy, M. P. (1951). Frequency of fronts in the Northern Hemisphere. *Arch. Meteorol., Geophys. Bioklimatol.* **A4**, 87–97.

Staley, D. O. (1960). Evaluation of potential-vorticity changes near the tropopause and the related vertical motions, vertical advection of vorticity, and transfer of radioactive debris from stratosphere to troposphere. *J. Meteorol.* **17**, 591–620.

Sutcliffe, R. C. (1940). Rapid development where cold and warm air masses move toward each other. Synop. Div., Tech. Mem. No. 12. Air Ministry, Great Britain.

van Loon, H. (1965). A climatological study of the atmospheric circulation in the Southern Hemisphere during the IGY. Part I. *J. Appl. Meteorol.* **4**, 479–491.

von Arx, W. S., Bumpus, D. F., and Richardson, W. S. (1955). On the fine structure of the Gulf Stream front. *Deep-Sea Res.* **3**, 46–65.

Vuorela, L. A. (1953). On the air flow connected with the invasion of upper tropical air over northwestern Europe. *Geophysica (Helsinki)* **4**, 105–130.

Welander, P. (1963). Steady plane fronts in a rotating fluid. *Tellus* **15**, 33–43.

10

THREE-DIMENSIONAL FLOW PATTERNS IN
EXTRATROPICAL DISTURBANCES

Cyclonic disturbances appear in such diverse forms that it is impossible to give a description that is uniformly applicable to all cyclones. The same is true of anticyclones and, in general, of "wave" disturbances. We must therefore limit our presentation to a few categories of disturbances that can be considered as typical or common. When using synoptic examples to describe these, it is natural to select cases that illustrate as best as possible the phenomena of interest. For this reason we shall generally show cases of rather strong development. Because there exist intermediate types of disturbances, types with very different origins, and cyclones that are weak both with respect to frontal systems and intensity of development, it would be easy to find synoptic cases that deviate from the structures of the particular examples we have selected.

In this chapter we shall be concerned essentially with frontal cyclones and the associated wave disturbances, excluding other types of which the most important is the tropical cyclone (Chapter 15). The expression "frontal cyclone" should not be taken too literally. As will be illustrated in Chapter 11, clearly defined fronts do not necessarily characterize the whole cyclone, but in a general sense the term *frontal cyclone* applies to any cyclone forming in the baroclinic westerlies, regardless of whether the baroclinity is well concentrated into fronts.

As noted in Sections 4.1 and 6.5, the actual trajectories taken by air particles are often very different from those that would be judged from the flow pattern at a given level, say, in the middle troposphere. For this reason, while the bulk movements of the main air masses are portrayed by changes in configuration of the frontal zone at a given level, these give an incomplete idea of the actual exchange processes in the general circulation. Nevertheless, to provide a framework for a more detailed discussion, it is convenient to begin by describing the broad forms assumed by some typical disturbances.

10.1   FORMATION OF UPPER-TROPOSPHERIC CYCLONES AND ANTICYCLONES

The deformation of the middle- and upper-tropospheric flow connected with the deepening of waves in the westerlies often results in the formation of closed lows equatorward (and closed highs poleward) of the main belt of zonally averaged westerlies. These lows and highs may then persist for quite a long time and have a profound influence on the weather. Such an upper low forms from a preexisting cold trough that is ultimately cut off from its connection with the polar-source region in the upper troposphere (Crocker *et al.*, 1947; Schwerdtfeger, 1948; Palmén, 1949). Similarly, upper highs are formed from warm ridges that extrude from the warm source region in the south.

An extreme example of the end product of this process is shown in Fig. 10.1, taken from Palmén and Nagler (1949). The thermal and contour fields at 500 mb show how the cold cyclonic vortex has been cut off from the mother cold region in the north, with which it is still united by an "umbilical cord" in the form of a shear line. Also, the northern warm high is already almost completely separated from the warm-air mass in the south, from which it originated. The process leading to such a deformation is suggested qualitatively by Fig. 10.2. As discussed in Section 10.4, polar-air masses, in moving equatorward generally undergo sinking. As a result of strong subsidence of the cold-air mass in the trough, vertical stretching and horizontal convergence in the upper troposphere intensifies the vorticity, deforming the current from the "sinusoidal" shape it would normally have. East of the trough, the jet stream assumes a more meridional orientation; with approximate conservation of absolute vorticity in the middle troposphere, the current must again bend anticyclonically in higher latitudes where the coriolis parameter is larger.

Such a process often takes place, in a repetitive manner, in "blocking" situations wherein the westerly waves break down into strongly meridional flow with vortices. The characteristic features of such a situation are illustrated in Fig. 10.3. Berggren *et al.* (1949) indicated that such deformations

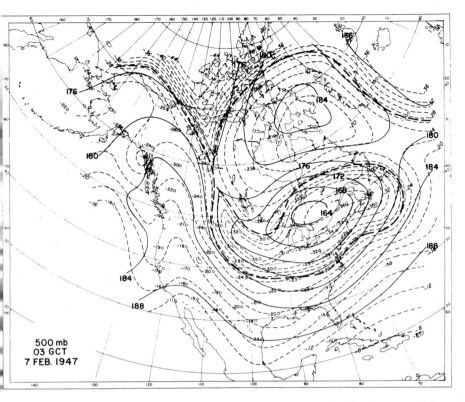

FIG. 10.1   Chart for 500 mb at 03 GCT Feb. 7, 1947, showing isotherms at 2°C intervals and isobaric contours at 200-ft intervals. Heavy line is warm-air boundary of frontal layer at this level. (After Palmén and Nagler, 1949.)

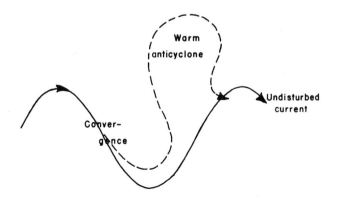

FIG. 10.2   Schematic diagram showing the changes from the normal "sinusoidal" air movement in the upper troposphere, resulting from convergence above a sinking cold-air mass in the trough. (After Palmén and Nagler, 1949.)

occur when the zonal current is strong well upstream from the region where
the block forms, but decreases in intensity downstream with a marked
diffluence of the current. Troughs that approach the blocking region decrease in eastward speed, the wavelength between successive troughs diminishing downstream where the waves also grow in amplitude. Strongly de-

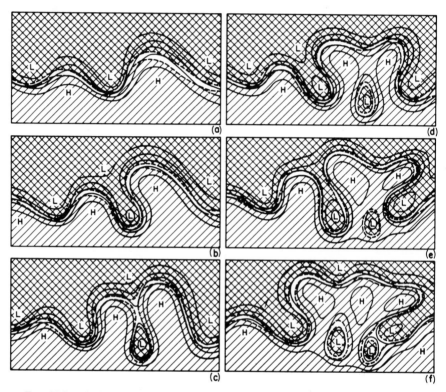

FIG. 10.3   Idealized sketches of the development of unstable waves at the 500-mb
level, in association with the establishment of a blocking anticyclone in high latitudes
and cutoff cyclones in low latitudes. Warm air hatched and cold air cross-hatched;
broken lines are frontal boundaries and solid lines are streamlines. (After Berggren
et al., 1949.)

formed perturbations of the type of Fig. 10.2 grow in succession from each
wave approaching the block, resulting in an accumulation of cutoff lows
in lower latitudes and of cutoff highs in higher latitudes as in the later stages
of Fig. 10.3. Rex (1950, 1951) has provided a comprehensive discussion of
this phenomenon along with statistical evidence that relates to the geographical and seasonal occurrences of blocking situations.

Both the formation of warm upper highs poleward of the main strong westerlies and the formation of cold upper lows on their equatorward side are processes expressing the bulk meridional exchange of air masses. That the large bodies of warm air transported poleward in this way should appear as anticyclonic vortices, and the large bodies of cold air as cyclonic vortices, follows from their meridional displacement and from the shrinking and stretching that occur in the upper troposphere during this displacement (Section 10.4).

Synoptic analyses show that the shapes of cold troughs can change in greatly differing ways. In Fig. 10.4, some characteristic types are sketched. Figure 10.4a shows the result of degeneration of an upper cold trough into

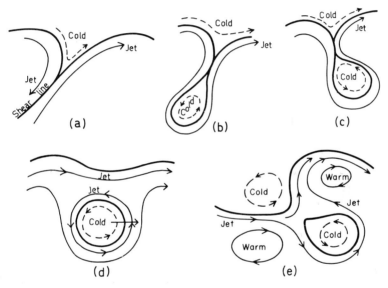

Fig. 10.4 Five characteristic types of disturbances resulting from the extreme growth of upper-level waves. Heavy lines represent warm-air boundaries of fronts; solid thin lines show streamlines in warm air and dashed arrows show streamlines in cold air.

a shear line with high cyclonic vorticity between the jet streams. Figs. 10.4b and 10.4c represent a shear line combined with a cold upper cyclone to the south. In Fig. 10.4d, a new jet stream has formed north of the cold upper cyclone, which now has the character of an almost circular vortex. The "Viererdruckfeld" pattern in Fig. 10.4e, whose characteristics have been described thoroughly by Raethjen (1949), is especially common over the east Atlantic and west European region. Cold lows and warm highs are paired meridionally, with a strong barocline field between the "upstream" ones and an overall reversal of the normal meridional arrangement of air masses in the "downstream" pair.[1]

These typical end products of a strong change in the upper flow generally have a tendency to move only slowly eastward. In connection with a slow sinking in the cold air, the temperature of such a cold vortex gradually increases, and ultimately the formation weakens and becomes absorbed by the surrounding warmer atmosphere. In Fig. 10.4e, with a warm high on the poleward side, a cold vortex can on occasion be steered westward by a strong upper easterly current and can survive for a considerable time. It may be noted, however, that (as in Fig. 10.3) *the westward movement characterizing "blocking highs" is generally accomplished by the successive generation of new warm high cells on the west side*; the component high cells of the blocking complex may themselves drift slowly eastward.

The formation of upper cold cyclones seems to occur in preferred regions of the westerlies. One such region is the western United States; another is southwestern Europe. Both regions are characterized by the common existence of a northwesterly jet stream and a tendency for the lower fronts to become indistinct. As Kerr (1953) and van Loon (1956) have pointed out, in the Southern Hemisphere there is a strong preference for cutoff lows, of the type in Fig. 10.4e, to form in the Australia-New Zealand region in connection with blocking situations. As in the Northern Hemisphere, there is a maximum frequency of blocking in late winter and early spring.

10.2 OVERALL STRUCTURES OF CUTOFF CYCLONES

Hsieh (1949, 1950) has studied upper-level developments of the types in Figs. 10.4a, b, and d, and has described the weather associated with them. A typical example of the type in Fig. 10.4c is given in Fig. 10.5, showing the frontal topography during its development. Such cases are characterized by cloudiness and precipitation on the east side and clear weather on the west side. Figure 10.6, after Hsieh (1949), shows schematically the distri-

---

[1] The different types of disturbances may to a certain extent perform different functions in the meridional transfer of physical properties. Types b to e in Fig. 10.4 are clearly associated with the equatorward transfer of large masses of cold air and a compensating poleward flow of warm air outside the disturbances. The northward and southward currents in type a of Fig. 10.4 (see cross sections in Newton *et al.*, 1951) often do not have appreciable differences in temperature and thus do not transfer much heat. At the same time, such shear lines are almost always tilted as shown so that, through the correlation between the u- and v-components of the wind, they are very effective in transferring angular momentum poleward.

bution of precipitation related to such a cold dome; obviously, this has a close connection with the distribution of upper divergence, in accord with the general rules presented in Chapter 6. Synoptic examples of cutoff lows over Europe were presented by Scherhag (1948) and Schwerdtfeger (1948), who also studied the characteristic weather distribution accompanying this type of atmospheric disturbance.

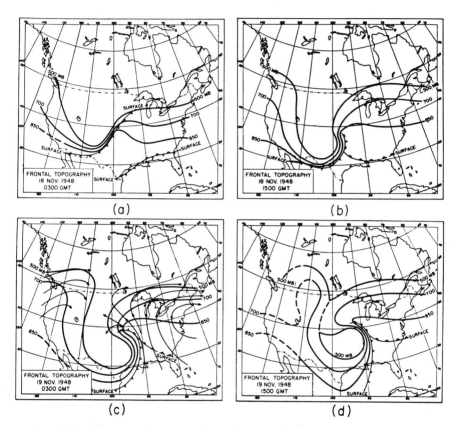

Fɪɢ. 10.5 Frontal contour charts, somewhat idealized, at 12-hr intervals on Nov. 18–19, 1948. In (c), streamlines at warm-air boundaries of frontal surfaces, based on actual winds. (From Palmén, 1951.)

Peltonen (1963) has carried out a case study of a high-level cyclone over Europe, which displayed an unusual behavior and offers an interesting comparison with other types. This cyclone formed from an upper cold trough around Nov. 8, 1959, over western Siberia and moved along the path shown in Fig. 10.7a, ultimately disappearing about 12 days later. It

reached full development on Nov. 14, moving generally westward under
the influence of the steering current south of a warm high over northern
Russia and Scandinavia. On the surface map (Fig. 10.7a) it appeared as a
weak wave-shaped disturbance in a general easterly current, with a quite
intense precipitation area on the east side of the surface trough. In that
respect it was similar to an easterly wave in the tropics (Section 14.7). No
strong temperature contrasts or front can be seen on the surface map.

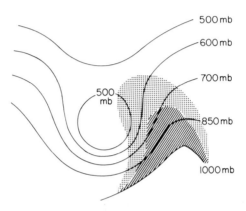

Fɪɢ. 10.6 Schematic representation of precipitation relative to the contours of the
surface of a cold dome. Heavier precipitation hatched; light precipitation stippled. (After
Hsieh, 1949.)

In Fig. 10.7 we reproduce charts for 850 mb, 500 mb and 300 mb, with
isotherms drawn for every degree centigrade. At 850 mb the thermal pattern
is already clear, with very low temperatures somewhat west of the trough
line. The mean westward movement of the "easterly wave" between Nov.
16 and 17 was 9 m/sec, compared with a gradient wind speed of about
13 m/sec north of the wave. Hence, the air above the friction layer moved
faster than the disturbance; the isotherms indicate low-level convergence
with ascent on the east side and divergence and descent on the west side.
If we suppose that the air during this movement ascends from about 950 mb
(top of frictional layer) to 600 mb along a condensation adiabat, the temperature along the three-dimensional path would decrease from $-6°C$ to
$-34°C$, which closely corresponds to the lowest observed temperature at
600 mb (Fig. 10.8). The cold core of the disturbance in the lower part of
the troposphere can be therefore explained by the vertical movement; this
is also in agreement with the observed precipitation pattern on its east side.
    The 500- and 300-mb charts (Figs. 10.7c-d) show the high-level cyclone
fully developed, with a closer approach to thermal symmetry. The heavy

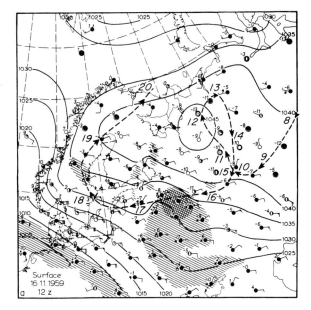

Fig. 10.7a

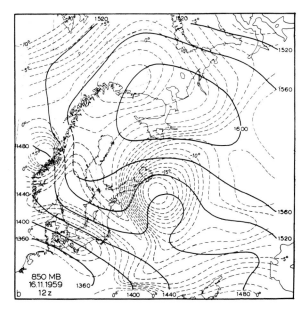

Fig. 10.7b

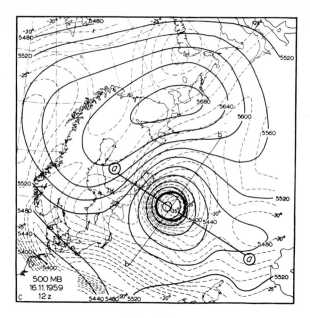

FIG. 10.7c

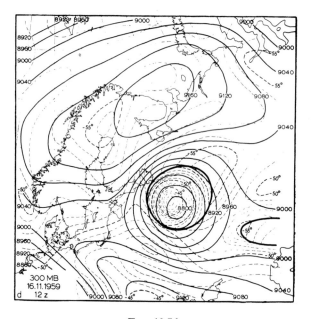

FIG. 10.7d

FIG. 10.7   (a) surface, (b) 850-mb, (c) 500-mb, and (d) 300-mb charts, 12 GCT Nov. 16, 1959. In (a), temperatures are in degrees centigrade; precipitation areas are hatched; cross-hatching indicates water-equivalent amounts exceeding 1 mm in 12 hr. In other charts, isotherms are at 1°C intervals, and contours at 40-m intervals. Heavy line in (c) and (d) is "tropopause" intersection. The path of the 500-mb low center is shown in (a), the arrowheads indicating its location at 00 GCT on the dates given. (After Peltonen, 1963.)

lines in these charts show the intersection with the relatively warm upper core (interpreted as a "tropopause funnel" in the cross section of Fig. 10.8, along line *a-a* in Fig. 10.7c). This illustrates in beautiful form the characteristic thermal structure of a high-level cyclone. The heavy line can be inter-

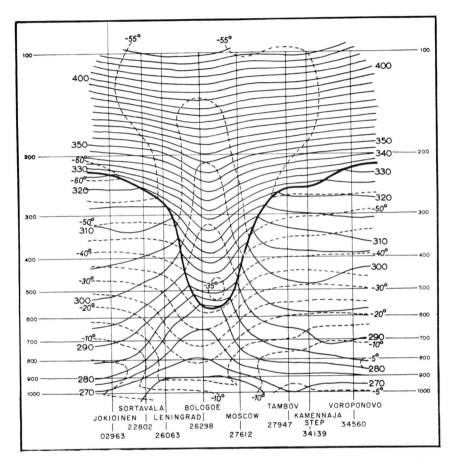

FIG. 10.8   Vertical section along line *a-a* in Fig. 10.7c. Heavy line corresponds to tropopause; dashed lines are isotherms at 5°C intervals; solid lines, isentropes (°K). (After Peltonen, 1963.)

preted as some kind of "tropopause" separating the outer part of the vortex, where the circulation increases with height, from the central parts where the circulation decreases upward. This surface represents a discontinuity of first order, the properties of which were discussed in Section 7.3. In that respect it is similar to a real tropopause. However, *the air just above this surface could not have originated in the stratosphere* and moved down to 570 mb because in that case the temperature of the tropopause should have been very much higher. "False tropopauses" of this kind often form in connection with the strong deformation of upper cold troughs. They must be regarded as a result of the tendency for the atmosphere to establish a hydrostatic equilibrium, expressed by the thermal wind equation, through a descent in the cyclone core of air that came originally from the upper troposphere. The "tropopause" in Fig. 10.8 is in some respects comparable to the boundary between the eye of a tropical hurricane and the surrounding moist and ascending air (Section 15.4).

The structure illustrated here was preserved for a long time; for example, a cross section 36 hr later, when the cyclone was centered over Stockholm, was virtually identical to Fig. 10.8. The persistence of cold upper lows of this type shows a remarkable similarity with that of tropical cyclones. However, in mature tropical cyclones, the region just outside the eye boundary is considerably warmer than the atmosphere farther outside, whereas in the type of cyclone discussed here, the coldest air appears at the "tropopause" in the middle and upper troposphere and the coldest air in the lower troposphere appears near the center of the cyclone. Hence the tropical cyclone is in this respect a warm-core vortex and the extratropical cyclone is a cold-core vortex. The persistency of tropical cyclones has generally been explained as a result of continuous generation of kinetic energy resulting from the warm-core structure (Chapters 15 and 16). The same explanation cannot be used for the maintenance of the extratropical cold-core vortices, the energetics of which is still obscure.

Most cyclones of the type discussed above have a remarkable thermal symmetry about the central axis. By contrast, *the weather shows a markedly asymmetric distribution, with precipitation on the east and northeast side and a tendency for clear skies on the west and southwest side, whether the vortex is moving eastward or westward.* This general distribution of precipitation and vertical motion also characterizes waves in both westerlies and easterlies, although waves (unlike closed vortices) generally have marked asymmetries in thermal structure.

Although thermal symmetry is generally characteristic, cold lows of this kind may take on an asymmetrical structure when they combine with an independent current system. A very complete analysis of the evolution and structure of a cold low and of its three-dimensional motion field has been presented by Omoto (1966). Figure 10.9 shows a north-south cross section at a time when the cyclone had become cut off from the polar westerlies and was centered near latitude 30°. An east-west section indicated almost complete symmetry, with a structure very like that in Fig. 10.8 and a "tropopause funnel" down to the 500-mb level. By contrast, although the meridional section in Fig. 10.9 shows a quite symmetrical structure in the isentropes below $\theta \approx 294°$K, there is marked asymmetry higher up. On the north side, the "tropopause" appears essentially continuous with the "middle tropopause." At the low latitude involved, however, the south side of the vortex was contiguous with the barocline zone of the subtropical jet stream,

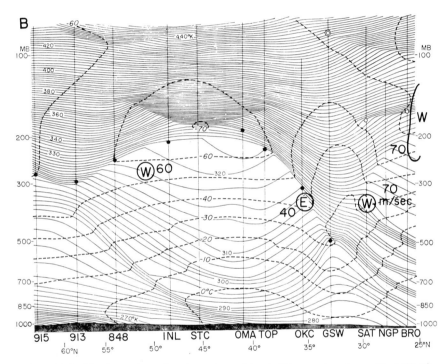

FIG. 10.9   North-south section through a cold cutoff low centered near Fort Worth, Texas (GSW), at 12 GCT Feb. 5, 1964. Dashed lines, isotherms at 10°C interval; solid lines, isentropes at 2°K intervals. Isotachs (m/sec) indicate locations of jet streams (W, westerly; E, easterly current). During the preceding three days, the cold vortex at the right became separated from the main body of polar air at left of the section. (After Omoto, 1966.)

so that real tropical air was present in the upper troposphere on that side. The thick stratum with several stable layers in the middle troposphere on the south side is similar to the structure in Fig. 8.23, which results when polar air moves to such low latitudes that the polar-front zone combines with the subtropical front.

10.3   ROLE OF EXTRATROPICAL DISTURBANCES IN THE MERIDIONAL
AND VERTICAL EXCHANGE PROCESSES

The importance of extratropical disturbances for the transfer of physical properties has been emphasized several times in earlier chapters. It was shown in Chapters 1 and 2 that *an essential role of extratropical disturbances is to effect a large part of the necessary meridional and vertical exchange of angular momentum and heat,* the vertical exchange of heat also being identified with generation of kinetic energy. The vertical heat exchange by disturbances will be discussed in Section 10.6. As regards angular momentum, the process of meridional transfer is fairly well understood, but the quantitative mechanism of vertical transfer has not been firmly established. Only some qualitative comments on this can be offered here.

According to the interpretation in Section 1.3, the required vertical flux of absolute angular momentum in extratropical latitudes is mostly accomplished by the Ferrel circulation, in which downward motions take place at lower latitudes where the earth-angular (Ω) momentum is large, with ascent in latitudes where this is smaller (Fig. 1.11). By a further interpretation in Section 16.3, the Ferrel circulation may be viewed as a statistical result of the sinking of cold air in the lower-latitude parts and the rising of warm air in the higher-latitude parts of wave disturbances on the polar front. In this sense the downward transfer of Ω-momentum by the mean meridional circulation is ultimately an effect of a systematic component in the vertical motion fields of large disturbances.

From the circumstance that the required downward transfer apparently exceeds that provided by the mean circulation, it was concluded in Section 1.3 that an additional downward eddy transfer (about one-fourth of the total) is needed.[2] Direct evaluation of this would depend upon a knowledge of the $w'u'$ correlation expressed by the term J_z in Eqs. (1.5) and (1.6). Although a transfer in the needed sense is accomplished by cumulus convection (Section 13.6), it is unlikely that this mechanism is significant in extratropical latitudes during the cooler months. Hence a transfer by the broad-scale vertical motions in the disturbances must be called upon.

Figure 10.15 typifies the middle-tropospheric contour field above a wave disturbance in its earlier lifetime. In this case, descent in the cold air, and ascent in the warm air east of the trough, would imply $\overline{w'u'} > 0$ with an

---

[2] The amount of this depends on estimates of the surface torque and of the mass circulation in the Ferrel cell. In the original computation of this meridional circulation by Mintz and Lang, referred to in Section 1.1, the assumption was made that the downward flux by small eddies is canceled by an upward flux in synoptic-scale eddies.

upward transfer of relative angular momentum. By contrast, we may consider an occluded system at an advanced stage of development when the upper contours are strongly distorted (Fig. 10.10). In this case a substantial part of the ascent in the precipitation area may take place where there is a weak westerly or even an easterly component; and descent takes place where the westerlies are stronger, with a net downward transfer of angular momentum.

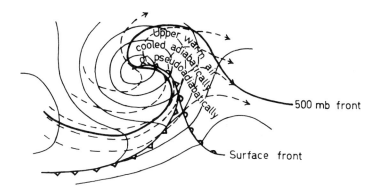

500 mb front

Surface front

FIG. 10.10 Schematic structure of a cyclone at an advanced stage of occlusion. Surface and 500-mb fronts are shown, with sea level isobars (solid) and 500-mb streamlines (dashed). At this stage, sinking in the cold air is most active on the south side of the sea-level cyclone center, and in upper levels the tongue of warm air is strongly distorted, with moist-adiabatic ascent largely in regions of weak westerly or even easterly wind component.

Moreover, it may be observed that when either an open-wave disturbance or an occluded system is superposed upon the east side of a long-wave trough, the westerly wind component is augmented in the rear of the cyclone-scale disturbance and is diminished ahead of it, as indicated by the contours of the short-wave troughs in Fig. 6.6b. Since the majority of cyclone waves have this broad relationship to larger-scale waves, and vertical motions are most intense in the short cyclone waves, this circumstance evidently contributes to a negative $u'w'$ correlation, with downward transport of westerly momentum. The reverse may be true in waves east of a major ridge, which are less frequent and usually less intense.

From these examples it can be seen that although synoptic disturbances in general transfer heat upward, the transfer of relative angular momentum is downward in some and upward in others. From an investigation of rawin soundings over the eastern United States during the period January–April 1949, White (1950) found that soundings during precipitation had mid-tropospheric zonal winds averaging 3 m/sec less than those in clear-sky regions, suggesting a net downward transfer of westerly momentum. Priestley (1967) indicates that the nongeostrophic component of the vertical momentum flux is very important for the transfer through the top of the friction layer and that the vertical circulations near fronts are a principal mechanism for this transfer.

The function performed by cyclones in the meridional and vertical exchange of air masses was emphasized in the original Norwegian polar-front theory (Section 5.4). From the examples in Figs. 10.1 and 10.5, a partial idea of this exchange in the form of discrete huge masses of cold and warm air moving across the latitude circles can be obtained. The conception on this basis is, however, incomplete, since the air moves vertically as well as horizontally through the synoptic systems. A superficial study of synoptic charts at upper levels suggests that the atmosphere is typified by a zonal circulation upon which wave-shaped disturbances are superposed. The quasi-geostrophic motion of the air gives the impression that air parcels at different latitudes are swaying north and south, but, on the whole, around a more or less fixed latitude. A detailed study of real air movement (e.g., with the aid of actual trajectories) gives an essentially different picture, showing that the amplitudes of air trajectories are much larger than the amplitudes of horizontal streamlines. It is therefore necessary to supplement the description of bulk movement of air masses with an examination of the actual three-dimensional motions.

Until recently, few attempts were made to study three-dimensional air trajectories, partly owing to the difficulty involved.[3] The simplest air trajectories to follow are those close to the earth's surface, where the vertical component of the air movement vanishes. Unfortunately, just at this level the thermal influence of the underlying earth's surface is so strong that temperature is very poorly conserved. In the free atmosphere the vertical component of the air motion cannot be neglected, but because of the character of the atmosphere as a "thin" layer in which vertical movement is small and not directly measurable, it is not easy to consider the vertical component accurately.

The problem is simplified if temperature changes can be considered

---

[3] This problem has been resolved (Danielsen, 1966) by the development of a machine program for constructing trajectories, largely by Bleck (1967). This makes use of observed wind and temperature fields and of the consistency of the trajectories with energy principles, according to Eq. (8.5). Danielsen gives numerous examples of the vertical motion fields on isentropic surfaces around cyclones, which agree very well with the cloud and precipitation distribution.

adiabatic. Hence, true air trajectories can with satisfactory accuracy be constructed in horizontally moving or subsiding air masses in which potential temperature can be considered approximately conserved, at least in layers to which heating from the ground hardly penetrates. Where ascending motion is present, it is much more difficult to identify an individual air particle by the aid of its potential temperature, owing to the liberation of heat of vaporization as soon as the air has reached saturation. In a study of the vertical and meridional exchange of air masses, it is therefore convenient to start with the three-dimensional flow either in regions where condensation processes do not occur at all or where such processes are of minor importance. We shall therefore proceed to describe the three-dimensional flow in cold-air masses of polar origin after making some general comments about the relations between meridional and vertical displacements.

10.4   Changes in Depth of Air Masses during Meridional Displacement

Polar air originating in the arctic and subarctic regions (if we relate our discussion to the Northern Hemisphere) undergoes characteristic transformations when it moves out from its source regions. Aside from radiation, these transformations are effected not only by transfer of heat and moisture from the earth's surface and its upward transport by eddies, but also by adiabatic heating or cooling that results from descent or ascent of the air. It was pointed out earlier (e.g., in Section 6.5) that polar air generally has a tendency to subside when moving toward lower latitudes.

Although the processes are in reality much more complicated, it is instructive to examine the behavior of a hypothetical air column that is displaced equatorward from a higher latitude. We assume conservation of potential vorticity, defined by the relationship (Rossby, 1940)

$$\frac{\zeta + f}{\Delta p} = \text{const.} \tag{10.1}$$

where ζ is relative vorticity and Δp is the pressure depth of the column.

In the special case wherein the relative vorticity is initially (and remains) zero, Δp is simply proportional to the coriolis parameter f, and the air column must shrink vertically as it is displaced southward, in the manner shown in Table 10.1. Thus, if the top of a cold-air column were initially at the 300-mb level at latitude 60°, its displacement to latitude 30° would result in a shrinkage in depth by 42%, the top at the final latitude being at

TABLE 10.1

CHANGE IN DEPTH OF AN AIR COLUMN MOVING FROM NORTH TO SOUTH WITH ZERO
RELATIVE VORTICITY[a]

			Latitude			
	60°	50°	40°	30°	20°	10°
Ratio of final to initial depth (%)	100	88	73	58	39	20
Pressure at top of column (mb)	300	380	490	590	730	860

[a] Adapted from Palmén (1951). The pressure at the base of the column is assumed to be 1000 mb.

590 mb. In the process, if the temperature at the top of the column is initially, say, $-58°C$, adiabatic compression would result in a warming to $-12°C$ at 590 mb at the lower latitude. This corresponds (Fig. 4.4) to mid-latitude warm air. Thus, if polar air is to retain its identity on moving to lower latitudes, there must be some limit imposed on its degree of sinking. This would evidently depend on the original temperature; *a very cold air mass could move farther equatorward than a moderately cold one without losing its contrast of temperature with the surrounding air.*[4]

As a less restrictive assumption, we may examine the conditions resulting if the relative vorticity is allowed to change as the air column moves along. According to Eq. (10.1), the depth of an air column could remain constant only if (corresponding to the decrease of f as it moves equatorward) there is a like increase of relative vorticity. In a broad sense, increasing relative vorticity is largely expressed by increasing cyclonic curvature. An air parcel moving equatorward without shrinking therefore has a tendency to bend eastward until it reaches a lowest latitude and again starts moving poleward. Petterssen (1956) has given examples of the change of the depth of air masses moving from latitude 60°, with zero relative vorticity initially, in terms of the relative vorticity achieved on reaching latitude 30°. According

[4] Along with the adiabatic warming in upper levels, heat transfer from the earth's surface warms the lower layers. This influence is naturally greatest in cold air moving equatorward over the oceans, whose heat capacity is large. An air mass that has undergone subsidence is on these grounds more readily recognizable from its stability over land than over water. Correspondingly, the weather phenomena in cold outbreaks over water and over land are likely to be very different (Section 12.2).

to Eq. (10.1) the ratio of the depth at latitude 30° to that at latitude 60° would be

$$r = \frac{\zeta_{30} + f_{30}}{f_{60}} \tag{10.2}$$

This ratio is given in Table 10.2, where case B corresponds to Table 10.1. Comparison with the other cases indicates that an equatorward-moving column of cold air could retain its depth only if it moves along a cyclonically curved path, in which event it would gradually turn eastward and ultimately poleward. This reasoning shows why deep air masses of real

TABLE 10.2

RATIO OF FINAL TO INITIAL DEPTH OF COLD AIR, FOR AN AIR COLUMN WITH ZERO RELATIVE
VORTICITY AT 60° LATITUDE, WHICH ARRIVES AT LATITUDE 30° WITH A PRESCRIBED
RELATIVE VORTICITY[a]

Trajectory	Case			
	A	B	C	D
ζ at 30°	$-0.5f_{30}$	0	$f_{60} - f_{30}$	f_{30}
Ratio (r)	0.3	0.6	1.0	1.2

[a] After Petterssen (1956).

polar origin can be found in low latitudes only in connection with strong cyclonic circulation. Movement far equatorward can be achieved only with strong vertical shrinking; observations show that the lowest latitude at which genuine polar air can be found at the 500-mb level is about 25 to 30°.

From the scheme above the following general rule can be derived: *If a deep polar air mass has been advected to low latitudes, it must appear there as a mass with strong cyclonic vorticity; if the mass is characterized by anticyclonic vorticity, it must be shallow.* Conversely, an air column starting at low latitudes will undergo vertical stretching unless its relative vorticity is markedly decreased as it moves poleward.

In real situations, the air trajectories in different parts of a cold- (or warm-) air mass moving meridionally differ from one another. The pattern in Fig. 10.11a is usual in cold-air outbreaks. In accord with Table 10.2, *the polar air on the east side of an equatorward intrusion is typically much deeper than that on the west side, and the slope of the frontal boundary between the polar*

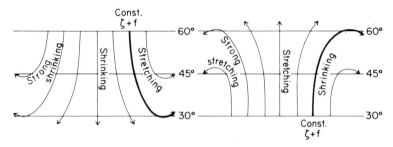

Fig. 10.11   Meridional motion with anticyclonic and cyclonic branches and the implied vertical stretching or shrinking. Along heavy streamlines, absolute vorticity is conserved, the change of coriolis parameter being compensated by an opposite change in relative vorticity expressed by curvature. (By permission from S. Petterssen, *Weather Analysis and Forecasting*, 2d ed., Vol. 1, McGraw-Hill Book Company, Inc., New York, 1956).

air and the warm air tends to be much steeper on the eastern than on the western flank of a cold tongue. The scheme in Fig. 10.11b is also quite characteristic of poleward surges of warm air ahead of cyclones (see, e.g., Fig. 5.6), in which the warm air near the cyclone center may curve cyclonically while that some distance ahead of the center typically has an anticyclonic curvature above the warm-front surface, where it undergoes horizontal divergence and vertical shrinking in the upper troposphere.

10.5   Computations of the Vertical Mass Transport in Cold Outbreaks

It is easy to visualize the general tendency for subsidence in an outbreak of polar air in middle latitudes, using a synoptic demonstration elaborated by Palmén and Newton (1951). This will be illustrated in connection with the situation in Fig. 10.12. This shows the sea-level chart and isotherm analyses at selected upper levels at 03 GCT Apr. 5, 1950. Over central North America, the isotherms are concentrated into a front that is clearly distinguishable in the middle troposphere (Figs. 10.12c and d), but only on the east side of the cold tongue at 850 mb (Fig. 10.12b). On the west side, at 850 mb, the broadness of the barocline zone reflects in part the thermal influence of the earth's surface and adiabatic heating in the subsiding air, and in part a shallower slope of the front. Comparison of the left- and right-hand cross sections (at a time 12 hr earlier) in Fig. 10.13 reveals that the front was twice as steep on the southeast side as on the southwest side of the cold tongue, in accord with the previous discussion.

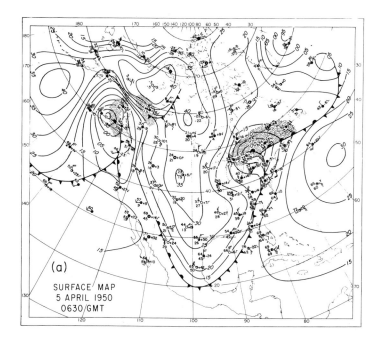

FIG. 10.12a

FIG. 10.12b

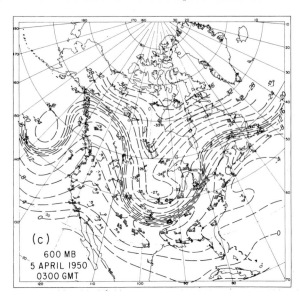

FIG. 10.12c

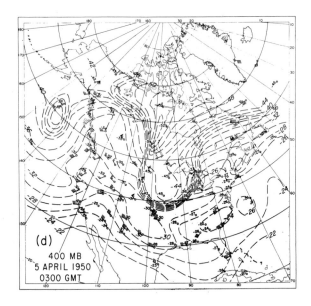

FIG. 10.12d

FIG. 10.12   (a) Surface map at 0630 GCT Apr. 5, 1950, with fronts and sea-level
isobars at 5-mb intervals; areas of continuous precipitation stippled. (b) 850-mb; (c)
600-mb, and (d) 400-mb charts at 03 GCT, with isotherms at 2°C intervals. (From
Palmén and Newton, 1951.)

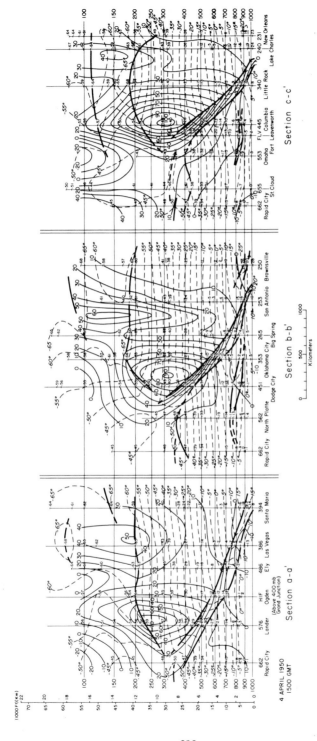

FIG. 10.13   Vertical sections along dotted lines in Fig. 10.14b. Heavy lines are tropopauses and frontal boundaries; dashed lines, isotherms at 5°C intervals; solid thin lines, isotachs of geostrophic wind component normal to sections, at 10 m/sec intervals. (From Palmén and Newton, 1951.)

295

The thermal structure at an earlier time during the formation of this same trough, in Fig. 9.8, showed that the upper-tropospheric front was strong on the west side but absent to the east. However, at the time of Fig. 10.12, and 12 hr earlier and later, the portion of the front south of 45°N was very distinct, as in Fig. 10.13. This makes it possible to use continuity considerations, as outlined below, to appraise the large-scale, three-dimensional air movements.

This method depends essentially upon an assumption that *the warm-air boundary of the front is an impermeable surface.* As indicated in Section 9.6, this assumption is not strictly valid in the very lowest layers, where there are frictional and diabatic influences. However, the entrainment of air into the front in the surface layers is likely to be insubstantial compared with the mass fluxes to be computed in higher levels. In the free atmosphere, the preceding assumption is justified by the observation that (as shown, e.g., by Fig. 7.4) there is a discontinuity of both horizontal (or isentropic) shear and stability across a frontal surface. These are related in such a way that a very large discontinuous change of potential vorticity would have to be undergone by an air parcel if it were to cross the front.

Frontal contours at selected times are shown in Fig. 10.14, in which Fig. 10.14c corresponds to Fig. 10.12. On the grounds discussed above, the volume of cold air enclosed between the frontal surface and the 45° parallel should change with time according to the influx or efflux of air through a vertical wall at 45°N. Also, if the expansion of an individual slice of that volume (say, between 400 and 500 mb) is not balanced by the influx in the same layer, the difference must be expressed by a corresponding net vertical mass flux through the bounding isobaric surfaces. These considerations form the basis for the computations that follow.

The method is illustrated by Fig. 10.15, showing the characteristic contour lines of an eastward-moving trough. The "front" corresponds to the warm-air boundary of a frontal layer, shown here as if it were a discontinuity of zero order. The lag between the pressure trough and the thermal trough (which is outlined by the front) is characteristic of developing disturbances. Letting L denote the distance between the intersections A and B of the frontal contour with a fixed latitude φ, the advective increase of the area C of polar air south of this latitude[5] during time t is given by

$$\Delta C_{\mathrm{II}} = -\,\bar{v}L \cdot t \approx \frac{g}{f}\,\Delta H \cdot t \qquad (10.3)$$

[5] Namely, the increase that would take place if the air moved horizontally without divergence within area C.

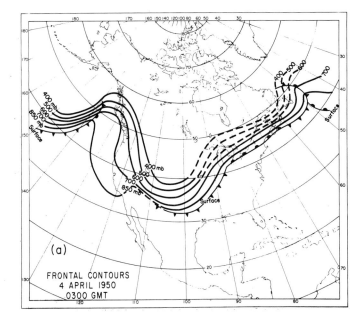

FIG. 10.14a

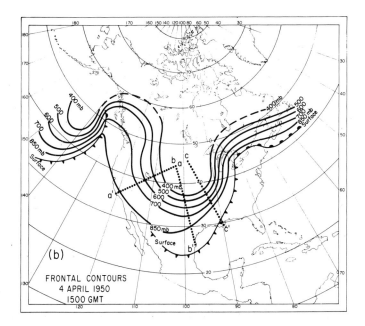

FIG. 10.14b

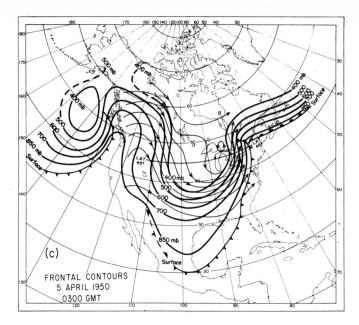

Fig. 10.14c

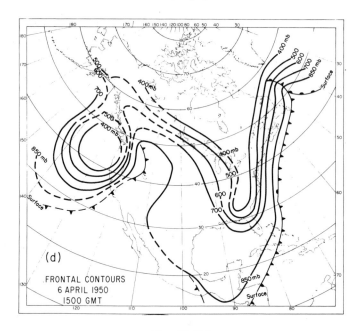

Fig. 10.14d

Fig. 10.14 Frontal contours (warm side of polar-front layer) at (a) 03 GCT April 4; (b) 15 GCT April 4; (c) 03 GCT April 5; and (d) 15 GCT April 6, 1950. Contours are dashed where front was indistinct. In (c), thin lines correspond to isobaric contours (200-ft interval) at 500 mb. Note that the time interval between the last two charts is 36 hr. (From Palmén and Newton, 1951.)

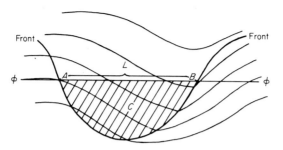

FIG. 10.15 Height contours (thin lines) and frontal contour at an isobaric surface in the upper troposphere. (See text.)

where $\Delta H = (H_A - H_B)$ is the difference in height of the isobaric surface between points A and B. Here $-\bar{v}$ denotes the mean north-wind component across L, approximated geostrophically. The total mass advection through the northward-bounding latitude wall between isobaric surfaces p_1 and p_2 ($p_1 > p_2$) is then given by

$$\frac{1}{g} \int_{p_2}^{p_1} \Delta C_{\text{II}} \, dp = -\frac{1}{g} \int_{p_2}^{p_1} \bar{v} L \cdot t \, dp \approx \int_{p_2}^{p_1} \frac{\Delta H}{f} \cdot t \, dp \qquad (10.4)$$

Denoting the observed change in area C by ΔC_{I}, the change in the mass of cold air south of latitude φ is

$$\frac{1}{g} \int_{p_2}^{p_1} \Delta C_{\text{I}} \, dp$$

Only at the level where there is no divergence within area C does $\Delta C_{\text{I}} = \Delta C_{\text{II}}$. If the integration is performed between the ground ($p = p_0$) and the pressure p_h at the top of the cold-air mass, continuity prescribes that

$$\int_{p_h}^{p_0} \Delta C_{\text{II}} \, dp = \int_{p_h}^{p_0} \Delta C_{\text{I}} \, dp \qquad (10.5)$$

Likewise, integrating from the top of the cold air to an arbitrary isobaric surface p, the vertical mass transport through that surface during the time considered is, according to continuity requirements,

$$(M)_p = \frac{1}{g} \int_{p_h}^{p} (\Delta C_{\text{II}} - \Delta C_{\text{I}}) \, dp \qquad (10.6)$$

A check on the validity of the method is that, using Eq. (10.5) in combination with Eq. (10.6), no vertical mass transport should occur through the pressure surface p_0 corresponding to the mean ground level. It should be stressed that the method is applicable only if the front between the cold air and the surrounding warmer air is so well marked that it can be considered a substantial surface through which there is no exchange of air.

Figure 10.16 illustrates at various levels the differences between the advective changes and the really observed changes of cold-mass area south of 45°N during the 24 hr centered on Fig. 10.14c. In the upper troposphere,

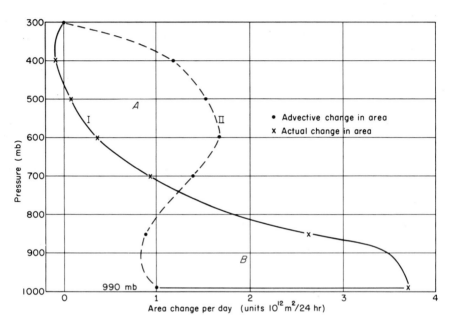

FIG. 10.16   Curve I shows actual change in area of cold air at isobaric surfaces south of 45°N during 24-hr period 15 GCT April 4 to 15 GCT April 5, 1950, and curve II shows the "advective change" (see text). (From Palmén and Newton, 1951.)

a systematic southward flow into the cold tongue was indicated as in Fig. 10.15. However, the areal expansion of the cold air was slight. In lower levels, the net southward flow was smaller; however, a considerable increase in the area of the cold air took place. These features are especially well emphasized by a comparison of the frontal contours at 400 mb and at the surface in Figs. 10.14a and 10.14d. During this period, *a vast expansion was observed in low levels; however, in the upper troposphere the cold tongue became narrower, with an appreciable decrease of the area south of 45°N at 700 mb and above.* The marked decrease in slope of the front in lower levels, especially on the west side of the cold tongue, is quite characteristic of the later stages of cold outbreaks and is in general accord with Fig. 10.11a.

The curves in Fig. 10.16 cut each other at 740 mb, which then represents the level of maximum downward mass transport of cold air according to Eq. (10.6). Since areas A and B are almost equal, the computed vertical mass transport at the ground (mean pressure 990 mb) is negligible, indicating that the assumption of no mass transport across the frontal surface is justified.[6]

The method was applied on four synoptic cases of cold outbreaks through the same latitude 45°N, with the results in Table 10.3. In all cases the computed vertical mass transport was negligible at the ground, indicating that the values are not too erroneous. According to the average for the four cases in the last columns, the strongest downward mass transport occurred at 760 mb, whereas the strongest mean downward velocity (average value about -3.5 cm/sec) is noted around 530 mb. The difference in these pressure levels depends partly on the downward-increasing area of cold air and partly on the vertical variation of density.

The strongest downward mass transport in Table 10.3 was computed for the case Nov. 24–25, 1950. This was the famous "Thanksgiving Storm" extensively studied by Charney and Phillips (1953) and by Phillips (1958). This case represents one of the most extreme baroclinic developments over the United States in recent years, a development that resulted in a very severe storm.

The preceding method could also be applied for computation of the vertical mass transport in the warm air, e.g., on the east side of a cold trough. However, we have not attempted this because such a computation in most cases must be extended farther north, where the network of stations is not so good as in lower latitudes.

10.6 FLUX OF HEAT IN SYNOPTIC DISTURBANCES

It is possible to make some general conclusions concerning the meridional and vertical flux of heat, with aid of the data in Table 10.3. For this purpose we use the synoptic case Apr. 4–5, 1950, which has been most thoroughly analyzed. In this case the total southward transport of cold air through 45°N amounted to 8.3×10^{12} tons in the 24-hr period studied. The length of the parallel was, in the middle troposphere, about 30° longitude. If we assume that the same mass of warm air is advected northward through a corresponding length of latitude circle (to maintain the mass north of the

---

[6] The additional assumption of mean geostrophic inflow, in Eq. (10.4), was verified by comparison with the mass inflow computed from actual winds.

TABLE 10.3

Downward Mass Transport $(M)_D$ and Mean Vertical Velocity w in Four Outbreaks of Polar Air South of Latitude 45° N[a]

Isobaric surface (mb)	April 4–5, 1950			Nov. 24–25, 1950			Jan. 21–22, 1951			Jan. 22–23, 1951			Mean values		
	C	$(M)_D$	w	C	$(M)_D$	w	C	$(M)_D$	w	C	$(M)_D$	w	C	$(M)_D$	w
400	1.1	0.8	−1.4	0.5	1.0	−4.0	0.6	0.1	−0.4	0.3	0.4	−2.6	0.6	0.6	−2.1
500	1.8	2.2	−2.0	1.0	2.9	−4.8	0.9	1.1	−1.9	0.5	1.6	−4.8	1.1	2.0	−3.4
600	2.3	3.6	−2.2	1.6	4.7	−4.2	1.4	2.3	−2.3	1.1	2.9	−3.8	1.6	3.4	−3.1
700	3.5	4.6	−1.6	2.4	6.1	−3.1	2.5	3.7	−1.9	2.4	4.1	−2.1	2.7	4.6	−2.2
800	5.0	4.5	−1.0	4.0	6.3	−1.7	4.2	4.5	−1.2	4.1	4.3	−1.2	4.3	4.9	−1.3
900	6.7	2.7	−0.4	5.8	4.4	−0.7	5.6	4.5	−0.8	6.4	3.2	−0.5	6.1	3.7	−0.6
Ground	8.2	0.0	0.0	6.6	−0.2	0.0	9.0	0.8	−0.1	8.9	−0.6	0.1	7.7	0.0	0.0

[a] Units: Area of cold air, C: 10^{12} m²; $(M)_D$: 10^{12} tons/day; w: cm/sec.

latitude), six such disturbances would extend around the globe. In that case, the total mass exchange would be 50×10^{12} tons/day. The total mass of air in the cap north of 45°N is about 7.5×10^{14} tons. On this basis the exchange of air across the whole latitude circle in one day would correspond to 6.6% of the air in the cap. If we further assume that the temperature difference between the two air masses participating in the mass exchange is 10°C (in the case above it was really somewhat larger), the total heat flux northward through latitude 45°N would amount to about 1.2×10^{20} cal/day, or 5.8×10^{12} kJ/sec.

This value compares favorably with, and is somewhat larger than, the estimated total eddy flux of heat during the winter season (Table 2.5). Since the case considered here involved a greater-than-average development with a strong temperature contrast, it is probably not very representative of the heat exchange in spring. However, it suffices to demonstrate that *the horizontal transport of heat by polar-front disturbances is of the magnitude required to offset the radiative imbalance between low and high latitudes.*

The vertical heat flux may also be estimated from the values in Table 10.3. Let us consider only the upward heat flux across the 500-mb surface. The mass flux $(M)_p$ in one cold-air outbreak is, from the averages in the last columns, 2.0×10^{12} tons/day at 500 mb. For six disturbances around the hemisphere, the downward mass flux would be 12×10^{12} tons/day. Assuming an equivalent upward mass flux in the warm air,[7] and again that there is a 10°C temperature difference between warm and cold air, the upward heat transport accomplished by the six disturbances would be 28.7×10^{18} cal/day, or 1.39×10^{12} kJ/sec. From Table 10.3, if the area covered by warm air is considered equivalent to that of the cold air, the area of the six disturbances is 13.2×10^{12} m², giving an upward heat flux averaged over the area of the disturbances of 0.105 kJ m⁻² sec⁻¹ (0.15 ly/min).

This heat transfer may be compared with the required eddy flux of heat through the 500-mb surface for the whole polar cap north of 32°N, in Fig. 2.10. The area covered by the portions of the disturbances considered is only about 12% of the area of the polar cap. Thus, if the heat flux (0.15 ly/min over the area of the disturbances alone) were distributed over the

---

[7] Since other values used here (e.g., for the temperature difference between cold and warm air) are only approximate, we have neglected the possibility that the upward and downward mass transports may differ within the latitude belt concerned. From Fig. 1.5, the total mass flux in the Ferrel circulation averages 2.6×10^{12} tons/day in winter, or about 20% of the vertical mass flux in the cold air in six disturbances like that considered here.

whole polar cap, the mean vertical heat flux would amount to 0.018 ly/min. The corresponding required heat flux, from Fig. 2.10, is 0.020 ly/min. Again, this comparison can be considered valid only in a semiquantitative sense. On the one hand, the disturbances in Table 10.3 may be stronger than average, while on the other hand, only a portion of each disturbance is encompassed by the computations and the contributions from disturbances in other parts of the polar cap may be significant. This example clearly indicates, however, that *polar-front disturbances provide the most essential mechanism for the vertical eddy flux of heat in extratropical regions.*

10.7   THREE-DIMENSIONAL AIR TRAJECTORIES IN COLD OUTBREAKS

The synoptic case discussed in Section 10.6 may also be used to illustrate the three-dimensional air trajectories in a typical outbreak of polar air. In the cold air, computation of these by use of isentropic charts is permitted in levels where the heating from below is not too strong. Figure 10.17 shows some trajectories, during the period 03 GCT Apr. 4, to 15 GCT Apr. 5, on the 290°K isentropic surface (within the polar air, near the lower boundary of the frontal layer). These clearly show the anticyclonic outflow in most of the cold-air region. During the 36-hr period, air parcels in the central part move down from about 630 mb to 900 mb at the coast of the Gulf of Mexico. Only in the northeast part are some trajectories cyclonic. The figure is in good qualitative agreement with the scheme in Fig. 10.11a, showing vertical shrinking in the central and western parts of the cold outbreak and stretching only in the region close to the cyclone center. It may also be noted that the vertical motions computed from isentropic trajectories, in this and other cases, show that the strongest descent characteristically takes place in the lower-latitude parts of cold outbreaks, while (as in Fig. 10.17) the parts farther poleward comprise air that is descending in a more restricted region, and in part often ascending.

This analysis, and several other analyses of similar cases, shows that the outflow of cold air close to the low-level cold front partly originates in the central (and largely also in the upper) part of the cold trough farther north. As noted earlier, by looking only at constant-pressure charts at upper levels, one could easily get the impression that the cold air sways alternately southward and northward, following more or less the wave-shaped contour pattern. In reality, both computations of the general subsidence in the cold air and analyses of three-dimensional air trajectories show that *the real amplitude of cold air trajectories is much larger than that suggested by the contour waves. Also, part of the cold upper air in a trough is finally lost as a*

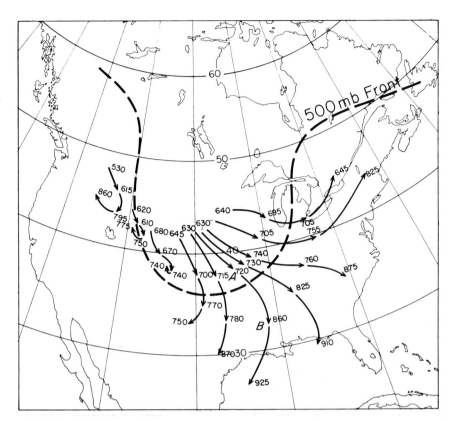

FIG. 10.17   Trajectories on isentropic surface 290°K, from 03 GCT April 4 to 15 GCT April 5, 1950. Pressures corresponding to positions of different air particles are indicated in millibars at 12-hr intervals. Dashed line shows 500-mb frontal contour, roughly corresponding to location of jet-stream axis, in the middle of the 36-hr period of the trajectories. (From Palmén and Newton, 1951.)

low-level anticyclonic current ultimately penetrating into the subtropics (see Fig. 6.6b). Some of the cold air on the western flank of a cold outbreak bends back poleward and becomes involved in other cyclones approaching from the west.

Similar conclusions concerning the three-dimensional flow of air in cold outbreaks have been drawn by Staley (1960), Danielsen (1961), and Reiter (1963). These authors have utilized isentropic trajectories to examine how radioactive debris is transferred from the stratosphere into the troposphere and from the poleward side of the jet stream to the equatorward side, where heavy ground depositions are observed. Such occurrences take place in connection with developments of the type described above, and trajectories have in some cases been traced from the lower polar stratosphere into the lower troposphere. These debris trajectories, in subsiding cold air in or near the polar front, cross from the poleward to the equatorward side of the jet stream. This is accomplished, however, not by the movement of air across the jet stream (i.e., through the "tropopause break") itself, but by its traversing in lower levels beneath the jet-stream axis, where the wind direction differs from that in higher levels, in the manner indicated by the trajectories of Fig. 10.17.

10.8   THREE-DIMENSIONAL FLOW OF WARM AIR IN CYCLONES

That the warm air east of a cold trough generally ascends follows from the model of the vertical distribution of divergence in westerly waves and is supported by the well-known distribution of precipitation in cyclones. Because of its poleward flow on the east side of a cold trough, the warm air generally acquires anticyclonic relative vorticity (Fig. 10.11b) if it is not subjected to vertical stretching. In the inner parts of the warm sector of a developing cyclone, the air is subjected to low-level convergence and upper-level divergence. In connection with this influence it gains cyclonic vorticity in low levels and anticyclonic vorticity in upper levels. At the same time, according to Eq. (10.1), because of the latitudinal variation of the coriolis parameter, the general poleward flow contributes to reduce the production of cyclonic relative vorticity in low levels and at the same time to augment the production of anticyclonic relative vorticity in upper levels.

Farther poleward, where the warm air ascends the warm-front surface, it experiences strong vertical shrinking in the upper troposphere. On the other hand, in the western part of a disturbance where the warm air flows down the cold-front surface, it undergoes vertical stretching in the upper troposphere and gains vorticity. The flow of air over sloping frontal surfaces is thus linked in a general way with the fields of divergence connected with upper waves (Chapter 6). The influence of these processes is described in the schematic Fig. 10.18, showing the types of three-dimensional trajectories (or, approximately, streamlines) that follow from the foregoing considerations. This scheme of the flow of warm air in a polar-front cyclone was verified by the studies mentioned in Section 5.5, soon after aerological observations became available.

It is obvious that the shapes of the three-dimensional trajectories or streamlines depend on the intensity of the vertical component of motion or on the strength of the divergence field. At the time when the cyclone undergoes strong development, one can therefore also expect a pronounced

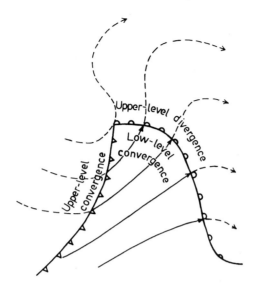

FIG. 10.18 Showing the general regions of convergence and divergence associated with air streaming over the cold-front surface, across the warm sector, and up over the warm-front surface of a wave cyclone.

change of the three-dimensional flow pattern with time. This change must generally appear as an increase of the amplitude of the upper baroclinic wave or partly as a shortening of its wavelength.

From the vorticity equation in the somewhat simplified form,

$$\mathbf{V} \cdot \nabla(\zeta + f) = -\frac{\partial \zeta}{\partial t} - \omega \frac{\partial \zeta}{\partial p} - (\zeta + f)\nabla \cdot \mathbf{V} \qquad (10.7)$$

the general shape of the streamlines in different levels may be illustrated. The l.h.s. term expresses vorticity advection by the flow at a fixed synoptic time. In the high troposphere the air generally flows through a disturbance at a speed much faster than the moment of the disturbance. In that case, $|\partial\zeta/\partial t|$ is much smaller than $|\mathbf{V} \cdot \nabla(\zeta + f)|$. To simplify the discussion, we consider a layer (at 300 to 200 mb) where $(\partial\zeta/\partial p) \approx 0$. With these special conditions, Eq. (10.7) becomes

$$\mathbf{V} \cdot \nabla(\zeta + f) \approx -(\zeta + f)\nabla \cdot \mathbf{V} \qquad (10.8)$$

It may then be inferred that $\nabla \cdot \mathbf{V} < 0$ on the west side of an upper trough, where $\mathbf{V} \cdot \nabla(\zeta + f) > 0$; and $\nabla \cdot \mathbf{V} > 0$ on the east side, where $\mathbf{V} \cdot \nabla(\zeta + f) < 0$. The change of absolute vorticity along streamlines is thus closely related to the upper divergence field; hence, the vorticity advection in the upper troposphere may be taken as a good indication of the distribution and intensity of the upper divergence over a cyclonic disturbance.[8]

As indicated in Chapter 6, since the net divergence through the whole depth of the atmosphere must be small compared with the divergence at individual levels, appreciable upper divergence of one sign must generally be compensated by lower-level divergence of the opposite sign. Consequently, the upper field of divergence and convergence is also a measure of the distribution of convergence and divergence in low levels, of the field of vertical motion in the middle troposphere, and ultimately also of the intensity of cyclonic development. On the same grounds, the air trajectories at different levels of a disturbance should be systematically interrelated, since their character is influenced by the divergence.

10.9   Composite Air Flow in Frontal Cyclones

Applying the previous results to frontal cyclones, we get the following general description, valid at the time when a cyclone still has a frontal system not yet destroyed by the occlusion process. This is illustrated in Fig. 10.19; a similar scheme has been discussed by Eliassen and Kleinschmidt (1957).

On the west side, the cold air descends and is subjected to vertical shrinking in lower levels except near the central part of the surface cyclone center where, in association with the westward tilt of the trough aloft, there is divergence aloft accompanied by convergence in low levels within the cold air. In lower levels, the warm air is subjected to prevailing convergence and vertical stretching. In the upper troposphere the cold air west of the trough is generally sinking, and the warm air on the east side rising. In these levels, the vertical motions decrease in magnitude upward; hence, the upper cold air is subjected to general vertical stretching and the warm air to vertical shrinking. An intense trough or cyclonic vortex therefore tends to form in the upper part of the cold air (if it reaches sufficiently high up), and an intense ridge tends to form in the upper warm air on the east side. While flowing over the cold air west of a surface cyclone, the warm air partly subsides with the cold air underneath, gaining cyclonic vorticity through

---

[8] This discussion, although cast in different terms, is closely analogous to that in Section 6.1, since the changes in curvature of the flow treated there are related to variations of relative vorticity; the latitudinal variation of the coriolis parameter is also considered in both treatments.

vertical stretching, but acquiring anticyclonic vorticity after entering the region of general ascending motion and upper divergence farther east.

In Fig. 10.19, the low-level streamlines indicate the predominance of convergence in the central parts of the cyclone and of divergence in the outskirts, whereas the changes of curvature along the upper-level stream-lines reflect the influence of convergence over the region of the cold out-break to the west and of divergence over the central parts of the cyclone. All synoptic analyses of polar-front disturbances give essentially the same characteristic streamline patterns, although the details vary in individual cases.

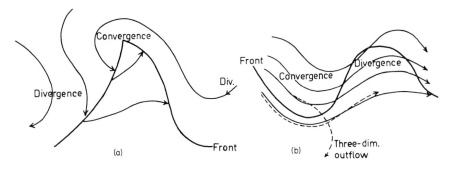

FIG. 10.19   The general character (a) of the surface streamlines, and (b) of the upper-tropospheric horizontal streamlines, in the vicinity of a frontal wave. In (b), the dashed lines represent three-dimensional trajectories.

The characteristic three-dimensional flow in connection with these pro-cesses is illustrated in the schematic Fig. 10.20, which embodies trajectories similar to those in Figs. 10.5c and 10.17. In discussing this figure, we use typical numerical values, which correspond approximately to the cyclone discussed in Section 16.3. In that case the maximum vertical velocity in the warm air at 600 mb is about $-8 \, \mu b/sec$, or 10 cm/sec, near the cyclone center (Figs. 16.2 and 16.4). With a wind speed of 40 m/sec, the trajectory at this level has an upward slope of 2.5 km in 1000-km horizontal distance, corresponding to about 6 hr travel. Thus we can visualize an air parcel moving along a three-dimensional path like that on the right side of Fig. 10.20, rising from lower levels in the warm sector where both the horizontal and vertical components of wind velocity are smaller. Assuming a wind speed of 20 m/sec at the 900-mb level, at the beginning point of the trajec-tory (where ω is assumed small) the average values of $\bar{\omega} = -4 \, \mu b/sec$ and $\bar{V} = 30$ m/sec would imply a rise from 900 mb to 600 mb in about 21 hr,

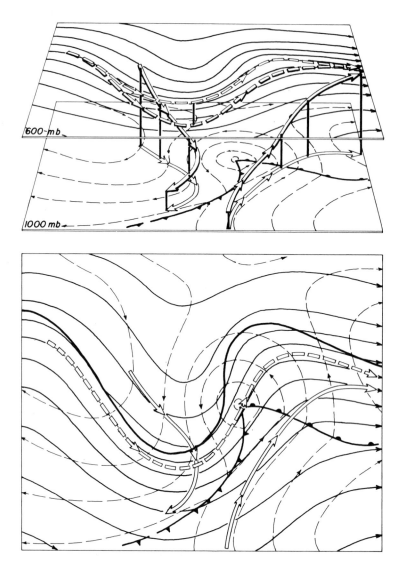

FIG. 10.20   Lower part shows schematic 1000-mb and 600-mb contours relative to a frontal wave; heavy lines are corresponding fronts. Upper part shows a perspective view of the same, with three-dimensional trajectories of air parcels originating in the central part of the cold air, in lower levels in the warm sector, and in the middle-tropospheric warm air near the polar front. Thin-line arrows in top figure, which correspond to those in lower diagram, show projections of three-dimensional trajectories onto the 1000-mb or 600-mb surface. The three portions of each trajectory are for approximately equal time intervals.

during which the horizontal movement would be 2250 km. In the cold air both ω and V at 600 mb in Fig. 16.2 are about half the values above; thus the average slope of the cold-air trajectory is about equivalent, as indicated on the left side of Fig. 10.20.

If, however, we consider an air parcel in the *warm* mass at 600 mb on the western side of the upper trough, but relatively close to the front at that level, this would move around the contour of the front and ultimately arrive at the eastern side of the cold trough. During this movement it would have descended somewhat on the west side of the trough and ascended on the east side. However, the slope of the trajectory would be small, owing partly to the large horizontal speed near the jet stream and partly to the vertical velocity in this region being weaker than it is in the cold and warm air at some distance from the front (cf. Figs. 16.2a and 16.2c). From this we can conclude that the warm air over and east of the cyclone in the middle troposphere partly originates from the strong meandering current entering this region from the west, but partly also from air rising from low levels in the warm sector, as on the right side of Fig. 10.20. Such an intermixing was discussed in Section 4.1 as part of the process of maintaining the properties of "middle-latitude air" in the upper troposphere.

If we assume that the warm air reaches saturation at the 900-mb level with a temperature of 13°C (corresponding to marine tropical air in the eastern United States in winter), it would reach the 500-mb level with a temperature of −15°C, or about the same temperature that characterizes the warm middle-latitude air at this level (e.g., Fig. 16.2a). Ascent to such a high level could not take place without the release of latent heat. Dry air would have cooled adiabatically to about −32°C at 500 mb and would therefore be almost as cold as genuine polar air at the same latitude. This shows that *the strong ascending motion regularly observed over deepening polar-front cyclones would not be possible without the additional heat released through condensation of water vapor.*

In both warm and cold air, the amplitudes and anticyclonic curvatures of the trajectories in Fig. 10.20 are more pronounced than would be suggested by the upper wave pattern. Anticyclonic curvature tends to dominate both in the lower-level portions of trajectories in cold air (Fig. 10.17) and the upper-level portions of those in warm air (Fig. 10.5c), both of which are located in regions of horizontal divergence and vertical shrinking. Also, as noted by Danielsen (1961), a preference for three-dimensional trajectories to be anticyclonically curved (compared with the horizontal streamlines) is a natural accompaniment of the fact that in regions of cold advection, the geostrophic wind backs with height, whereas in regions of warm advection the wind veers with height. Thus, to the extent that cold advection is identified with sinking motion and warm advection with ascent, vertical movement in both cases carries the air toward levels where the geostrophic wind is oriented clockwise from that at the level of origin.

Figure 10.20 and the preceding examples indicate that *descent takes place predominantly in the cold air and ascent mainly in the warm air*. This arrangement is an essential feature for the vertical eddy flux of heat and at the same time for the generation of kinetic energy in the atmosphere (Sections 2.6 and 16.1). Also, relative to the meandering jet stream and polar front, the cumulative effect of synoptic disturbances is evidently an average vertical circulation in a direct solenoidal sense (Section 16.3).

To satisfy continuity, the horizontal branches must also be in a direct sense, when averaged around the hemisphere. That is, when the motions are referred to a coordinate system centered on the polar-front zone, there must be a net flow in a "poleward" direction in the upper troposphere and in an "equatorward" direction in the lower troposphere. This is to be expected on the grounds discussed in Section 9.7. It was noted that, owing to the general increase of geostrophic wind with height, rising air tends to flow toward the left and sinking air toward the right of the contours. The generation of such transverse motions is especially favored in the precipitation regions of cyclones. When water is precipitated out of rising air, heating of the air by condensation is an irreversible process. Correspondingly, *the transverse movement resulting from upward motion in precipitation regions is also irreversible.*

As discussed above, the connected ascending branch depends essentially upon the release of latent heat, which takes place to a significant degree only in the organized precipitation regions of synoptic disturbances and preferentially in the warm-air mass. These precipitation regions at a given time occupy only a relatively small portion of the total area so that most of the middle-latitude air mass in upper levels must be viewed as meandering around the hemisphere, more or less in accord with the waves in the contour field. As pointed out in Section 9.7, however, the amount of heat released in precipitation regions and mixed into this middle-latitude air is very significant and contributes (being released mainly on the anticyclonic shear side of the jet stream) to maintaining the warmth of the middle-latitude air mass and the meridional temperature contrast across the polar-front zone.

REFERENCES

Berggren, R., Bolin, B., and Rossby, C.-G. (1949). An aerological study of zonal motion, its perturbation and breakdown. *Tellus* 1, No. 2, 14–37.

Bleck, R. (1967). Numerical methods for computing moist isentropic trajectories. Final Rept., Contr. No. AF 19(628)-4762, pp. 35–96. Dept. Meteorol., Pennsylvania State Univ. (AFCRL 67-0617).

Charney, J. G., and Phillips, N. A. (1953). Numerical integration of the quasi-geostrophic equations for barotropic and simple baroclinic flows. *J. Meteorol.* 10, 71–99.

Crocker, A. M., Godson, W. L., and Penner, C. M. (1947). Frontal contour charts. *J. Meteorol.* 4, 95–99.

Danielsen, E. F. (1961). Trajectories: Isobaric, isentropic and actual. *J. Meteorol.* 18, 479–486.

Danielsen, E. F. (1966). Research in four-dimensional diagnosis of cyclonic storm cloud systems. Sci. Rept. No. 2, Contr. No. AF 19(628)-4762. Dept. Meteorol., Pennsylvania State Univ.

Eliassen, A., and Kleinschmidt, E. (1957). Dynamic meteorology. *In* "Handbuch der Physik" (S. Flügge, ed.), Vol. 48, pp. 135–136. Springer, Berlin.

Hsieh, Y.-P. (1949). An investigation of a selected cold vortex over North America. *J. Meteorol.* 6, 401–410.

Hsieh, Y.-P. (1950). On the formation of shear lines in the upper atmosphere. *J. Meteorol.* 7, 382–387.

Kerr, I. S. (1953). Some features of upper level depressions. *Tech. Note. Meteorol. Serv. New Zealand* No. 106.

Newton, C. W., Phillips, N. A., Carson, J. E., and Bradbury, D. L. (1951). Structure of shear lines near the tropopause in summer. *Tellus* 3, 154–171.

Omoto, Y. (1966). On the structure of an intense upper cyclone (Part. I). *J. Meteorol. Soc. Japan* 44, 320–340.

Palmén, E. (1949). On the origin and structure of high-level cyclones south of the maximum westerlies. *Tellus* 1, No. 1, 22–31.

Palmén, E. (1951). The aerology of extratropical disturbances. *In* "Compendium of Meteorology" (T. F. Malone, ed.), pp. 599–620. Am. Meteorol. Soc., Boston, Massachusetts.

Palmén, E., and Nagler, K. M. (1949). The formation and structure of a large-scale disturbance in the westerlies. *J. Meteorol.* 6, 227–242.

Palmén, E., and Newton, C. W. (1951). On the three-dimensional motions in an outbreak of polar air. *J. Meteorol.* 8, 25–39.

Peltonen, T. (1963). A case study of an intense upper cyclone over eastern and northern Europe in November 1959. *Geophysica (Helsinki)* 8, 225–251.

Petterssen, S. (1956). "Weather Analysis and Forecasting," 2nd ed., Vol. 1, Chapter 12. McGraw-Hill, New York.

Phillips, N. A. (1958). Geostrophic errors in predicting the Appalachian storm of November 1950. *Geophysica (Helsinki)* 6, 389–405.

Priestley, C. H. B. (1967). On the importance of variability in the planetary boundary layer. Rept. *Study Conf. Global Atmospheric Res. Programme, Stockholm,* 1967 Appendix VI, 5 pp. ICSU/IUGG Comm. Atmos. Sci.

Raethjen, P. (1949). Zyklogenetische Probleme. *Arch. Meteorol., Geophys. Bioklimatol.* A1, 295–346.

Reiter, E. R. (1963). A case study of radioactive fallout. *J. Appl. Meteorol.* **2**, 691–705.

Rex, D. F. (1950). Blocking action in the middle troposphere and its effect upon regional climate. I. *Tellus* **2**, 196–211; II. pp. 275–301.

Rex, D. F. (1951). The effect of blocking action upon European climate. *Tellus* **3**, 100–111.

Rossby, C.-G. (1940). Planetary flow patterns in the atmosphere. *Quart. J. Roy. Meteorol. Soc.* **66**, Suppl. 68–87.

Scherhag, R. (1948). "Neue Methoden der Wetteranalyse und Wetterprognose," pp. 227–235. Springer, Berlin.

Schwerdtfeger, W. (1948). Untersuchungen über den Aufbau von Fronten und Kaltluft-tropfen. *Ber. Deut. Wetterdienstes U. S. Zone* No. 3, 1–35.

Staley, D. O. (1960). Evaluation of potential-vorticity changes near the tropopause and the related vertical motions, vertical advection of vorticity, and transfer of radioactive debris from stratosphere to troposphere. *J. Meteorol.* **17**, 591–620.

van Loon, H. (1956). Blocking action in the Southern Hemisphere. *Notos* **6**, 171–175.

White, R. M. (1950). A mechanism for the vertical transport of angular momentum in the atmosphere. *J. Meteorol.* **7**, 349–350.

11

DEVELOPMENT OF EXTRATROPICAL CYCLONES

In Chapter 10, emphasis was placed upon the role of synoptic disturbances in the meridional and vertical exchange of properties. The general nature of cyclones, in relation to the polar front and upper-level waves, was discussed in Chapters 5 to 7. In the present chapter, our objective is to outline some of the overall processes associated with the development of cyclones and the regular evolutions of their structures. To describe even a single case in detail, with presentation of the vorticity, divergence, vertical motion, and other fields, would require more space than is available. Hence only the gross features will be considered, with a few examples.

11.1   PERTURBATION THEORIES OF INSTABILITY

A very large literature exists concerning the theory of growth of small perturbations. We shall not attempt to interpret these theories, but shall mention only some of the broad aspects. For general discussions, the reader may refer to Eliassen (1956) and Thompson (1961).

In these theories it is generally agreed that atmospheric disturbances derive their kinetic energy from release of the available potential energy inherent in air-mass contrasts. Unlike tropical cyclones, which generate available potential energy as a result of their vertical circulation (Chapter 15), extratropical cyclones depend on a preexisting barocline structure of the basic current. The kinetic energy of the disturbance may be converted either from the kinetic energy of the basic current or from the potential energy, or from both. Instability has been studied from the viewpoint of amplifying wave disturbances on a frontal surface, the character of the horizontal velocity profile, and the concepts discussed in Chapter 6 relating to the vertical wind shear in a baroclinic current.

The general results of baroclinic instability theories indicate that (1) disturbances with a wavelength below certain values will not amplify regardless of the vertical shear; (2) growth of disturbances is most favored at certain intermediate wavelengths[1]; (3) the rate of intensification is, at the wavelength of maximum instability, roughly proportional to the vertical shear; (4) at longer wavelengths, a corresponding rate of growth requires greater vertical shear (baroclinity) with increasing wavelength. Árnason (1963) also finds that deviations from geostrophic flow are favored by weak static stability and large vertical shear, among other factors.

These general results agree with reality in the respect that they predict a most favored wavelength for instability, which broadly conforms with the most common dimensions of synoptic disturbances (see Section 17.4). They also indicate the importance of strong baroclinity, which is a basic feature of developing polar-front cyclones. The questionability of extrapolating the results of small-perturbation theory to finite perturbations is generally acknowledged as a limitation. It appears that significant perturbations are usually, and perhaps always, present in upper levels prior to the development of lower-tropospheric disturbances. Emphasis is placed upon this in the following discussion. However, it should be stressed that the various factors involved in cyclogenesis do not have the same weight in all cases, and the several illustrations that will be given do not exhaust the possibilities.

11.2 Upper-Level Divergence and Initiation of Lower-Level Cyclone Development

In Chapter 6 it was indicated that there is a general level of nondivergence in the middle troposphere, with nearly compensating mass divergences of opposite sign above and below. The development and maintenance of a cyclone, according to this concept, requires the presence over the cyclone of a region of upper-level divergence, which is characteristically found on the east side of a trough in the upper troposphere. The strength of this

---

[1] Between about 2500 and 5000 km, according to different theoretical treatments. According to Árnason (1963), the wavelength below which all waves are stable increases with decreasing latitude; for a given static stability, the corresponding wave number is largest at latitude 45°.

divergence depends mainly on the length and amplitude of the upper waves and on the strength of the upper wind, which in turn depends on the baroclinity in the troposphere.

The formulation by J. Bjerknes (Section 5.6) provided a physical basis for the relationships mentioned above, and made it clear that the behavior of lower-tropospheric disturbances has an intimate connection with the flow patterns of the upper troposphere and lower stratosphere. As stated by Sutcliffe (1939): "Neither the lower nor the upper member of the dynamical system can exist without the other and if a field of divergence (or convergence) can be recognized in the upper atmosphere the associated lower convergence (or divergence) completing the scheme of cyclonic (or anticyclonic) development must occur automatically."

In a series of papers, Sutcliffe formulated the problem of forecasting the development of synoptic systems in terms of the relationship between vorticity change and divergence. He demonstrated that the difference between the upper-level and lower-level vorticity changes, and correspondingly the magnitude of the compensating divergences in both upper and lower levels, could be assessed from the thickness pattern in combination with the wind field. Following an increase in upper-air coverage, it became possible to analyze high-tropospheric charts on a routine basis, and forecasting rules based directly upon the high-level wind (or vorticity) patterns were developed by Riehl *et al.* (1952) and by Petterssen (1955, 1956). The following discussion, in abbreviated form, is essentially founded upon the treatment by Petterssen, who acknowledges its evolution largely from the ideas of Sutcliffe and collaborators.

The requirement for appreciable divergence aloft, to generate a sea-level cyclone, is illustrated by Table 11.1. This gives some values of the "doubling time" (or the time required for the vorticity of a system to increase to twice its initial value) in response to convergence, according to the relationship

$$\frac{d\zeta_a}{dt} = -\zeta_a \nabla \cdot \mathbf{V} \tag{11.1}$$

where ζ_a is the absolute vorticity. According to this table, moderate to strong development, in which a cyclone doubles in intensity in a day or less, requires convergence of 1 to 3×10^{-5} sec^{-1}, or a nearly corresponding divergence in the upper troposphere, considering the requirement for near-compensation of mass in a vertical column.

In waves of moderate amplitude the distribution of geostrophic absolute vorticity in Fig. 11.4a is fairly typical. If we consider air motions near the level of maximum wind, where influences of vertical motions can be neglect-

TABLE 11.1

MAGNITUDE OF DIVERGENCE OR CONVERGENCE ASSOCIATED WITH SYNOPTIC DEVELOPMENT.
Δt CORRESPONDS TO THE TIME REQUIRED TO DOUBLE THE VORTICITY[a]

$\lvert \nabla \cdot \mathbf{V} \rvert$ (sec^{-1})	Δt	Motion system
1.9×10^{-4}	1 hr	Subsynoptic
3.2×10^{-5}	6 hr	Intense synoptic
0.8×10^{-5}	1 day	Medium synoptic
0.4×10^{-5}	2 days	Feeble synoptic
1.1×10^{-6}	1 week	Planetary waves

[a] After Petterssen (1956).

ed, the vorticity changes are given by the simplified Eq. (11.1), from which the divergence is

$$\nabla \cdot \mathbf{V} = -\frac{1}{\zeta_a}\left(\frac{\partial \zeta_a}{\partial t} + \mathbf{V} \cdot \nabla \zeta_a\right) \tag{11.2}$$

The local change can be expressed in terms of movement of the vorticity isopleths; thus, if at a given point such an isopleth moves with velocity $\mathbf{C}_\zeta$, Eq. (11.2) may be written as

$$\nabla \cdot \mathbf{V} = -\frac{1}{\zeta_a}(\mathbf{V} - \mathbf{C}_\zeta) \cdot \nabla \zeta_a \tag{11.3}$$

For convenience, $\mathbf{C}_\zeta$ may be taken along the wind direction so that if s denotes distance along a streamline,

$$\nabla \cdot \mathbf{V} = -\frac{1}{\zeta_a}(V - C_s)\frac{\partial \zeta_a}{\partial s} \tag{11.4}$$

Upper-level troughs and ridges, with their associated vorticity maxima and minima, move slowly in comparison with the high-tropospheric winds in the main part of the current. Hence, in this region, $C_s \ll V$, and the sign and approximate magnitude of the divergence may be inferred from the vorticity advection, $-V\,\partial \zeta_a/\partial s$. Since $(V - C_s)$ is greatest near the jet stream, where also the gradients of vorticity are typically large, divergence and convergence in upper levels attain greatest magnitude in the general vicinity of the jet stream. *Lower-tropospheric cyclones and anticyclones of any significant intensity, being also typified by regions of appreciable convergence and divergence, are for this reason characteristically found in close association with the baroclinic region of the jet stream* (Sutcliffe, 1939; Palmén, 1951).

At the level of nondivergence, the vorticity pattern according to Eq. (11.4) moves with the horizontal winds,[2] while in general $V > C_s$ above that level and $V < C_s$ below it. Hence, the nearly compensating divergences of opposite sign in the upper and lower troposphere are connected, through Eq. (11.4), with the vertical wind shear. The relation between the present discussion and that outlined in different terms in Section 6.2 is evident.

From an examination of a large number of cyclone developments over and near North America, Petterssen *et al.* (1955) found that the vast majority occurred in connection with the overtaking of a slow-moving cold front or stationary front at the surface by the region of "positive vorticity advection" in advance of an upper trough, in the manner illustrated by Fig. 11.1. The

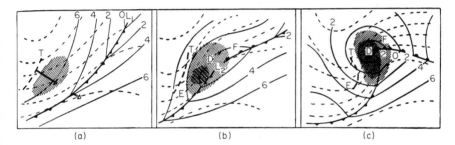

| (a) | (b) | (c) |

FIG. 11.1 Stages in the development of a cyclone. As the upper cold trough advances and the area of appreciable vorticity advection aloft (hatched area) spreads over the frontal zone, the imbalance created results in convergence at low levels. (After Petterssen, 1956.)

rapidity of cyclone development was well correlated with the intensity of vorticity advection or with the combined strength of the upper-tropospheric wind and the intensity of the upper trough. This sequence led Petterssen (1955) to formulate the rule that "cyclone development will occur when and where a high-level (e.g., 300 mb) positive-vorticity advection region becomes superposed over a low-level baroclinic (frontal) zone along which the thermal advection is discontinuous." The reason for most rapid cyclonic development when the upper divergence is superposed over the frontal trough, and not before, is obviously in part related to the circumstance that the increase of vorticity due to convergence is directly proportional to the

[2] This is approximately true, but not strictly so because at that level the vertical motions are strongest, and their influence is neglected here.

vorticity itself or to the preexisting circulation that is always cyclonic at any surface front.[3]

Relative movements between upper- and lower-tropospheric systems (which are essential to the process described above) must, of course, be linked in a consistent way to the changes of thermal structure in the intervening layer. If h is the thickness of an isobaric layer, then (see Sutcliffe and Forsdyke, 1950)

$$\frac{\partial h}{\partial t} = R \int_{p_1}^{p_2} \left\{ -\mathbf{V} \cdot \nabla T + \omega(\Gamma_a - \Gamma) + \frac{1}{c_p} \frac{dq}{dt} \right\} d \ln p \qquad (11.5)$$

where Γ_a is the adiabatic lapse rate (dT/dp), moist or dry depending upon whether the air is saturated; Γ is the actual lapse rate $(\partial T/\partial p)$; and dq/dt is the heat-energy change per unit mass caused by processes other than condensation. If ζ_{go} is the geostrophic relative vorticity at a lower isobaric surface (say, 1000 mb, or near the ground) and ζ_{gu} is that at a upper-level isobaric surface, changes of the two are linked by the relationship

$$\frac{\partial \zeta_{go}}{\partial t} = \frac{\partial \zeta_{gu}}{\partial t} - \frac{g}{f} \nabla^2 \frac{\partial h}{\partial t} \qquad (11.6)$$

Based on Eqs. (11.5) and (11.6), the development of lower-tropospheric circulation systems may be viewed in different ways that are interrelated. Since these are elaborated fully by Petterssen (1956), only a few broad aspects will be outlined here. We shall confine the discussion essentially to the vicinity of an incipient or developing cyclone where, since the low-level vorticity advection is generally small, the l.h.s. term of Eq. (11.6) represents essentially an intensification associated with convergence.

If level u in Eq. (11.6) is considered to be the level of nondivergence L, then according to Eq. (11.2), the first r.h.s. term may be approximated by $(-\mathbf{V}_L \cdot \nabla \zeta_{aL})$. Consequently, cyclonic development at level p_0 is fostered by a local increase of vorticity reflecting a similar change aloft, in harmony with the advection of vorticity downstream from an upper trough. This is pronounced in the case of troughs with appreciable amplitude and short wavelength, in which case the variations of relative vorticity dominate the

---

[3] The essential synoptic features embodied in Petterssen's formulation were described earlier by Saucier (1949) in an instructive discussion of cyclone formation in the southern states or Gulf of Mexico east of the Rockies. Saucier observed that cyclogenesis is especially favored in this region when a cold anticyclone covers the continental region down to the coast, and when an upper-level trough (often moving out of the southwestern United States) moves eastward over the region of strong lower-tropospheric baroclinity near the Gulf coast.

variations of coriolis parameter. This influence on the sea-level system, however, may be either enhanced or suppressed, according to Eq. (11.6), by changes of the thermal vorticity (or thickness distribution) in the intervening layer. In the stage of Fig. 11.1a, when the advancing upper trough is well to the rear of a surface front, the region of positive vorticity advection aloft is likely to be superposed over a region of pronounced cold advection, with a distribution of $\partial h/\partial t$ which may oppose development according to the last term of Eq. (11.6).

At the later stage of Fig. 11.1b, when the region of positive vorticity advection has further advanced so that the advective cooling is weaker beneath it than in the environs to the rear, this contribution favors rather than retards development, according to the last term of Eq. (11.6). At the same time, the influence of vertical motions expressed by the middle r.h.s. term of Eq. (11.5) ordinarily acts to suppress development, since the atmosphere is generally stable and upward motion with adiabatic cooling is greatest near the center of cyclonic development. Finally, the latter damping influence is minimized with the onset of condensation and precipitation, with a corresponding release of heat that enhances development, according to the last terms of Eqs. (11.5) and (11.6). The exchange of heat with the earth's surface must, of course, also be taken into account, especially over oceans where this is large. As observed by Petterssen, the influence of this is mainly to retard the movement of a cyclone through heating of the cold air to its rear (Fig. 12.5).

In summary, during the advance of an upper-level trough relative to a surface frontal system, the sequence is as follows: When the upper trough is well to the rear (Fig. 11.1a) the tendency for cyclonic development beneath the region of advective vorticity increase aloft is likely to be opposed both by the distribution of horizontal advective cooling and by the influence of induced vertical motions. When the trough has advanced so that the vorticity increases over the surface front, the distribution of thermal advection becomes favorable, but that of vertical motions is unfavorable, the damping influence of the latter being largely offset by the effects of condensation heating when this sets in.

The sequence described above is, of course, at the same time linked with the field of divergence in the upper troposphere. Above the middle troposphere, the tilt of the wave system is small. Owing to the thermal structure (cold troughs and warm ridges), the extreme values of vorticity are magnified in the upper compared with the middle troposphere, but the general patterns correspond closely at different levels. Considering also that the wind increases generally with height, *a region of positive vorticity advec-*

tion at the level of nondivergence, related in the discussion above to sea-level vorticity development, also corresponds closely to a region of upper-tropospheric divergence or ascending motion in the troposphere.

A characteristic example of a "wave cyclone" over the North Atlantic and northwestern Europe is illustrated by Fig. 11.2, analyzed by L. Oredsson. Although at the earliest time shown here a deep cyclone was already present, the general configuration of the flow patterns is quite typical of the earlier stages of wave-cyclone formation. Such cyclones form on slow-moving cold fronts trailing behind older well-developed cyclones, upon superposition of the upper-divergence region in advance of a minor trough, steered by the larger-scale circulation, which enters the preexisting major trough on its west side.

An example of an incipient wave of this sort is seen southwest of Iceland in the lower right-hand chart, the upper trough having approached from northwest. In the middle chart, it is seen that the upper divergence region ahead of this trough was superposed over a region of cyclonically curved flow, owing to the presence of the cyclone farther east. In such a case, some cyclonic development can take place in the cold air even before the strong upper divergence region spreads over the surface front, at which time more rapid development is likely to ensue. Although this sea-level wave was feeble at the last time shown, there was already a strong pattern of thermal advection (indicated by the intersections of sea-level isobars and 500-mb contours). This favors amplification of the thermal wave and operation of the "self-development" process discussed below. At the same time, according to the last term of Eq. (11.6), rapid movement of the cyclone is favored by the large differences of cold and warm advection, which is identified with the thickness gradient and thus with the strength of the winds aloft, as indicated in the "thermal steering" principle discussed by Sutcliffe (1947).

In this example the cross sections before, during, and after passage of the cyclone suggest that the polar front was well developed on both east and west sides. It has not been established to what extent such a frontal symmetry characterizes oceanic cyclones in their earlier stages of development. Over the vicinity of North America, a well-developed polar front is often confined to the lower troposphere in the immediate locale of cyclogenesis, although a frontal layer extending through the whole troposphere may be present upstream from the cyclone.

A physical interpretation of the latter structure is given in Section 11.6. It seems likely that the process discussed, which leads to the development of a strong upper trough prior to significant sea-level cyclogenesis, is more

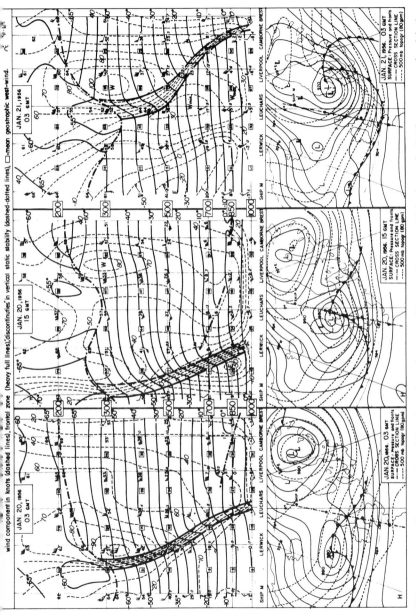

FIG. 11.2   A wave cyclone over the Atlantic and northwestern Europe, shown at 12-hr intervals. In the maps, sea-level isobars (solid) at 5-mb interval and 500-mb contours (dashed) at 80-m interval. Cross sections along indicated lines show frontal boundaries and tropopause, isotherms (solid) at 5°C intervals, and west-wind component (dashed) at 10-knot intervals. (Analysis by L. Oredsson, University of Stockholm; reproduced by his permission.)

important in some areas than in others. This is suggested by a comparison of cyclone developments over North America and over the north Atlantic Ocean, made by Petterssen *et al.* (1962). These authors observed that rapid cyclogenesis over North America commonly followed the overtaking of a surface front by a strong preexisting trough in upper levels. By contrast, they found that the corresponding troughs leading to maritime cyclogenesis were comparatively weak, with initially weaker upper-level divergence, and that the thermal advection processes incorporated in the last term of Eq. (11.6) played a more dominant role in the early cyclone development.

11.3   Development of Thermal Field, and Limiting Processes

It is typical that when a cyclone achieves a significant circulation, the upper-level contours acquire a deformation, tuned to the scale of the cyclone, of the sort indicated in Fig. 11.1c. This change of configuration, evident in the growing amplitude of the upper-level contours in Fig. 11.2, evidently represents a progressive increase of the vorticity advection, which favors intensified development of the surface cyclone. The process has been discussed by Sutcliffe and Forsdyke (1950), who called it "self-development." Self-development is a natural consequence of the initiation of a cyclone, but once it has begun, other processes are immediately initiated and these eventually offset the tendency of the cyclone to develop further.

The self-development process may be described in terms of the change in structure of the upper-level flow pattern, as affected by changes in the thermal structure of the troposphere. The local temperature change at a given level may be written as

$$\frac{\partial T}{\partial t} = - \left(\mathbf{V} \cdot \nabla T + \omega \frac{\partial T}{\partial p} \right) + \frac{dT}{dt} \qquad (11.7)$$

Replacing dT/dt by $(dT/dp)(dp/dt)$ and taking the average between the 1000-mb surface and an isobaric surface (say, 300 mb) in the upper troposphere,

$$\frac{\partial \bar{T}}{\partial t} = - \overline{\mathbf{V} \cdot \nabla T} + \overline{\omega(\Gamma_a - \Gamma)} \qquad (11.8)$$

To the extent that the thermal wind approximation is valid, the wind velocity at any upper level H may be expressed as

$$\mathbf{V}_H = \mathbf{V}_0 + \mathbf{V}_T$$

where $\mathbf{V}_0$ is the 1000-mb wind that may be approximated by the geostrophic

wind. Sutcliffe observed that since the thermal wind $\mathbf{V}_T$ parallels the isotherms, it does not contribute to the thermal advection.[4] Consequently,
Eq. (11.8) may be written as

$$\frac{\partial \bar{T}}{\partial t} \approx - \underbrace{\mathbf{V}_0 \cdot \nabla \bar{T}}_{\text{A}} + \underbrace{\overline{\omega(\Gamma_a - \Gamma)}}_{\text{S}} \tag{11.9}$$

which is equivalent to Eq. (11.5) if heat sources and sinks other than condensation are neglected.

In broad terms, the "self-development" process may be visualized as a
progressive deformation of the upper current in the vicinity of a cyclone,
in the manner shown schematically in Fig. 11.3. If we consider two locations
upstream and downstream from the cyclone center, on the same 1000-mb
contour, the height of an upper-level isobaric contour can be identified
with the thickness. Deformation of the upper-level contour field can correspondingly be identified with a deformation of the mean isotherms through
the layer considered. When a cyclonic circulation has been initiated, the
effect of horizontal temperature advection, term A in Eq. (11.9), is obviously
to contribute to a deformation of the isotherms and of the upper-level flow
pattern, in the manner shown. *The increased amplitude of the upper wave
corresponds to an enhanced variation of vorticity (as expressed by the curvature) along the current. This in turn implies an increase of upper-level
divergence, which favors progressively stronger low-level convergence and
intensification of the cyclone.*

If only the influence of horizontal advection were important, this process
would imply an unbounded deformation of the thermal field and of the
upper flow. It is observed, however, that after a certain amplification has
been achieved, the process comes to a stop and the thickness field of a
cyclone is not further distorted.

This is due to a damping influence of term S in Eq. (11.9). Since Γ_a is
ordinarily larger than Γ, the effect of stability is generally to contribute
to local cooling in regions of ascent and to warming in regions of descent
(an exception being in areas of convective precipitation, which are typically
small compared with the size of a cyclone). According to the discussion in
Section 6.2, the divergence in an upper wave is (for a given wavelength)
proportional to its amplitude. Thus, the vertical motion ω becomes larger
as the wave amplifies, and correspondingly the damping influence of static
stability increases. Eventually, the influence of horizontal advection is bal-

---

[4] When a layer mean is taken, this statement is strictly true only when the hodograph
is linear throughout the depth of the layer.

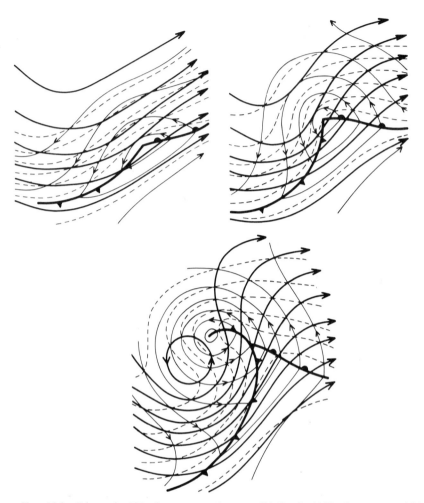

Fig. 11.3   Schematic 500-mb contours (heavy solid lines), 1000-mb contours (thin lines), and 1000-500 mb thickness (dashed), illustrating the "self-development" process during growth of a cyclone (see text).

anced by the opposing influence of vertical motion. In this way it can be seen that *"self-development" also calls into play a "self-limiting" process that prevents the development from progressing without limit.* Since $(\Gamma_a - \Gamma)$ is much smaller for moist-adiabatic ascent than for dry ascent, the damping term S in Eq. (11.9) is correspondingly smaller for a given vertical motion when there is extensive condensation in a cyclone. Thus, from the preceding viewpoint, "self-development" can proceed further and the cyclone can achieve greater intensity than it could if precipitation were absent.

11.4 AN EXAMPLE OF CYCLONE EVOLUTION

The charts in Figs. 11.4 through 11.6 illustrate a cyclone in its early, mature, and dissipative stages. In this particular example the initial cyclogenesis took place in the lee of the Rocky Mountains, by processes outlined in Section 11.8. At the time shown in Fig. 11.4 a cold front, originating from a cyclone that had perished on entering the Pacific coast, had entered the lower-level cyclonic circulation east of the mountains. As seen from the pattern of vorticity advection in Fig. 11.4a, upper-tropospheric divergence was present over the cyclone center.

Two days later (Fig. 11.5), the upper-level trough had amplified appreciably and the cyclone was fully developed. Note that the considerable development of the upper-level trough goes hand in hand with a corresponding amplification of the mean isotherms, in accord with the scheme discussed in the preceding section. At this time there was still an appreciable phase lag between the 1000-mb contour field and the mean isotherm field, with a corresponding westward displacement of the upper-level trough. This phase lag diminished during the period between Figs. 11.4 and 11.5, during which time the advance of the strongest upper-level divergence region relative to the surface cyclone, together with intensification at the upper level,[5] increasingly favored low-level development. Continuance of the same trend of relative movement, however, eventually brought the cold air and upper trough into phase with the low-level cyclone center (Fig. 11.6). Being no longer supported by upper divergence over the center, this cyclone filled during the next 24 hr. A new cyclone formed farther east on the warm front over which the upper divergence region ahead of the trough became superposed.

The rapid eastward movement of the upper-level trough, causing it to overtake the surface cyclone, was undoubtedly influenced by the diminishing wavelength between it and the next trough upstream. At the same time, a striking change in the thermal pattern was observed in the form of a marked decrease in amplitude of the mean isotherms west of the cyclone. This was essentially associated with a collapse of the cold dome west of the cyclone. The influence of this rapid subsidence upon heating the cold air is seen by comparing the thickness lines at successive times. For example, the 18,200-ft

[5] This intensification is somewhat exaggerated by portrayal in terms of geostrophic vorticity. In an example comparing the geostrophic and actual vorticity fields, given by Newton and Palmén (1963), striking differences are observed. Experience has nevertheless shown that indications from the geostrophic vorticity field, or that computed using the balance equation, are qualitatively useful.

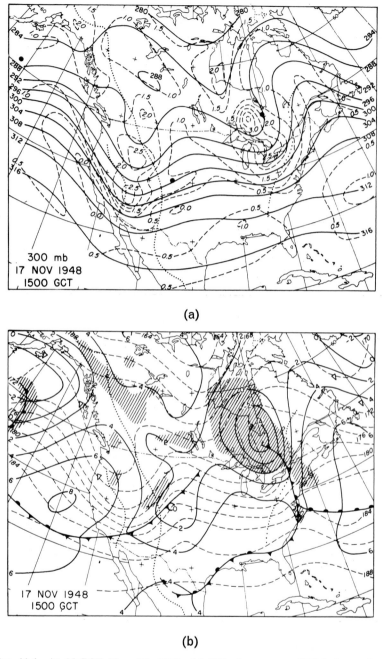

(a)

(b)

FIG. 11.4 At 15 GCT Nov. 17, 1948: (a) 300-mb contours (400-ft interval) and geostrophic absolute vorticity (10^{-4} sec^{-1}); (b) surface fronts and 1000-mb contours (200-ft interval), with dashed 500-1000 mb thickness isopleths (200-ft interval). In (a), heavy dots show locations of cyclone centers; in (b), precipitation areas are shaded. Dotted lines are ridges of principal mountain ranges. (From Newton, 1956.)

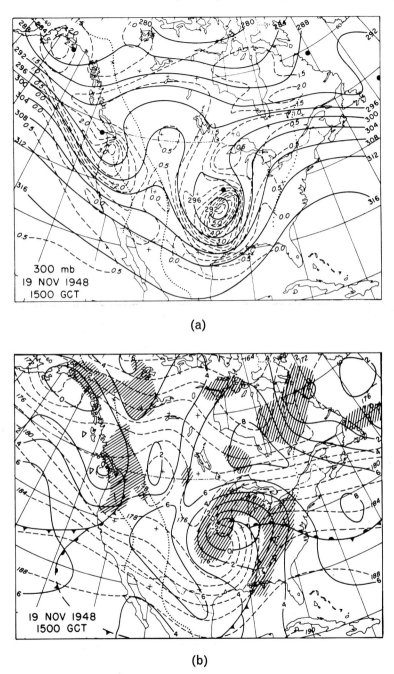

(a)

(b)

FIG. 11.5   Charts at 300 and 1000 mb at 15 GCT Nov. 19, 1948. Legend as in Fig. 11.4.

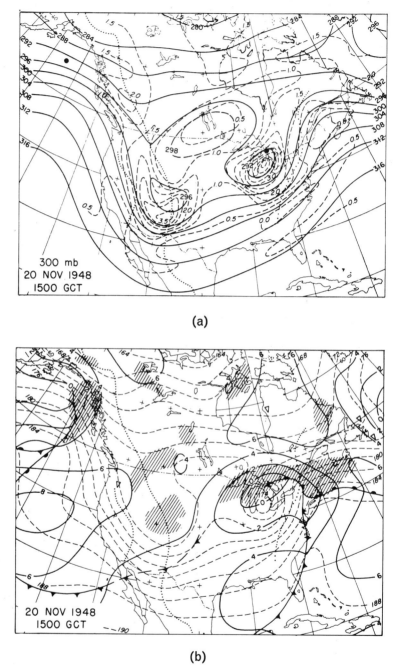

(a)

(b)

FIG. 11.6   Charts at 300 and 1000 mb at 15 GCT Nov. 20, 1948. Legend as in Fig. 11.4.

thickness isopleth, located as far south as the Gulf Coast in Fig. 11.5b, is considerably farther north in Fig. 11.6b despite the southward advection indicated by the 1000-mb contours during the intervening time.

Maximum cyclone intensity was achieved near the time of Fig. 11.5. At that time, as shown by the frontal contours in Fig. 10.5d, the top of the polar-air dome southwest of the cyclone still extended above the 500-mb level. By the time of Fig. 11.6, the top of the polar-air dome had descended well below that level, showing that intense sinking in the cold air accounted for the retreat of the mean isotherms noted above. The equatorward extrusion of cold tongues from the main polar-air source in the manner illustrated by this example, in which the cold air maintains its depth during the earlier stages of cyclone formation and later collapses, is typical. This behavior is partly connected with the changes of wind structure discussed in Sections 11.6 and 11.7.

11.5 Cyclone Development with Change in Planetary Waves

In Section 6.4 it was mentioned that major cyclone developments are commonly associated with changes of the long-wave pattern, and in particular that this occurs with the establishment of a new major wave in the upper-tropospheric westerlies. An event of this kind is illustrated in Figs. 11.7 to 11.11.

For many days previously, the hemispheric flow pattern had been dominated by four strongly developed long waves that remained essentially fixed, as shown by Fig. 6.9 for an earlier time in this same situation. The first intimation of a significant change was evident on Jan. 11, 1959 (Fig. 11.7), when a cold minor trough (identified by letter A) appeared west of the major trough in the eastern Pacific. As shown by Fig. 11.8, two days later, this minor trough developed strongly, establishing a new major trough west of the old major trough position. This event, a discontinuous retrogression, was followed by similar retrogressions downstream, and eventually by an increase of the hemispheric wave number from 4 to 5.

At the time of Fig. 11.8, the old major trough is evident as a minor trough (B) near the west coast. At sea level, an anticyclone has built up over the eastern Pacific, a region earlier occupied persistently by cyclones. Comparing the 1000- and 500-mb contours, it is seen that there is strong warm advection between this anticyclone and the rapidly developing cyclone on its west side, and that further intensification of the sea-level high (current indications of which were given by the pressure tendencies) would lead to increased cold advection in the vicinity of the minor trough on its east side.

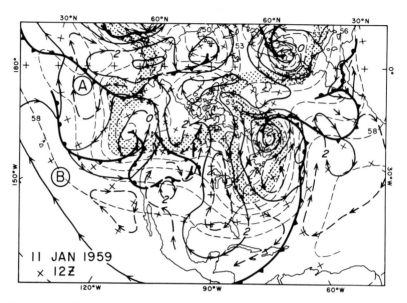

FIG. 11.7　Surface fronts and 1000-mb (solid) and 500-mb (dashed) contours, 12 GCT Jan. 11, 1959. Contours are at 100-m intervals; portions of intermediate 50-m contours at 1000 mb shown by long dashes with arrowheads. Stippled where 500-mb height is less than 5400 m. (Analysis taken from U.S. Weather Bureau, 1959.)

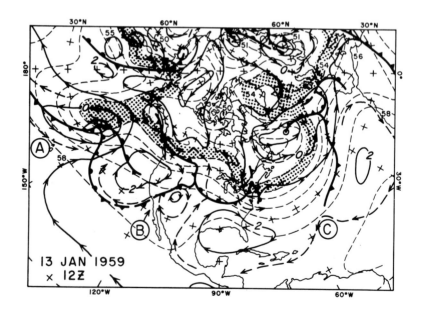

FIG. 11.8　Same as Fig. 11.7, at 12 GCT Jan. 13.

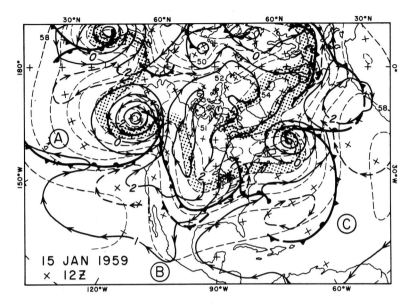

Fig. 11.9   Same as Fig. 11.7, at 12 GCT Jan. 15.

These features, combined with the presence of very cold air to the north of the minor trough (500-mb temperatures as low as −37°C in southern British Columbia) and especially with the circumstance that the wavelength between existing major troughs *A* and *C* was excessively long, all favored development of minor trough *B* into a new major trough, in the manner described by Cressman (1948). As shown by Fig. 11.9, two days later, a very marked intensification of both the upstream ridge and this trough had taken place.

At the time of Fig. 11.9, the upper-level trough was already quite sharp, with pronounced vorticity advection indicating upper-level divergence on its downstream side. This was, however, confined to the near-vicinity of the trough and did not extend significantly over the feeble cyclone to the east. Rather, the strong upper-level divergence was located over the region where cold advection was most pronounced in low levels. This is a circumstance that, as indicated in Section 11.2 (see Fig. 11.1a), is not yet favorable for low-level development.

Figure 11.10 shows the situation a day later. During the intervening time, the upper-level trough had moved significantly faster than the surface system, and the cyclone had intensified somewhat. At this time, the strongest vorticity advection aloft was situated directly over the cold front, at the location marked *CG*. At this location the divergence at 300 mb, estimated from

Eq. (11.2), was 6×10^{-5} sec^{-1}. This corresponds to "intense synoptic development," according to Table 11.1. Surface-pressure falls were most rapid at this location, southwest of the existing cyclone center; a new cyclone developed there and deepened 30 mb in the next 24 hr while moving north northeast (Fig. 11.11).

The sequence of adjustments in the upper-wave pattern, beginning with cyclogenesis in the central Pacific and eventuating in strong cyclogenesis over eastern North America, was quite similar to that described in another case by Hughes *et al.* (1955). During the development of an upper-level trough in this manner, a substantial increase of vorticity takes place within it (compare trough *B* in Figs. 11.8 and 11.9) through a mechanism discussed in the next section.

The significance of such an upper-level increase of vorticity for cyclogenesis is evident, since it represents an enhancement of vorticity advection and correspondingly of upper divergence downstream from the trough. The pattern that will be described corresponds to the "diffluent upper trough," whose importance for the deepening of cyclones was recognized by Scherhag (1937). The emphasis placed upon this upper-air configuration by Scherhag, and later by other authors, has been amply justified by synoptic experience. Polster (1960) examined a large number of cyclone developments and found that although initial wave formation took place in about

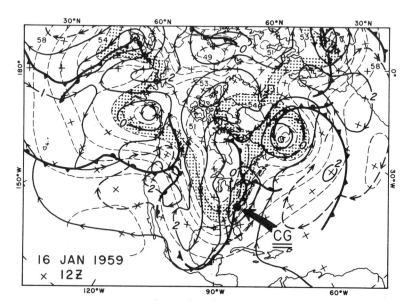

FIG. 11.10   Same as Fig. 11.7, at 12 GCT Jan. 16.

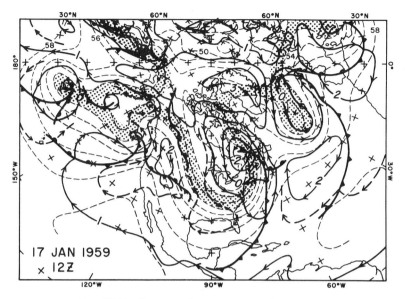

FIG. 11.11   Same as Fig. 11.7, at 12 GCT Jan. 17.

one-fifth of the cases under confluent flow aloft, deepening under such flow was rare. By contrast, 73% of the cases of deepening cyclones were found under diffluent contours, predominantly ahead of upper troughs, while most of the remainder were beneath "parallel upper flow."

11.6   ASYMMETRICAL STRUCTURE AND ITS IMPLICATIONS

Forecasters have long observed the rule that when an intense ridge builds up on the upstream side of a trough, the trough is likely to deepen, especially if the equatorward upper current on its upwind side is much stronger than the poleward current on its downwind side. These two features tend to be interrelated, as stressed by Bjerknes (1951).

The rapid development of trough B in this example (Figs. 11.8 and 11.9) followed the scheme described by Bjerknes. He pointed out that the maximum possible anticyclonic curvature that can be achieved by an air parcel in gradient flow is given by $r_{min} = 2V_g/f$ in the case where $V = 2V_g$. If $V > 2V_g$, the sum of centrifugal and pressure-gradient forces is greater than the coriolis force, and the curvature of the flow will be smaller than the curvature of the contours.[6] In that case, an air particle downstream from

[6] The criterion given here is for a stationary contour field; more general criteria, including moving fields, are given by Bjerknes (1954).

a ridge cannot follow the contours and must deviate toward lower geopotential. "This must imply a forward acceleration of the particle leading up to maximum speed at the end of the anticyclonic sweep. If such fast-moving air is fed directly into a pressure trough downstream, in which the pressure gradients were adapted to smaller wind speeds, an intensification of the trough should follow", since the fast-moving air entering the trough must deviate across the contours to a lower latitude.

The criterion given for a "dynamically unstable" ridge is commonly achieved at jet-stream level. The 250-mb chart corresponding to the time of Fig. 11.9 is given in Fig. 8.19. From the latter figure, the maximum wind in the upstream crest was 60 m/sec, compared with a geostrophic wind speed of about 27 m/sec. Thus $V > 2V_g$; the isotach field in Fig. 8.19 corresponds to the condition described by Bjerknes, with strong northwest winds on the upstream side of the trough, which was rapidly increasing in amplitude at the time.

Portions of the 500-mb charts at this time and 36 hr later, when the trough had achieved full amplitude, are shown in Fig. 11.12. At the earlier

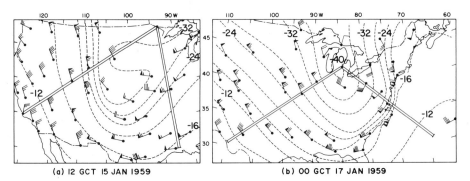

FIG. 11.12   Portions of 500-mb charts at (a) 12 GCT Jan. 15; (b) 00 GCT Jan. 17, showing winds and isotherms at 4°C interval.

time, the asymmetry noted above for the wind field and a corresponding asymmetry in the thermal gradient was evident at this level. At the later time, however, the character of the asymmetry had essentially reversed, with strongest winds and temperature gradient on the east side. This change in structure with time is very characteristic; other examples are given by Newton (1958).

Vertical sections across the trough at these two times are shown in Fig. 11.13. At the earlier time when the trough was still amplifying, a well-developed frontal layer was present throughout the troposphere on the

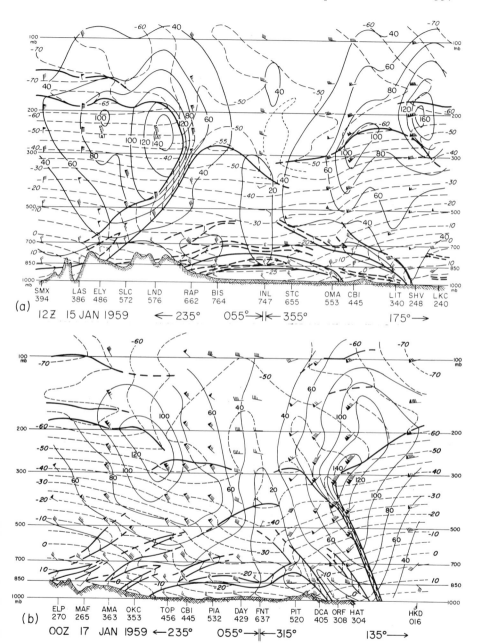

FIG. 11.13 Vertical sections (a) along lines in Fig. 11.12a; (b) along lines in Fig. 11.12b. In each case the left-hand portion is approximately normal to upper-tropospheric flow on upwind side of trough; right-hand portion, to that on downwind side. Isotachs at 20-knot interval; isotherms (dashed) at 5°C interval.

west side. On the east side, however, a distinct polar front was evident only in lower levels, with a diffuse baroclinic field in the upper troposphere. During the evolution of an upper trough, the upper-tropospheric frontal layer advances downstream at a rate faster than the movement of the trough, so that around the time when the trough is most fully developed, a more-or-less symmetrical structure is characteristic, as in Fig. 7.9. Later on, the strong upper-tropospheric front is typically found on the east side of the trough, as in Fig. 11.13b, with a diffuse thermal structure on the west side.

Strong cyclogenesis in this case began just prior to the time of Fig. 11.10 in a locale where vertical sections still showed a distinct polar front only in the lower troposphere. The onset of cyclogenesis thus occurred not where a deep-tropospheric front was present, but rather in advance of the "exit" region of the high-tropospheric frontal layer. This was also true in the examples of November, 1948 (Section 11.4), and April, 1950 (Section 10.5; see also Figs. 9.8 and 9.9). In both cases the upper-tropospheric frontal layer eventually moved forward and combined with lower-tropospheric fronts on the east side of the trough, but only after the cyclones had attained considerable development.

Thus, *a frontal layer extending through the entire troposphere is, at least in some cases, a characteristic acquired by a cyclone during, rather than prior to, its development.* Nyberg (1945) was apparently the first to recognize that cyclogenesis often takes place in a region where the polar front is relatively shallow. Based on his analyses of disturbances over Europe, he stated: "The existence of distinct fronts also at higher air levels is in many cases undeniable, but there is good reason to believe that their existence is neither a necessary nor a sufficient condition for cyclogenesis or for the formation of cyclonic rain areas."

In the preceding discussion, a commonly observed sequence was described in which the development of a sea-level cyclone is preceded by development of a strong trough in upper levels. The requisite increase of vorticity (and its decrease later on) may be examined in light of the asymmetry of the wind-velocity field connected with the frontal structure. Equation (11.2) may be written as

$$\frac{\partial \zeta_a}{\partial t} = - \zeta_a \nabla \cdot \mathbf{V} - \mathbf{V} \cdot \nabla \zeta_a = - \nabla \cdot (\zeta_a \mathbf{V})$$

or, from Gauss' theorem,

$$\frac{\partial [\zeta_a]}{\partial t} = - \frac{1}{A} \oint_L \zeta_a v_n \, \delta L = - \frac{1}{A} (\underbrace{\hat{\zeta}_a \hat{v}_n L}_{(D)} + \underbrace{\widehat{\zeta_a' v_n' L}}_{(E)}) \quad (11.10)$$

where v_n is the outward velocity component normal to an arbitrarily chosen boundary L; the circumflex accents denote mean values along the boundary, and the primes indicate deviations from the mean. The brackets on the l.h.s. denote an average over the area A included within L.

Since $\hat{v}_n L/A$ is the mean divergence, term D represents the contribution to vorticity change due to net divergence over area A. Term E then corresponds to a change due to the eddy flux of vorticity through the boundary, and depends only upon a correlation between the vorticity and the normal velocity component v_n at the boundary.[7] It was observed in Section 7.4 that the distinguishing feature of a front at the level of maximum wind is

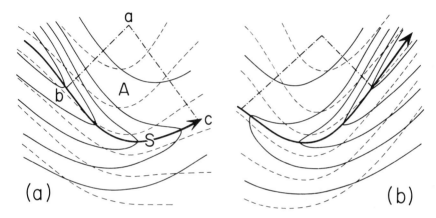

FIG. 11.14   Schematic contours (dashed), jet-stream axis, and isotachs near tropopause level, at (a) early and (b) late stages in evolution of upper trough. (See text.)

a concentration of the velocity gradient. Thus the asymmetry of frontal strength in the cross sections of Fig. 11.13 corresponds to an asymmetry of vorticity, implied by the isotachs in the schematic Fig. 11.14 at that level.

In Fig. 11.14a, the area A is bounded by the jet stream axis and lines ab and ac, which are chosen normal to the flow and some distance upstream and downstream from the trough where the curvature is small; then the relative vorticity is essentially expressed by the shear. In this case a large value of vorticity is, on the west side, associated with the region near the

---

[7] In Eq. (11.10), and the following expressions, $\partial[\zeta_a]/\partial t$ is (by strict interpretation) the change within a fixed area, and not in an area bound to a moving system. In the latter case, v_n would have to be modified to take into account the movements of the boundaries. Since a trough moves slowly compared with the wind near jet-stream level, the refinement introduced by a more complicated treatment is likely to be relatively small.

jet stream where the wind speed (or v_n) is strong, while on the east side this correlation is less marked. Hence, according to term E of Eq. (11.10), the eddy flux contributes to an increase of vorticity in the trough.[8] Similarly, in Fig. 11.14b, the eddy flux contributes to a decrease.

The foregoing discussion may be stated in a different way, which brings out a relationship between trough development and kinetic-energy variations along the current. If boundaries ab and ac are chosen as indicated above, the total wind speed V is equivalent to v_n, and the shear to the relative vorticity. The advection of vorticity across the jet-stream axis is nil. Then, expanding the middle expression of Eq. (11.10), this expression becomes

$$\frac{\partial [\zeta_a]}{\partial t} = -\frac{1}{A} \oint_L f v_n \, \delta L + \frac{1}{A} \int_a^b \frac{\partial V}{\partial L} \cdot V \, \delta L - \frac{1}{A} \int_a^c \frac{\partial V}{\partial L} \cdot V \, \delta L$$

The sum of the last two integrals is $(V^2/2)_b - (V^2/2)_c$, so that

$$\frac{\partial [\zeta_a]}{\partial t} = -\frac{1}{A} \oint_L f v_n \, \delta L - \frac{1}{A} \frac{\partial}{\partial s} \left(\frac{V^2}{2} \right) S \qquad (11.11a)$$

where S is the distance bc along the jet-stream axis that approximates a streamline. Denoting the mean radius of curvature by r_s, $A = r_s S/2$. Also, the mean value of $f v_n$ along side ab is approximately $-(f_b V_b + f_a V_a)/2$, while that along ca is $(f_c V_c + f_a V_a)/2$. Making the appropriate substitutions into Eq. (11.11a), we have

$$\frac{\partial [\zeta_a]}{\partial t} \approx -\frac{\partial (fV)}{\partial s} - 2k_s \frac{\partial}{\partial s} \left(\frac{V^2}{2} \right) \qquad (11.11b)$$

where k_s is the curvature of a streamline in the vicinity of the jet-stream

---

[8] This "eddy" influence is, of course, implicitly contained in numerical prediction models to the extent that the finite-difference approximations describe the details of the velocity and vorticity fields. The size of the grid used is obviously relevant to the fidelity achieved.

The principle outlined here was earlier expressed by Bjerknes (1954): "If the new trough formation takes place just north of an intense jet stream the pre-existing, concentrated, cyclonic shear-vorticity pours into the diffluent trough from the upwind side while none is leaving the trough downwind. A strong boost is thus being given to the vorticity budget of the upper trough and excess cyclonic vorticity is supplied also to the surface cyclone by the mechanism of downward vorticity flux." A very complete numerical diagnostic analysis of the three-dimensional motion field in a strongly developing cyclone has been carried out by Krishnamurti (1966). This cyclone was similar to those discussed here, the formation being connected with a diffluent upper trough that later transformed to a confluent trough.

axis, along which $\partial/\partial s$ is measured as a mean value between locations upstream and downstream from the trough.

Considering a trough centered on a meridian, Eq. (11.11b) may be written as

$$\frac{\partial[\zeta_a]}{\partial t} \approx -(f + 2k_s V)\frac{\partial V}{\partial s}$$

This expression implies that *the vorticity in the interior of a trough will increase as long as the wind speed decreases downstream along the jet axis*, as in Fig. 11.14a or, vice versa, in Fig. 11.14b. If the variation of f is taken into account, the first r.h.s. term of Eq. (11.11b) suggests that the growth of a trough is enhanced if it is superposed on a equatorward current (e.g., a minor trough entering a major trough) and is suppressed if the basic current has a component poleward. This aspect of the development of waves in the jet stream was brought out by Petterssen (1952).

11.7 AIR-MASS EXCHANGE IN RELATION TO EVOLUTION OF DISTURBANCES

Based upon the previous discussions, an attempt can be made to summarize in broad terms the way in which upper-wave and cyclone developments are related to the meridional and vertical movements of air masses discussed in Chapter 10. As elaborated in Chapter 16, in extratropical latitudes a regular conversion of energy takes place, from the "mean available potential energy" arising from the meridional distribution of heat sources and sinks to "eddy available potential energy" expressed by the waves in the westerlies. This conversion is accomplished by the horizontal advection of cold and warm air in disturbances, as in Fig. 11.3. Kinetic energy is generated mainly in the disturbances and depends upon organized vertical movements in which cold air sinks and warm air rises, representing a conversion from eddy-available potential energy that is destroyed in the process.

In an individual disturbance, as illustrated by the example in Figs. 11.4 to 11.6, there is evidence that the processes of horizontal and vertical exchange, and of energy conversion, do not proceed at a uniform rate. Calculations of the energy budgets of fully developed cyclones (Section 16.3) indicate that large amounts of kinetic energy are generated and at times almost comparable quantities are exported from their vicinity. At this stage, the kinetic energy generated comes largely from a localized readjustment of the air masses, in which there is a decrease of available potential energy in the neighborhood of the cyclone.

Early in the development, however, it appears that there must be a net import of energy into the region surrounding the cyclone, at least when the types illustrated in Sections 11.4 to 11.6 are considered. For a time, the winds in both lower and upper levels increase in intensity, and at the same time the frontal contrasts increase locally; thus there is a local increase of both kinetic and available potential energy. Hence, as emphasized in Section 16.1, a cyclone cannot be considered as an isolated system in which the kinetic energy of the circulation is derived purely from readjustment of the locally available potential energy expressed by the air-mass contrasts.

Figure 11.15 shows the thermal structure at 500 mb and at selected times during the development of the major cyclone of November 1948. Figure

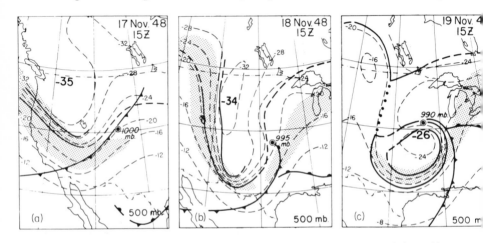

FIG. 11.15   Surface fronts, 500-mb isotherms, and boundaries of 500-mb frontal layer, at (a) 15 GCT Nov. 17, 1948; (b) 24 hr later; and (c) 48 hr later. Dots show location and central pressure of sea-level cyclone; shading, region where 300-mb wind speed exceeds 120 knots. (From Newton, 1958.)

11.15a corresponds to the time of Fig. 11.4, and Fig. 11.15c to Fig. 11.5. The regions of strongest wind at 300 mb (shaded) show an evolution similar to that described in Section 11.6. On Nov. 17, the upper-tropospheric wind was much stronger on the west than on the east side of the trough; at the intermediate time, when the trough reached full development, the wind field was fairly symmetrical; and on Nov. 19 the strongest winds were on the east side.

During the time when the trough was intensifying, as shown by Figs. 11.15a and b, the lowest temperature at the 500-mb level remained essentially unchanged. This suggests that, as seems typical, the cold air retained its depth during its extension southward and underwent massive subsidence only after the trough reached full amplitude. Cressman (1948) cited evidence that during the deepening of an upper trough, the cold air in the central portion actually ascended.

As discussed by Hsieh (1949), this behavior is qualitatively in accord with the structure of the wind field at the various times. According to the principles discussed in Chapter 7, the frontal zone bounding the cold air can maintain an equilibrium slope as long as the solenoids in a vertical plane across the flow, which tend always to cause a direct circulation, are balanced by an opposite coriolis plus centrifugal acceleration associated with an upward increase of wind speed. During the trough development (Fig. 11.15a), when there is strong kinetic-energy advection into the neighborhood of the trough in upper levels and an increase of vertical shear south of the cold-air dome, it is prevented from collapsing. When at a later time (Fig. 11.15c) there is a kinetic-energy export in upper levels and the vertical shear decreases, the cold dome must collapse, since the solenoidal circulation is no longer balanced. One of the results of this process is that a deep mass of cold air is permitted to move quite far equatorward during the time when horizontal advection dominates over the effects of vertical movement, so that the sinking that evolves later takes place mainly in the lower-latitude part of the cold trough.

As noted in Section 11.4, the sea-level cyclone was best developed near the time of Fig. 11.15c. The increase of 500-mb temperature during the preceding 24 hr indicates that during this period of cyclogenesis, the cold tongue was already actively subsiding, especially in the central part connecting it with the main body of polar air. During the next 24 hr, the southern cold dome sank well below the 500-mb surface. The overall process, as discussed by Palmén (1951), is one in which characteristically the development of an occluded cyclone, with warm air being excluded from the cyclone center at low levels and lifted to progressively higher levels, goes hand in hand with a seclusion of cold air and its eventual removal downward from the upper levels. With the available potential energy largely exhausted by the processes mentioned, there is no longer an appreciable source for production of kinetic energy. This is therefore the dying stage in the life history of a cyclone, as emphasized by Bjerknes and Solberg (Section 5.2).

In relation to the overall process of cyclogenesis, the various features described above may be summarized as follows: During the developing stage of an upper-level trough, when the front with its corresponding stronger jet stream and high cyclonic vorticity are present on the west side, the vorticity increases in the trough essentially as an effect of the eddy influx of vorticity. At the same time, cold air can move equatorward in the trough without subsiding bodily, due to an increase of the vertical shear as kinetic energy is imported in upper levels. As a result of increased vorticity in the trough, the upper-level vorticity gradient and correspondingly the upper divergence are augmented on the downstream side. Upon advance of the trough and superposition of the connected upper divergence over a preexisting region of relatively high vorticity (generally a front), significant low-level cyclogenesis ensues according to the scheme outlined in Section 11.2. As the cyclone develops, vertical motions increase in its vicinity according to the pattern described in earlier chapters, with ascent mainly in advance and sinking motions in the cold air to the rear. Vorticity is generated both in the low-level convergence region and in the upper-level trough where there is convergence above the sinking cold dome behind the cyclone.

When the strong winds have migrated to the east side, the cold-air dome sinks bodily at an increased rate as the upper-level winds diminish in the trough. The upper-level vorticity tends to decrease, owing to an eddy export of vorticity, but this tendency is to some extent counteracted by the influence of vertical stretching and horizontal convergence above the subsiding cold dome. The massive sinking of cold air in the middle and later stages of this sequence, together with the ascent of warm air with release of latent heat in other parts of the accompanying cyclone, represents a conversion from available potential energy to kinetic energy. This is partly consumed in increasing or maintaining the cyclone circulation against dissipative forces, but is largely exported from the vicinity of the cyclone as discussed in Chapter 16. Successive contributions from many cyclones, in the vicinity of which the most intense energy conversion takes place, furnish a large part of the energy required to maintain the extratropical westerly circulation.

11.8   OROGRAPHIC INFLUENCES

The influence of orography upon cyclone development will, of course, vary with geographic location and with the general character of the large-scale flow pattern. In this section we show broad-scale aspects of only two characteristic types of orographic development. In one type, cyclogenesis occurs in the lee of the meridionally extensive Rocky Mountain massif, with general westerly flow across it; in the other, a somewhat different type results from deformation of the thermal and flow patterns by the Alps, which are high but much less extensive, and are oriented mainly west to east.

The basic features of cyclogenesis in the lee of the Rocky Mountains were described by Hess and Wagner (1948). Their analyses showed that, in general, under conditions of westerly flow in winter, the isentropic surfaces in the lower troposphere are elevated over the plateau region and depressed in the lee, with a nodal surface in the middle troposphere and an opposite configuration in the upper troposphere. Thus, the isentropes in a vertical section superficially resemble a pattern expected from mountain-wave theory, and suggest vertical shrinking of air columns during passage over the mountains, with vertical stretching on the lee side where the static stability is weakest some distance east of the mountain crest.

Composite maps prepared by Hess and Wagner showed that cyclogenesis on the lee side "depends upon the twofold effect of a Pacific low entering the west coast. First, the low increases the flow over the mountains with consequent intensification of the standing lee trough. Second, the low then passes over the mountains and produces additional lee pressure falls." The general features found by Hess and Wagner have been borne out by later studies.

Distinct frequency maxima of cyclogenesis are found, in both winter and summer, to the east of the highest parts of the Rockies in Alberta and Colorado (Petterssen, 1956). Formation in either of these favored regions depends in general on the latitude of the maximum westerlies at the outset. The details of the development differ, but the basic mechanisms are much the same. In the following discussion we employ schematic diagrams that embody structures consistent with case studies by Newton (1956), McClain (1960), and Bonner (1961). Reference is made to these papers for detailed discussions of the processes, which are more complex than in the simplified discussion below. A somewhat different type of orographically influenced cyclogenesis over Alberta has been described by Hage (1961).

Figure 11.16d represents a time when the lee cyclogenesis process has essentially been completed, after which the cyclone moves off and behaves in a manner essentially similar to a cyclone over flat terrain. Figures 11.16a-c correspond to 500-mb and 1000-mb contours approximately 36, 24, and 12 hr earlier. The shaded area corresponds to the general region of high terrain.

Typically, at the stage in Fig. 11.16a, a well-developed frontal cyclone is approaching the west coast; an anticyclone overlies the plateau region. Some time before the frontal system enters the west coast (at which time the Pacific cyclone rapidly fills), surface pressures will usually have begun falling in the lee of the mountains. This occurs in response to adiabatic warming, owing to westerly or southwesterly winds with a component down

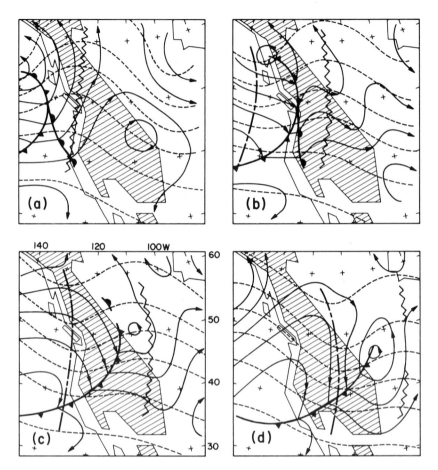

FIG. 11.16   At approximately 12-hr intervals, schematic fronts and 1000-mb contours (solid) and 500-mb contours (dashed) during a common type of orographic cyclogenesis. Shading represents, in simplified form, general region where terrain is higher than 1500 m, including interior basins where the terrain is lower. Upper trough line dash-dotted; ridge line, zigzag.

the lee slope. At the time of Fig. 11.16b, a thermal ridge will normally have begun to form in the lee of the mountains, with a deepening sea-level trough that, in some cases, is already of moderate intensity at this time.

Formation of this trough is commonly well under way when the upper-level ridge is still somewhat to the west. Thus, during the initial stages at least, the vorticity change in upper levels may act against sea-level development. In terms of Eq. (11.6), the increase of sea-level geostrophic vorticity is due to generation of the warm tongue in the lower troposphere east of

the mountains. This is in harmony with an increase in the vorticity of the
real wind in low levels, due to vertical stretching as the air descends the
slope at the surface, with weaker descent higher up.[9]

The process of generation of a lee trough continues as long as the low-
level winds have a component down the mountain slope. East of the trough,
there is warm advection (indicated by the crossings between 1000-mb and
500-mb contours). Hence the air warmed by descent spreads eastward and
the lee trough broadens (Fig. 11.16c), while the orographically caused
pressure falls eventually diminish when cold advection to the rear compensates the warming due to descent. At this stage, the upper-level ridge will
have moved east of the lee trough, and positive vorticity advection aloft
will have started to contribute to further cyclonic development. As the
upper-level trough advances eastward, low-level convergence in response
to increasing upper-level divergence intensifies the vorticity in low levels.
Rapid low-level development normally continues only until the surface
cold front has moved into the lee trough. At this time, the orographic
contribution to development stops abruptly, or reverses, owing to a shifting
of surface winds to northwest with cessation of surface downslope motion.
As a result of the development itself, the region of orographic production
of low-level vorticity is transferred to the south side. This largely accounts
for the commonly observed movement of such cyclones with a component
equatorward across the upper-level "steering" flow, until they have moved
far enough eastward to be free of orographic influences.

The overall process is one in which *vorticity is generated in the low-level
lee trough, which is held fixed to the eastern slope, and is finally overtaken
by the divergence region aloft as the upper-level trough approaches.* The rapidity of development normally observed is then due to the availability
of a region of large vorticity in low levels. This plays the same role as
(over flat terrain) a sea-level front with its preexisting vorticity plays when

---

[9] Over flat terrain, Eq. (5.1) shows that a net horizontal mass divergence is necessary
for the pressure to fall. Over sloping terrain, as shown by Godson (1948), this is not
required because, with downslope motion, air is in effect removed vertically through the
base of an atmospheric column. Even with a modest descending motion of $\omega_0 = 1\ \mu$b/sec
($w_0 \approx -1$ cm/sec), this effect alone would cause a pressure fall of 3.6 mb/hr. Orographic
descent commonly exceeds this magnitude, and such pressure falls are rare, showing that
there must be overall convergence in the air column. Observed temperature fields (see,
e.g., McClain, 1960) indicate maximum warming in low levels and little effect at 700 mb,
suggesting that the convergence takes place in a relatively shallow layer. With the pressure
falling at the ground and no corresponding fall at an upper level during this phase of
the development, hydrostatic consistency is maintained by warming of the air column
through adiabatic descent.

overtaken by a trough in upper levels. Since the belt of westerly winds in low levels is usually quite broad, the orographic influence that causes formation of a lee trough may be correspondingly extensive in the meridional direction. It is the combination of this influence and of divergence aloft that leads to the final intense cyclogenesis. This consequently takes place in a location primarily governed by the upper-level divergence, being most intense somewhat north of the jet stream.

After the stage in Fig. 11.16d has been reached, the frontal cyclone advances generally eastward and developes further by normal processes, as in the example of Figs. 11.4 to 11.6 in which the cyclone was initially an orographic disturbance. Many "Colorado cyclones," as in this example, undergo two distinct periods of intensification, weakening somewhat between them. The first of these takes place by the processes discussed above, when development is favored both by orographic influences and divergence aloft; a slackening occurs when the cyclone has moved eastward and the orographic influence ceases; and further development ensues when the upper-level trough has moved forward relative to the sea-level cyclone, with intensified upper divergence over it, as discussed in Section 11.4.

The orographic influence is obvious not only in the cyclone development associated with disturbances moving in from the west, but also in the changes of the structure of fronts associated with such disturbances. The orographic effect is twofold. First there is the direct adiabatic influence of the forced vertical motions, which tend to distort the regular thermal structure of the fronts in lower levels. More important, however, is the influence of the mountains on the moisture content of the air. Since a large part of the regular moisture content of the warm air is dispersed on the western slope of the mountains as orographic rain, and only weak moisture sources are available in the mountain region, the fronts on the eastern slope have more of the character of fronts in a dry atmosphere. As was pointed out before in connection with frontogenesis, the lack of moisture tends to exaggerate the frontolytic effect of the vertical circulation associated with fronts.

As a result of this, the characteristic cloud sequence and precipitation pattern associated with warm fronts is very much distorted over large parts of North America. On the other hand, cold fronts moving eastward over the eastern slope of the mountains tend to become weakened in their lower parts, especially when they encounter a cool-air residual from a preceding disturbance. Later on, when the accompanying cyclone moves farther eastward and moist warm air is drawn into the circulation from the Gulf source region, frontogenesis again ensues and the frontal structure gradually becomes more similar to structures observed in other parts of the westerlies. As stressed earlier, condensation resulting from the renewed latent heat source in the warm air contributes both to frontogenesis and to the general development of the cyclone.

In the geographical area discussed above, the mountains are so extensive in the north-south direction that the main belt of westerlies flows over them, and they are so broad that massive lifting and sinking of the air take place. Orographic cyclogenesis is common in the Mediterranean region where, as described by Radinović (1965), the dominant physical processes are of a different nature. Radinović (1960) gives convincing arguments

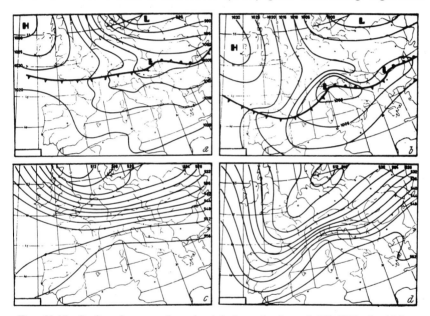

FIG. 11.17   Surface fronts and sea-level isobars (top), and 500-1000 mb thickness (decameters) during formation of an orographic cyclone over northern Italy. Left-hand charts are for 12 GCT Feb. 17, 1958, and right-hand charts are 24 hr later. (After Radinović, 1965.)

that although the influence of surface heating is important in the general Mediterranean region, the primary influence upon cyclogenesis is orographic. The frequency of cyclogenesis is greatest in the Gulf of Genoa, not over the warmest waters of the Mediterranean in winter; it occurs in summer as well as winter, and cyclones typically form before the cold air moves over the warm water. Moreover, in a large majority of cases, the cyclonic development occurs prior to the inception of significant precipitation so that release of latent heat cannot be a primary factor in the beginning, although it becomes important later on.

Cyclogenesis in this region occurs typically with cold outbreaks from the north or northwest. An example is given in Fig. 11.17. The Alps form a semicircular barrier, convex toward north northwest, with a mean elevation of 2.5 to 3 km. This restricts the southward movement of cold air; however, this air can flow southward around the mountains, principally in the Rhône Valley to the west. Thus the thickness pattern is distorted, as in Fig. 11.17d, with the formation of a surface-cyclone center in the lee of the Alps. The surface cyclone reflects a ridge in the thickness field, with maximum thermal vorticity in the cold air to west and east.

REFERENCES

Árnason, G. (1963). The stability of nongeostrophic perturbations in a baroclinic zonal flow. *Tellus* **15**, 205–229.

Bjerknes, J. (1951). Extratropical cyclones. *In* "Compendium of Meteorology" (T. F. Malone, ed.), pp. 577–598. Am. Meteorol. Soc., Boston, Massachusetts.

Bjerknes, J. (1954). The diffluent upper trough. *Arch. Meteorol., Geophys. Bioklimatol.* A7, 41–46.

Bonner, W. D. (1961). Development processes associated with the formation and movement of an Alberta cyclone. Tech. Rept. No. 4, Grant NSF-G6353, 29 pp. plus figs. Dept. Meteorol., Univ. Chicago.

Cressman, G. P. (1948). On the forecasting of long waves in the upper westerlies. *J. Meteorol.* **5**, 44–57.

Eliassen, A. (1956). Instability theories of cyclone formation. Chapter 15 in Petterssen (1956).

Godson, W. L. (1948). A new tendency equation and its application to the analysis of surface pressure changes. *J. Meteorol.* **5**, 227–235.

Hage, K. D. (1961). On summer cyclogenesis in the lee of the Rocky Mountains. *Bull. Am. Meteorol. Soc.* **42**, 20–33.

Hess, S. L., and Wagner, H. (1948). Atmospheric waves in the northwestern United States. *J. Meteorol.* **5**, 1–19.

Hughes, L. A., Baer, F., Birchfield, G. E., and Kaylor, R. E. (1955). Hurricane Hazel and a long-wave outlook. *Bull. Am. Meteorol. Soc.* **36**, 528–533.

Hsieh, Y.-P. (1949). An investigation of a selected cold vortex over North America. *J. Meteorol.* **6**, 401–410.

Krishnamurti, T. N. (1966). A study of a wave cyclone development. Final Rept., Contr. AF 19(628)-4777, Sect. II. Dept. Meteorol., Univ. Calif., Los Angeles.

McClain, E. P. (1960). Some effects of the Western Cordillera of North America on cyclonic activity. *J. Meteorol.* **17**, 104–115.

Newton, C. W. (1956). Mechanisms of circulation change during a lee cyclogenesis. *J. Meteorol.* **13**, 528–539.

Newton, C. W. (1958). Variations in frontal structure of upper level troughs. *Geophysica (Helsinki)* **6**, 357–375.

Newton, C. W., and Palmén, E. (1963). Kinematic and thermal properties of a large-amplitude wave in the westerlies. *Tellus* **15**, 99–119.

Nyberg, A. (1945). Synoptic-aerological investigation of weather conditions in Europe, 17–24 April 1939. *Statens Meteorol.-Hydrogr. Anstalt, Medd., Ser. Uppsater* No. 48, 1–122.

Palmén, E. (1951). The aerology of extratropical disturbances. *In* "Compendium of Meteorology" (T. F. Malone, ed.), pp. 599–620. Am. Meteorol. Soc., Boston, Massachusetts.

Petterssen, S. (1952). On the propagation and growth of jet-stream waves. *Quart. J. Roy. Meteorol. Soc.* **78**, 337–353.

Petterssen, S. (1955). A general survey of factors influencing development at sea level. *J. Meteorol.* **12**, 36–42.

Petterssen, S. (1956). "Weather Analysis and Forecasting," 2nd ed., Vol. I, Chapters 13, 14, and 17. McGraw-Hill, New York.

Petterssen, S., Dunn, G. E., and Means, L. L. (1955). Report of an experiment in forecasting of cyclone development. *J. Meteorol.* **12**, 58–67.

Petterssen, S., Bradbury, D. L., and Pedersen, K. (1962). The Norwegian cyclone models in relation to heat and cold sources. *Geofys. Publikasjoner, Norske Videnskaps-Akad. Oslo* **24**, 243–280.

Polster, G. (1960). Über die Bildung und Vertiefung von Zyklonen und Frontwellenentwicklungen am konfluenten Höhentrog. *Meteorol. Abhandl., Inst. Meteorol. Geophys. Freien Univ. Berlin* **14**, No. 1, 1–70.

Radinović, D. (1960). Analysis of the cyclogenetic effects in the west Mediterranean. *6th Intern. Meeting Alpine Meteorol., 1960*, Bled, Yugoslavia p. 33–40.

Radinović, D. (1965). On forecasting of cyclogenesis in the western Mediterranean and other areas bounded by mountain ranges by baroclinic model. *Arch. Meteorol., Geophys. Bioklimatol.* **A14**, 279–299.

Riehl, H., La Seur, N. E. *et al.* (1952). Forecasting in middle latitudes. *Meteorol. Monographs* **1**, No. 5, 1–80.

Saucier, W. J. (1949). Texas-West Gulf cyclones. *Monthly Weather Rev.* **77**, 219–231.

Scherhag, R. (1937). Bermerkungen über die Bedeutung der Konvergenzen und Divergenzen des Geschwindigkeitsfeldes für die Druckänderungen. *Beitr. Physik Atmosphäre* **24**, 122–129.

Sutcliffe, R. C. (1939). Cyclonic and anticyclonic development. *Quart. J. Roy. Meteorol. Soc.* **65**, 518–524.

Sutcliffe, R. C. (1947). A contribution to the problem of development. *Quart. J. Roy. Meteorol. Soc.* **73**, 370–383.

Sutcliffe, R. C., and Forsdyke, A. G. (1950). The theory and use of upper air thickness patterns in forecasting. *Quart. J. Roy. Meteorol. Soc.* **76**, 189–217.

Thompson, P. D. (1961). "Numerical Weather Analysis and Prediction," Chapter 10. Macmillan, New York.

U. S. Weather Bureau. (1959). Daily series, synoptic weather maps, Part I, January 1959. U. S. Weather Bur., Washington, D. C.

12

WEATHER IN RELATION TO DISTURBANCES

The fact that synoptic disturbances with essentially similar flow patterns may have very different weather distributions is well known to all forecasters. Nevertheless, the practical use of models to relate cloud and weather distributions to various kinds of circulation systems suggests that such models are considered to have enough validity to provide a broad framework for the forecast. Modifications are, of course, necessary to take account of deviations such as those that arise from topographic influences, variations in the moisture content, and stability of the air masses that enter the circulation system. Hence it is not surprising that a model developed in a given region, such as northwestern Europe, may be found inadequate when applied in a physiographically dissimilar region.

In this chapter we deal only with a few aspects of cloud and precipitation as related to fronts and synoptic disturbances, insofar as these can be described in very general terms. We shall not discuss the application of numerical techniques for diagnosis and prognosis, since justice cannot be done to the subject in the space of a few pages. For extended discussions of the general weather forecasting problem, the reader is referred to the books by George (1960) and by Petterssen (1956), and to the appropriate chapters in Godske et al. (1957). Bergeron (1960) gives an overall summary of the important influences of orographic and other local effects upon the detailed distribution of precipitation. Organized convective systems in extratropical disturbances are dealt with in Chapter 13.

12.1   BROAD-SCALE WEATHER DISTRIBUTION

The general weather distribution around cyclones, as deduced by the Bergen school, was illustrated in Figs. 5.1 and 5.2. Bergeron (1951) has synthesized the main features of the weather distribution in occluded cyclones; the scheme is shown in Fig. 12.1. This especially typifies cyclones over the North Atlantic Ocean. In the map, "upper polar front" refers to a mid-tropospheric level, say, 500 mb. The upper waves are broadly related so that in the west-east cross section *B-B* there is a wind component into the page over the cold front and out of the page over the warm front.

The weather and cloud symbols in Fig. 12.1 are self-explanatory, and will not be discussed in detail. Three principal species of cloud and precipitation areas are represented: (1) "frontal upgliding" types that result from broad-scale lifting; (2) showers in the southward-moving cold air, which is heated from below; and (3) stratus and drizzle in the northeastward-moving warm air, which is cooled from below. The latter two types tend to be especially prominent over oceanic regions where there is a significant north-south gradient of sea-surface temperature, but are less common over interior land areas (although on occasion large areas of type 3 weather are observed in the cooler months). For thorough descriptions of the classical weather models, including cross sections showing the characteristic hydrometeor distributions in relation to different types of frontal systems, see Godske *et al.* (1957). Borovikov *et al.* (1961) provide an excellent summary of the combined microphysical and synoptic aspects of frontal clouds.

The advent of weather satellites made it possible for the first time to see directly the entire cloud pattern on the scale of a cyclone. No attempt will be made to summarize the extensive literature on satellite observations.[1] The general interpretations of satellite-observed cloud patterns have been reviewed by Widger (1964), who elaborates in Fig. 12.2 upon a model by Boucher and Newcomb (1962) of the cloud structure of a mature occluded cyclone over the North Atlantic Ocean. The typical cloud distributions agree remarkably well with Fig. 12.1, which was based on other kinds of observations.

The overall configuration in Fig. 12.2 is very characteristic of mature *oceanic* cyclones in both hemispheres. Extended cloud bands are normally present along both warm and cold fronts, with broader cloud sheets in

---

[1] For articles related to different aspects of satellite meteorology, see, e.g., the proceedings edited by van de Boogaard (1966).

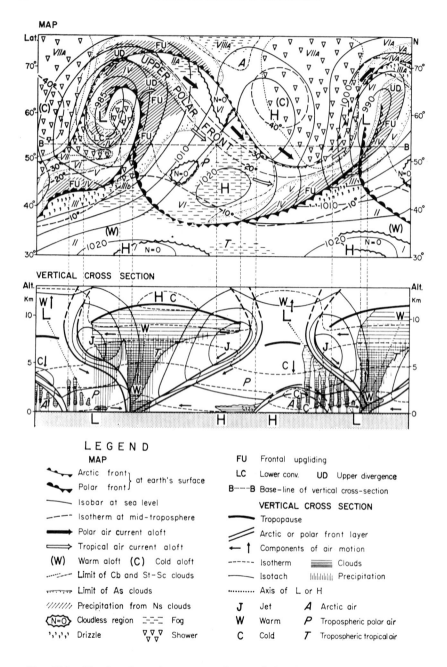

FIG. 12.1   Cloud and weather systems characteristic of occluded cyclones over the north Atlantic Ocean and western Europe. (After Bergeron, 1951.)

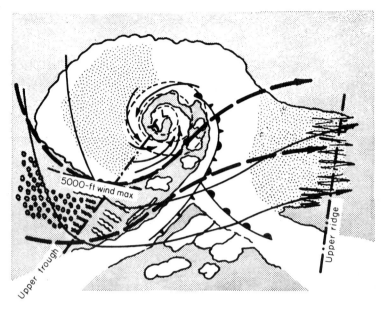

FIG. 12.2 Cloud structure of an occluded cyclone just prior to fullest maturity, from TIROS V photograph, near 56°N, 75°W, 1250 GCT Sept. 7, 1962. Arrows near cyclone center show lower-tropospheric flow; solid thin arrows, flow in middle troposphere; heavy dashed arrows, flow at tropopause level. (After Widger, 1964.)

advance of cyclone centers and extensive spiral-cloud systems around them. Near the cyclone center it is common to observe a tongue of relatively clear air, which may to different degrees be partially filled with patches of cloud. This is interpreted as *air that has become dry through subsidence west of the cyclone, and then advected by the inward-spiraling winds.* The degree to which this "clear" tongue has intruded into and around its center is considered to be an indication of the age and intensity of circulation of a cyclone.

The striking tendency for convective clouds in cold outbreaks behind cyclones to be arranged in "cloud streets" almost parallel to the direction of low-tropospheric winds (see, e.g., Rogers, 1965) was earlier seen from high-altitude aircraft photos, and is discussed by Kuettner (1959). An apparently universal accompaniment is a curved wind profile, similar to a low-level jet, in the cloud and subcloud layer. Kuettner observes that most extratropical cloud streets are found in vigorous outbreaks of polar air heated from below, the characteristic low-level wind maximum being due to "a frictional wind increase with height in its lower portion and to a thermal wind decrease in the upper portion" (i.e., the wind has an equatorward component, and with warm air to the east, the geostrophic wind decreases with height in lower levels).

Although regularly aligned rows predominate, Nagle and Serebreny (1962) give examples of a vortical structure of the rows in regions of apparently uninterrupted northwesterly flow in the low levels. Winston and Tourville (1961) suggested that these are decayed vortices that show the past history of the air motions. Oliver and Ferguson (1966) have also found vortical or "comma-shaped" sheet-cloud systems (with longest dimensions up to 1000 km) in the cold air well behind cold fronts, which they connect with the positive-vorticity-advection regions ahead of minor upper troughs.

The fibrous eastern edge of the cloud system in Fig. 12.2 is clearly a characteristic of high cloud. This boundary between dissipating cirrostratus and the clear area downstream is, according to Serebreny *et al.* (1962) and others, likely to be found near the upper ridge downstream from the cyclone. The overall picture must be therefore interpreted in the light of clouds responding to the air motions at different levels. *Only the low clouds really spiral inward, whereas those typically seen at a distance poleward of the warm front are upper clouds streaming away from the cyclone.* This is consistent with the three-dimensional flow as described in Chapter 10, due to which the vertical development of clouds is suppressed on the west side while those on the east side can penetrate into the upper-tropospheric flow.

Although the general locations of cold fronts over the ocean are apparent from long bands of more-or-less continuous cloud in the pictures, the bulk of the cloud may at a given place be either pre- or postfrontal (see, e.g., Winston and Tourville, 1961). Concerning warm fronts, Widger (1964) states: "Seldom does there appear to be a clearly discernible boundary between the convective cloudiness in the warm air and the more stratiform cloudiness due to over-running poleward of the surface warm sector."

According to satellite and other evidence presented by several authors, stratiform cloud tends for the most part to be confined to the anticyclonic-shear side of the jet stream, which in some locations almost coincides with the cloud limit. A general preference for the anticyclonic-shear region accords with the consideration that lateral upgliding motions can take place most readily where the dynamic stability is weak (Section 9.6). In the vicinity of well-developed cyclones, however, this general rule does not apply. A typical variation in structure during the life of a cyclone is illustrated in Fig. 12.3. In Fig. 12.3a, corresponding to a young cyclone, the jet-stream core is indicated vertically above the 500-mb front, according to the scheme discussed in Section 7.4. This correspondence, approximately

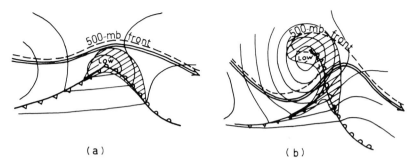

Fig. 12.3   Schematic wave cyclone (a) and occluded cyclone (b), precipitation areas hatched. Dashed line is 500-mb front; double arrow is jet stream.

valid for weak disturbances on the polar front, is no longer valid during a later stage of cyclone development, owing to the pronounced change of the thermal structure resulting from the occlusion process.

During this process, the thermal structure gradually becomes diffuse as a result of sinking cold air and rising warm air; thus the thermal contrast is weak for some distance equatorward of the cyclone center. The polar front in upper levels (at 500 mb, say) is strongly deformed, as indicated in Fig. 12.3b, with a tongue of ascending warm air displaced far poleward. The jet core is now separated from its former intimate connection with this frontal contour and remains equatorward of it, particularly on the east side where the front is strongly perturbed.

This behavior of the jet stream is supported by aerological experience (Pogosian, 1960; Boyden, 1963), and follows from hydrostatic considerations. With low-level isobars of the type in Fig. 12.3b and the reduced baroclinity resulting from the frontolytic effect of the occlusion process, the integrated thermal wind can no longer support a wind maximum above the 500-mb front in the area near the center of the cyclone. Hence, in this region, the jet core occupies a position farther south, where the low-level westerlies combined with the greater thermal contrast south of the cyclone imply the strongest geostrophic wind in upper levels. Thus, after occlusion of the surface cyclone, the jet core is south of the center and also south of a large part of the precipitation area. There have been many discussions of the relationship between the jet stream and rainfall (Riehl, 1948). Although some statistical patterns have emerged (Starrett, 1949), if we consider the manifold combinations that can appear among the low-level frontal system, the upper wind system, and the vertical motion pattern, it does not seem possible to express any simple rules about this connection in individual cases, except the broad sense depicted by Fig. 12.3.

12.2   Weather Systems in Relation to Heat Transfer

In a study encompassing 51 cases of cyclone development over the North Atlantic, and mainly in the cold season, Petterssen *et al.* (1962) determined the characteristic weather distributions. These are illustrated by Fig. 12.4 for a well-developed wave cyclone, and by Fig. 12.5 for a fully occluded cyclone. "Predominant weather" is represented in five categories: heavy convective activity, shower symbol with bar; moderate convective activity, shower symbol; weak convection, cumulus symbol; fog and stratus, dashed hatching; and frontal activity, continuous hatching. The numbers in Figs.

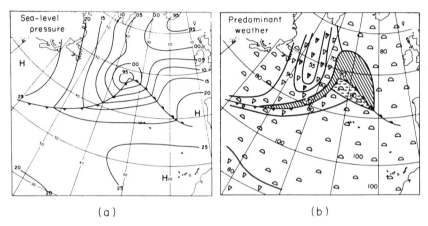

(a)   (b)

Fig. 12.4   For an open-wave cyclone over the north Atlantic Ocean, the typical (a) sea-level pressure and (b) weather distribution. For explanation of symbols, see text. (After Petterssen *et al.*, 1962.)

12.4b and 12.5b correspond to the percentage of cases in which the indicated weather types were reported in a given region relative to the synoptic system. Where mixed symbols are shown, each type occurred in less than 50% of the cases, and the number refers to the frequency of the combined types. Thus, within the warm-sector areas of mixed "fog and stratus" and "light convective activity," the former was present in somewhat more than 30% and the latter in some 30 to 50% of the observations.

In both wave and occluded cyclones, the distribution of the frontal type of weather (altostratus genera and nonshowery rain or snow) agrees well with the classical models. In each case, a band centered about 1000 km west to southwest of the cyclone center is characterized by heavy convective activity. Through the region 500 to 1000 km farther southwest, this shades off to light convection (cumulus of weak vertical development;

no showers). Within the cold air, there is everywhere a transfer of both sensible and latent heat from the sea to the atmosphere (Figs. 12.5c and d). However, *the pattern of this transfer is related only in a broad sense to the pattern of convection, which is most intense between the cyclone center and the region of maximum transfer of both sensible and latent heat.* As observed

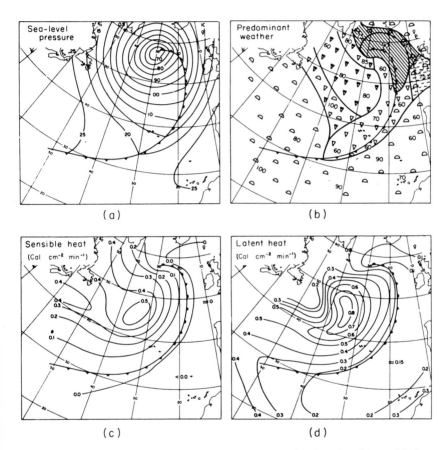

FIG. 12.5   For fully occluded cyclones: (a) and (b) as in Fig. 12.4, (c) sensible heat transfer, and (d) latent heat transfer from sea to atmosphere. (After Petterssen *et al.,* 1962.)

by Petterssen *et al.* (1962), "heating from the underlying surface always results in some convective activity; however, unless the air takes part in a cyclonic circulation, moderate and heavy convection (i.e., convection with precipitation) will not occur even if the surface heating is very intense; as far as the state of the sky is concerned, there is often a larger difference between *cyclonic-* and *anticyclonic*-air masses than there is between *cold-* and *warm*-air masses in the meaning used in customary classifications."[2]

This distinction is brought out further by Fig. 12.6, showing conditions typical of outbreaks from the subarctic region. The cyclonically curved branch in the northeast Atlantic is characterized by showery weather. Farther southwest, where the streamlines are more nearly straight or anti-cyclonically curved, light convection is typical, even though the surface heating is on the whole stronger there than in the region of heavy convection. This is in accord with Section 10.4 (Fig. 10.11), where it was shown that anticyclonically curved or even weakly cyclonic equatorward flow is

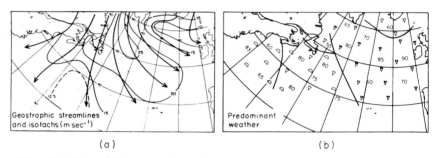

(a)　　　　　　　　　　　　　　　　　(b)

FIG. 12.6　For cold air streaming southward over the north Atlantic Ocean: (a) geostrophic streamlines and isotachs, and (b) typical weather distribution. (After Petterssen *et al.*, 1962.)

subsident. Adiabatic heating in such a case so stabilizes the air mass some distance above the surface that the depth of convection resulting from surface heating is greatly restricted. Similarly, Petterssen *et al.* (1945) found that the tops of convective clouds over England are appreciably higher with cyclonic curvature; and Crutcher *et al.* (1950) established a high correlation between the 700-mb vorticity and cloud tops on the Washington-Bermuda airway.

The observations summarized in Figs. 12.4 and 12.5 indicate that although convective clouds are present in very large regions of synoptic disturbances, where they transfer heat and water vapor upward into the lower troposphere, the regions occupied by clouds of considerable vertical development are comparatively small. Moreover, especially in high latitudes, many shower clouds do not extend much higher than middle-tropospheric levels. This was considered in Section 2.6 (see Fig. 2.12), where it was stressed that the

---

[2] In Bergeron's classification a "cold" air mass is one that is heated from below (showing unstable cloud types), while a "warm" air mass loses heat to the surface.

effect of "small eddies" including cumulus clouds decays rapidly with height so that the upward energy transfer through the middle and upper troposphere is accomplished essentially by synoptic-scale systems. However, even though the energy gained in the region of suppressed convection is not immediately realized and transferred to high levels, this takes place eventually. A significant part of the air modified in this manner moves in an anticyclonic trajectory, and in returning poleward in advance of the next approaching cyclone it rises in the ascending branch of its circulation.

While extensive areas of showers are the rule in the cold air west of oceanic cyclones, significant shower activity (if any) is generally confined to the region close to the center of a cyclone over land. Exceptions are found where cold air streams over large and relatively warm water surfaces such as the Great Lakes and the Baltic, Caspian, Black, and Mediterranean seas, especially in autumn and winter. The effect is not only to add water vapor and heat, but also to modify the circulation in a way that favors lower-level convergence, producing heavy showers especially in the lee of the water bodies (Petterssen and Calabrese, 1959).

The modification of cold-air masses passing over warmer surfaces has been dealt with in theory and observation by Burke (1945; see also Jarvis, 1964), Craddock (1951), Bunker (1956), and others. No attempt will be made to review these studies, but a few aspects will be mentioned. Figure 12.7, taken from Priestley (1959), illustrates the rate of heat gain by the air from various kinds of underlying surfaces (not considering insolational heating) and at varying distances behind the front of an outbreak of cold air. We may examine first the solid curves, which correspond to an air mass with neutral stability.

Curve *A* shows that after the passage of a cold front over *water*, heat is received at a rapid rate, which diminishes slowly with time. Passage over an *ice or land* surface results in heating that is initially smaller, and which diminishes rapidly with time. This difference in behavior, especially pronounced in the case of dry soil or snow, is due to the low thermal conductivity; after the uppermost layer has been chilled, the rate at which heat is made available from the substrata is slow. Consequently, a cold-air mass is heated only slowly over a solid surface (except for a short distance behind the front), and the air mass may retain the characteristics acquired in its source region for a long time. Hence, air-mass contrasts tend to be rapidly effaced in the lower atmospheric layers over water, but there is a much smaller tendency for frontolysis from this effect over land.

The dashed curves in Fig. 12.7 correspond to a very unstable air mass, the difference from the solid curves evincing a greater capability for upward transfer of the heat gained from the surface. Thus, the decay of heating rate after cold-front passage is proportionately greater than it is in a stable air mass. As Priestley observes, "over land the temperature response of the soil becomes the dominating factor, and a highly unstable air mass gains little more heat than one which is only moderately unstable, except immediately behind the front."

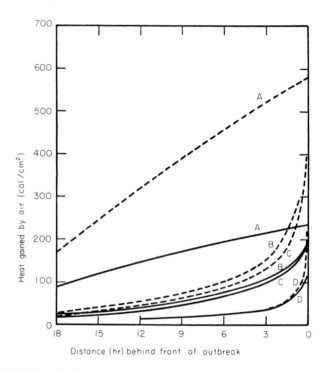

Fig. 12.7   Heat gained per 24 hr by air in a cold outbreak over different underlying surfaces: *A*, water; *B*, ice; *C*, wet soils; *D*, dry sand. Full curves represent a stable, and dashed curves, an unstable air mass. (Reprinted from *Turbulent Transfer in the Lower Atmosphere*, by C. H. B. Priestley, by permission of the University of Chicago Press, copyright 1965 by the University of Chicago). See original for numerical parameters used.

The gradient of temperature at the surface over which the cold air passes is naturally very important in determining the rate of energy transfer. Thus, cold-air masses moving over the waters off the east coast of North America (Fig. 12.8) acquire heat and water vapor much more rapidly than do those moving southward over the eastern Atlantic Ocean. This follows because (considering that the heat gained at the surface must be transferred upward in the atmosphere in order to warm it, and that for corresponding wind speeds the advective contribution to local temperature change depends on the horizontal temperature gradient) large sea-air temperature differences characterize the westernmost oceans, whereas a closer equilibrium typifies the eastern parts. Where polar air moves off the east coasts in winter, air-sea temperature differences of the order of 20°C may be observed, as in the example analyzed by Ichiye and Zipser (1967).

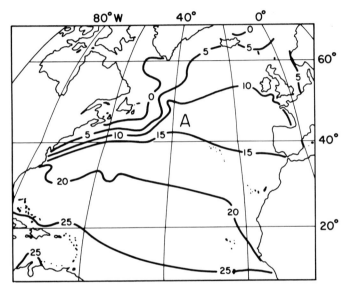

FIG. 12.8   Mean north Atlantic sea-surface isotherms in winter. (After Sverdrup, 1942.)

With regard to the vertical stability and the corresponding degrees of convective development, it is necessary to consider both the intensity of diabatic heating in lower levels and heating associated with vertical movements in higher levels. In subsiding air, these effects tend to oppose one another in varying degrees. We may consider an air column moving east southeast at 30-knot speed in the region of strongest sea-air energy transfer in Fig. 12.5. Suppose that a moist-adiabatic lapse rate is maintained in the layer between 950 and 500 mb and that the air-sea temperature difference is maintained constant. In the appropriate region A of Fig. 12.8, the air column would, in 12 hr, experience a temperature increase of 3°C at 1000 mb and 5.2°C at 500 mb. The heating required is $c_p(\delta\bar{T} + \delta\bar{T}')\,\Delta p/g$, where $\delta\bar{T}' \approx 1°$ is added to account for the 12-hr radiative cooling, and would be 0.85 ly/min. Part of the heating is, however, accomplished by subsidence so that roughly 0.5 ly/min heating from below would be needed, as suggested by the schematic Fig. 12.9.

In addition, an increase of the water-vapor content should be considered. If the column is maintained near saturation at the 950-mb level, with relative humidities of 50% at 500 mb and 80% at 1000 mb, the specific humidity would have to increase by an average amount of 1.1 g/kg, equivalent to 0.45 ly/min (about half the maximum evaporation in Fig. 12.5d). Con-

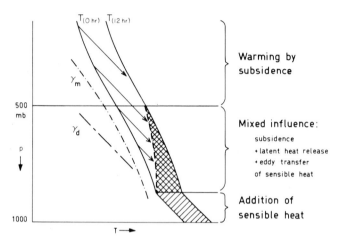

FIG. 12.9   Schematic soundings in an air column moving over warmer water, illustrating the influences of surface heating and convection combined with heating through subsidence in upper levels.

sidering both latent and sensible heat change in the column, about 0.95 ly/min energy increase would be required. In Figs. 12.5c and d, maximum transfers of about 0.6 ly/min (sensible heat) and 0.9 ly/min (latent heat) take place from the sea surface. The total of 1.5 ly/min considerably exceeds the amount required for heating and moistening the air column, showing that a lateral export of energy may be expected from this region.

The foregoing estimates are given only to illustrate roughly how the energy is partitioned in a particular region. In certain parts of disturbances over the ocean, diabatic and adiabatic influences may conspire to maintain the lapse rate approximately. However, this special condition would depend on a balance between the influences of heating at the various levels. *In the western part of a cold outbreak where anticyclonic or straight trajectories indicate subsidence, the heating in higher levels dominates; the lapse rate is stabilized, and convective development is suppressed. In the eastern part, where subsidence is weaker or there may be actual ascending motions near the cyclone center, the air mass would be more unstable; a greater proportion of the heat is realized from condensation, and convective clouds transfer heat to a higher level.* Over a land area where less heat is given off from the surface, the depth of the layer maintained in a state of neutral or unstable lapse rate by the processes mentioned must obviously be smaller. This is mainly so during the colder seasons.

Although cyclones passing inland over the British Isles and northwestern Europe tend regularly to have the weather structure depicted in Figs. 12.4 and 12.5, such a distribution is not to be expected everywhere, even in cyclones having similar frontal and isobaric configurations. The similitude of western European cyclones depends largely upon their entering the continent from an oceanic region and thus having considerable regularity in their air-mass properties, at least in lower levels.

Thus, the air behind a cold front entering a west-coast region is likely to be moist and unstable so that daytime heating (during the warmer season) would be more productive of showers in the *cold* air than in the continental interior, where air masses are dryer and more stable as a result of their prehistory. By contrast, convection due to daytime heating is more likely in the *warm* sectors of continental cyclones, particularly in lower latitudes during the warmer seasons, if these are in a location where moist air can be drawn into their circulations. For obvious reasons, showery precipitation in oceanic cyclones does not vary greatly from day to night because surface heating does not depend on insolation, while a strong diurnal variation is characteristic over land.

There is likewise great variability of warm-air weather in different localities. Warm air entering the circulation of a cyclone over northwestern Europe is likely to have traversed cool surface waters (Fig. 12.8) so that frontal clouds and precipitation are mostly of the stable type. Also fog and low stratus are common warm-sector phenomena (Fig. 12.5b). *In inland North American cyclones, on the other hand, the moisture availability is often uncertain, and the stability characteristics may be very different from those in European cyclones, with corresponding differences in weather phenomena.* Also, cyclones tend not to be so regular in structure after entering the west coast of North America because the mountains inhibit cold air from being drawn into the circulation on their east sides, or modify it through Föhn influences. Hence, as noted in Section 11.8, warm fronts analyzed from ascents near the coast often do not have the simple and regular structures observed near the European west coast; also, the difference in latitude affects the stability characteristics.

12.3 Stable "Upglide" Precipitation

Frontal cloud and precipitation was earlier viewed in terms of upgliding above sloping discontinuity surfaces. The weather distribution must, of course, also be considered in light of the ascent and descent connected with the large-scale convergence and divergence fields of synoptic systems, outlined in earlier chapters. In Chapter 9 it was emphasized that the formation and maintenance of fronts should be considered in the context of a three-dimensional circulation system in which frontogenesis at the surface requires convergence as well as deformation and is thus connected with organized vertical circulations in the overlying layers.

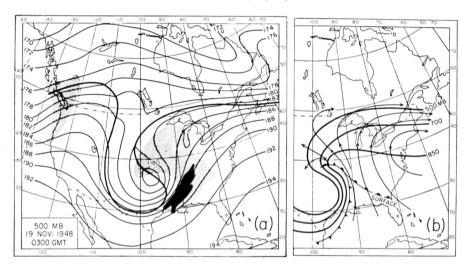

Fig. 12.10 (a) 500-mb contours (hundreds of feet) and front (heavy line), and surface fronts, 03 GCT Nov. 19, 1948; (b) frontal contours and streamlines at corresponding levels. Precipitation areas stippled; solid area, region where 24-hr rainfall exceeded 2 in. (After Palmén, 1951.)

Figure 12.10a shows the precipitation shield of a major cyclone over North America. The superimposed 500-mb contours show, as discussed by Palmén (1951), that the precipitation occurred in the general region where the cyclonic curvature and generally the vorticity decreased downstream. In accord with the principles in Section 10.8, this is the region of divergence aloft and rising motion east of the upper wave. A portion of the frontal-contour chart is reproduced in Fig. 12.10b, in which the "streamlines" correspond to the observed wind directions at the warm boundary of the frontal layer at various levels. Whereas in lower levels the air motions were predominantly normal to the front, the strong anticyclonic turning, in accord with divergence in the warm air, results in air movements almost parallel to the frontal contours in higher levels except near the cyclone center.

Since in upper levels the frontal contours north of the warm front were advancing at 10 to 15 knots (see Figs. 10.5c and d), the impression of upgliding there is somewhat exaggerated. Thus the forward edge of the precipitation shield falls somewhat short of the 500-mb frontal contour where the air flows nearly parallel to the front. Although "upglide" motion is also suggested by the streamlines some distance east of the precipitation area, the southerly wind component was weak compared with the northward advance of the frontal contours in this region. The broad agreement between the upper divergence judged from the isobaric contours, the frontal upglide, and the precipitation region in Fig. 12.10 depends, of course, not only on the motion field but also upon the water-vapor content of the air.

Some general aspects of precipitation production in the vicinity of a warm front can be seen from the schematic Fig. 12.11. This shows hypothetical motions in a vertical section normal to the front in the region where the southerly flow is strongest. In constructing the streamlines, an approximately parabolic distribution of ω in the vertical is assumed, with $\omega_{max} = -10\,\mu b/\text{sec}$ aoove the surface warm front, point C. Over B and D, 500 km to either side, ω is assumed half as large. At the 1000- and 100-mb surfaces, and over A and above the 600-mb intersection of the front, ω is assumed to vanish. The horizontal velocity component V in direction A–E is taken as 25 m/sec, constant with height. This is not necessarily realistic, but is broadly so, since the thermal wind is mainly parallel to the front and thus normal to the section. The assumption of a uniform V-component between A and E is obviously inconsistent with reality, but suffices for the present purpose.

The solid streamlines in Fig. 12.11 correspond to relative trajectories if the system is not moving. An air parcel starting at 900 mb over point A (trajectory I) would, after traveling 1000 km to point C over the surface front, rise to the 780-mb level, and in moving 500 km farther would rise to 640 mb. However, if the front and the attendant vertical velocity field move in direction A–E with speed c, the slope of the streamlines relative to the moving system would increase by the ratio $V/(V - c)$. Because the horizontal movement relative to the system is diminished, an air parcel would for a longer time be subjected to the field of vertical movement.

If, for example, $c = V/2$, in a relative framework the air would take twice as long to move from A to C, or C to D, than if $c = 0$. Correspondingly,

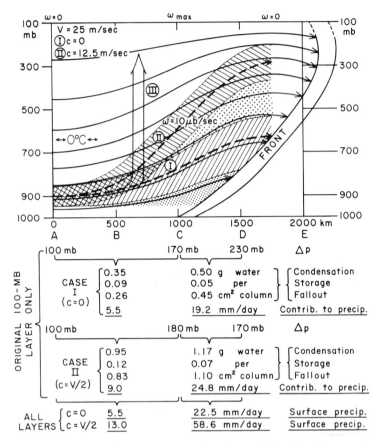

FIG. 12.11 A warm front and streamlines in vertical section. (See text.)

as indicated by relative trajectory II, an air parcel starting at 900 mb over
A would reach 590 mb above the surface front and 330 mb over *D*. Comparing the two situations, it is evident that *the vertical development of cloud
systems associated with fronts, and the corresponding precipitation, may vary
considerably from one case to another even though the thermal structures
and overall flow patterns are quite similar.* These differences depend on variations in the intensity of vertical motion and in the movements of the
synoptic systems as well as the moisture content and stability of the air.

Let us now assume that the air above point *A* is just saturated in the
layer 950 to 850 mb, but is relatively dry higher up. In case I the moist
layer would then be lifted to between 870 and 700 mb over *C*, allowing
for the convergence within the layer. Assuming a saturated-adiabatic process

and $T = 15°C$ at 900 mb, then 2.6 g/kg of water would be condensed, or considering the mean depth (135 mb) of the column between A and C, about 0.35 g/cm² horizontal area. The volume of a 1-cm² column over C (1.7 km deep) being 0.17 m³, if 0.5 g/m³ of the condensed water goes into cloud-droplet storage, 0.26 g/cm² would be precipitated out.[3] Between C and D the same column would rise from the 700- to 870-mb layer to the 550- to 780-mb layer, with further condensation of 2.5 g/kg, producing 0.50 g/cm² water, considering the 200-mb mean depth of the layer. Of this, 0.05 g/cm² would go into cloud storage (the vertical column having been stretched by 1 km), leaving 0.45 g/cm² for precipitation between C and D over the warm front.

These amounts of 0.26 g/cm² between A and C and 0.45 g/cm² between C and D refer to the quantity of precipitation shed from the base of an air column moving with the wind. Considering the time required for this movement at 25 m/sec (11 hr from A to C and 5.5 hr from C to D), the average precipitation rate would be 5.5 mm/day between A and C and 19.2 mm/day between C and D in the "upglide" region over the warm front. Thus, although the water condensed per kilogram of air between C and D is about the same as that between A and C (in a symmetrical ω-field), the precipitation would be much heavier over the warm front than in the warm sector. The difference, by a factor of 3.5, is due to the combined influences of greater vertical velocity as the sheet of air rises from the lower toward the middle troposphere, an increasing depth of the condensing layer as it experiences horizontal convergence, and the decreasing effect of water storage after cloud has formed in the layer.

Corresponding calculations for trajectory II, considering a moving system, are shown beneath Fig. 12.11. In this case, although the total amount of water condensed out of the corresponding layer of air is somewhat more than doubled, the ascent (in moving from A to D relative to the moving system) requires twice as long a time. Hence the surface precipitation rate contributed by this layer alone is not greatly different in the moving and stationary systems, considering the region A to D as a whole. It would be more realistic, however, to assume that all air rising from the boundary layer becomes saturated upon reaching the 950-mb level. If this is taken into account, calculations by the same method give the precipitation rates in the last two lines under Fig. 12.11, indicating a large augmentation for

---

[3] According to Burkovskaya (1959, quoted in Borovikov *et al.* 1961), liquid-water contents average 0.32 g/m³ in a layer of limited depth above warm fronts over Russia. A somewhat greater value is used here in consideration of the high temperature assumed (see Bannon, 1948).

the deep-cloud system. In terms of the water-budget considerations in Section 12.6, although the mass divergence field is the same in the two cases, the water-vapor convergence expressed by the last term of Eq. (12.6) is greater in the moving system with its deeper moist layer.

The detailed distribution of precipitation may, of course, be influenced by the microphysical processes in the clouds, such as the Bergeron-Findeisen mechanism for droplet growth. Since ice nucleation does not normally proceed until the temperature is well below 0°C, under the assumed conditions of Fig. 12.11 this process would not be significant in case I. In case II, however, the formation of ice clouds might begin somewhat in the warm sector; correspondingly, the rate of precipitation near the surface front might exceed the local condensation because part of the liquid water condensed earlier is caused to fall out when seeded by ice crystals.

The specific conditions in Fig. 12.11 are hypothetical, although they agree broadly with calculations of the ω-field in real cases. The actual rainfall would depend upon the nearness to saturation in the warm sector, the location of maximum ascent relative to the front, and the air temperature and maximum ω. The values assumed for the latter typify stronger-than-average disturbances and temperatures equivalent to real tropical air. Also the frontal cloud is generally in layers, indicating that part of the air entering the frontal circulation is too dry to yield precipitation.

The preceding discussion is valid only in the case of stable upgliding motions. If the air is potentially unstable, a moderate amount of lifting generally suffices to produce buoyant convection; in some cyclones (Chapter 13) most of the precipitation is convective. In that case, the moist low-level air ascends very rapidly after reaching condensation, with vertical velocities of meters per second rather than centimeters per second. As a result, the air rises essentially in the vertical (III in Fig. 12.11); a considerable part of the water content of the warm air is precipitated in the warm sector and less water vapor is left for frontal rain.

As noted earlier, characteristic differences in precipitation patterns between cyclones in the United States and in northern Europe can thus be explained largely by the differences of stability. In northern Europe, especially during the cold season, the warm air is essentially stable. Frontal precipitation of nonshowery type therefore predominates, with little real rain in the warm sector except for the innermost parts of a cyclone. By contrast, cyclones over the eastern United States often draw in air that has recently passed over the Gulf of Mexico, where the sea-surface temperature in winter is 10 to 15°C higher than the waters adjoining northern Europe (Fig. 12.8). Correspondingly, potential instability is more likely (also considering the greater insolation at the lower latitude), and many cyclones over the United States have substantial convective precipitation even during the cooler months.[4] In the warmer months, convective precipitation mainly in the warm sector is also observed in many European cyclones.

12.4   Clouds in Relation to Fronts

The extensive observations from British reconnaisance flights during the period 1939 to 1945 have been analyzed, especially with regard to warm fronts, by Matthewman (1955) and by Sawyer and Dinsdale (1955). They sought to relate the extent of development of cloud (in categories ranging from less than 25% of the space between the lower limit of the warm-front zone and the 400-mb level up to continuous cloud through the whole layer) to a large number of synoptic parameters. The only one found to be of real significance was the difference between the observed wind component normal to the upper boundary of the front, and its movement. The percentage of cloud-filled space on the average increased with this difference, which would be a measure of the upslope motion (for a corresponding conclusion regarding cold fronts, see Section 7.8). However, the correlation was not high, and the variability between cases was large. This led Matthewman to suggest that "the cloud in frontal regions may be associated more closely with the general upward motion over a large volume of air, including the front, rather than with any specified upgliding at the frontal surface."

Only on rare occasions did warm-front clouds (in the cooler months) extend through the whole layer from the front up to 400 mb. Rather, layers separated by clear spaces of significant depth predominated. Sawyer and Dinsdale found that within 100 nmi of the surface front, there was about 90% probability of cloud in the upper part of the frontal layer, decreasing to 60% at 4 km above the upper surface of the front. By contrast, at distances greater than 200 nmi, the probability was 30% in the upper part of the front (being nil 600 m below the lower frontal surface), increasing to about 75% at a height 4 km above the upper frontal surface. Thus the cloudy region generally has a slope exceeding that of the front, as in the example of Fig. 12.12, taken from Freeman (1961). This shows a west southwest to east

---

[4] Omoto (1965), however, has analyzed several cases with abundant warm-sector precipitation that is mainly nonconvective. The precipitation zones are in some cases superficially similar to squall lines, but more diffuse, and the circulation in a vertical plane is also similar. Omoto concludes that they are not self-propagating, as is a squall line (Chapter 13), rather being induced by the divergence fields associated with small-scale troughs or shear zones that are most distinct in the middle-tropospheric wind field.

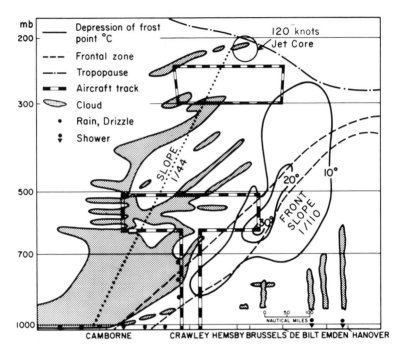

FIG. 12.12   Vertical section along aircraft tracks normal to a warm front over southern Britain on Oct. 7, 1955, showing clouds and saturation depression relative to frontal layer and jet-stream core. (After Freeman, 1961.)

northeast section normal to a warm front, 1000 km from the point of occlusion at the surface and 1500 km southeast of a 988-mb low center.

The slope of the cloud systems may be examined if we consider the mechanics of the "upgliding" motion in the following way: Let u be the component of air motion parallel to the front (positive with warm air to the right) and v the component toward the cold air. Then

$$\frac{du}{dt} = f(v - v_g)$$

Expanding the left-hand term (assuming uniformity in the x-direction so that $u\,\partial u/\partial x = 0$) and rearranging,

$$\frac{w}{v} = \frac{f\left(1 - \dfrac{v_g}{v}\right) - \dfrac{1}{v}\dfrac{\partial u}{\partial t} - \dfrac{\partial u}{\partial y}}{\partial u/\partial z} \tag{12.1}$$

Matthewman (1955) found that warm fronts moved on the average with the normal component of wind v_0 at 900 mb in the cold air. If we assume that this corresponds to the geostrophic wind component[5] and that the isotherms are parallel to the frontal zone (as shown by the 500 to 1000 mb thickness lines in this example), v_0 can replace v_g. Furthermore, if the structure of the whole wind system is conserved, $\partial u/\partial t = -v_0 \partial u/\partial y$. If β is the angle between the plane in which the air rises and the horizontal, then this angle (relative to the moving system) is given by $\tan \beta = w/(v - v_0)$ Making the appropriate substitutions into Eq. (12.1), this reduces simply to

$$\tan \beta = \frac{f - (\partial u/\partial y)}{\partial u/\partial z} \qquad (12.2)$$

Freeman also gives the wind field corresponding to Fig. 12.12. As an average value, $\partial u/\partial y \approx 10$ knots/150 nmi $= 0.18 \times 10^{-4}$ sec^{-1} in the cloud region, where $\partial u/\partial z \approx 4.1 \times 10^{-3}$ sec^{-1}. Using these values and $f = 1.12 \times 10^{-4}$ sec^{-1}, $\tan \beta \approx 1/44$, which is 2.5 times the frontal slope and corresponds roughly with the inclination of an envelope drawn around the whole cloud system.

It may be an oversimplification to view the process in terms of a continuous relative streamline extending through the whole depth. However, it must be considered that the layered structure in Fig. 12.12 is at a particular location, while the three-dimensional trajectories have a large component normal to the section. Other examples show a more continuous cloud system, broad in low levels and tapering to thin cloud in the upper troposphere. In the cases observed, *the slope of the cloud system seems typically to be about twice that of the front*, although variations may be expected because the horizontal and vertical shears differ from case to case. The general correspondence between the lower part of the cloudy sheet and the surface front, shown by the observations, is in accord with this being the region of strongest convergence (Chapter 9).

The research flights and the related soundings revealed that both warm and cold fronts are characterized by a pronounced saturation depression in the middle and upper troposphere, in and near the frontal layer. This feature, which fits the scant cloudiness in and near the frontal layer in the upper levels, is connected with the sinking motions involved in the process of upper-tropospheric frontogenesis (Section 9.4). Observations show that the tops of cold-air cumulus are lowest near the front and rise at increasing distances from it, compatible with Figs. 12.4b and 12.5b. The frontal in-

---

[5] This assumption is not quite justified; Sawyer and Dinsdale (1955) indicate that warm fronts move with about two-thirds of the geostrophic normal component at that level.

version would set a limit to the penetration of unstable clouds; in addition, however, entrainment of the dry air in the vicinity of the frontal layer into cumulus clouds would discourage their reaching even to the lower frontal surface. In the case of "active" cold fronts, or anafronts, the slopes of cloud systems in the warm air tend (as with warm fronts) to be greater than the frontal slope. Thus, in the neighborhood of an active cold front, there may be two separated regions with clouds extending to considerable elevations, one in the warm air above the front and the other well within the cold air where convective clouds can attain their greatest development, with a minimum of cloudiness near the upper part of the frontal layer itself.

Petterssen *et al.* (1963) have investigated the cloud structure of several winter and spring cyclones over the United States east of the Rockies, utilizing specially collected aircraft observations. Cyclones of moderate intensity were generally involved, with fronts that in no case reached above the lower troposphere. In the cyclone of Fig. 12.13, the fronts were quite distinct at the earth's surface, but did not extend above 700 to 800 mb. Despite the shallow slope of the warm front, a deep cloud mass was present over the east side of the cyclone (Fig. 12.14c), with a broad precipitation region. Over most of the area surrounding the cyclone there was an extensive

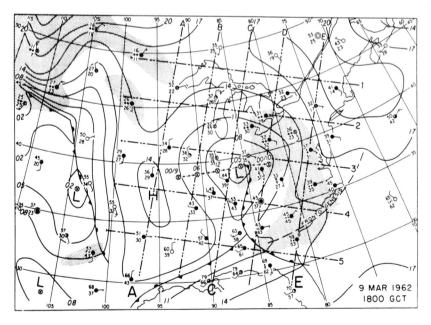

FIG. 12.13   Sea-level isobars, surface fronts, and (stippled) continuous precipitation areas, 18 GCT Mar. 9, 1962. (After E. P. McClain, in Petterssen *et al.*, 1963.)

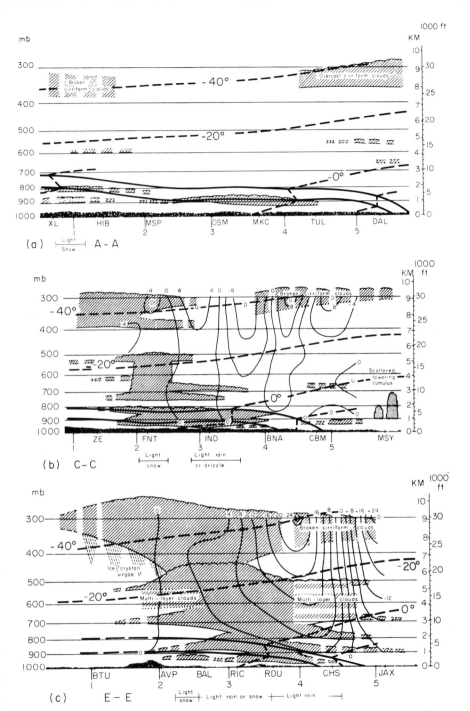

FIG. 12.14 Vertical sections along (a) line *A-A*, (b) line *C-C*, and (c) line *E-E* in Fig. 12.13. Clouds hatched; thick lines are boundaries of front; dashed lines, isotherms (°C). (After E. P. McClain, in Petterssen *et al.*, 1963.)

sheet of cirrostratus clouds, with wide breaks in the portion west of the cyclone. In that region, where the air was dry above 800 to 900 mb, middle clouds were generally absent, although a widespread low-cloud deck was present in the cold air. In other examples, distinct cloud layers dominated, often separated by clear layers of appreciable depth. In the weaker cyclones, the highest altostratus was generally below the 500 mb level.

Comparing the regular weather distributions in oceanic cyclones with those over land, Petterssen states: "Although many aspects of cyclone development became clarified and useful models were developed for the North Atlantic region, the results are not directly applicable to continental structures. Over the North America Continent, particularly to the east of the Continental Divide, the saturation deficit varies within wide limits, with consequent variability in the structure of cloud and weather systems that accompany fronts and cyclones...the variability of the vertical structures was found to be so large that the construction of representative models could not be attempted." For other examples illustrating this variability and the marked changes in weather patterns that can occur in a given cyclone, see Petterssen (1956).

12.5   Cloud Systems in Occlusions

Kreitzberg (1964) has made a systematic study of occluded systems as they passed over Seattle, Washington. These were analyzed by means of serial rawinsondes, vertical-pointing radar, and pilot reports as well as reports from regular surface stations. An example of a warm-front type occlusion is shown in Fig. 12.15. Here the thermodynamic structure is portrayed in terms of the quantity $(c_p T + gz + Lq)$, which Kreitzberg calls "static energy." This distribution is essentially related to that of wet-bulb potential temperature.

In the tongue of warmest air aloft, deep cloud penetrated from a lower-cloud deck into the middle troposphere, where the horizontal extent of clouds was least, and extensive decks of clouds were present in the upper troposphere. Rising motions as large as 30 to 40 cm/sec were computed[6] in parts of the cloud regions, while descent was usually indicated in the clear spaces between the cloud layers. The bases of the extensive cloud systems sloped upward toward east, bearing some relationship to the slopes

---

[6] By a single-station technique similar to that introduced by Árnason (1942), wherein the thermal advection is based on the wind hodograph. Kreitzberg's method retains part of the accelerations rather than being based purely on the geostrophic thermal-wind assumption.

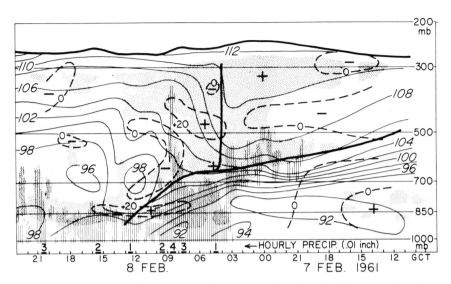

FIG. 12.15 Fronts and tropopause (heavy lines), cloud (stippled), and precipitation detected by vertical-pointing radar, in a time section during the passage of a warm-front occlusion over Seattle, Washington, Feb. 7–8, 1961. Dashed lines, vertical motion (centimeters per second); thin solid lines, static energy (joules per gram). (Redrawn from Kreitzberg, 1964.)

of the fronts overlying the coldest air masses. The "static energy" isopleths showed configurations broadly similar to the distribution of potential temperature in cross sections through occlusions by McClain and Danielsen (1955), who stressed their indistinct frontal structures.[7]

From a statistical analysis of the cloud distribution in such time sections, Kreitzberg concluded that in a given situation even the best generalized model has a small chance of verifying in all major respects, but that Bergeron's classical occlusion model is as good as any generalization that might be devised. The variability can again be laid partly to the variety of air masses with different moisture contents and stabilities, which might be drawn into the circulation of a cyclone (in the locality of Kreitzberg's studies, varying orographic influences could also be significant).

An extensive study of occlusions entering the southwestern United States, where the air masses are likely to be less stable, has been carried out by Elliott *et al.* (1961), and by Elliott and Hovind (1964). Some of the results

---

[7] The vague structure of Pacific Coast occlusions was earlier noted by A. K. Showalter, who called them "moist synclines." Particularly in Canada, some forecasters use the acronym "TROWAL" (TRough Of Warm air ALoft) to describe this characteristic.

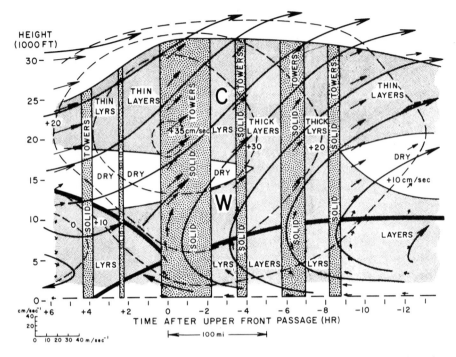

FIG. 12.16 Idealized time section during passage of an occlusion into California coast on Jan. 25–26, 1961. Arrows along verticals represent wind relative to moving system (scale in lower left), from which streamlines were drawn; dashed lines are isotachs (cm/sec) of vertical velocity. Heavy lines, frontal surfaces; light stippling, layer cloud (base unknown); heavy stippling, towers of convective bands. *C* and *W* indicate general regions of advective cooling and warming. (After Elliott *et al.*, 1961; streamlines redrawn from data in Elliott and Hovind, 1964. See text.)

are summarized in Fig. 12.16, using the occlusion of Jan. 26, 1961, as an example. At 00 GCT, a 990-mb cyclone was centered about 1000 km west of San Francisco. Figure 12.16 describes the main features of the occlusion southeast of this low as it passed inland near 34°N. Serial soundings were made at a triangle of stations (Point Arguello, Santa Monica, and San Nicolas Island, 100 km offshore), at 1- to 6-hr intervals. The stations are 130 to 180 km apart so that vertical motions computed from the winds[8]

---

[8] By integrating the kinematic divergence, generally by a "triangle method", but when winds from only two stations were available by a "rectangle method," assuming conservation of the structure in time. The vertical motions in Fig. 12.16 are taken from a figure by Elliott *et al.* (1961), giving preference to the more reliable triangle-method computations.

represent a scale between the "synoptic-scale" and "mesoscale." In Fig. 12.16 the vertical motions are combined with the horizontal motions normal to the occluded front, after subtracting its movement, to give streamlines relative to the frontal system. The horizontal and vertical scales of the relative velocity vectors are commensurate with the vertical exaggeration (42:1) in the figure.

Throughout the time represented, there was a southerly component, amounting to about 5 knots on the right and 25 knots on the left in the middle troposphere. Thus there was a flow partly into the figure, and the streamlines on the right side should be visualized as a half-helical motion. Also, the frontal surfaces sloped upward toward north. Maximum upward motions of 35 cm/sec were computed at 20,000 ft, near the time of passage of the cold front aloft. Weak downward motions in the cold air followed passage of the surface front. The frontal layers were diffuse, although the passage at the surface was marked by a distinct wind shift and discontinuity of barometric tendency. Temperature contrasts across the whole system were greatest in the middle troposphere, where there was an anomaly from the mean temperature, of $+2°C$ on the right, and $-2°C$ on the left side.

The schematic cloud system typifies other cold-season occlusions passing over this area. In this case, six bursts of moderate to heavy rainfall were observed, occurring at all stations and indicating distinct lines of convection oriented north northwest to south southeast. These traveled with a mean speed of 19 knots, approximately the 700-mb wind component normal to the lines, and also nearly the speed of the occluded front. The convective lines were embedded in stratiform layers, as outlined in Fig. 12.16. In the middle troposphere, substantial unsaturated layers were present. The general configurations of the cloud and dry layers are similar to those in Fig. 12.15. The streamlines in Fig. 12.16 pertain to average vertical motions over areas considerably wider than the convective lines. Hence the proper interpretation is that ascent takes place predominantly in the convective lines, with only feeble vertical motion in the stratiform cloud or the dry layers between them. For an illustration of the circulation and discussion of the precipitation mechanisms in a cold-front type of occlusion, see Elliott (1958).

12.6   Precipitation and the Budget of Water Vapor

Relationships between the vertical motions and water-vapor distribution in an atmospheric column and the rate of rainfall were developed by Fulks (1935) and also by Bannon (1948), who showed that the rainfall could be used to estimate the vertical motion. If liquid-water storage is neglected, the precipitation contributed per unit mass of air is

$$P_i = -\frac{dq}{dt} = -\omega \frac{dq}{dp} \tag{12.3}$$

where dq/dp means the change of specific humidity q along the three-dimensional path of an air parcel that, with a saturated adiabatic process, implies a decrease of water vapor during its movement to lower pressure, the condensed water falling out as precipitation. Since dq/dp is not readily determined, it is (in Fulks' method) replaced by $\partial q/\partial p$, which can be evaluated from a local sounding.

Full expansion of Eq. (12.3) gives

$$P_i = -\frac{\partial q}{\partial t} - \mathbf{V} \cdot \nabla q - \omega \frac{\partial q}{\partial p} \tag{12.4}$$

Thus, utilizing only the last term, the method is valid if the moisture content does not change locally and the advection of moisture is insignificant, as in a uniform air mass, or if the local change balances the advection. By use of the continuity equation $\nabla \cdot \mathbf{V} = -\partial \omega/\partial p$, Eq. (12.4) may be replaced by the identity

$$P_i = -\frac{\partial q}{\partial t} - \nabla \cdot q\mathbf{V} - \frac{\partial (q\omega)}{\partial p} \tag{12.5}$$

In Eqs. (12.3) through (12.5), P_i is the water condensed in the atmosphere; for total water balance, evaporation must be considered. Hence,

$$P - E = \frac{1}{g} \int_0^{p_0} P_i \, dp$$

where P and E are the precipitation and evaporation at the earth's surface. Then, upon integration of Eq. (12.5) through the whole vertical column,

$$P = E - \frac{1}{g} \int_0^{p_0} \frac{\partial q}{\partial t} \, dp - \frac{1}{g} \int_0^{p_0} \nabla \cdot q\mathbf{V} \, dp \tag{12.6}$$

since $(q\omega)$ in the last term of Eq. (12.5) vanishes at the top of the atmosphere (or at a lower level where $\omega = 0$), and in most cases has a small value at the earth's surface where ω is generally small.

The formulation in Eq. (12.6), although equivalent to Eq. (12.4), can be considered more reliable because it does not require an evaluation of $\partial q/\partial p$. The last term of Eq. (12.4) gives a contribution to the rainfall only if the air is saturated. Hence, in its evaluation, all layers have to be excluded where the air is not saturated. The interpretation of soundings in this regard becomes to some extent arbitrary because a radiosonde, even when ascending through a cloud layer, often does not register saturation. Moreover, ambiguity arises in cases of showery precipitation because the measured value of $\partial q/\partial p$ can be very different if the sounding balloon does or does not rise through an active cloud. In the use of an individual sounding to represent the mean condition over a large surrounding area, it could happen, for example, that in the region of intense showers no balloon rises through a cloud; in this case, the air at practically no level would appear to be saturated.

In the form of Eq. (12.6), on the other hand, the rate of saturation does not appear. Also, the computations involve evaluation of q and $\nabla \cdot q\mathbf{V}$ over relatively large regions but need not take account of the detailed distributions of these quantities, as would be necessary for a strict evaluation of $\omega\, \partial q/\partial p$ integrated over a corresponding area if the precipitation were uneven. The good agreement obtained between computations based on Eq. (12.6) and the observed precipitation suggests that, despite the considerable distance between sounding stations, the measured values at their locations are representative enough to allow a reliable computation of the horizontal moisture flux and its divergence.

An example (Palmén and Holopainen, 1962) is illustrated in Fig. 12.17, which shows: the surface fronts and 1000-mb contours of a developing cyclone with the 500-mb contours superimposed; the kinematically computed horizontal divergence at the 950-mb level; and the ω-field at 500 mb from the integrated divergence. Figure 12.18a shows $[\nabla \cdot \mathbf{V}]$ and $[\omega]$, area averages over the outlined region A ($0.96 \times 10^{12}\ \mathrm{m^2}$). The variations with height are fairly typical of the central part of a cyclone in advance of a well-developed upper trough. The corresponding mean divergence of water vapor, multiplied by the area A, is shown in Fig. 12.18b. When integrated over pressure, this gives the total mass of precipitation within the area per unit time, contributed by the last term of Eq. (12.6). In this case no attempt was made to evaluate the local change (second r.h.s. term). The evaporation must also have been small, since it was nighttime, the region was generally overcast, and the dew-point deficit was small. Based on Fig. 12.18b, the water-vapor accumulation had a mean value of

$$-\frac{1}{g} \int_{400\ \mathrm{mb}}^{980\ \mathrm{mb}} [\nabla \cdot q\mathbf{V}]\, dp = 7.4\ \mathrm{mm/6\ hr}$$

which agrees closely with the observed area-mean rainfall of 7.6 mm during a 6-hr period centered on the time of the computation.

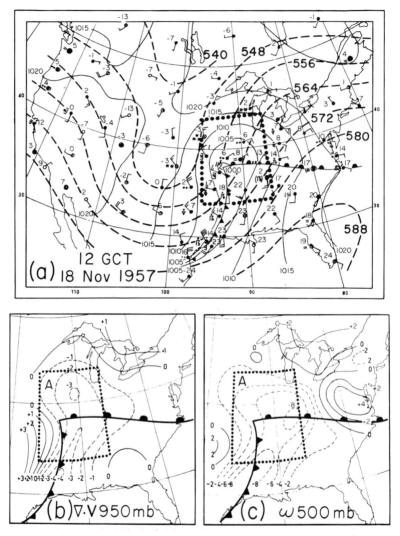

Fig. 12.17   For 12 GCT Nov. 18, 1957: (a) surface fronts, sea-level isobars, and 500-mb contours (tens of decameters); (b) horizontal divergence (10^{-5} sec^{-1}) at 950 mb; and (c) vertical motion (μb/sec) at 500 mb. (After Palmén and Holopainen, 1962.)

Considering the decrease of q with height, the curve of water-vapor divergence in Fig. 12.18b is very similar to the curve of horizontal velocity divergence in Fig. 12.18a. If the last term of Eq. (12.6) is expanded,

$$P = E - \frac{1}{g} \int_0^{p_0} \frac{\partial q}{\partial t}\, dp - \frac{1}{g} \int_0^{p_0} \mathbf{V}\cdot\nabla q\, dp - \frac{1}{g} \int_0^{p_0} q\nabla\cdot\mathbf{V}\, dp \quad (12.7)$$

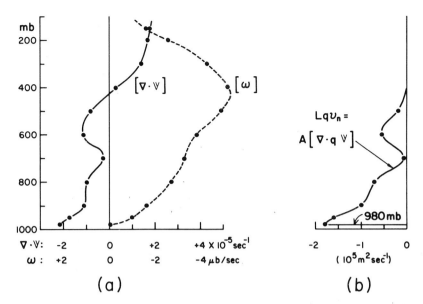

FIG. 12.18   (a) Mean divergence and vertical velocity, and (b) water-vapor accumulation, over area A outlined in Fig. 12.17. (After Palmén and Holopainen, 1962.)

When only the last term of Eq. (12.7) was considered, this gave

$$-\frac{1}{g}\int_{400\text{ mb}}^{980\text{ mb}}[q\nabla\cdot\mathbf{V}]\,dp = 6.2\text{ mm/6 hr}$$

Thus, *in this cyclone, whose central portion was occupied by moist air, the overwhelming part of the water-vapor accumulation was controlled by the actual velocity convergence*, the contributions from the other terms taken together being small. In the case of no divergence and no precipitation, Eq. (12.7) reduces to

$$\frac{1}{g}\int_0^{p_0}\frac{\partial q}{\partial t}\,dp = E - \frac{1}{g}\int_0^{p_0}\mathbf{V}\cdot\nabla q\,dp$$

Although the last term may be quite large (when the air flowing into a region has significantly greater q than that flowing out), a close balance can generally be expected between the vapor accumulation and the evaporation from the earth's surface plus the advection of water vapor. It may also be noted that in the preceding calculation, extended to the 400-mb level only, it is tacitly assumed that water or water vapor carried upward through that level falls back as precipitation. The good agreement between computed and observed rainfall suggests that change of water storage and the transport of water across the lateral boundaries were relatively unimportant. This is a reasonable deduction, since at the time considered the cloud system of the cyclone was already well developed, and it is likely that the volume considered was largely filled with cloud at the outset.

Equation (12.6) may be alternatively written as

$$P = E - \frac{1}{g} \int_0^{p_0} \frac{\partial q}{\partial t} \, dp - \frac{1}{Ag} \oint_L \int_0^{p_0} q v_n \, dp \, dL \qquad (12.8)$$

where v_n denotes the outward velocity component normal to the boundary L of area A. When it is desired only to compute the water budget of a large area, without regard to the detailed distribution within it, this expression is convenient because the distribution of $\nabla \cdot q\mathbf{V}$ need not be evaluated. For example, Fletcher (1951) used this formulation in discussing hydrologic applications, such as the calculation of runoff from a watershed. Computations based on Eq. (12.8) are discussed in Sections 13.4 and 13.5 for individual synoptic systems and convective storms and in Chapter 2 for the water budgets over latitudinal zones.

The method was applied by Benton and Estoque (1954) to determine the monthly water budget over North America during 1949; this was in good agreement with hydrologic data and empirical estimates of evapotranspiration. Regional water-budget studies using the same principles have been presented by Hutchings (1957) for southern England, Hutchings (1961) for Australia, Nyberg (1965) for southern Sweden, Palmén and Söderman (1966) for the Baltic Sea, and Rasmusson (1967) for North America and vicinity. These papers may be referred to for data concerning the geographical variations, partitions between zonal and meridional, mean and eddy fluxes, and the distributions of vapor flux with height. It may be remarked that in some cases the water-vapor flux was computed on the basis of geostrophic winds. This is an approximation that may be permissible for large areas and extended periods. However, it is questionable in individual cyclones wherein the wind divergence is strong, since divergence is largely associated with ageostrophic motions.

In the example given above, there was already abundant water vapor in the neighborhood of the cyclone at the time discussed. Thus, as was pointed out, the velocity convergence near the cyclone center was closely related to the rainfall intensity. This would be typical of cyclones over oceanic regions or those (e.g.) in winter over western Europe where the region is predominantly occupied by humid air from the Atlantic Ocean. However, in such regions as central North America or eastern Asia, the moisture source is more uncertain.

The cyclone described in Section 11.4 offers a striking example of the influence of moisture availability upon the weather distribution. As shown in Fig. 12.10, this cyclone when fully developed had a well-developed precipitation shield. However, a day earlier there was no precipitation in advance of the cyclone, although there was appreciable ascent in southerly flow. Backward trajectories (Newton, 1958) showed that the air in the center of upward motion had earlier descended behind the preceding cyclone, following a path over the continent.

As the cyclone moved eastward, it came into a location where southerly flow advanced moist air from the Gulf of Mexico. The ensuing rapid increase of water vapor is illustrated in Fig. 12.19, a section approximately along the axis of the moist tongue east of the cyclone. Local increases of water vapor took place partly by the northward advection of moist air by the low-level jet and partly a simultaneous increase in the depth of the moist layer. During this time, when precipitation was absent and evaporation negligible compared with the large local changes of water-vapor content, the balance was essentially expressed by Eq. (12.7) in the form

$$\frac{1}{g} \int_0^{p_0} \frac{\partial q}{\partial t} \, dp \approx -\frac{1}{g} \int_0^{p_0} \mathbf{V} \cdot \nabla q \, dp + \frac{1}{g} \int_0^{p_0} q \frac{\partial \omega}{\partial p} \, dp$$

Considering the horizontal winds and moisture gradient together with the upward motions, both r.h.s. terms were large. *The rapid northward spreading of the moist tongue, with accompanying low cloudiness, resulted in a transformation of this cyclone from an almost completely dry vortex to one with abundant precipitation in less than 24 hr.* Although at the fully developed stage there was a widespread precipitation shield characteristic of occluded cyclones, most of the precipitation occurred as heavy rain in the restricted region shown in Fig. 12.10, largely within the warm sector where some stations reported over 20 cm of rain. The rapid development of convection in this area resulted from processes outlined in Section 13.2.

Such a marked change in weather characteristics was clearly due to geographical circumstances. Fawcett and Saylor (1965) give an instructive description of the weather patterns accompanying cyclones that form in the Colorado region and move northeastward from there. This is based on 5-year composites of the sea-level pressure and 500-mb contours, thickness fields, vorticity and vertical motion patterns, and precipitation distribution for 28 cyclones in February through April. Their charts show that although the likelihood of precipitation is greatest on the general northwest or north side of the cyclone throughout its lifetime, the probability of rainfall on the

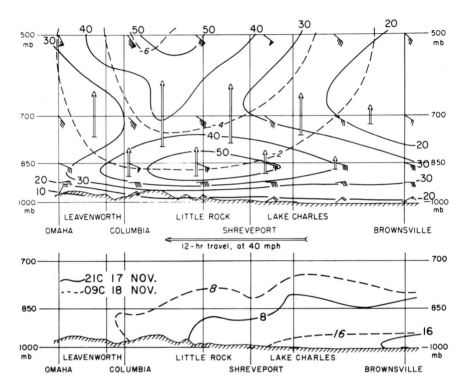

FIG. 12.19   Vertical section (approximately north-south) along moist tongue east of cyclone, 12 and 24 hr prior to the time shown in Fig. 12.10. In upper diagram, mean conditions for the period: wind symbols (pennant is 50 mph) are plotted so that wind from right is in plane of section; solid lines, isotachs; dashed lines, vertical motions (μb/sec), computed from vorticity equation. Vertical movement in 12 hr shown by double arrows. In lower section, specific-humidity isopleths at beginning (solid) and end (dashed) of 12-hr period. (From Newton, 1958.)

east and southeast sides increases markedly during the period 24 hr after cyclogenesis. Fawcett and Saylor observe: "During the early history of the storm, the moisture source is the Pacific Ocean while in later stages the Gulf of Mexico serves as the source." During eastward movement of the cyclones, severe weather also increased. This reached a maximum 24 hr after cyclogenesis, when 84% of the storms had heavy snow (most frequent north northeast of the cyclone center) and 81% were accompanied by severe thunderstorms (southeast of the center). These changes of weather patterns clearly show the influence of moisture importation into cyclones that were occupied by relatively dry air in their early lifetimes.

REFERENCES

Árnason, G. (1942). Distribution of mass variations in atmospheric air columns. *Meteorol. Ann.* **10**, 255–279.

Bannon, J. K. (1948). The estimation of large-scale vertical currents from the rate of rainfall. *Quart. J. Roy. Meteorol. Soc.* **74**, 57–66.

Benton, G. S., and Estoque, M. A. (1954). Water-vapor transfer over the North-American continent. *J. Meteorol.* **11**, 462–477.

Bergeron, T. (1951). A general survey in the field of cloud physics. *Intern. Union Geod. Geophys., Assoc. Meteorol. Brussels, 1951*, Ninth Gen. Assembly Mem., pp. 120–134.

Bergeron, T. (1960). Problems and method of rainfall investigation. *In* "Physics of Precipitation" (H. K. Weickmann, ed.), Geophys. Monograph No. 5, pp. 5–30. Am. Geophys. Union, Washington, D. C.

Borovikov, A. M., Gaivoronskii, I. I., Zak, E. G., Kostarev, V. V., Mazin, I. P., Minervin, V. E., Khrgian, A. Kh., and Shmeter, S. M. (1961). "Fizika oblakov," Chapter 7. Gidrometeorologicheskoe Izdatel'stvo, Leningrad. (Transl., "Cloud Physics," 392 pp. Israel Progr. Sci. Transl., Ltd.; Office Tech. Serv., OTS-63-11141, Washington, D. C.)

Boucher, R. J., and Newcomb, R. J. (1962). Synoptic interpretation of some TIROS vortex patterns: A preliminary cyclone model. *J. Appl. Meteorol.* **1**, 127–136.

Boyden, C. J. (1963). Jet streams in relation to fronts and the flow at low levels. *Meteorol. Mag.* **92**, 319–328.

Bunker, A. F. (1956). Measurement of counter-gradient heat flows in the atmosphere. *Australian J. Phys.* **9**, 133–143.

Burke, C. J. (1945). Transformation of polar continental air to polar maritime air. *J. Meteorol.* **2**, 94–112.

Burkovskaya, S. N. (1959). Oraspredelenii vodnosti v oblakakh teplogo fronta (The distribution of water content in warm-front clouds.) *Tr. Tsentr. Aerolog. Observ.* No. 28.

Craddock, J. M. (1951). The warming of arctic air masses over the eastern North Atlantic. *Quart. J. Roy. Meteorol. Soc.* **77**, 355–364.

Crutcher, H. L., Hunter, J. C., Sanders, R. A., and Price, S. (1950). Forecasting the heights of cumulus cloud-tops on the Washington-Bermuda airways route. *Bull. Am. Meteorol. Soc.* **31**, 1–7.

Elliott, R. D. (1958). California storm characteristics and weather modification. *J. Meteorol.* **15**, 486–493.

Elliott, R. D., and Hovind, E. L. (1964). On convection bands within Pacific Coast storms and their relation to storm structure. *J. Appl. Meteorol.* **3**, 143–154.

Elliott, R. D., Hovind, E. L., and Flavin, J. W., Jr. (1961). Investigation of cloud-water budget of Pacific storms. Final Rept., Natl. Sci. Found. Contr. C-104, 64 pp. Aerometric Res. Inc., Goleta, Calif.

Fawcett, E. B., and Saylor, H. K. (1965). A study of the distribution of weather accompanying Colorado cyclogenesis. *Monthly Weather Rev.* **93**, 359–367.

Fletcher, R. D. (1951). Hydrometeorology in the United States. *In* "Compendium of Meteorology" (T. F. Malone, ed.), pp. 1033–1047. Am. Meteorol. Soc., Boston, Massachusetts.

Freeman, M. H. (1961). Fronts investigated by the Meteorological Research Flight. *Meteorol. Mag.* **90**, 189–203.

Fulks, J. R. (1935). Rate of precipitation from adiabatically ascending air. *Monthly Weather Rev.* **63**, 291–294.

George, J. J. (1960). "Weather Forecasting for Aeronautics," 673 pp. Academic Press, New York.

Godske, C. L., Bergeron, T., Bjerknes, J., and Bundgaard, R. C. (1957). "Dynamic Meteorology and Weather Forecasting," Chapter 14. Am. Meteorol. Soc., Boston, Massachusetts.

Hutchings, J. W. (1957). Water-vapour flux-divergence over southern England: Summer 1954. Quart. J. Roy. Meteorol. Soc. 83, 30–48.

Hutchings, J. W. (1961). Water-vapor transfer over the Australian continent. J. Meteorol. 18, 615–634.

Ichiye, T., and Zipser, E. J. (1967). An example of heat transfer at the air-sea boundary over the Gulf Stream during a cold air outbreak. J. Meteorol. Soc. Japan 45, 261–270.

Jarvis, E. C. (1964). An adaptation of Burke's graphs of air mass modification for operational use. J. Appl. Meteorol. 3, 744–749.

Kreitzberg, C. W. (1964). The structure of occlusions as determined from serial ascents and vertically-directed radar. AFCRL-64-26. Air Force Cambridge Res. Labs., L. G. Hanscom Field, Massachusetts.

Kuettner, J. (1959). The band structure of the atmosphere. Tellus 11, 267–295.

McClain, E. P., and Danielsen, E. F. (1955). Zonal distribution of baroclinity for three Pacific storms. J. Meteorol. 12, 314–323.

Matthewman, A. G. (1955). A study of warm fronts. Prof. Notes No. 114, 23 pp. Gr. Brit. Meteorol. Office, London.

Nagle, R. E., and Serebreny, S. M. (1962). Radar precipitation echo and satellite cloud observations of a maritime cyclone. J. Appl. Meteorol. 1, 279–295.

Newton, C. W. (1958). Convective precipitation in an occluded cyclone. J. Meteorol. Soc. Japan 75th Anniv. Vol., pp. 243–255.

Nyberg, A. (1965). A computation of the evaporation in Southern Sweden during 1957. Tellus 17, 473–483.

Oliver, V. J., and Ferguson, E. W. (1966). The use of satellite data in weather analysis. In "Satellite Data in Meteorological Research" (H. M. E. van de Boogaard, ed.), pp. 85–101, NCAR-TN-11.

Omoto, Y. (1965). On pre-frontal precipitation zones in the United States. J. Meteorol. Soc. Japan 43, 310–330.

Palmén, E. (1951). The aerology of extratropical disturbances. In "Compendium of Meteorology" (T. F. Malone, ed.), pp. 599–620. Am. Meteorol. Soc., Boston, Massachusetts.

Palmén, E., and Holopainen, E. O. (1962). Divergence, vertical velocity and conversion between potential and kinetic energy in an extratropical disturbance. Geophysica (Helsinki) 8, 89–113.

Palmén, E., and Söderman, D. (1966). Computation of the evaporation from the Baltic Sea from the flux of water vapor in the atmosphere. Geophysica (Helsinki) 8, 261–279.

Petterssen, S. (1956). "Weather Analysis and Forecasting," 2nd ed., Vol. II, Chapter 26. McGraw-Hill, New York.

Petterssen, S., and Calabrese, P. A. (1959). On some weather influences due to warming of the air by the Great Lakes in winter. J. Meteorol. 16, 646–652.

Petterssen, S., Knighting, E., James, R. W., and Herlofson, N. (1945). Convection in theory and practice. Geofys. Publikasjoner, Norske Videnskaps-Akad. Oslo 16, No. 10, 1–44.

Petterssen, S., Bradbury, D. L., and Pedersen, K. (1962). The Norwegian cyclone models in relation to heat and cold sources. *Geofys. Publikasjoner, Norske Videnskaps-Akad. Oslo* **24**, 243–280.

Petterssen, S., Bradbury, D. L., McClain, E. P., Omoto, Y., Rao, G. V., Ternes, E. R., and Watson, G. F. (1963). An investigation of the structure of cloud and weather systems associated with cyclones in the United States. Final Rept., Contr. AF 19(604)-7230, Article H. Dept. Geophys. Sci., Univ. Chicago.

Pogosian, Kh. P. (1960). "Struinye techeniia v atmosfere," Chapter 4. Gidrometeorologicheskoe Izdatel'stvo, Moscow. (Transl. by R. M. Holden, "Jet Streams in the Atmosphere," Mimeograph, 187 pp. Am. Meteorol. Soc., Boston, Massachusetts.)

Priestley, C. H. B. (1959). "Turbulent Transfer in the Lower Atmosphere," 130 pp. Univ. of Chicago Press, Chicago, Illinois.

Rasmusson, E. M. (1967). Atmospheric water vapor transport and the water balance of North America: Part 1. Characteristics of the water vapor flux field. *Monthly Weather Rev.* **95**, 403–426.

Riehl, H. (1948). Jet stream in upper troposphere and cyclone formation. *Trans. Am. Geophys. Union* **29**, 175–186.

Rogers, C. W. C. (1965). A technique for estimating low-level wind velocity from satellite photographs of cellular convection. *J. Appl. Meteorol.* **4**, 387–393.

Sawyer, J. S., and Dinsdale, F. E. (1955). Cloud in relation to active warm fronts, Bircham Newton, 1942–46. Prof. Notes, No. 115, 10 pp. Gr. Brit. Meteorol. Office, London.

Serebreny, S. M., Wiegman, E. J., and Hadfield, R. G. (1962). Investigation of the operational use of cloud photographs from weather satellites in the North Pacific. Final Rept., Contr. Cwb-10238, 93 pp. Stanford Res. Inst., Menlo Park, Calif.

Starrett, L. G. (1949). The relation of precipitation patterns in North America to certain types of jet streams at the 300-mb level. *J. Meteorol.* **6**, 347–352.

Sverdrup, H. U. (1942). "Oceanography for Meteorologists," 246 pp. Prentice-Hall, Englewood Cliffs, New Jersey.

van de Boogaard, H. M. E., ed. (1966). "Satellite Data in Meteorological Research," NCAR-TN-11, 349 pp. Natl. Center Atmospheric Res., Boulder, Colorado.

Widger, W. K. (1964). A synthesis of interpretations of extratropical vortex patterns as seen by TIROS. *Monthly Weather Rev.* **92**, 263–282.

Winston, J. S., and Tourville, L. (1961). Cloud structure of an occluded cyclone over the Gulf of Alaska as viewed by TIROS I. *Bull. Am. Meteorol. Soc.* **42**, 151–165.

13

ORGANIZED CONVECTIVE SYSTEMS
IN MIDDLE LATITUDES

Over many regions of the middle latitudes as well as the tropics, a large portion of the total annual precipitation is derived from convective showers and thunderstorms, which also account for the greatest overall weather damage in many areas. Convective storms are also of particular interest because they manifest perhaps the most obvious interaction between circulations on large and small (or medium) scales. The occurrence of significant convection is strongly influenced by synoptic-scale disturbances, and convective clouds in turn are very effective agents for the vertical transfer of energy and horizontal momentum in amounts significant for the general circulation.

In this chapter we treat mainly a characteristic type of organized convective system connected with synoptic disturbances. The type of "air mass" convection initiated by surface heating will not be discussed, since the principles involved are described in many textbooks. It should, however, be stressed that the degree of "air mass" convection is also influenced by the larger-scale synoptic systems. This is well known to all forecasters, in terms of the flow patterns that determine the air masses over a given region, and in terms of the influence of general ascent or subsidence upon the stability and the depth of the moist layer. It also follows from the principles of the atmospheric water budget (Section 12.6). Since the total mean precipitation over a large region of shower activity is determined by the evaporation from the earth's surface and the net influx of water vapor, in the absence of such an influx the mean precipitation will correspond to the mean evaporation. Thus "air mass" showers can occur in a quiescent synoptic situation, but they may be significantly enhanced or suppressed by relatively feeble organized convergence or divergence in the lower layers.

13.1   DISTRIBUTION OF CONVECTIVE STORMS AND OF SEVERE STORMS

The frequency of thunderstorm occurrence varies greatly, owing to (among other influences) the character of the earth's surface. Over the eastern oceans in middle latitudes, 5 to 10 thunderstorm days per year are characteristic, while in some parts of tropical Africa and South America, thunderstorms occur on 180 to 200 days per year (World Meteorological Organization, 1956).

There is, however, no close relationship between the frequency of occurrence of thunderstorms and the damage they produce. In Fig. 13.1, Ludlam

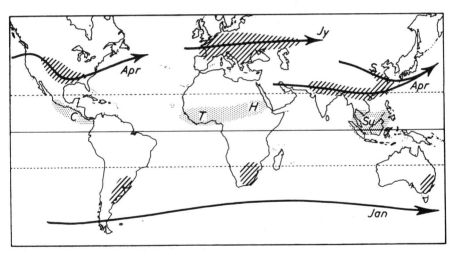

FIG. 13.1   According to Ludlam (1963): "Distribution of areas subject to severe squall-thunderstorms. In middle latitudes these areas (hatched) are closely related to the mean position of the jet-stream axis at the 500-mb level for the appropriate season... In midsummer the jet over N. India shifts (to north of the Tibetan plateau), and the severe hailstorms [and squalls] there become infrequent. In the tropics the squall-storms are not accompanied by large hail. They mostly occur over regions (stippled) where monsoonal circulations produce well-defined intertropical fronts near which there is a lower-tropospheric wind shear not closely related to a middle-level jet stream." The regions *T-H* and *Su* are, however, overlain by strong mean easterlies at the 200-mb level (see Fig. 3.13b) in summer. Also, although the strongest winds in the Southern Hemisphere are at the latitude shown, a secondary maximum of mean westerlies passes directly over the squall areas.

(1963) has summarized the regions where severe squall thunderstorms are most common. In middle latitudes, these are in general the localities that most commonly come under the influence of the jet stream. Since cyclones are also associated with the jet-stream region, this is one indication of the frequent connection between synoptic-scale disturbances and severe thunderstorms. This association results from two influences: the action of synoptic-scale circulations in creating an air-mass structure favorable for convective overturning; and invigoration of the energy processes of the storm itself, due to the structure of the wind system in which it is embedded.

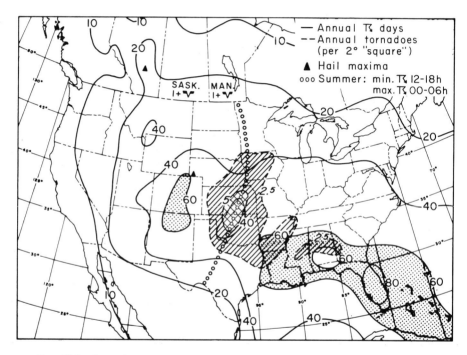

FIG. 13.2 Annual number of thunderstorm days (solid lines, taken with some alterations from World Meteorological Organization, 1956) and annual number of tornadoes per 2° latitude-longitude tessera (dashed lines, from House, 1963). Along the line of circles there is a nocturnal maximum and afternoon minimum in thunderstorm frequency.

The lack of any clear correlation between the frequency of thunderstorms and their severity is emphasized by Fig. 13.2. Although tornadoes (which occur with thunderstorms that are in other respects severe) occur in the broad region of frequent thunderstorms, the *maximum* number of *severe* storms coincides nearly with a relative *minimum* (< 40) of the total thun-

derstorm days per year. Most thunderstorms do not reach destructive proportions, for which special environmental conditions seem to be required.

The western maximum in Fig. 13.2 is due to the frequent formation of
heat thunderstorms over mountains; the southeastern one, to the direct influence of surface heating, the effect of which is greatly enhanced by the
convergence of sea breezes during daytime over the Florida peninsula
(Byers and Rodebush, 1949). In the Great Plains region, where there are
fewer thunderstorms, the air is not normally so rich in water vapor as in
the area farther east. However, this region is peculiarly suited, by virtue

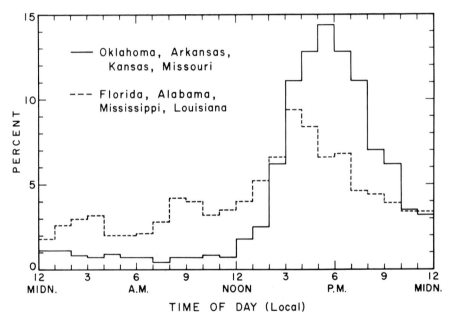

FIG. 13.3   Percentage frequency of tornado occurrences (by hours) for regions in the
central United States (solid) and for Gulf Coast states (dashed). (After House, 1963.)

of orographic features, for the rapid northward flux of water vapor on the
approach of a cyclone from the west and for the generation of a state of
considerable potential instability on such occasions. This accounts for the
intensity of the more limited number of storms which do occur.

Figure 13.3 indicates that the diurnal variation of surface heating is
important, but that its effect relative to other influences varies in different
regions. The solid curve corresponds to the central states, where the maximum frequency of severe storms is in late spring and early summer. In
the southeastern states, the maximum occurrence is in early spring when

surface heating is not so pronounced, the influence of insolation and outgoing radiation also frequently being suppressed by extensive cloud cover in the marine-air mass. Dynamical influences due to synoptic disturbances assume a relatively large importance, and these may initiate thunderstorms at any time of day or night.

In the central states, Fig. 13.3 also shows that the maximum occurrence is not in early afternoon when the surface temperature is usually highest, but is displaced toward late afternoon and evening. As indicated in Fig. 13.2, in parts of this region in the warmer part of the year, thunderstorms are more frequent in the early morning hours than in the afternoon. Means (1944) has connected this behavior with the diurnal variation in the strength of the low-level southerly winds (see Fig. 8.25), in its effect on changes of stability through horizontal advection. Sangster (1958) demonstrated the influence of this wind variation in causing maximum low-level convergence in the evening over the Great Plains. Bleeker and Andre (1951), who had earlier shown that a significant diurnal variation existed (with maximum convergence over the Great Plains in the early morning), interpreted this as the result of circulations set up by differential heating and cooling over the Rocky Mountains and the plains to the east. Brown (1962) has also demonstrated that one of the principal factors is that squall lines, generated in the afternoon over the western plains and moving eastward with ordinary speeds, arrive at night in the region concerned. These viewpoints should not be considered as alternatives; rather they are complementary.

13.2   GENERATION AND RELEASE OF INSTABILITY

Figure 13.4 illustrates the diversity of air-mass structures observed prior to the occurrence of destructive thunderstorms. The "Oklahoma tornado" sounding in Fig. 13.4a is also representative of situations that produce medium-sized hail; on occasion, appreciably greater instability may be present, producing very large hailstones. Studies in different regions by Donaldson (1958), Douglas and Hitschfeld (1959), and Schleusener and Grant (1961) indicate that the higher the top of a thunderstorm, especially when it penetrates the tropopause, the greater is the probability of hail from it. This association is due to the fact that strong upward currents are required both to carry the air in the updraft to levels where it becomes colder than the surroundings (and thus negatively buoyant) and to suspend or recirculate hailstones, whose terminal velocities increase with their sizes, long enough for them to grow. Other manifestations of storm severity are in a general way correlated with the conditions favorable for hail. Thus, for example, tornadoes may occur with either of the sounding types shown in Fig. 13.4, but are more likely and more severe with the type in Fig. 13.4a.

The common feature of these soundings, considering the overall depth of the troposphere, is the presence of both convective and conditional

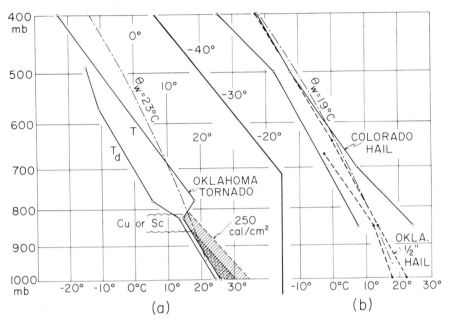

FIG. 13.4   (a) Mean temperature (T) and dewpoint (T_d) sounding shortly before the occurrence of tornadoes, after Fawbush and Miller (1953). (b) Dashed curves show mean sounding for 1/2-in. diameter hail in Oklahoma area (Fawbush and Miller). Solid curves show mean sounding for hail situations at Denver (Beckwith, 1960). Dash-dotted curves are moist adiabats corresponding to θ_w in lowest 100 mb, not considering the increase that would result from solar heating of this layer (the curve $\theta_w = 19°C$ is valid for both soundings on the right).

instability. The *degree* of instability, however, may vary greatly, and the conditions required to *release* the instability may also differ.

This may be appreciated by comparison of the two Oklahoma soundings. The tornado sounding is characterized by great potential instability,[1] but the upward penetration of air from the moist layer is inhibited by the

[1] In our usage, "potential instability" will refer to a combined condition of "convective instability" ($\partial\theta_w/\partial z < 0$) and "conditional instability" ($\gamma > \gamma_m$) when a deep layer is considered. In general, either of these latter types of instability may exist without the other. The terms *potential* and *convective instability* are sometimes used synonymously.

inversion near 800 mb. In spring, for which season this sounding is most representative, the potential insolation (heat accumulation from incoming minus outgoing radiation, under clear skies, between the coolest and warmest part of the day) is roughly 250 cal/cm² at the latitude concerned. If a layer Δp in thickness is warmed by an amount $\overline{\Delta T}$, the heating required per unit area is $c_p \overline{\Delta T} \Delta p/g$. The maximum possible heating (hatched area) would be inadequate to eliminate the inversion. Moreover, with this type of sounding about one-third of this amount of heating would suffice to produce shallow convective clouds in the layer indicated, and these reduce the insolation.

With this thermal structure, then, it is not possible to realize deep convection purely through the agency of solar heating unless the inversion is considerably weaker than that shown. By contrast, in the case of the Oklahoma small-hail sounding, when the convective temperature is reached (requiring three-tenths of the potential insolation), the ensuing convective clouds are free to penetrate to high levels, since there is no inversion to inhibit their vertical growth.

Hailstorms are most frequent in the High Plains regions marked by triangles in Fig. 13.2, which is seldom invaded by marine tropical air in low levels. The typical air-mass structure in the Alberta region (Longley and Thompson, 1965) is also very similar to the "Colorado hail" sounding in Fig. 13.4b. The relative humidity is characteristically low in the layers near the ground; however, a realizable amount of surface heating will start convection. At Denver (Beckwith, 1960), hailstorms commonly occur some time after a cold-front passage; northeasterly winds increase the moisture content in low levels, and orographic lifting is also influential.

Longley and Thompson found it difficult to differentiate clearly between the stabilities and lower-level moisture contents on major hail days and days without hail in Alberta. The synoptic parameters indicated, however, that afternoon hail is most likely in Alberta when there is an upper-tropospheric trough over British Columbia in the morning, with an approaching cold front and a low-level warm tongue ahead of it. These are conditions under which general ascending motions are to be expected; it may be noted that a lifting of the High Plains sounding by even as little as 50 mb makes this air mass much more susceptible to thunderstorm formation from surface heating.

The foregoing considerations indicate that it is necessary to consider the dynamical factors associated with synoptic systems as well as the vertical structure of the air mass observed at a given time. This is reflected in the procedures for forecasting severe local storms, described by Showalter and

Fulks (1943), Fawbush *et al.* (1951), and the staff of the Severe Local Storm Forecast Center (U.S. Weather Bureau, 1956). The importance of synoptic-scale processes for generation of severe convective storms is also recognized in tropical regions (Section 14.10).

The conditions that contribute to generate convective instability and that lead to its release were in general terms outlined rather completely by Normand (1938). Figure 13.5 illustrates these conditions schematically for

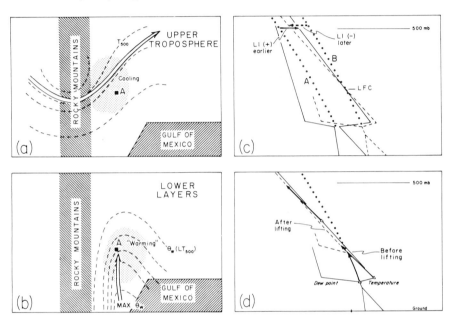

FIG. 13.5   Schematic illustration of processes in formation and modification of potentially unstable air mass. In (a) and (b), arrows are axes of high-tropospheric and low-level jets. In (a), 500-mb isotherms are indicated; in (b), wet-bulb potential temperature in moist layer, which when lifted adiabatically reaches temperature LT_{500} at the 500-mb level. In (c) and (d), temperature and dewpoint curves; lines of circles are moist adiabats corresponding to θ_w of moist layer. In (c), the lifted index (LI) indicates potential stability at first, changing to potential instability later. In (d), sounding curves before (solid) and after (dashed) lifting. (From Newton, 1963.)

a common situation east of the Rocky Mountains. For examples of real situations, see Sugg and Foster (1954) and U.S. Weather Bureau (1956). One of the essential processes is differential horizontal advection in the lower and upper troposphere, the importance of which was emphasized by Means (1944) and which has been discussed further by Miller (1955). It should be noted that horizontal advective processes do not *release* the instability, although the upward motions leading to such release tend to be associated with the regions where differential advection contributes to growth of potential instability.

On the approach of a cold trough aloft from the west (Fig. 13.5a), there is usually an increase in the southerly current over the Great Plains region, often accompanying cyclogenesis east of the Rocky Mountains. Air with high-moisture content is thus drawn northward from the Gulf of Mexico, as shown in Fig. 13.5b, where the wet-bulb potential temperature (θ_w) of the air in the moist layer is identified with its temperature when lifted unmixed to the middle troposphere. Comparison of this "lifted temperature" (which would characterize the innermost part of a convective cloud) with the environment temperature at a corresponding level gives a useful index of the possible intensity of convection (Showalter, 1953; U.S. Weather Bureau, 1956).

Figure 13.5c represents a sounding at point A in Figs. 13.5a and b. Increase of θ_w in lower levels, in itself or combined with cooling aloft, results in destabilization. Although the lower-level changes are dominated by horizontal advection (modified by diabatic heating or cooling due to proximity to the earth's surface), cooling in the upper troposphere reflects the combined effects of vertical motion and horizontal advection. The advent of a synoptic disturbance may occur night or day. Also, advection in low levels tends to be strongest in early morning (Means, 1944; Blackadar, 1957) when the low-level jet stream is best developed. Thus, the generation of potential instability, and on occasion its release, may take place at night even though the additional influence of insolation favors daytime onset of convection. The importance of synoptic circulations varies with the situation. Occasionally, extreme instability may be created in a previously stable region, mainly by differential advection, in 12 hr or so. On other occasions, particularly in the warmer months, an area may be occupied for extended periods by marine tropical air, in which case differential advection is unimportant and insolation dominates the onset of convection.

As discussed in connection with Fig. 13.4a, when a stable layer is present it may not be possible to realize deep convection through surface heating alone. Beebe and Bates (1955) have demonstrated that such a stable layer may be eliminated by organized vertical motions, in the manner shown by Fig. 13.5d. If, as is usual, the inversion layer is convectively unstable, lifting results in the air beneath it becoming saturated before the air above (Rossby, 1932) and thus cooling at a different rate. The lifting required to vitiate a strong inversion may be of the order of 100 mb. Hence, ordinary cyclone-scale vertical motions (of, say, 5 μb/sec) can eliminate the inversion in 6 hr.[2]

As pointed out by Fulks (1951), the inversion plays an important role in the generation of potential instability. If a strong inversion is present, initially it prevents deep convective overturning. Thus the potential instability may increase to a high degree in the manner of Fig. 13.5c if the differential advection and related processes are pronounced and go on long enough. Finally, when a mechanism is imposed to eliminate the inversion, the stratocumulus or stunted cumulus clouds, which may earlier have been formed in the marine layer, are free to penetrate upward and form cumulonimbus.

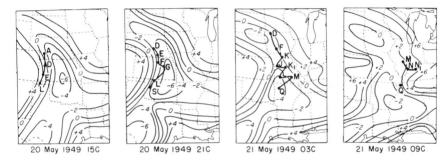

FIG. 13.6   Showalter stability index (500-mb temperature minus temperature of parcel lifted to 500-mb level from 850 mb; negative values indicate potential instability). Dots on lines, identified by letters, show successive locations of heaviest precipitation in individual rainstorms associated with squall line. Stability isopleths in first and third charts were interpolated; there were no upper-air observations at those times. Potential instability was present in some regions outside the zero isopleth, where conditions at 850 mb did not represent the maximum θ_w of the low-level moist air. (From Newton and Newton, 1959.)

The mechanism initiating convection may be, as has been suggested, simply the organized slow vertical motions connected with synoptic disturbances, combined with insolation. In some cases the first convection is clearly associated with a cold front.[3] Figure 13.6 illustrates the behavior of a squall line initiated by a cold front that, moving more rapidly eastward

[2] Horizontal advection is often an additional factor in removing the inversion locally. Advection may first contribute to warming in the lower middle troposphere and then to cooling after passage of a warm tongue at 850 to 700 mb (U.S. Weather Bureau, 1956.)

[3] Analyses of the circulations in the vicinity of cold fronts, by Clarke (1961) and Brundidge (1965), indicate upward motions of 50 to 70 cm/sec at 500- to 1000-m height in a relatively narrow band preceding them.

than the tongue of potentially unstable air, swept into its western edge and set off a line of thunderstorms there. These progressed eastward through the tongue of instability, first increasing in intensity and then mostly dying out as they encountered the stable air farther east. This is a fairly common pattern of behavior. Note that since the tongue of instability may itself move eastward with appreciable speed, the extent of the swath swept by a squall line of this sort is likely to be considerably greater than the breadth of the unstable tongue at a given time.

The western boundary of the tropical marine layer is often very sharp, as in the example analyzed by L. D. Sanders in Fig. 13.7 (U.S. Weather

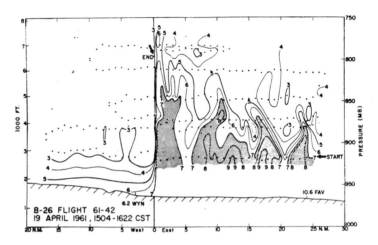

Fig. 13.7 Isopleths of mixing ratio (g/kg) in vicinity of boundary between continental and maritime tropical-air masses over northwest Oklahoma, Apr. 19, 1961. Constructed from stepwise aircraft traverses shown by dotted lines. Analysis by L. D. Sanders, from U.S. Weather Bureau (1963).

Bureau, 1963). In this case the moisture transition took place within about 1 km in the horizontal; the varying depth of the moist air to the east was probably associated with gravity waves set up at the western boundary. That boundary, which is normally characterized by a shift from southerly winds in the marine air to southwesterly winds in the dry air, is in the warmer months a semipermanent zone of convergence. This has long been known as a favorite place for thunderstorms to form, even without the intervention of cold fronts, as brought out by the statistics of Rhea (1966).

It will be noted in Fig. 13.6 that the potential instability was also quite pronounced in the southern region where no thunderstorms developed. This is because the stronger ascending motions were confined to the northern portion, where there was a well-developed, upper-level wave and jet stream (Section 6.2). The inversion capping the moist layer was more or less undisturbed farther south so that penetrative convection could not take place there.

Fulks (1951), Fawbush *et al.* (1951), and others have emphasized the role of orography in the climatological distribution of severe storms. The reason for their great frequency in the Oklahoma-Kansas region (Fig. 13.2) is that this region is conveniently close to a source of moist air, in the Gulf of Mexico, and just east of a high plateau. Tropical marine air moving northward from the Gulf is constrained by the Rocky Mountains from spreading westward. The low-level jet stream, advecting air with high θ_w northward (Fig. 13.5b), tends to achieve strongest development a few hundred kilometers east of the Rockies. This, along with the circumstance that cold upper troughs frequently move in from the Pacific Ocean, makes this region particularly susceptible to the generation and release of strong potential instability.

13.3 SYNOPTIC STRUCTURE OF SQUALL LINES

Although severe weather is on many occasions associated with isolated large thunderstorms, it usually reaches most destructive proportions with squall lines that sweep over large areas during their lifetimes. The general structure of an extensive squall line is well illustrated by Fig. 13.8, taken from Bergeron (1954).

An outstanding feature of the thunderstorm region behind the squall line is its general association with an area of high pressure. This is attributed to cooling in the lower layers, owing primarily to the evaporation of rain and to some extent also to melting of snow or hail or both. Surface winds diverge with a strong component across the isobars (Tepper, 1950) because the air does not remain long enough in the region of strong pressure gradient for the coriolis acceleration to be effective. With the usual pressure rise of about 2 to 5 mb, the approximate Bernoulli relationship $\Delta p/\varrho = \Delta V^2/2$ suggests that air parcels traveling from the high to the region of lower pressure should attain a speed of about 18 to 28 m/sec. On the average, this is somewhat diminished as a result of surface friction, but locally stronger winds may occur.

At the leading edge of the thunderstorm system, a sharp wind shift is generally observed so that, as in Fig. 13.8, the squall line has the appearance of a cold front. A rapid drop of temperature takes place near the wind-shift line, along with a decrease in θ_w, despite a general increase of relative

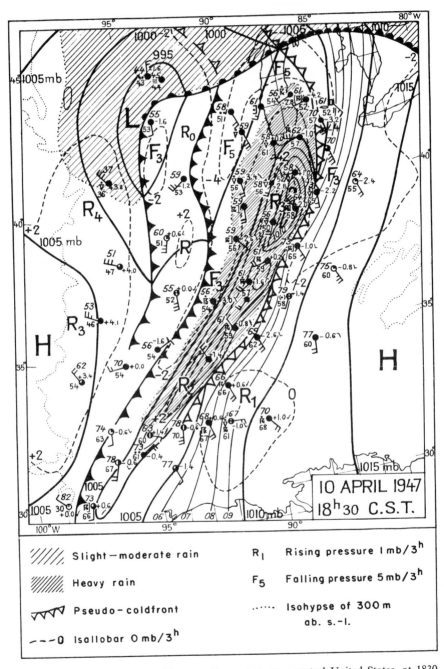

FIG. 13.8 Weather map with convective system over central United States, at 1830 CST Apr. 10, 1947. (After Bergeron, 1954).

humidity. Although squall lines most commonly occur in the warm sectors of cyclones (or in tropical air well removed from frontal systems), they often extend far north of warm fronts and occasionally form some distance behind cold fronts: in both cases the thunderstorms are in the warm air above the front.

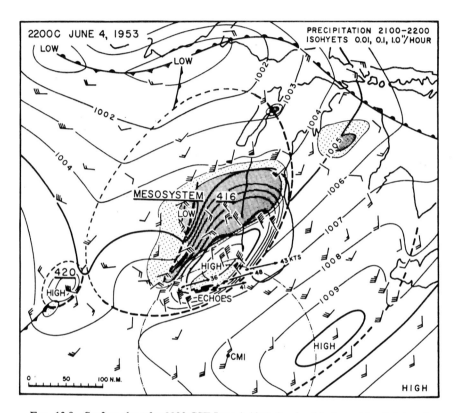

FIG. 13.9   Surface chart for 2200 CST June 4, 1953. Sea-level isobars at 1-mb intervals. In wind symbols, a full barb is 5 knots and a flag is 25 knots. Dashed lines are boundaries of cold-air domes of mesosystems. Isohyets are for precipitation (stippled) during previous hour. (After Fujita and Brown, 1958.)

Some squall lines appear as single outbreaks of thunderstorms that achieve great lengths, as in Fig. 13.8. In other instances, several discrete "mesosystems" (Fig. 13.9) may appear successively. At the stage shown, the cold air produced by the thunderstorms has spread out over a much larger area than is covered by the precipitation itself. Fujita (1959) has demonstrated that the mass of cold air produced is proportional to the cumulative mass of rain that has fallen from the convective system and to the amount that has been evaporated into the subcloud layer. Figure 13.10 shows how, as this cold air is continuously augmented and diverges from the region of active thunderstorms, the area covered by the mesosystem expands with time.

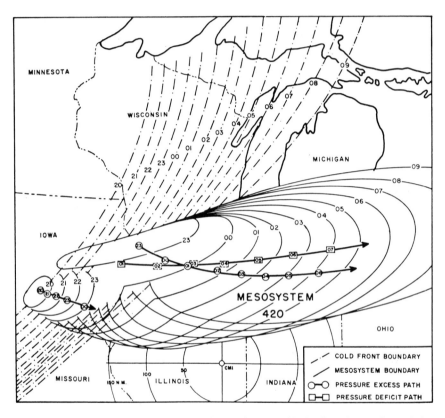

FIG. 13.10   Isochrones (hours in CST) of cold front (dashed) and boundary of the smaller mesosystem (420) in Fig. 13.9, on June 4–5. Arrows indicate tracks of high center (circles) and of a low center to its rear, these features being analogous to those observed in mesosystem 416 of Fig. 13.9. (After Fujita and Brown, 1958.)

In Fig. 13.9 a small-scale low is seen to the west of the precipitating system. This feature, whose cause has not been conclusively explained, is characteristic of the mature stage of a mesosystem (Brunk, 1953; Williams, 1953; Fujita, 1955); it has also been observed in England (Pedgley, 1962). Intense pressure dips and strong winds may occur with its passage (Brunk, 1949).

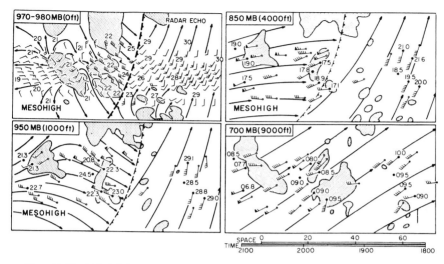

FIG. 13.11   Wind field of a squall-line system passing over the Thunderstorm Project network in Ohio, on June 29, 1947. Dashed line is squall line; shaded areas are radar echoes from precipitation. The numerous observations represent a combined analysis of space and time data from a dense surface network and serial radiosonde ascents from several stations. (After Fujita, 1959.)

The wind shift observed at the ground on passage of a squall line decreases in intensity upward. This is illustrated by Fig. 13.11, where the shift is seen to be relatively weak already at 850 mb and undetectable at 700 mb. Observations show (Byers and Braham, 1949) that the temperature contrast between the rain-cooled air and the undisturbed air is greatest near the ground and becomes quite small near cloud base. Thus the excess pressure beneath the storm (Figs. 13.8 and 13.9), which reflects the weight of the cold air above, rapidly decreases in magnitude upward, as does the disturbance of the wind field associated with it.

13.4   GENERAL ASPECTS OF STORM CIRCULATION

Because of the difficulties and hazards of making direct measurements in intense convective systems, aerological descriptions of them are fragmentary, and the character of their circulations must be largely inferred from indirect evidence.

This evidence suggests that there is a variety of structures as well as a wide range of storm sizes and intensities; the larger and more severe storms have in general more highly organized, more asymmetrical, and more persistent circulations. The "air mass" thunderstorm, which has been extensively explored and which was described in detail by Byers and Braham (1949), is characterized by sporadic development of relatively small "cells." These pass through an updraft phase, to mixed updraft and downdraft, and finally to downdraft, an individual cell perishing after about a half-hour from its inception. On the other end of the scale, some large severe storms are apparently dominated by a single huge "supercell" (Browning, 1964), which may persist for hours. Generally, these severe storms are embedded in a baroclinic environment, which (Section 13.2) plays an important role in the generation and release of convection; in addition, the associated vertical shear has a profound influence on the structure and mechanics of the storm itself.

Here we shall deal only with certain features that we consider most relevant to interactions with the synoptic-scale circulations. The reader interested in a comprehensive discussion of the structures of individual storms and groups of storms is referred particularly to Browning et al. (1965), who provide the most complete and detailed description up to the time of this writing.

The broad relationship between a large, organized, convective system and the associated synoptic disturbance is illustrated by Fig. 13.12, taken from Fankhauser (1965). This shows that the general lower-tropospheric confluence and upper-tropospheric diffluence associated with the cyclone and upper trough were concentrated in the vicinity of the convective system. Computations of the air flow across the dotted boundary in Fig. 13.12 give the vertical distribution of mean divergence over the enclosed area, shown by curve D_p in Fig. 13.13. This indicates marked convergence in the lowest layers, weak convergence through most of the troposphere, and increasing divergence above 400 mb to 200 mb, near the tropopause level that corresponded approximately to the general top of the cloud system. Fankhauser's further analysis shows that weak and indifferent vertical motions were present in the middle troposphere outside the near-vicinity of the squall line. Hence, it may be concluded that *nearly all ascent of air associated with the cyclone was taking place in the convective system.* The computed water-vapor influx, represented by

$$\oint_L \int_{250\text{ mb}}^{p_0} v_n q \, \frac{dp}{g} \, dL$$

where v_n is the inward component of the wind across the boundary L and q is specific humidity, is represented by curve M for 50-mb layers. The vertically integrated vapor accumulation was within 5% of the mass rate of

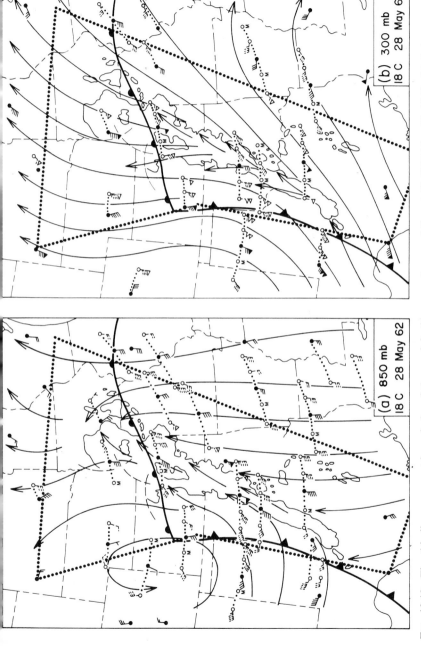

FIG. 13.12   The thin lines represent boundaries of the precipitation area of a squall line at 18 CST May 28, 1962, this being a composite of the radar echoes observed from several stations. The wind observations plotted on open circles along dotted lines were from serial ascents made 3 to 6 hr before and after the time of the figure, displaced according to the speed of movement of the system. Surface fronts and streamlines at (a) 850 mb, and (b) 300 mb. (After Fankhauser, 1965.)

rainfall estimated from analyses of hourly rainfall data. A fairly close agreement was obtained earlier in similar computations by Bradbury (1957).

On a finer scale, the general character of the circulation in a convective
system can be deduced from the wind field in its vicinity together with
applications of thermodynamic principles. The result of one such analysis
is shown in Fig. 13.14, a time section through a large thunderstorm in a
squall line that passed over Oklahoma City in late afternoon on May 21,

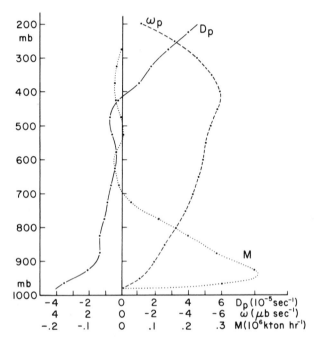

FIG. 13.13   The mean, over the area enclosed by dotted lines in Fig. 13.12, of the
horizontal divergence (D_p) and the vertical velocity (ω_p); scales at bottom. Curve M
shows the horizontal flux of water vapor per 50-mb layer into the region enclosed by
this boundary. This value goes to zero at the lowest point because the volume considered
is over sloping terrain. (After Fankhauser, 1965.)

1961. The cloud outline is schematic, being based on surface observations
and on radar profiles observed at earlier times (the limited upward tilt of
the radar antenna precluded scanning the whole storm at this time). Other
storms occurring in the vicinity prior to squall-line passage are omitted for
the sake of clarity. Isotach analyses indicated that the general winds near
tropopause level were from west 50 to 60 knots. Thus, there was a pronounced veering and an increase of wind speed with height, characteristic

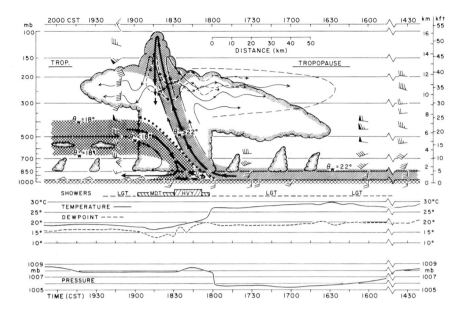

FIG. 13.14 Partly schematic cross section through squall line of May 21, 1961, as it passed Oklahoma City. The vertical exaggeration is fivefold. Winds plotted with directions relative to north at top of figure. Hatching indicates depth of air, with θ_w in excess of 22°C ahead of squall line, and its probable extent in updraft and upper portion of cloud. Crosshatching indicates extent of air, with θ_w less than 18°C, based on sounding behind storm and on surface observations. A similar layer at closely corresponding heights was present before squall-line passage; this is omitted for the sake of clarity. Shown at bottom are periods of precipitation, temperature, dewpoint, and sea-level pressure traces (somewhat smoothed). Heavy arrows, axes of main drafts; light arrows, relative streamlines, dashed where air emanates from core of stratospheric tower. At right, long dashes suggest outline of mass of *air* plume originating in storm and spreading out essentially horizontally; the radar-detected *cloud* plume, at lower elevations, consists of small precipitation particles that have partly fallen out of the air plume. The dimensions of the stratospheric tower are about the maximum attained and may at a given time be considerably smaller in a pulsating storm. Anvil length is typical for a storm about 2 hr old. (From Newton, 1966.)

of the environment of most severe storms. In the discussion to follow, the circulations are, for simplicity, treated only in two dimensions.[4]

In this particular squall line, the cloud profile indicated in Fig. 13.14

[4] The important influences of wind veer with height have been discussed by Browning and Ludlam (1962) and Browning (1964) in relation to the three-dimensional configurations of severe storms and the distribution of hail and rain, and by Newton and Fankhauser (1964) in regard to the direction in which an individual storm moves, which partly controls its intake of water vapor to balance the rainout.

was characteristic. A distinguishing feature was the presence on the up-shear side of the main storm columns (i.e., the side upwind with respect to the shear vector from cloud base to tropopause) of towers that pulsated in height but remained persistently in this location (the mean height of the towers of ten storms in a 200-km segment of the squall line being 1.5 km above the tropopause, with a standard deviation of 1.5 km). The anvils, although most extensive on the downshear side, also protruded as a heavily mammatiform shelf up to 50 km upshear from the storm cores.

This squall line, oriented northeast to southwest, advanced southeastward at a speed of 11 m/sec against southerly winds in the lower levels; thus the moist air feeding the updrafts entered the storms on their forward sides. The cloud towers on the rear sides are indicative of updrafts with high kinetic energy at the tropopause level, implying that the updrafts tilt in an upshear sense (Fig. 13.14) as they traverse through the storm. As indicated by Bates (1961; see also Newton, 1966), a configuration of this sort is consistent with physical principles.[5] We shall outline here in some detail the principles underlying the structure of such a storm because these govern not only the behavior of the storm itself but also, as will be seen in Section 13.6, the way in which the convective cloud acts to transfer properties significant for the general atmospheric circulation.

At the considerable heights to which air parcels rise above the tropopause, they become colder than the ambient air and must subsequently acquire large downward velocities. Thermodynamic analysis suggests, however, that the downdraft must diminish rapidly in intensity below the tropopause and that *no appreciable amount of the air originating in the updraft is likely to return to the lower troposphere.* The considerations involved are illustrated in Fig. 13.15, in which the thin lines correspond to a composite sounding representative of the undisturbed environment of the storm of Fig. 13.14.

Normand (1946), Stommel (1947), and Schmidt (1947) introduced the important concept of entrainment; the drawing in of air from the environment of a convective cloud and its mixing into the saturated air of an up-draft or downdraft modify both thermal properties and vertical momentum of the draft. Because the surface-to-volume ratio is greater for a small-than for a large-diameter draft, a small cloud is more influenced by the mixing-in of air through its sides than is a large one. From cloud observa-

---

[5] The upshear canting of the draft is due in part to a strong tendency of the air in a large-diameter draft to conserve its horizontal momentum despite drag forces exerted by the ambient winds, and in part to the movement of the storm as a whole, which causes the air elements during their rise to lag the foot of the updraft on the advancing side of the storm.

tions and theory, Malkus and Williams (1963) concluded that

$$\frac{1}{M}\frac{dM}{dz} \approx \frac{1}{D} \qquad (13.1)$$

where D is cloud diameter and M is the vertical mass flux at a given level. "Thus trade cumulus sized clouds ($D \sim 1$ km) entrain a mass flux comparable to their own initial mass flux in 1 km, while thunderstorm sized towers (D up to 10 km) can shoot through most of the troposphere before experiencing one-to-one dilution." In an illuminating numerical experiment, Squires

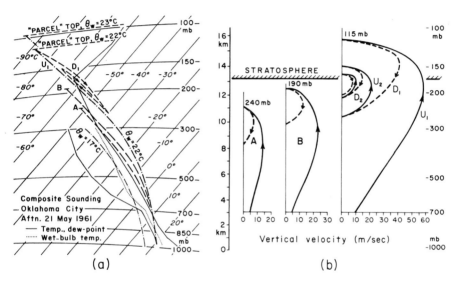

(a)   (b)

FIG. 13.15 (a) Representative sounding at Oklahoma City, afternoon of May 21, 1961 (thin solid lines, temperature and dewpoint; dotted line, wet-bulb temperature); plotted on a tephigram. For meaning of other curves, see text. (b) Vertical velocities in updrafts (solid lines) and downdrafts (dashed lines) under different assumptions described in text. (From Newton, 1966.)

and Turner (1962) demonstrated the influence of varying cloud diameter and environmental relative humidity upon the vigor of the updraft, its condensed water content, and the vertical development achieved by the cloud.

Since the towers of large cumulonimbus clouds often penetrate to maximum heights predicted from "parcel" theory, the air in the *cores* of their updrafts is evidently at times undiluted by mixing with the environment. However, since the diameters of drafts even in large storms are not likely to exceed about 10 km, Eq. (13.1) suggests that significant entrainment is to be expected. Consequently, the outer sheaths of drafts must undergo strong mixing. Then, under the influence of buoyant accelerations, the air in an updraft core would rise to the greatest height, whereas air parcels in the outer portions of the draft, having decreased buoyancy due to chilling, would not rise so far.

Curves U_1 in Fig. 13.15 show the characteristics of an unmixed updraft parcel, having $\theta_w = 22°C$ corresponding to the average in the moist layer below 700 mb. Such a parcel would rise to about 16 km. With a draft diameter of 7.5 km, the overall entrainment into the draft would be 1/7.5 km, according to Eq. (13.1). If, as a crude assumption, the inner protected core occupied half the draft diameter and all the entrained air were mixed into the outer sheath, this would correspond to a rate of 1/6 km. To account for the mass transport determined later, an updraft is required whose diameter is much smaller than the overall diameter of the cloud column. This suggests that the bulk of the cloud through which the updraft rises (Fig. 13.14) has only weak buoyancy, and thus that the cloud air outside the vigorous draft has a temperature near that of the storm environment. Curves B in Fig. 13.15 show the temperature and vertical velocity profiles resulting from entrainment (at the rate of 1/6 km) into the outer sheath of the updraft, of cloudy (saturated) air having these characteristics. For comparison, curves A show the results of a corresponding entrainment, but from the unsaturated environment. In these computations the entrainment of inert air (i.e., of air that has no initial vertical momentum) was taken into account, using the modified buoyancy equation

$$\frac{d}{dt}(wm) = mg\frac{\Delta T}{T} \tag{13.2}$$

where ΔT is temperature excess over the environment and m is the mass of an air parcel, initially a unit mass, that is augmented according to the entrainment rate.

At the height of its maximum upward penetration, an air parcel is colder than its environment and is subjected to a downward buoyancy acceleration. Analyses of the motions of cloud elements (e.g., by Anderson, 1960; Saunders, 1961) suggest that air rising in the cores of cloud towers subsequently diverges outward, making it susceptible to mixing with the environment. Thus, although some air may rise undiluted, it is very unlikely that the returning downdraft is not subjected to some degree of mixing. For the case U_1, the further behavior of an air parcel entraining at the rate of 1/6 km is indicated in Fig. 13.15. In this case, it undergoes an oscillation that is damped partly because it gradually approaches the environment temperature in the process of entrainment and partly because inert air is incorporated, according to Eq. (13.2). Even with the very small mixing rate of only 1/20 km, a downdraft parcel could descend no lower than to 360 mb, and after damped oscillation would approach equilibrium near the 250-mb level.

The conclusion from this analysis is that *essentially all air in the updraft must, after undergoing oscillations, remain in the upper troposphere, eventually passing into the expanding anvil of the storm.* This behavior, and the rising of air in different parts of the draft to varying heights depending on their different degrees of mixing, is indicated schematically by the streamlines in Fig. 13.14. It is obvious, of course, that all these oscillations cannot coexist; rather they would interfere with each other and be expressed as "turbulence."

The spreading out of air in the surface layers beneath thunderstorms (Section 13.3) indicates that there must be a substantial downward mass flux in the lower part of the cloud system. In this example, the lowest θ_w in the surface layers was $17°C$. It is apparent from Fig. 13.15a that the cold air beneath the thunderstorm must have originated in the middle troposphere where the environmental θ_w was correspondingly low. It may then be inferred that there is a separate downdraft circulation in the lower part of the cumulonimbus (Fig. 13.14), as indicated in other analyses by Newton (1950), Bates (1961), Fulks (1962), and Browning and Ludlam (1962).

On entering the storm, the air taking part in the lower downdraft circulation has moderately high momentum characteristic of the environment in middle levels. Its momentum enables it to move forward through the storm, converging with and undercutting the warm air in advance of it. *The mechanical lifting provided by the cold downdraft thus continually regenerates the updraft, prolonging the storm circulation.* The vertical circulation in the lower downdraft is caused by gravitational sinking, its coldness being due to chilling by partial evaporation of water condensed in the sloping updraft and falling through the downdraft. Normand (1946) recognized the desirability of some such process, pointing out that the energy is greatly augmented if a thunderstorm is "organized to take in potentially cool air at high levels as well as potentially warm air at the lower levels."

The principle was discussed earlier by Normand (1938). Desai and Mal (1938) recognized its applicability to the severe springtime thunderstorms of northeast India. They pointed out that the fall of both atmospheric temperature and wet-bulb temperature, observed at the ground with the onset of these storms, could be explained by evaporative cooling of dry continental air descending from levels above the moist layer. Harrison and Orendorff (1941) were the first to describe the essential structure of squall lines over the United States, observing that the squall line is a self-regenerative mechanism of the character described above.

The present conception of a highly organized, severe storm circulation has been most strongly influenced by the analysis carried out by Browning and Ludlam (1962) of a giant hailstorm over England. Through an ingenious and thorough interpretation of radar and other observations, they were able to deduce the main features of the storm circulation and trace its influence on the growth and redistribution of hail and other hydrometeors. They showed also that in its early and late stages, the storm was composed of numerous cells that propagated in a systematic manner related to the vertical shear, as suggested by Newton and Newton (1959). During the time when the storm was best developed and disgorged huge quantities of hail, however, it was dominated by a persistent steady circulation, basically similar to that in Fig. 13.14, but with a more erect updraft in upper levels.

13.5   MASS AND WATER BUDGETS

The mass budget of a squall-line thunderstorm may be estimated from the winds in its environment together with the considerations mentioned in Section 13.4. For a "solid" squall line[6] moving with speed c, the fluxes of air and of water vapor into or out of an individual storm are given by

$$\int (c - v_n) D \frac{dp}{g}; \qquad \int (c - v_n) q D \frac{dp}{g} \qquad (13.3)$$

where v_n is the wind component along the direction toward which the squall line is advancing, q is specific humidity, and D is distance along the line between thunderstorms, taken to be the width of an individual storm. The vertical profiles in Fig. 13.16 indicate a flux of air (and water vapor) into the forward side of the storms below 700 mb, a flux outward from the rear side below 810 mb, a weaker flux into the rear side between 360 and 810 mb, and a strong efflux above 360 mb where the balloon passed through the anvil.

---

[6] Squall lines are ordinarily composed of discrete thunderstorms, which as individuals conserve their identities for long periods of time. In the present example, thunderstorms were spaced at average intervals of 20 km along the line. Although the radar-detected cores of heaviest precipitation were distinct, the cloud mass between these was continuous, as was the squall front at the surface.

Based on expressions (13.3), Fig. 13.16, and the scheme of Fig. 13.14, the mass fluxes in various branches of a single large thunderstorm are shown in Fig. 13.17. Of the 700 kton/sec of air entering the updraft in low levels, 20 to 60% circulates through the tower above the tropopause,[7] the remainder rising to more limited heights in the strongly entraining outskirts of the updraft and spreading out directly into the anvil. An updraft mass

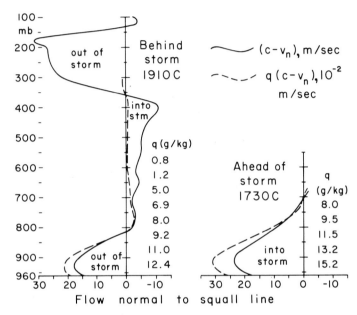

FIG. 13.16   Vertical profiles of relative velocity components normal to squall line (solid) and of the product of relative velocity and specific humidity (dashed), based on soundings at 1730 and 1910 CST in Fig. 13.14. (From Newton, 1966.)

flux of this magnitude would lead to an expansion of the anvil volume at the rate of 2 km³/sec (consistent with observed rates of anvil growth). The mass flux in the lower-tropospheric downdraft, 400 kton/sec, is about 60% of that in the updraft.

The flux of water vapor entering the foot of the updraft is 8.8 kton/sec. With a mean mixing ratio of 0.9 g/kg and an assumed water substance

---

[7] This estimate was made by an indirect method described by Newton (1966). For an average storm with strongly pulsating towers, a mean flux of 150 kton/sec upward and downward through the tropopause is realistic. For a "steady state" storm whose tower is broad and remains persistently high above the tropopause, as much as 400 kton/sec of air may circulate through the stratospheric tower.

density of 0.3 g/m³, 0.6 kton/sec of vapor and 0.6 kton/sec of water are exported in the anvil expansion, leaving 7.6 kton/sec for precipitation aloft in the main storm column. From Fig. 13.16, the vapor flow aloft into the rear of the storm is 0.7 kton/sec, and the outflow toward the rear in low levels is 4.3 kton/sec, requiring 3.6 kton/sec of this rain to be evaporated into the downdraft. This leaves a residual of 4.0 kton/sec water available for precipitation at the earth's surface. Allowing for uncertainties in representativeness of the various observations, this compares fairly well with the observed rainfall rate, 4.7 kton/sec, determined by averaging the precipi-

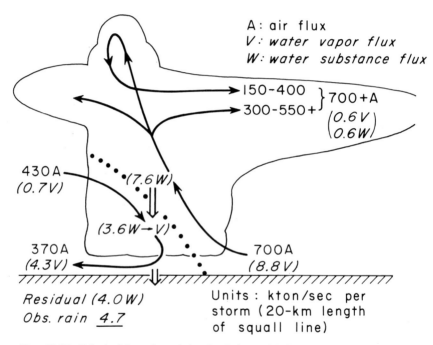

FIG. 13.17   Principal branches of the circulation, with fluxes of air (A), water vapor (V), and water (W), corresponding to a single thunderstorm in squall line. (From Newton, 1966.)

tation over the U.S. Weather Bureau "Beta" mesometeorological network south of Oklahoma City, over which the squall line passed.

It should be noted that the mass fluxes in Fig. 13.17 are representative of a large intense thunderstorm; they are more than an order of magnitude greater than the fluxes determined by Braham (1952) for typical "air mass" storms. Also, Braham found in the latter that the surface rain was only about 10% of the vapor flux through the storm, while in the large storm discussed above, the rain reaching the ground was over 50% of the vapor flux entering the updraft. It may also be noted that since the influx and efflux almost balance within the upper troposphere, the amount of rain deposited is essentially equivalent to the moisture convergence in the moist layer below 700 mb (compare Fig. 13.13).

13.6   VERTICAL TRANSFER OF PROPERTIES

The vertical transports of heat, angular momentum, and water vapor in convective systems have never been evaluated in relation to the general circulation as a whole. As mentioned in Section 2.5 and discussed further in Sections 14.4 and 17.2, the heat transfer from the low to the high troposphere in the Hadley circulation is largely accomplished by high-reaching cumulonimbus clouds covering only a small portion of the tropics.

According to the "parcel" theory of convection, the temperature of an unmixed parcel would be prescribed by the moist adiabat from cloud base. The change of its kinetic energy from a lower level z_0 to an upper level z would be

$$\Delta \left(\frac{w^2}{2} \right) = \int_{z_0}^{z} g \frac{\Delta T}{T} \, dz = - \int_{p_0}^{p} R \, \Delta T \, d \ln p \qquad (13.4)$$

With a sounding such as that in the left side of Fig. 13.18, the maximum kinetic energy should be achieved near the tropopause level. Petterssen *et al.* (1945), in a study of cumulus clouds over the vicinity of the British Isles, found that the tops rose on the average only to a level (corresponding to A in Fig. 13.18) where, according to the parcel concept, the buoyancy should be greatest. This was so also in the sample shown in this figure, in which only a very small fraction of the clouds grew to stratospheric heights predicted by parcel theory.

The "slice" concept (Bjerknes, 1938; Petterssen, 1939), wherein the environment is warmed by dry-adiabatic descent of the air between clouds (thereby diminishing their excess temperature), can partially account for the limited developments of cumuli. Stommel's findings (1947) indicate that the diminution of their buoyancies through entrainment has a great additional influence; Eq. (13.1) indicates that only a favored few clouds that manage to grow to large diameter can survive the erosive influences of entrainment and penetrate to high levels.

Thus, in any approach to the appraisal of heat transfer by cumulus convection, it will be essential to determine the spectra of their sizes under varying environmental conditions and to take into account the differences

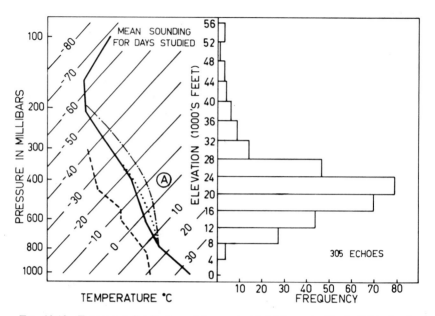

FIG. 13.18   Frequency distribution of the greatest heights reached by individual radar-echo tops in convective clouds over Texas. The mean sounding (temperature, solid; dewpoint, dashed) for the ten warm-season days studied is shown at left. Added dotted line is schematic; shows qualitative influence of strong entrainment. (After Clark, 1960.)

of thermal and kinematic properties with cloud size.[8] In a crude sense, the vertical eddy-heat transport over a given horizontal area A is given by $\int \varrho w c_p \, \Delta T \, dA$. Since, according to Eq. (13.4), the vertical velocity is proportional to $\int \Delta T^{1/2} \, dz$, the vertical heat flux (say, through the 500-mb level) should be roughly proportional to

$$\int \int \Delta T^{3/2} \, dA \, dz$$

Thus the large clouds, whose extensive horizontal areas tend to be associated with large buoyancies and great vertical development, must predominate in the mechanism of vertical heat transport between the lower and upper troposphere.

[8] A first attack on this problem has been made by Kasahara and Asai (1967) on the basis of a population of clouds of uniform size. The size of an individual cloud can be varied, and both entrainment and the compensating descent between clouds are taken into account in their model. This model was devised to study the two-way interaction between the clouds, as a mechanism for transporting and generating energy, and the influence of the large-scale motion field upon the production of clouds.

As far as the whole atmosphere is concerned, the heat gain *from a single large storm,* given by the latent heat of condensation times the observed mass rate of precipitation at the ground, would (for the case in Fig. 13.17) be 2.8×10^{12} cal/sec. From data derived from radar measurements for eight squall lines as given by Austin *et al.* (1961), an equivalent average value of 2.45×10^{12} cal/sec (per 20-km length of squall line) was determined, which is closely comparable. The vertical energy transfer may alternatively be estimated from the expression

$$\int \varrho w(c_p T' + Lq' + gz') \, dA \approx \Sigma \, F_z(c_p \bar{T}' + L\bar{q}' + g\bar{z}') \qquad (13.5)$$

where the primes indicate deviations of temperature, specific humidity, and height at an isobaric surface within the storm from the values in the environment,[9] and F_z is the mass flux (positive upward) in a branch of the circulation. The value of the height deviation z' is unknown, but inclusion of values as large as 50 m influence the result by less than 10%.

Assuming properties prescribed by Figs. 13.17 and 13.15a, with saturation in both updraft (at $\theta_w = 22°C$) and downdraft (at $\theta_w = 17°C$), the upward energy flux is found to be 1.6×10^{12} cal/sec through the 700-mb surface (where both updraft and downdraft fluxes must be considered) and 2.2×10^{12} cal/sec through the 450-mb surface (decreasing at higher levels). Allowing for the discrepancy between the results of the moisture-flux budget and the observed rainfall in Fig. 13.17, the foregoing evaluation suggests that *the heat transported upward in the middle troposphere is essentially equivalent to the total heat realized from condensation, this being transferred into and remaining in the upper troposphere.* In the preceding computations the heat of fusion has not been considered. The freezing and subsequent sublimation of small ice particles entirely in the upper troposphere would evidently not influence the heat transfer. On the other hand, if all the precipitation fell as hail (a condition approached in some storms), the heat transfer would be augmented by 13%, since freezing occurs in the upper troposphere with melting in the lower troposphere or at the ground.

Fankhauser (1965) found that over a 6-hr period, the squall line in Figs. 13.12 and 13.13 precipitated 8.4×10^{12} kg of water. The corresponding release of condensation heat, 23×10^{13} cal/sec, may be compared with the

---

[9] In expressions (13.5), the upward mass transport through an isobaric surface within the storm is assumed to be compensated by an equal downward mass transport outside it of air having the properties of the environment. Since this compensation may take place distant from the storm, the latter condition may not be satisfied exactly. On the right side, the use of mean (barred) values implies a neglect of small eddies and of systematic variations of properties across the drafts.

value of 20×10^{13} cal/sec found by Palmén and Holopainen (1962) for the partially nonconvective rainfall shield of a winter cyclone. Since, as indicated earlier, virtually all upward mass flux in the cyclone of Fig. 13.12 took place in the convective system, it is likely that this was also true of the vertical heat flux. In such a cyclone with a well-developed convective system, it is often observed that little ordinary frontal precipitation occurs, presumably because (Fig. 13.12a) the warm, moist air is intercepted and overturned before it reaches the fronts.[10]

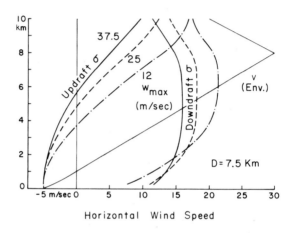

FIG. 13.19   Horizontal wind speed (σ) within the updraft of a thunderstorm, embedded in an environment wind field V, for cases wherein the maximum vertical velocity (at tropopause level) has the three indicated values. The horizontal velocities indicated for the downdraft were calculated on the supposition that the downdraft extends through the whole depth of the cloud. As indicated in Figs. 13.14 and 13.15, the downdraft must in reality consist of two separate branches, one in the upper part of the storm and the other entering the storm in the middle troposphere. Thus, the downdraft wind profile is qualitatively valid only in the uppermost levels and in the lower troposphere. (After Bates and Newton, 1965.)

Only a crude estimate can be made to illustrate the effect of thunderstorms in transferring momentum vertically. Figure 13.19 shows a computation (Bates and Newton, 1965) of the horizontal speed within an updraft, which would result if a draft of a 7.5-km diameter were subjected to drag forces

---

[10] In the cyclone model developed by Bjerknes and Solberg (Fig. 5.1) from observations in northwestern Europe, the precipitation was well organized along fronts, and very little rain was assumed to occur in the warm sector. Their classical model is in this respect not applicable to some cyclones in lower latitudes (e.g., in the United States) because these often contain potentially unstable air not characteristic of cyclones in higher latitudes, particularly in the cooler seasons.

while rising through an environment with vertical shear fairly typical of severe storm situations (an updraft of this diameter would, with a vertical velocity of 25 m/sec at 500 mb, carry a mass flux of the magnitude in Fig. 13.17).

If we assume that the upward mass flux in the storm is compensated by a downward mass flux in the environment *at the same latitude*, the contribution to the vertical flux of earth angular momentum may be neglected, assuming also that the downward flux of relative angular momentum is characterized by the wind velocity in the immediate storm environment. The upward flux of angular momentum may then be approximated by

$$\int \varrho wua \cos \varphi \, dA \approx \Sigma \, F_z \bar{u}'a \cos \varphi \qquad (13.6)$$

where the conventions are as in Eq. (13.5).

In the middle troposphere (say, at 500 mb), $\bar{u}' \approx 0$ in the downdraft; in the updraft $\bar{u}' \approx -15$ m/sec according to Fig. 13.19, and $F_z = 700$ kton/ sec from Fig. 13.17. The corresponding angular-momentum flux for a single storm of this magnitude is then -5.5×10^{23} gm cm^2 sec^{-2} through the 500-mb surface at latitude 35°. In the lower troposphere, momentum is carried downward by the downdraft and air with deficient momentum (though by a smaller magnitude than that above) is carried upward, the computed net transfer at 700 mb being nearly the same as that at 500 mb.

In the squall line of Fig. 13.12 (nearly 1500 km in length), the total rainfall rate evaluated by Fankhauser was about 80 times that of the single storm represented in Fig. 13.17. The vertical shear of the west-wind component was about half as great as in Fig. 13.19. Correspondingly, assuming the vertical mass flux to be proportional to the mass rainfall rate, the vertical flux of angular momentum through the middle troposphere, accomplished by the entire squall line, would be -2.2×10^{25} gm cm^2 sec^{-2}. It must be emphasized that this is a raw estimate.[11] While it can probably be taken as typical for a well-developed squall line, it should also be noted that a squall line of this magnitude is a special event.

---

[11] Note also the uncertainties arising from the assumptions stated before Eq. (13.6). We may, however, compare a cyclone in which convective ascent occurs with one in which the ascending air is stable and rises slowly. In the latter case, the horizontal velocity accommodates nearly to the general wind field. With a wind profile as in Fig. 13.19, there would in this case be a large upward transfer of momentum through the middle troposphere, whereas the corresponding upward transfer by convective clouds would be small. This comparison concerns only the transfer within the precipitation region; the overall vertical transfer of momentum depends in either case upon the properties of the compensating air that descends elsewhere.

A complete evaluation of the momentum transfer by thunderstorms would require a knowledge of their populations and consideration of the different conditions under which they occur. However, it may be pointed out that the average torque required to balance the angular-momentum flux for the whole latitude belt between 30° and 50°N (Table 1.1) in summer is -6×10^{25} gm cm^2 sec^{-2}; in winter -31×10^{25} gm cm^2 sec^{-2}. The value given above for a single squall line is 37% of the required amount in summer and 7% in winter, suggesting that *vertical angular-momentum transfer by convective storms is very significant during the warmer months*; it is *probably of little significance compared with other processes in winter* when the frequency of large convective systems is comparatively low.

References

Anderson, C. E. (1960). A technique for classifying cumulus clouds based on photogrammetry. *In* "Cumulus Dynamics" (C. E. Anderson, ed.), pp. 50–59. Pergamon Press, Oxford.

Austin, P. M., Cochran, H. B., and Patrick, G. O. (1961). Investigations concerning the internal structure of New England squall lines. *Proc. 9th Weather Radar Conf., Kansas City, Oct. 1961*, pp. 193–198.

Bates, F. C. (1961). "The Great-Plains Squall-Line Thunderstorm—A Model," Ph. D. dissertation, 164 pp. St. Louis Univ. (Univ. Microfilms, Ann Arbor, Michigan).

Bates, F. C., and Newton, C. W. (1965). The forms of updrafts and downdrafts in cumulonimbus in a sheared environment. *Natl. Meeting Am. Meteorol. Soc. Cloud Phys. Severe Storms, Reno, Nev.*, 1965 9 pp. plus figs.

Beckwith, W. B. (1960). Analysis of hailstorms in the Denver network, 1949–1958. *In* "Physics of Precipitation" (H. K. Weickmann, ed.), Geophys. Monograph No. 5, pp. 348–353. Am. Geophys. Union, Washington, D. C.

Beebe, R. G., and Bates, F. C. (1955). A mechanism for assisting in the release of convective instability. *Monthly Weather Rev.* **83**, 1–10.

Bergeron, T. (1954). The problem of tropical hurricanes. *Quart. J. Roy. Meteorol. Soc.* **80**, 131–164.

Bjerknes, J. (1938). Saturated-adiabatic ascent of air through dry-adiabatically descending environment. *Quart. J. Roy. Meteorol. Soc.* **64**, 325–330.

Blackadar, A. K. (1957). Boundary layer wind maxima and their significance for the growth of nocturnal inversions. *Bull. Am. Meteorol. Soc.* **38**, 283–290.

Bleeker, W., and Andre, M. J. (1951). On the diurnal variation of precipitation, particularly over central U.S.A., and its relationship to large-scale orographic circulation systems. *Quart. J. Roy. Meteorol. Soc.* **77**, 260–271.

Bradbury, D. L. (1957). Moisture analysis and water budget in three different types of storms. *J. Meteorol.* **14**, 559–565.

Braham, R. R., Jr. (1952). The water and energy budgets of the thunderstorm and their relation to thunderstorm development. *J. Meteorol.* **9**, 227–242.

Brown, H. A. (1962). Unpublished manuscript.

Browning, K. A. (1964). Airflow and precipitation trajectories within severe local storms which travel to the right of the winds. *J. Atmospheric Sci.* **21**, 634–639.

Browning, K. A., and Ludlam, F. H. (1962). Airflow in convective storms. *Quart. J. Roy. Meteorol. Soc.* **88**, 117–135.

Browning, K. A. *et al.* (1965). A family outbreak of severe local storms—a comprehensive study of the storms in Oklahoma on 26 May 1963. Part I. Spec. Rept., No. 32, 346 pp., AFCRL-65-695(1). Air Force Cambridge Res. Labs., Bedford, Massachusetts.

Brundidge, K. C. (1965). The wind and temperature structure of nocturnal cold fronts in the first 1,420 feet. *Monthly Weather Rev.* **93**, 587–603.

Brunk, I. W. (1949). The pressure pulsation of 11 April 1944. *J. Meteorol.* **6**, 181–187.

Brunk, I. W. (1953). Squall lines. *Bull. Am. Meteorol. Soc.* **34**, 1–9.

Byers, H. R., and Braham, R. R., Jr. (1949). "The Thunderstorm," 287 pp. U.S. Govt. Printing Office, Washington, D. C.

Byers, H. R., and Rodebush, H. R. (1949). Causes of thunderstorms of the Florida peninsula. *J. Meteorol.* **5**, 275–280.

Clark, R. A. (1960). A study of convective precipitation as revealed by radar observation, Texas, 1958–59. *J. Meteorol.* **17**, 415–425.

Clarke, R. H. (1961). Mesostructure of dry cold fronts over featureless terrain. *J. Meteorol.* **18**, 715–735.

Desai, B. N., and Mal, S. (1938). Thundersqualls of Bengal. *Beitr. Geophys.* **53**, 285–304.

Donaldson, R. J., Jr. (1958). Analysis of severe convective storms observed by radar. *J. Meteorol.* **15**, 44–50.

Douglas, R. H., and Hitschfeld, W. (1959). Patterns of hail storms in Alberta. *Quart. J. Roy. Meteorol. Soc.* **85**, 105–119.

Fankhauser, J. C. (1965). Water budget considerations in an extensive squall-line development. Tech. Note 4-NSSL-25, 28 pp. U. S. Weather Bur., Washington, D. C.

Fawbush, E. J., and Miller, R. C. (1953). A method for forecasting hailstone size at the earth's surface. *Bull. Am. Meteorol. Soc.* **34**, 235–244.

Fawbush, E. J., Miller, R. C., and Starrett, L. G. (1951). An empirical method of forecasting tornado development. *Bull. Am. Meteorol. Soc.* **32**, 1–9.

Fujita, T. (1955). Results of detailed synoptic studies of squall lines. *Tellus* **7**, 405–436.

Fujita, T. (1959). Precipitation and cold air production in mesoscale thunderstorm systems. *J. Meteorol.* **16**, 454–466.

Fujita, T., and Brown, H. A. (1958). A study of mesosystems and their radar echoes. *Bull. Am. Meteorol. Soc.* **39**, 538–554.

Fulks, J. R. (1951). The instability line. *In* "Compendium of Meteorology" (T. F. Malone, ed.), pp. 647–652. Am. Meteorol. Soc., Boston, Massachusetts.

Fulks, J. R. (1962). On the mechanics of the tornado. Natl. Severe Storms Proj. Rept. No. 4, 33 pp. U. S. Weather Bur., Washington, D. C.

Harrison, H. T., and Orendorff, W. K. (1941). Pre-frontal squall lines. United Air Lines Meteorol. Circ. No. 16, Mimeograph, 12 pp.

House, D. C. (1963). Forecasting tornadoes and severe thunderstorms. *Meteorol. Monographs* **5**, No. 27, 141–155.

Kasahara, A., and Asai, T. (1967). Effects of an ensemble of convective elements on the large-scale motions of the atmosphere. *J. Meteorol. Soc. Japan* **45**, 280–290.

Longley, R. W., and Thompson, C. E. (1965). A study of the causes of hail. *J. Appl. Meteorol.* **4**, 69–82.

Ludlam, F. H. (1963). Severe local storms—a review. *Meteorol. Monographs* **5**, No. 27, 1–30.

Malkus, J. S., and Williams, R. T. (1963). On the interaction between severe storms and large cumulus clouds. *Meteorol. Monographs* **5**, No. 27, 59–64.

Means, L. L. (1944). The nocturnal maximum occurrence of thunderstorms in the Midwestern States. *Misc. Rept.* No. 16, 38 pp. Dept. Meteorol., Univ. Chicago.

Miller, J. E. (1955). Intensification of precipitation by differential advection. *J. Meteorol.* **12**, 472–477.

Newton, C. W. (1950). Structure and mechanism of the prefrontal squall line. *J. Meteorol.* **7**, 210–222.

Newton, C. W. (1963). Dynamics of severe convective storms. *Meteorol. Monographs* **5**, No. 27, 33–58.

Newton, C. W. (1966). Circulations in large sheared cumulonimbus. *Tellus* **18**, 699–713.

Newton, C. W., and Fankhauser, J. C. (1964). On the movements of convective storms, with emphasis on size discrimination in relation to water-budget requirements. *J. Appl. Meteorol.* **3**, 651–668.

Newton, C. W., and Newton, H. R. (1959). Dynamical interactions between large convective clouds and environment with vertical shear. *J. Meteorol.* **16**, 483–496.

Normand, C. W. B. (1938). On instability from water vapour. *Quart. J. Roy. Meteorol. Soc.* **64**, 47–69.

Normand, C. W. B. (1946). Energy in the atmosphere. *Quart. J. Roy. Meteorol. Soc.* **72**, 145–167.

Palmén, E., and Holopainen, E. O. (1962). Divergence, vertical velocity and conversion between potential and kinetic energy in an extratropical disturbance. *Geophysica (Helsinki)* **8**, 89–113.

Pedgley, D. E. (1962). A meso-synoptic analysis of the thunderstorms on 28 August 1958. *Geophys. Mem.* **14**, No. 106, 1–74.

Petterssen, S. (1939). Contribution to the theory of convection. *Geofys. Publikasjoner, Norske Videnskaps-Akad. Oslo* **12**, No. 9, 1–23.

Petterssen, S., Knighting, E., James, R. W., and Herlofson, N. (1945). Convection in theory and practice. *Geofys. Publikasjoner, Norske Videnskaps-Akad. Oslo* **16**, No. 10, 1–44.

Rhea, J. O. (1966). A study of thunderstorm formation along dry lines. *J. Appl. Meteorol.* **5**, 58–63.

Rossby, C.-G. (1932). Thermodynamics applied to air-mass analysis. *Mass. Inst. Technol., Meteorol. Papers* **1**, No. 3.

Sangster, W. E. (1958). An investigation of nighttime thunderstorms in the central United States. *Tech. Rept.* No. 5, Contr. AF 19(604)-2179, 37 pp. Dept. Meteorol., Univ. Chicago.

Saunders, P. M. (1961). An observational study of cumulus. *J. Meteorol.* **18**, 451–467.

Schleusener, R. A., and Grant, L. O. (1961). Characteristics of hailstorms in the Colorado State University Network, 1960–61. *Proc. 9th Weather Radar Conf., Kansas City, Oct. 1961*, pp. 140–145.

Schmidt, F. H. (1947). Some speculations on the resistance to the motions of cumuliform clouds. *Koninkl. Ned. Meteorol. Inst., Mededel. Verhandel.* **B1**, No. 8, 1–55.

Showalter, A. K. (1953). A stability index for thunderstorm forecasting. *Bull. Am. Meteorol. Soc.* **34**, 250–252.

Showalter, A. K., and Fulks, J. R. (1943). "Preliminary Report on Tornadoes," 162 pp. U. S. Weather Bur., Washington, D. C.

Squires, P., and Turner, J. S. (1962). An entraining jet model for cumulonimbus updraughts. *Tellus* **14**, 422–434.

Stommel, H. (1947). Entraining of air into a cumulus cloud. *J. Meteorol.* **4**, 91–94.

Sugg, A. L., and Foster, D. S. (1954). Oklahoma tornadoes May 1, 1954. *Monthly Weather Rev.* **82**, 131–140.

Tepper, M. (1950). A proposed mechanism of squall lines; the pressure jump line. *J. Meteorol.* **7**, 21–29.

U. S. Weather Bureau, Staff Members of Severe Local Storm Forecast Center, Kansas City. (1956). Forecasting tornadoes and severe thunderstorms. Forecasting Guide No. 1, 34 pp. U. S. Weather Bur., Washington, D. C.

U. S. Weather Bureau, Staff Members, National Severe Storms Project. (1963). Environmental and thunderstorm structures as shown by National Severe Storms Project observations in Spring 1960 and 1961. *Monthly Weather Rev.* **91**, 271–292.

Williams, D. T. (1953). Pressure wave observations in the Central Midwest, 1952. *Monthly Weather Rev.* **81**, 278–289.

World Meteorological Organization. (1956). World distribution of thunderstorm days, Part 2: Tables of marine data and world maps. *World Meteorol. Organ., Tech. Publ.* **21**, 1–71.

14

CIRCULATION AND DISTURBANCES OF THE TROPICS

The "tropics" is generally taken to comprise the region between the latitudes of the subtropical highs, considering their axes between easterlies and westerlies at the earth's surface. Relative to the global circulation, the tropical belt is the half of the earth over which the atmosphere gains both angular momentum from the earth's surface and heat energy in excess of the requirements to offset the outgoing radiation, and from which these properties are exported poleward to compensate the losses in extratropical latitudes (Chapters 1 and 2). Cressman (1948) described how wave disturbances of the tropics are affected in both their movements and intensities by interactions with troughs in the middle-latitude westerlies. Riehl (1950) observed that extended troughs often reach from very high to very low latitudes, and suggested that the poleward flow of heat must take place mainly in the restricted regions of these troughs. These and later studies have increasingly emphasized that the tropical and higher latitude circulations cannot be considered as isolated from each other.

Riehl (1954) characterizes the tropics as "that part of the world where most of the time the weather sequences differ distinctly from those of middle latitudes." It will become evident that, as in higher latitudes, the diverse natures of tropical disturbances can be understood only when these are viewed in the context of their broad-scale environments. It is therefore appropriate to begin by discussing the mean circulation and then to describe some kinds of typical disturbances. Considering the vast literature available, only a summary of selected aspects can be given here.[1]

---

[1] Reference may be made to the textbook by Riehl (1954). Also, several proceedings of tropical meteorology symposia have been published, which contain discussions ranging

14.1   MEAN STRUCTURE AND CIRCULATION OF THE TROPICAL BELT

To a greater extent than in middle latitudes, the broad-scale tropical circulation pattern on a given day is likely to resemble the seasonal mean circulation. The mean sea-level isobars and streamlines over the tropical belt are shown in Fig. 14.1 for the solstitial seasons. The main features are the subtropical highs, the trade-wind belts on their equatorward sides, and the zone where the trades meet, corresponding to a low-pressure belt. In the following discussion, the latter feature will be referred to as the "trade confluence" (TC).[2] Remarks on its structure are given in Section 14.11.

As emphasized in Chapter 2, the TC represents a belt across which no appreciable atmospheric energy flux occurs. In regions covered by oceans, the seasonal change of heat storage closely compensates the hemispheric radiative excess or deficit. Hence the TC in predominantly oceanic longitudes should be close to the geographic Equator and show only slight seasonal variations. In longitudes of large continents, whose heat storage is slight, the TC should undergo a strong seasonal migration to permit an appreciable heat transfer across the Equator. In these longitudes, a large flux of water vapor takes place into the summer hemisphere, in the low-level monsoon current. The corresponding latent heat, together with sensible and latent heat gained in the summer hemisphere, is realized and carried upward there, and partially exported (not necessarily in the same longitudes) into the winter hemisphere. As observed in Section 2.3, the overall transequatorial fluxes of latent and realized heat are in opposite directions and nearly compensate.

As shown by Fig. 14.1, around half the equatorial belt (over the eastern Pacific, the Atlantic, and west Africa) the TC lies north of the Equator the whole year. Over east Africa and through the longitudes of Asia and Aus-

---

from the large-scale aspects to the disturbances and weather regimes of special locations in the tropics. Concise and instructive overall views of the nature of the disturbances and of the philosophy and problems of analysis in the tropics are given in papers by LaSeur (1960) and Forsdyke (1960). See also Johnson and Mörth (1960) for an illuminating discussion of high-tropospheric flow patterns in low latitudes and their relation to the divergence field. The book by Watts (1955) describes the circulations of low latitudes (with special emphasis on southeast Asia), and gives an appreciation of the complexities of weather in tropical regions, especially as these are influenced by local effects.

[2] Also known as intertropical convergence zone (ITCZ), intertropical front (ITF), equatorial front, equatorial trough, and monsoon trough. Any general term applied to the zone around the whole earth is likely to connote attributes that are not valid in some regions. Alternatively, we sometimes use the term *thermal equator* in contexts relating to the release of energy in this zone.

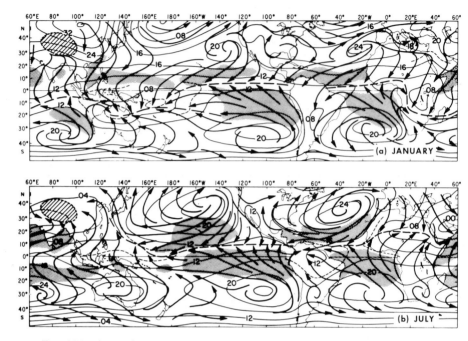

Fɪɢ. 14.1   Approximate mean streamlines at surface and mean sea-level isobars (mb) in (a) January and (b) July. Over oceans, based on analysis by McDonald (1938) of predominant wind direction; over land, a compromise between streamlines by Mintz and Dean (1952) and by Riehl (1954). The isobars are also a compromise between their analyses. Streamlines have been somewhat simplified. Stippling indicates regions where wind blows more than 80% of time from a direction within 45° of that given by the median streamline.

tralia, a large migration between the summer hemispheres is observed, in accord with the preceding remarks. In this region, the trades originating in the winter hemisphere turn on crossing the Equator and appear as westerlies in the summer hemisphere. A similar turning of the southern hemispheric trades takes place in a separate monsoon branch over the west Africa region, and in the eastern Pacific in summer.

Where the trades cross the Equator, they flow continuously from high pressure in one hemisphere toward low pressure in the other, and the turning is consistent with the condition that northward-decreasing pressure (or vice versa) corresponds to an easterly geostrophic wind in one hemisphere and a westerly geostrophic wind in the other. South of Asia (Fig. 14.1), the pressure is lower on the east than on the west side of the monsoon current. This is consistent with the sense of curvature of the flow in both

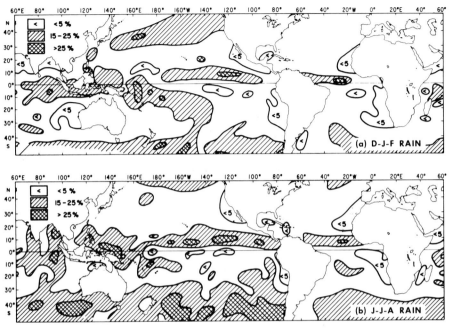

FIG. 14.2 Percent of observations in which rain in any form is observed in (a) December-January-February and (b) June-July-August. (After McDonald, 1938.)

seasons and with the turning of the winds to westerlies slightly before crossing the Equator. Over the central Pacific the pressure along the Equator is lower to the west, and (Fig. 14.1b) such a curvature is not evident. Schmidt (1951) has demonstrated that the observed configurations of the streamlines in the equatorial region are in accord with the principle of conservation of absolute vorticity.

Figure 14.2 illustrates the pronounced tendency for rain to occur near the TC, showing that in the mean it is a region of convergence. When averaged around the globe in a coordinate system relative to the "equatorial trough" (Fig. 14.3), the expected relationship between heaviest rain and maximum mean convergence appears in a broad sense. Not only in subtropical latitudes where there is mean low-level divergence, but also in the trade-confluence zone, *most of the rainfall is derived from synoptic disturbances deviating from the mean conditions.*

The importance of deviations from long-term mean conditions was recognized by Brooks and Braby (1921), in their study of the TC in the central Pacific. They pointed out the significant difference between the

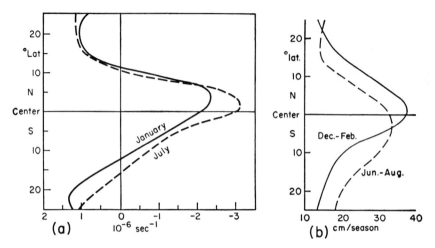

FIG. 14.3   (a) Profiles of mean surface divergence versus latitude in January and July; (b) profiles of seasonal rainfall. Curves are relative to equatorial trough. (By permission from H. Riehl, *Tropical Meteorology*, McGraw-Hill Book Company, Inc., 1954.)

"meetings of the trades" at a small angle east of 180° longitude and at a large angle farther west where there is a marked seasonal migration (see Fig. 14.1). In seasons when narrow-angle confluence dominated, the region 170°E to 155°W tended to be deficient in rainfall. Excessive rainfall in this region was associated with a displacement of the mid-Pacific equatorial low eastward of 170°E.[3] Also, other data presented by Brooks and Braby on the relation between wind direction and excessive rain very clearly indicated the importance of deviations from the mean tropical-flow patterns. Rainfall statistics (Riehl, 1954) for the tropics show that in most places about 50% of the total rainfall occurs on 10% of the days with rain, while 50% of the days with rain account for only 10% of the precipitation.

14.2   WATER-VAPOR TRANSPORT

In connection with the rainfall distribution and the global energy generation and transfer, it is of interest to examine the character of the water-vapor transport in the tropics. As indicated in Chapter 2, most of the heat imparted to the atmosphere is derived from condensation and precipitation.

[3] On the basis of rather small differences of temperature with wind direction, Brooks and Braby concluded that east of 180°, the southeast trades overran the northeast trades, and introduced the term "equatorial front" by analogy with the polar front. Although this concept enjoyed some acceptance, it is now generally discounted.

In the lowest latitudes, the heating from this source, together with the much smaller contribution of sensible heat transfer from the earth's surface, greatly exceeds the atmospheric loss by net radiation. Within the tropics, the excess of realized energy is transported poleward mainly by the mean (Hadley) circulation; however, eddies assume increasing importance poleward and account for most of the energy export from the tropics into extratropical latitudes (Fig. 2.7).

In addition, it was emphasized in Section 2.6 that the upward heat transport within extratropical regions is accomplished mainly by cyclone-scale eddies. In Section 10.9 it was pointed out that the movement of air from lower levels into the higher troposphere, required as part of this process, could not readily take place without the realization of latent heat in the rising currents. A significant part of this latent heat is derived from water vapor exported from the tropics in lower levels. Thus the upper levels of higher latitudes (block IV in Fig. 2.10) derive a large part of their heat energy, via two different routes, from sources that can be traced back to vapor evaporated within the tropical belt.

The essential aspects of the meridional transfer of water vapor in subtropical latitudes can be seen from Fig. 14.4. This pattern was derived by van de Boogaard (1964) from a detailed analysis of data for a single day in Northern Hemisphere winter, and can probably be considered typical.[4] *In low latitudes, the vapor transport is predominantly toward the thermal equator, being dominated by the mean flow in the lower trade-wind branch of the Hadley cell where the vapor density is greatest* (Fig. 14.4a). *The transport by eddies is almost exclusively poleward* (Fig. 14.4b), being almost nil at the Equator and strongest at the latitude of the subtropical highs. The vertically integrated vapor transport (Fig. 14.5a) shows that the flux accomplished by eddies (mainly wave number 6 on this day) is dominant poleward of 22°N. The equatorward boundary of the region of vapor-flux divergence (indicating an excess of evaporation over precipitation) in Fig. 14.5b corresponds approximately to the latitude separating mean low-level mass divergence from convergence (Fig. 14.3a), or roughly to the latitude of the strongest trade winds.

The meridional fluxes of water vapor in the Northern Hemisphere for the year 1958, computed by Peixoto and Crisi (1965), are given in Table 14.1. These are given separately for the flux by the mean circulation and by

---

[4] The mass circulation in the Hadley cell on this day corresponded closely to the winter mean in Fig. 1.5; however, van de Boogaard indicated that the eddy activity was well above normal.

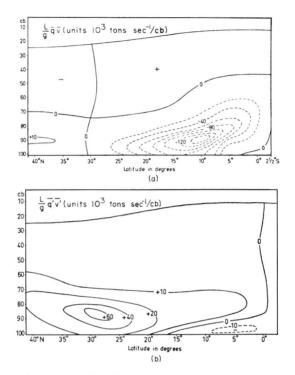

FIG. 14.4   Hemispheric meridional flux of water vapor (a) by the mean circulation and (b) by eddies on Dec. 12, 1957. (After van de Boogaard, 1964.)

the eddies, and for the cooler half-year (October-March) and the warmer half-year (April–September). The principal features are in accord with Fig. 14.5, in which the greater values can be ascribed to their being for a

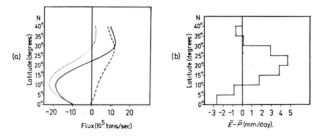

FIG. 14.5   (a) From data in Fig. 14.4, latitudinal profiles of the pressure-integrated total (full line), mean circulation (dotted line), and eddy (dashed line) meridional fluxes of water vapor. (b) The zonally averaged divergence of water vapor in belts of 5° latitude, expressed as evaporation minus precipitation (mm/day). (After van de Boogaard, 1964.)

TABLE 14.1

NORTHWARD FLUX OF WATER VAPOR IN NORTHERN HEMISPHERE (1958)[a]

Season	Latitude				
	40°	30°	20°	10°	0°
Cooler Months (Oct.-Mar.)					
Circ. flux	1.4	−1.7	−5.9	−16.4	−10.1
Eddy flux	5.8	5.9	3.5	2.1	1.0
Total	7.2	4.3	−2.4	−14.3	− 9.1
Warmer Months (Apr.-Sep.)					
Circ. flux	1.9	0.9	−3.1	− 0.4	8.7
Eddy flux	5.2	3.2	2.2	1.6	0.1
Total	7.1	4.1	−0.9	1.3	8.8

[a] Units are 10^5 ton/sec; negative sign corresponds to southward vapor flux. (From Tables 5 and 7 in Peixoto and Crisi, 1965.)

midwinter day, whereas the mean values in the upper part of Table 14.1 embrace parts of the autumn and spring seasons. The same features are evident for the warmer months, considering the northward displacement of the Hadley cell.[5] Corresponding computations for the year 1950 have been presented by Starr and Peixoto (1964), with broadly similar results.

In both seasons the eddy flux is directed poleward. In the tropics, the circulation flux is directed toward the TC, which on the average lies south of the Equator during northern winter and north of the Equator during northern summer. While the eddy flux differs relatively little between the seasons, the circulation flux varies strongly. This is consistent with the corresponding seasonal variations of the mass circulation of the Hadley cell (Figs. 1.3 and 1.4). In part, the weak equatorward flux in the warmer half-year is due to the zonal averaging process, since (for example) near latitude 20°N the southward transport of air with high vapor content around most of the hemisphere is largely compensated by a northward flux in the region south of Asia (Fig. 14.1b). On an annual basis (Table 2.2), both

[5] Peixoto and Crisi indicate, as an average for the year, a weak southward flux across the Equator. In the long-term mean (Table 2.2), there must be a northward flux. Based on a vertical profile using the values in their Table 4, the total warm-season northward flux across the Equator is estimated to be at least 13×10^5 ton/sec rather than 9×10^5 ton/sec as indicated in Table 14.1. This is consistent with the strong low-level northward mass flux across the Equator in northern summer (Fig. 1.4).

aerologically computed flux and the flux required by the water balance in latitude belts approach zero near latitude 5°N. This corresponds closely to the mean annual position of the TC or of the zone of heaviest annual rainfall (Fig. 2.3), where the excess of precipitation over evaporation is greatest.

14.3 Characteristics of the Trade-Wind Region

Over most of the tropical ocean the air near the surface is slightly cooler than the water. Owing to vertical mixing, which sustains an approach to a dry-adiabatic lapse rate and constant specific humidity in the lower levels, the relative humidity at the surface is maintained generally at about 80%. Thus, over practically the whole of the tropical waters, the atmosphere continually receives both heat and water vapor from the earth's surface.

Although in low latitudes only a minor portion of this energy transfer is in the form of sensible heat (Table 2.7), this portion is very significant with regard to the mechanism of upward energy transfer (Riehl, 1954). This is because a small amount of heating at the surface can maintain an unstable lapse rate in the subcloud layer, and slightly heated buoyant eddies can carry upward significant quantities of latent heat (observations at ship's deck level show a positive correlation between temperature and specific humidity).

The marine layer above the convection condensation level (about 600 m over much of the tropics) is somewhat more stable than is the subcloud layer, and the specific humidity decreases with height. The trade cumuli that partially fill this layer have tops usually no higher than 1 to 3 km. The typical trade cumulus is of small horizontal extent, and due to the erosive influences of strong entrainment and vertical shear, it has an average lifetime of less than a half-hour. The successive upward penetration, breaking off, and evaporation of cumulus towers give visible evidence of the transfer of water vapor and heat from the subcloud layer into the upper marine layer. Although, because of their limited depth, most trade cumuli can transfer surface-acquired properties only through a correspondingly shallow layer, this process is important in determining the air-mass structure of the tropics as a whole.

In most regions the marine layer is capped by a "trade inversion," which (where it is present) limits the height of maximum development of low clouds. Observations of the *Meteor* expedition (Ficker, 1936) disclosed that *the marine layer over the Atlantic is shallowest near latitude 15° and increases in depth both toward the western parts of the ocean and toward the trade confluence zone* (Fig. 14.6). A similar westward rise of the inversion over the eastern area of the north Pacific was confirmed in the extensive investigations by Neiburger, Beer, and Leopold (1945) and by Neiburger, Johnson, and Chien (1961). Computations by Neiburger *et al.* (1961) indicated on the average a strong low-level divergence in the eastern part of the subtropical high near the coast, with a marked decrease in magnitude of the divergence

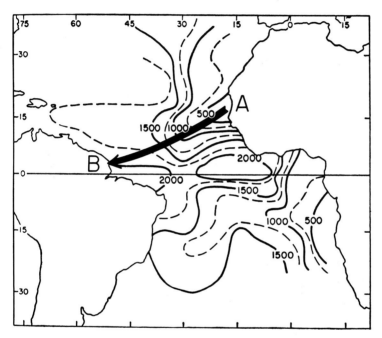

FIG. 14.6   Height of the base of the trade-wind inversion (m). Line *A–B* lies approximately along a mean streamline in the lower levels. (From Ficker, 1936.)

toward the central Pacific. Their charts demonstrate, consistent with Ficker's results for the Atlantic, that upward temperature increase, humidity decrease through the inversion, and frequency of inversion occurrence diminish westward, while the inversion height rises westward (Fig. 14.7). Data from successive cruises showed that trade inversion characteristics varied greatly in both time and space, although the forementioned conditions prevailed, as shown by the averaged data.

The typical low-level streamline *AB* in Fig. 14.6 suggests that as an air column moves from the African coast near 15°N toward the coast of equatorial South America, its depth is approximately quadrupled. While this

cannot be taken as an indication of vertical stretching alone, it is of interest to examine how much convergence would be required to account for
it. The distance *AB* is about 4000 km. With an average wind speed of 4 m/sec,
12 days would be required for the air column to traverse this region. A
mean upward motion of only 1.6 mm/sec at the inversion base would suffice to account for its rise along *AB*. The corresponding mean convergence

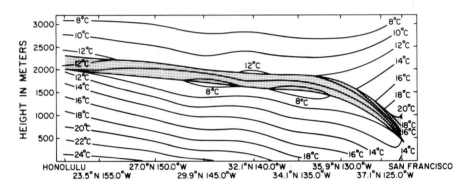

FIG. 14.7   Cross section showing average temperature (°C) between San Francisco
and Honolulu in summer; inversion layer shaded. (After Neiburger *et al.*, 1961.)

is slightly more than 10^{-6} sec^{-1}, roughly comparable to the average magnitude in this latitude belt in Fig. 14.3a.

A study by Riehl *et al.* (1951) demonstrated that lower-level convergence
or divergence is not the only process involved in changing the inversion
height. Poleward of approximately the latitudes of strongest trade winds,
there is surface divergence (Mintz and Dean, 1952), despite which the
inversion height increases westward. Based on ship and island observations
approximately along a mean streamline, Riehl *et al.* deduced a structure
over the eastern Pacific in summer, shown in Fig. 14.8. (The depth of the
inversion is considerably exaggerated, and its intensity vitiated, by the use
of mean soundings.) Accompanying a rise and weakening of the trade
inversion downstream, the cloud layer increased in depth downstream.
Since there was general divergence and subsidence, as suggested by the
streamlines in Fig. 14.8, it must be concluded that air is incorporated from
above into the inversion layer, and in turn from the inversion layer into
the cloud layer.

The mechanism is indicated schematically in Fig. 14.8. Trade-wind cumuli,
gaining upward momentum within the conditionally unstable cloud layer,
overshoot into the inversion layer. The cloud tops are potentially cooler

and have moisture contents higher than those of their surroundings. Their mixing into the lower part of the inversion layer, accompanied by evaporation of cloud water, contributes both to cooling and to enrichment of its moisture content. This in effect causes a continuous erosion of the lower part of the inversion layer, which gradually takes on the characteristics of the cloud layer.

The base of the inversion layer is consequently elevated by this process, which Riehl *et al.* call upon to explain the observed downstream increase in depth of the marine layer. Along the three-dimensional trajectories, there is an increase of both potential temperature and specific humidity in the cloud and subcloud layers, due to turbulent transfer as the air streams over

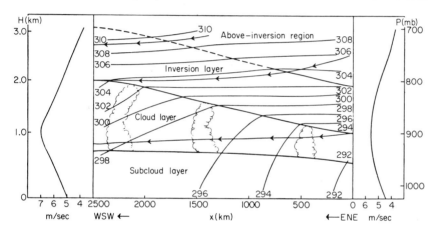

FIG. 14.8 Mean cross section approximately along low-level trade-wind trajectory between ship at 32°N, 136°W (right side) and Honolulu, for July–October 1945. Thin lines, potential temperature (°K); arrows, mean trajectories inferred from divergence; heavier lines delineate layers of interest. On right and left, vertical wind profiles at up-stream and downstream ends. (After Malkus, 1956.)

warmer water. A consequence of this heating is that it maintains a solenoid field such that the downstream pressure decrease along a streamline is appreciable in low levels, although it virtually vanishes at 3 km. A thorough discussion is given by Malkus (1956), who demonstrates the interrelations among the pressure-gradient accelerations, frictional stresses, and heat exchange. She concludes that an important factor governing the steadiness of the trades arises from the condition that the air must choose an equatorward trajectory in which the various balance conditions are satisfied, an essential constraint being imposed by the vertical transports in cumulus clouds.

14.4   Vertical Transfer in Tropical Zone

The Hadley cells are, as emphasized in Sections 1.1 and 2.5, solenoidally direct circulations. They are maintained through the addition of heat in the rising branch in the trade-confluence zone, combined with the radiative loss of heat in the sinking subtropical branches. In both meridional drifts connecting these vertical branches (i.e., the equatorward trades of the lower levels and the poleward branches of the upper troposphere), the flow is toward lower pressure. Hence, kinetic energy of the horizontal winds is generated, the major part of which is dissipated by friction in the lower branch, with a smaller loss in the upper. Thus the Hadley cells, viewed as mean circulations, appear to be relatively simple heat engines.

Although this straightforward concept of the meridional circulation cells is acceptable in its broad features, Riehl and Malkus (1958) have demonstrated that a steady circulation of the kind in Figs. 1.5 and 1.6, with uniform gradual rising motions through most of the troposphere over a broad area in the equatorial region, would be impossible. At the same time it is plainly necessary for both mass and energy to be transferred from low levels to the high troposphere in the equatorial zone because observations show that the strongest poleward mass flow and heat transfer in the Hadley cell take place near the 200-mb level.

The difficulty in accomplishing such a transport to high levels by a steady circulation is brought out by Fig. 14.9, showing the variation of static energy with pressure and latitude relative to the equatorial trough. The vertical variation in Fig. 14.10 is representative not only of mean conditions but also of individual situations. A steady upward movement in the tropics would imply a transfer down the energy gradient in the lower troposphere, but up the gradient in the upper troposphere. In the higher levels, it is therefore necessary that the vertical transfer be accomplished by eddy motions.

Riehl and Malkus concluded that the agent for this is large cumulus clouds with "protected cores." Small-diameter clouds (Section 13.4) entrain strongly from their environments; thus, air parcels rising from the lower moist layer with high static energy would have their energies diluted by mixing with the low-energy environment, as indicated by curve A in Fig. 14.10. On the other hand, the core of a large-diameter updraft may be essentially undiluted; an air parcel in such a draft might thus have the properties of line U, preserving its energy throughout its rise to the high troposphere.[6] In this manner *the required vertical transports can be accom-*

---

[6] On a thermodynamic diagram these curves correspond to A and U in Fig. 13.15a.

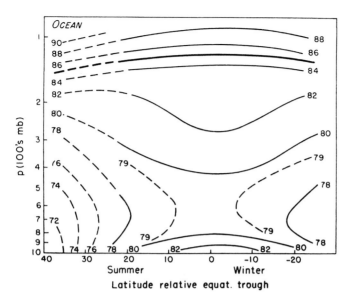

FIG. 14.9   Vertical cross section of the "static energy" $(gz + c_p T + Lq)$ relative to equatorial trough; ocean areas. Units are cal/gm. (After Riehl and Malkus, 1958.)

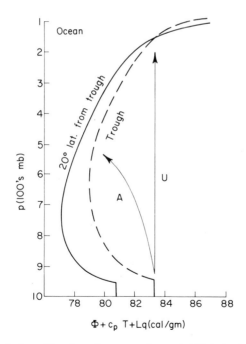

FIG. 14.10   Vertical profile of static energy in, and 20° latitude from, equatorial trough. (After Riehl and Malkus, 1958.) Lines U and A have been added (see text).

plished by tall cumulonimbus clouds. Their inner parts comprise insulated channels capable of conveying upward, through the lower-energy environment of the middle levels, large masses of air with high-energy content gained from the earth's surface.

On the basis of a heat budget for the equatorial zone, Riehl and Malkus estimate that the "protected cores" of cumulonimbi need occupy only about one-thousandth the area of the equatorial belt (considering this to be 10° latitude in width). Although the area covered by cumulus clouds in the tropics is very much larger, these are predominantly small trade-wind cumuli whose tops reach only to 700 mb, or lower, or to the levels of minimum static energy in Fig. 14.10. While the large clouds with protected cores transfer energy to the high troposphere, the lower trade cumuli along with those reaching intermediate heights carry heat to the middle troposphere.

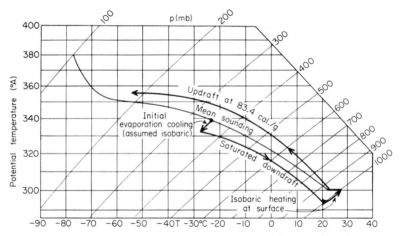

Fig. 14.11   Illustrating heat transfer from sea to air by means of downdraft cycle. (After Riehl and Malkus, 1958.)

Cumulonimbus clouds (as well as smaller shower clouds) are considered not only to be effective in transferring heat upward, but also to be a mechanism for enhancing the transfer of heat from the sea to the atmosphere, at the interface. The principle is illustrated in Fig. 14.11. Air in a downdraft, arriving at the surface with a temperature appreciably lower than that of the sea, must be correspondingly warmed. As discussed in Section 13.4, the coldness of the downdraft depends upon ingestion into the cloud of middle-tropospheric air with low-energy content.

It must be realized that, owing to variations of the ocean temperature, which to a large extent determines the energy content of the lower moist layer, the deepest convective clouds in different parts of the tropics can be expected to rise to varying heights. Thus, for the mean conditions in Fig. 14.10, undilute ascent would imply that a cloud parcel could rise somewhat above the 200-mb level in the equatorial trough, but only to about the 300-mb level at 20° distance from the trough. Moreover, although deep convective clouds are observed in the TC and in disturbances over the ocean, they show a distinct preference for land (see, e.g., Palmer *et al.*, 1955). Analyses of satellite radiation data by Saha (1966) demonstrate very clearly that, southeast of Asia, convective clouds with high tops are not uniformly distributed along an "equatorial trough zone." Rather, they show a strong preference to form over large islands such as Sumatra and Borneo, the neighboring oceans being relatively free of cloud except when synoptic disturbances are present. Marine air heated during the day over such land masses (as well as over tropical continents) has its θ_w considerably augmented. Consequently, cumulonimbus cloud tops may be expected to be much higher over the land areas than over the waters of the tropics. Thus the poleward branches of the Hadley circulations draw upon air derived from clouds extending to varying heights in different regions.[7]

14.5   Circulation during Asiatic Summer Monsoon

The principles of Section 14.4 apply also to the monsoon circulations. Figure 14.12, taken from Normand and Rao (1946), shows the approximate distribution of wet-bulb potential temperature (θ_w) in the Northern Hemisphere in summer and winter. In tropical latitudes, this section is most representative of the longitudes of India. The θ_w-isopleths are nearly equivalent to those of static energy. It may be noted that the maximum surface θ_w near latitude 30° in summer corresponds to a static energy of about 87 cal/g, appreciably greater than the mean of about 83 cal/g in the equatorial trough (Fig. 14.9).

Normand and Rao indicate that, on the average, convective instability is present below envelope A, while B, corresponding to the surface θ_w,

---

[7] In contrast to the 12- to 15-km cloud tops commonly observed in some tropical regions, during a two-month series of observations by the Line Islands Expedition in the early spring of 1967, only a single cloud was observed that extended to 11 km. The locale, in the central equatorial Pacific near 155°W, is influenced by cool equatorial waters over which the southeast trades must pass. Brooks and Braby (1921) noted that heavy rains occur in this region much more frequently with northeast winds than with east or southeast winds.

limits the layer of "convective liability" beyond which convection is not likely to penetrate. In terms of the argument by Riehl and Malkus, the transfer of energy upward through layer *A–B* requires eddy motions (hot convective towers).

The similarity between Figs. 14.12 and 14.9 is evident; in summer the transfer of energy to high levels is most favored, in the longitudes of south-

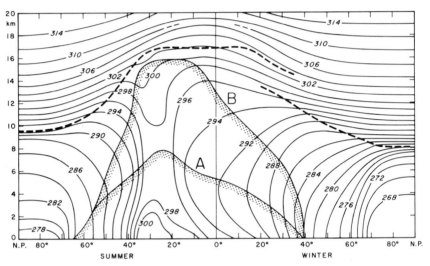

FIG. 14.12   Distribution of wet-bulb potential temperature (°K) with height and latitude, according to Normand and Rao (1946).

east Asia, over latitudes considerably removed from the Equator. As noted in Section 8.7, during the summer monsoon the warmest air of the globe, at levels up to 200 mb, is found in an east-west belt over northern India. That *the warmth is maintained essentially by the release of latent heat* is clearly indicated. In July, according to data given by Thompson (1965), the highest temperatures at levels between 700 and 300 mb lie along the moist adiabat defined by $\theta_w = 26°C$. This corresponds (with a relative humidity of 80%) to the mean sea-surface temperature (28 to 29°C) of the eastern Arabian Sea and the Bay of Bengal, over which the low-level monsoon currents pass before entering India.[8]

[8] The suggestion that the monsoon circulation is driven by differential heating of land and water masses was made by Halley in 1686. Ramage (1964) quotes Dallas (1887), who indicated that even though India is cooler during the monsoon than in the premonsoon period, "...the condensation of vapours and consequent liberation of latent heat suffice to prolong the existence of the indraught toward India."

As in the trades elsewhere, the available evidence indicates that the moist layer of the monsoon current is quite shallow over the ocean (from a limited number of soundings, about 1 to 1.5 km deep). In the longitudes of the Arabian Sea, where the southwest monsoon is strongest at the surface, the mean charts of Raman and Dixit (1964) and Frost and Stephenson (1964) show that the southerly flow is appreciably weakened at 850 mb and gives over to northwesterly flow at 700 mb. A shallow moist layer is apparently characteristic where the monsoon current crosses the Equator (Pisharoty, 1964), as well as farther north.

An inversion is normally present between the marine layer and the over-riding warm and dry air flowing from the Africa-Arabia region. Colón (1964) states: "The presence of this layer is of great importance to the rain-producing potential of the monsoon current, since once the inversion is destroyed there is a favorable stratification for rapid release upward of the moisture, leading to condensation and precipitation." The principle was discussed in Section 13.2. Indian meteorologists have increasingly empha-sized that the monsoon rains cannot be considered as a more-or-less steady phenomenon. Rather, the day-to-day and regional variations are consider-able, and synoptic disturbances must be held primarily responsible for the occurrence of deep convection and significant precipitation. The natures of some of these disturbances are outlined in Sections 14.8 and 14.10.

An estimate of the water-vapor budget by Pisharoty (1964) suggests that the mass of water precipitated over India may be 2.5 times or more as great as the vapor flux in the monsoon current as it crosses the Equator. The required augmentation of vapor is clearly due mostly to intense evapora-tion over the north Indian Ocean. Colón (1964) has studied the heat budget of the Arabian Sea. This behaves unlike other water bodies in that its highest temperature is reached in May, with a minimum in August and September and a secondary maximum in October and November. Colón concludes that "during spring, conditions are such that most of the incoming solar radia-tion is expended in increasing the temperature of the water.... With the establishment of the SW monsoon [with mean surface winds in June in excess of 14 m/sec near the African-Arabian coast] widespread cooling takes place on account of cold advection [of air from the cooler waters to the south] and heat flux [mostly by evaporation] to the atmosphere."

Thus the seasonal presence of the "thermal equator" over the latitude of northern India can evidently be ascribed to the release of latent heat over the subcontinent, the primary origin of this being evaporation from the warm-water bodies north of the Equator. The influence of this tempera-ture distribution on the upper-level wind system (namely, the presence of an easterly jet stream in the high troposphere south of Asia in latitudes 10 to 15°N) was discussed in Section 8.7. As suggested by Koteswaram (1958), this can be viewed as the consequence of a meridional circulation, in a sense reversed from that of the Hadley circulation but similarly driven. In the equatorward branch south of the belt of highest temperature, the tendency for conservation of absolute angular momentum results in generation of high-level easterly winds.

14.6   Varying Structures of Environments of Disturbances

Especially when the Hadley cells are contrasted with the summer monsoon, it is evident that in different regions of the tropics the structures of the basic currents are very dissimilar. As Riehl (1950) has observed, this fact is reflected in a greater variety in the character of synoptic disturbances in the tropics than in higher latitudes.

The middle latitudes are dominated by a westerly current increasing with height. Contrasted with this, Riehl notes that any of the four types of vertical wind profile in Fig. 14.13 may be found in various regions of the

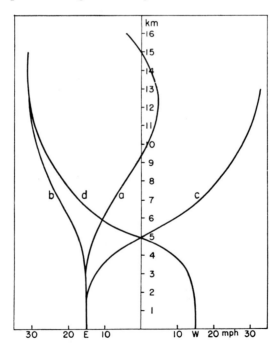

Fig. 14.13   Schematic representation of types of vertical profiles of the zonal wind in low latitudes. (After Riehl, 1950.)

tropics at various times. Over most of the trade-wind region in winter, lower-level easterlies give over to westerlies in the upper troposphere (curve *c*). The base of the westerlies slopes upward toward the Equator from the latitude of the subtropical highs so that both depth of the easterlies and strength of the westerlies vary with location. In summer, the average baroclinity in the trade-wind region is weaker; in various places and times the meridional temperature gradient may be directed either equatorward or poleward so that either profile *a* or *b* may be observed in the trades. Over the regions influenced by the summer monsoons, profile *d* is characteristic.

The nonuniformity of the flow patterns in the tropical belt is strikingly illustrated in Fig. 14.14, analyzed by Sadler (1965) for a day in northern winter. In the Western Hemisphere at 200 mb, the middle-latitude westerlies extended for the most part right across the equator, while at 700 mb, equatorial easterlies were flanked by regular belts of anticyclones near 15°N and S. In the Eastern Hemisphere, by contrast, at 200 mb there was a belt of equatorial easterlies with anticyclonic vortices both north and south. Beneath this, a narrow belt of westerlies was present near the Equator, with an irregular pattern of vortices in subtropical latitudes. Around half the equatorial belt, a vertical wind profile like *c* in Fig. 14.13 was present, with a profile like *d* or *b* over most of the remainder of the equatorial belt. Figure 3.13a shows that even on the average, equatorial westerlies appear in the Western Hemisphere at 200 mb, with easterlies in the Eastern Hemisphere. Hence, this feature of Fig. 14.14 is not anomalous.

These considerations and those in Section 14.5 indicate that there are marked variations in the structure of the basic currents in the tropics. Consequently, as emphasized by Riehl, it may be expected that *the kinds of disturbances also vary with geographical location and season, depending largely upon the regional barocline structure.* Some of the more common types of disturbances are briefly described below.

14.7   WAVES IN THE EASTERLIES

The regular presence of perturbations moving from the east over the tropical North Atlantic was discovered by Dunn (1940) through the analysis of 24-hr pressure-change patterns and their day-to-day movements. Utilizing airplane soundings in the Virgin Islands, Dunn found that these disturbances were connected with systematic fluctuations in the depth of the moist layer and of the temperatures aloft. There is increasing evidence that many of the perturbations observed in the western Atlantic have traveled all the way across the ocean; Riehl (1945) suggested a possible

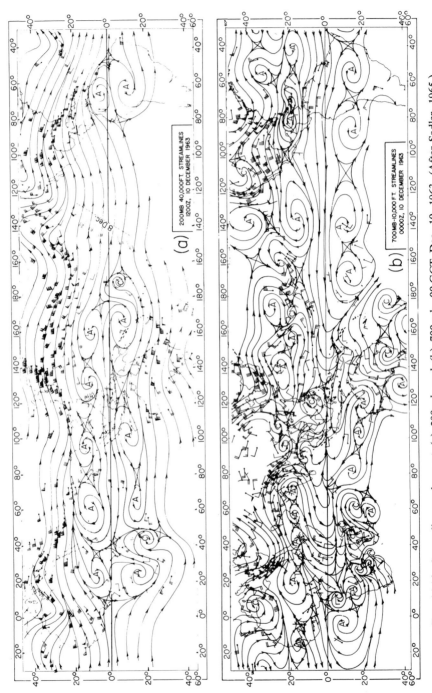

FIG. 14.14   Streamline analyses at (a) 200 mb and (b) 700 mb, 00 GCT Dec. 10, 1963. (After Sadler, 1965.)

connection with the westward-moving disturbances over north Africa, described by Regula (1936). The regular progression of vortices across the Atlantic, from near the west African coast, is illustrated in the daily analyses for August, 1963, published by Aspliden *et al.* (1966). Most vortices remained in the trade-confluence zone and did not develop significantly, while some, moving away from the equatorial zone, eventually developed into tropical cyclones in the west Atlantic.

The easterly perturbations in the Caribbean were found to have a predominantly wavelike character, although many of them contain vortices. The waves exist, according to Riehl, only when the easterly current is at least 6 to 8 km deep. For this and other reasons they are common in subtropical latitudes only in summer, in particular longitudes (generally westward of the subtropical highs, in the western Atlantic and mid-Pacific, mainly in the Northern Hemisphere), and they should not be expected to be a phenomenon typical of the whole tropical belt.

An example of an easterly wave in the mid-Pacific, taken from Malkus and Riehl (1964), is shown in Fig. 14.15. According to Riehl (1945, 1954) the leading dynamical characteristics may be deduced from the principle of conservation of potential vorticity. Considering the region between a wave trough and the upstream or downstream crest, we may write

$$\frac{\Delta(f + (E/r))}{\bar{f}} \approx \frac{\Delta d}{\bar{d}} \approx -\bar{D}\,\Delta t \tag{14.1}$$

where E is the easterly wind, d is the pressure depth of an air column, r is radius of streamline curvature (positive when cyclonic), and Δ represents the change along a trajectory between crest and trough of a wave. The time required to move between crest and trough is Δt, and the mean divergence experienced by the air column during this time is denoted by $\bar{D}$. Horizontal shear is neglected.

To illustrate general magnitudes, we may consider a wave with a length of 3000 km and a streamline amplitude of 2° latitude, as in Fig. 14.15. If $E = 9$ m/sec and the wave moves westward with a celerity $c = 6$ m/sec, the trajectory amplitude, greater by the ratio $E/(E - c)$, would be 6° latitude. With $|r| = 1000$ km at trough and crest, and a wave centered on latitude 20°, an air column would approximately triple in depth as it passes from the upstream crest into the wave trough, shrinking correspondingly as it moves into the downstream crest. If the base of the air column is at the earth's surface, this gives a rough idea of the variation to be expected in the depth of the moist layer. With the given values, about 10 days would be required for a particle to travel through an entire wavelength, and $|\bar{D}|$ would be about 2×10^{-6} sec^{-1}. A detailed kinematic analysis of an easterly wave has been carried out by Yanai and Nitta (1967). This involved a wave of much greater amplitude than the one considered above, and the maximum upward motion (near 500 mb about 500 km east of the trough) was $\omega \approx -3\,\mu$b/sec.

A distinguishing feature of the easterly wave configuration is that the troughs are in higher latitudes; therefore Δf and $\Delta(E/r)$ in Eq. (14.1) have

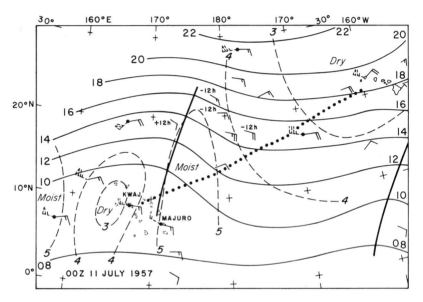

FIG. 14.15   Sea-level isobars and isopleths of vertically integrated water vapor (g/cm²), at 00 GCT July 11, 1957, over central tropical Pacific. Solid wind symbols, surface winds; dashed symbols, the shear from top of trade-wind layer to 40,000 ft. (By permission from J. S. Malkus and H. Riehl, *Cloud Structure and Distributions over the Tropical Pacific Ocean*, University of California Press, 1964).

the same sign. Consequently, as long as the wave moves slower than the wind in lower levels, convergence can be inferred to take place on the east side. With a vertical wind profile like *a* in Fig. 14.13, the air movement in upper levels is eastward relative to the westward-moving waves, with divergence of the opposite sort. The general pattern of divergence relative to the ideal easterly wave is thus similar to that in middle-latitude westerly waves, in the respect that there is lower-level convergence and upper-level divergence on its east side. This is an expression of the similar baroclinic structures of the basic currents. Although the basic current is from the east, the thermal wind is from west, as in middle latitudes.

In the example of Fig. 14.15 the vertically integrated water-vapor content increases westward toward the trough, with a substantial decrease on its west side, largely reflecting the moistness or dryness of the troposphere above the normal level of the undisturbed moist layer. The pronounced difference in air-mass structure across the trough is seen in the soundings of Fig. 14.16 and in the vertical section of Fig. 14.17.

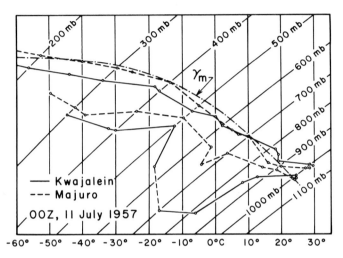

FIG. 14.16   Temperature and dewpoint soundings from Majuro and Kwajalein (see locations on Fig. 14.15). Dash-dotted line, moist adiabat corresponding to θ_w in lowest 100 mb. (By permission from J. S. Malkus and H. Riehl, *Cloud Structure and Distributions over the Tropical Pacific Ocean*, University of California Press, 1964).

Both because lifting vitiates the trade inversion and because the height to which a cumulus cloud can grow is strongly influenced by the environmental relative humidity in the middle troposphere (Section 13.4), the clouds tower to progressively greater heights as the wave is approached from the east side. Shower activity is generally greatest on that side of the trough. With a symmetrical wave pattern, Eq. (14.1) would imply greatest depth of the moist layer just at the trough, the depth being the same at given distances to each side. Thus the observation that the moist layer is deeper on the east side must be accounted for by additional processes.

In examining this question, one may consider that an essential function performed by cumulus convection is to transfer water vapor out of the subcloud layer. As pointed out by Zipser (1966), the need for such upward transfer is quite sensitive to the small values of divergence that characterize most of the tropical region. On the east side of the wave trough, vapor is accumulated in the subcloud layer both by evaporation from the surface and through horizontal convergence of water vapor, and it must therefore be transferred upward in large quantities. On the west side, vapor is added by surface evaporation, but removed through horizontal divergence. Based

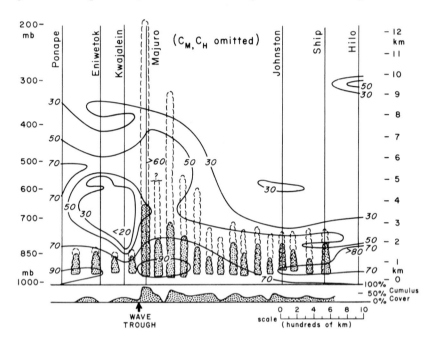

FIG. 14.17   Vertical section along dotted line in Fig. 14.15. Solid lines, relative humidity. Middle and high clouds were present, but have been omitted here. Stippled cloud symbols indicate average heights of bases and tops of cumulus clouds, determined from aircraft photographs, and dashed outlines show maximum vertical development of any cloud for each half-hour along flight leg. Widths of cloud symbols have no significance; the percent coverage of cumulus clouds is graphed at bottom. (By permission, adapted in simplified form from J. S. Malkus and H. Riehl, *Cloud Structure and Distribution over the Tropical Pacific Ocean*, University of California Press, 1964).

on water-budget calculations, Zipser found that (for example) with an evaporation rate of 0.65 cm/day, the likely precipitation rate would be 0 to 0.11 cm/day for divergence of 3×10^{-6} sec^{-1}, and 0.82 to 1.05 cm/day for convergence of 3×10^{-6} sec^{-1} (the range in precipitation amounts depends on whether the divergence or convergence extends up to 850 or to 700 mb). For a smaller evaporation rate, the magnitude of divergence required to offset evaporation without rainfall would be correspondingly smaller.[9] The example above illustrates how, even with the small divergence magnitudes of 2 to 3 $\times$ 10^{-6} sec^{-1} in relatively weak disturbances, the precipitation may vary greatly in different parts.

From this it can be seen that the need for upward convective transport of water vapor is small on the west side but large on the east side of an easterly wave trough, consistent with the difference in cumulus cloudiness in Fig. 14.17. In the middle troposphere, variations of humidity are influenced by the fact that the cumulus clouds carry up condensed water, an appreciable part of which is evaporated into their surroundings. This probably accounts in part for the asymmetry commented upon above.

The example in Figs. 14.15 through 14.17 was chosen because it is a clear-cut illustration, and it should not be supposed that all easterly waves have this ideal appearance. Cumulonimbi are often present west of wave troughs, although they are generally more abundant to the east. Various investigators have found from satellite data that cloud systems commonly deviate from the idealized scheme.[10] Merritt (1964) recognizes the easterly wave as one of a hierarchy of perturbations, noting that the disillusionment of some tropical meteorologists arises from a "tendency to attempt to force an easterly wave configuration into the analysis whenever any perturbation less intense than a tropical cyclone is detected in the tropical easterlies."

The energy processes of the type of easterly wave described above have remained obscure. By contrast with practically all kinds of synoptic disturbances that have been identified, the motions (on a synoptic scale) appear to be upward in the cooler air east of the trough and downward in the warmer air to the west. This would suggest that the disturbance consumes rather than generates kinetic energy, and it has been supposed that the circulation is somehow maintained by an energy input from the surroundings.

---

[9] The zonally averaged annual value of evaporation at 20°N is about 0.35 cm/day (Fig. 2.3). In the western parts of the oceans, Budyko's atlas indicates annual averages of 0.5 to 0.55 cm/day. In synoptic disturbances or stronger-than-average trades, this rate may be considerably enhanced, as shown by Garstang (1967). Thus the value selected above is not excessive.

[10] It may also be noted that in the Caribbean region, J. Simpson et al. (1967) have analyzed situations in which extensive cloud and precipitation systems appeared in latitudes 10 to 15°, in areas where there was no recognizable disturbance, or only weak perturbations, in the low-level streamline pattern. The upper tropospheric flow was highly disturbed, and Simpson et al. suggest that the formation and disappearance of the weather systems "are related to relatively small and subtle changes in the wind field at and above 500 mb."

Some light is shed upon this question by Riehl's analysis (1965) of aircraft observations in an easterly wave. The system was anomalous in that the wave trough moved westward with a speed exceeding that of the lower-tropospheric easterlies. The vorticity (due to shear) was greatest east of the wave trough. Considering this distribution and the wave movement, air columns on the west side and for some distance to the east of the trough must have experienced an increase of vorticity, with lower-level convergence. Correspondingly, convection started west of and became strong close to the trough. On the east side the air was more stable and there was extensive altostratus (with a base at 650 to 700 mb) and light rain. Riehl concluded that the coolness of the lower levels on the east side was due to the evaporation of rain, with a descent of air with low θ_w from middle levels on that side. About 50% of the condensed moisture was required for evaporation; the evaporation equaled the condensation in the system below 600 mb (considering both sides of the trough). There was a net heating from condensation above that level. East of the trough, the air was warmer in the upper troposphere than it was on the west side.

In this example, there was clear evidence of rising warm air to which heat was imparted by condensation, and of sinking cold air from which heat was abstracted by evaporation. Thus, this was an energy-generating system in the sense outlined in Section 16.3. Although this kind of system is more widespread and diffuse, the overall thermodynamic mechanism is superficially similar to that of a squall line (Section 13.4), which heats the upper and cools the lower troposphere in the same fashion. It is conceivable that the same general process is significant for the more common types of easterly waves, since convective showers on their east sides release heat, which is carried upward, and produce cold downdrafts. Thus the appearance of "rising" cold air on the synoptic scale, on their east sides, may be deceptive with regard to the actual energy processes. This suggestion is, however, speculative.

Palmer (1952) has given a detailed description of waves in the easterlies over the equatorial Pacific. Their wavelength is around 15° longitude, and they move westward with phase speeds of 10 to 15 knots, broadly similar to waves embedded in the trades. In the equatorial waves, crests in one hemisphere correspond to troughs in the other. Thus, along a given meridian, there may be divergence in one hemisphere and convergence in the other in a pattern similar to that related to the changes of streamline curvature in waves in the easterlies.

Rosenthal (1960) has provided a dynamical analysis that agrees very well in most essentials with Palmer's empirical description. This includes the features that the streamline amplitude is greatest near the Equator and damps poleward and that the strongest divergence and convergence (with a magnitude of about 2×10^{-6} sec^{-1} at the surface) are located at a distance from the Equator where the zonal easterly wind equals the phase speed of the waves. Again, this is a disturbance type that can be expected only in restricted regions where there are easterlies on both sides of the TC.

A different type of disturbance in the equatorial easterlies has been described by Freeman (1948). In this, there is no perturbation in the streamlines, but a discontinuity of wind speed is found, accompanied by an abrupt change in the height of the trade-wind inversion. The westward movement is at a speed roughly in accord with the speed of a gravity wave, considering the depth of the moist layer and the density change through the inversion; the physical analogy is to a hydraulic jump in flow through a channel.

14.8   Cyclonic Vortices

At least three distinctly different kinds of cyclones recognized in tropical latitudes may attain significant size and strength. The most common type is the "tropical cyclone," which has a warm core due to the release of latent heat and is embedded in a nearly barotropic basic current. The second is the cold-core "subtropical cyclone," which in some respects (some of the time) resembles a middle-latitude cyclone. The third, observed over India in summer, is a disturbance in a baroclinic current with basic westerlies in lower and easterlies in upper levels. Since tropical cyclones are treated in Chapter 15, only the latter two types will be discussed here.

The common existence of high-level ("distal") cyclones in latitudes as low as 10° to 15° was recognized by Palmer (1951). According to Palmer, these cyclones first appear at levels above 10 km and gradually extend downward; during much of their lives the lower-level trade-wind circulation may be undisturbed. Palmer et al. (1956) indicate that there is a general relation between the levels of associated cloudiness and the levels at which the cyclonic circulation exists. Palmer emphasizes that while the circulation at times may appear similar to the "cutoff low" (Section 10.1), a distal cyclone is not the same phenomenon because it appears first in tropical latitudes and does not involve polar air at any stage.

Particularly in late winter and spring, some cyclones in subtropical latitudes achieve considerable intensity at the surface as well as aloft. The heavy rain associated with two or three such cyclones may furnish the bulk of the annual rainfall in such places as the arid parts of Hawaii. A comprehensive examination of these "subtropical cyclones" was made by

R. H. Simpson (1952), who distinguishes between two modes of formation. He indicated that about two-thirds originate as waves on cold fronts trailing from middle-latitude systems when an upper-level trough deepens far southward and forms a cutoff low (Fig. 14.18a). While the low stagnates, the trough from which it separated moves rapidly eastward in higher latitudes. The migratory high originally on its west side also moves rapidly eastward, passing north of the cyclone and eventually merging with the subtropical anticyclone of the eastern Pacific. The result is that the influx of cold air is shut off, the vortex becoming embedded in surface easterlies, as in Fig. 14.18b. Although originating as a frontal disturbance, the cyclone is at this stage essentially fed by marine tropical air, having lost its frontal characteristics. The vortex is quite symmetrical and only slightly colder in the center than in the surroundings. As stressed in Section 10.4, it is hardly possible for polar air of considerable depth to move to the latitude of the closed low in Fig. 14.18a. Thus the upper-level air in such a subtropical cyclone should not be interpreted as polar air but as "middle-latitude air" (Section 4.1), which is cooler than real tropical air.

In a different mode of formation, a preexisting, low-latitude, upper-level cyclone overlies undisturbed easterlies. On the passage of an upper trough in the polar westerlies, this temporarily joins the lower-latitude upper cyclone, a surface circulation similar to that in Fig. 14.18 develops, and the upper cyclone is again cut off as the polar trough moves eastward. In this case the cyclogenesis takes place entirely in the low-level easterlies, and fronts are not involved. Cyclones of this type move slowly and erratically, and are observed not only in the Pacific, but also (less often) over the mid-Atlantic Ocean.

After reaching the stage shown in Fig. 14.18, such cyclones may (according to Simpson) have varying futures. In many cases a fresh intrusion of cold air from a passing polar-front trough results in the incorporation of a new frontal system, and the subtropical cyclone moves northeastward into middle latitudes. The lapse rate is characteristically near the moist adiabatic, and occasionally a relatively small increase of upper-tropospheric temperature, due to the release of latent heat, suffices to transform the vortex into a warm-core system resembling a tropical cyclone.

From an analysis of observations in Hawaii, Simpson found that the bands of heaviest rain and strongest surface wind (both of which attain maximum intensity on the east side) are located several hundred kilometers from the center of a subtropical cyclone. The same general distribution was found by Ramage (1962), using satellite pictures and regular observations in an analysis of a well-developed spring cyclone. Deep clouds, largely

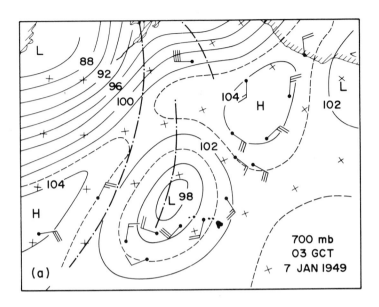

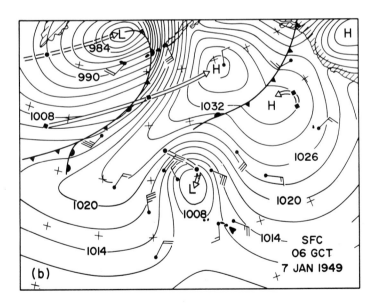

FIG. 14.18 Flow patterns over Pacific Ocean on Jan. 7, 1949; (a) 700-mb contours (hundreds of feet) at 03 GCT; (b) sea-level isobars at 06 GCT. Broad arrows show past 48-hr movements of sea-level pressure systems; positions of troughs on upper chart are shown for 24 and 48 hr earlier. (After R. H. Simpson, 1952.)

convective but also with broken middle and high clouds, were confined to a region 100 to 400 mi from the circulation center, with shallow clouds capped by a subsidence inversion near 800 mb in the outer region and evidence of an "eye" in the storm center. Noting that subtropical cyclones are very persistent, Ramage concludes that "old subtropical cyclones never die; they are only caught up by troughs in the polar westerlies."

Ramage (1962, 1964) gives evidence that some subtropical cyclones, both over the Pacific and the Indian Ocean (in the southeast Asia region) have maximum circulation intensity (and indraft) in the lower-middle troposphere, beneath which the disturbance has a cold core and above which it has a warm core. Ramage (1968) cites evidence that during the southwest monsoon season, disturbances of this type cause heavy rains in the vicinity of Bombay, the cyclonic circulation seldom extending to the surface.

A type of large cyclone with entirely different characteristics is the "monsoon depression" that forms at the head of the Bay of Bengal and moves west northwest over northern India. Its occurrence is confined to the summer monsoon season. The "cyclones" of the Bay of Bengal and Arabian Sea, which form in the pre- and postmonsoon seasons, but not when the monsoon current traverses these regions, are of a different (warm core) character, described in Chapter 15. Although monsoon depressions form on the average only about twice a month, according to Malurkar and Desai (1943) they account for most of the rainfall in upper India-Pakistan during the southwest monsoon.

An example is shown in Fig. 14.19. During the southwest monsoon season, the peninsula and neighboring waters are dominated by low-level westerlies. These are overlain by the easterly upper-level jet stream. Thus the vertical wind profile is like curve d in Fig. 14.13, the warmest air being to the north. As pointed out by Koteswaram and George (1958a), the monsoon depression is in many respects dynamically similar to a disturbance in the baroclinic westerlies, turned upside down with respect to directions.

Koteswaram and George found that the intensification and westward movement of such cyclones are associated with the approach of a trough in the high-tropospheric easterlies (or, in a minority of cases, of an easterly jet streak). Since the wind speed in the upper troposphere exceeds the westward movement of the wave system, it may be inferred from Eq. (14.1) that there is a divergence aloft on the west side of such a trough. In the example of Fig. 14.19, the amplitude of the wave at the 200-mb level is unimpressive. However, the vorticity variation along a streamline is appreciable because in an easterly current a trough represents not only a relative vorticity maximum but also the point on a streamline where the

coriolis parameter is greatest. Also, at such a low latitude, the variation of
f with latitude is large. Thus, in relating the divergence field to the flow
pattern, it must be considered that the amplitude of the upper wave is not
required to be so large as that which is characteristic of upper-level waves
associated with middle-latitude cyclones.

In the example of Fig. 14.19, a low-level depression had been nearly
stationary near the head of the Bay of Bengal. On the approach of the
upper-level easterly trough, this depression intensified appreciably (Fig.
14.19a), in response to superposition of the upper-level divergence ahead
(west) of the trough. As this trough moved westward, so did the surface
cyclone. Three days after it began to move (Fig. 14.19b), the monsoon

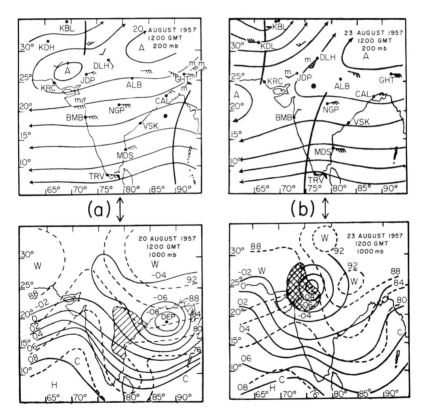

Fig. 14.19   200-mb winds and streamlines (upper) and 1000-mb contours (lower),
in tens of meters, for (a) 12 GCT Aug. 20, and (b) 12 GCT Aug. 23, 1957. On lower
charts, dashed lines are 1000 to 500 mb thickness (tens of meters over 5000 m); hatched
area indicates continuous rain and cross-hatching rainfall in excess of 5 cm/day. (After
Koteswaram and George, 1958b.)

depression was over northwest India where, coming under the influence of the southwesterly winds in advance of an approaching trough in the upper westerlies, it followed a path curving northward and then northeastward. At times, the upper-level easterly perturbation overtakes and passes the surface low, in which event the latter weakens before it has moved far westward.

Koteswaram and George give the general rule that "cyclone development at sea level occurs when and where an area of positive vorticity advection in the upper troposphere becomes superimposed upon a pre-existing trough at sea level [in the northern Bay of Bengal]." This is analogous to Petterssen's rule for extratropical cyclone development (Section 11.2). It is noteworthy that *although both upper trough passages and extensions of the monsoon trough into the Bay are much more frequent than cyclone developments, the latter took place in all cases (and only) when these circumstances coincided.*

Characteristically, the precipitation is mostly in the southwest quadrant of a monsoon depression. Considering that the cyclone moves westward, with warm air to the north rather than to the south and thus with a reversed vertical shear, the main precipitation region is in a position analogous to the warm-frontal precipitation in a middle-latitude cyclone, the whole system being inverted. The low-level moist air from the west undergoes an increase in vorticity as it approaches the trough south of the cyclone so that lower-level convergence must be present on the southwest side. Also (Koteswaram and George, 1958b), a local change of vorticity, contributing in the same sense, can be inferred from the isallobaric pattern connected with the westward movement of the cyclone.

The air masses entering the circulation are predominantly marine tropical, potentially unstable and with high moisture content, and in their regions of heaviest precipitation monsoon depressions typically disgorge rainfalls of 10 to 15 cm/day. When the cyclones move far westward, the easterly winds on their north sides advance marine air into normally dry regions. On occasion, heavy thunderstorms are observed on the north side, which are not necessarily connected with the organized precipitation regions characterizing the southwest sector but are more like the squall-line type of thunderstorm observed in warm sectors of middle-latitude cyclones.

14.9   HIGH-LEVEL FLOW PATTERNS IN SUMMER

In the winter season the high-tropospheric flow is dominated by westerlies that extend to low latitudes. This is also true around much of the Southern Hemisphere, even in summer. In the Northern Hemisphere in summer, the most striking feature of the high-level flow pattern in low latitudes is the presence of the easterly jet stream around half the subtropical belt (Fig. 3.13b). A salient characteristic of this system is its great steadiness. Although small-amplitude waves do pass westward and there are significant variations of wind speed from day to day, the high-tropospheric wind has a steadiness comparable to that of the low-level oceanic trade winds (Section 8.7).

By contrast, great unsteadiness prevails in the high troposphere over much of the remainder of the tropics in the warmer months. Riehl (1948) and Hubert (1949) found that in the tropical western Pacific, large anti-cyclonic vortices were present at 200 mb during the typhoon seasons of 1945 and 1946. These moved slowly westward at about the speed (12 to 15 knots) characteristic of lower-tropospheric easterly perturbations. The dimensions of the latter are much smaller, and on the average two easterly waves corresponded to an upper-level anticyclone. Figure 14.20, taken from Yanai (1964), illustrates this feature. The period between passages of anticyclones (with easterly troughs on their south sides) over the western Pacific corresponds nearly to the average period of 6 days (Koteswaram and George, 1958a) between the passages of high-level easterly waves over Calcutta in the height of the monsoon season, and a connection between these events is plausible.

Sadler (1963) has established that in summer there is in the high tropo-sphere a semipermanent trough extending east northeast to west southwest from about 35°N in the east Pacific to about 15 to 20°N in the central west Pacific. A mean ridge line (near 30°N in the west Pacific) corresponds to the latitude of the westward-moving anticyclonic vortices discussed by Riehl and by Yanai, and another mean ridge line is present south of the mean trough. This trough (which is not evident on most climatological charts such as Fig. 3.13b) has appeared in nearly the same location on charts analyzed for four individual summer seasons. Sadler indicates a similar structure over the Atlantic Ocean, where the mean trough extends from Cuba toward Spain. The features mentioned above appeared very regularly on the daily 200-mb charts over the Atlantic during August, 1963, analyzed by Aspliden *et al.* (1966.)

Figure 14.21 shows an example of the 200-mb flow pattern over the Pacific, superimposed upon the sea-level pressure pattern. The main features of the upper-tropospheric flow are analyzed from wind observations, al-though the suggested locations of cyclonic vortices are partly inferred from the organized cloud systems (shaded) shown by satellite pictures. In the easternmost vortex *M*, which was weakly reflected at sea level, there were

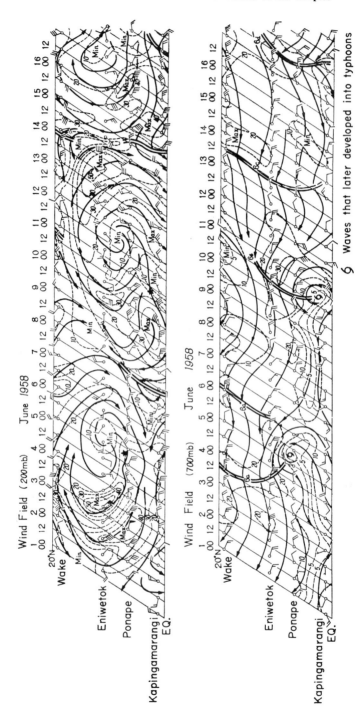

Fig. 14.20   Time sections showing passages of migratory vortices at 200 mb (top) and of waves at 700 mb (bottom) over a line of stations in the Marshall Islands, June 1958. Solid lines, streamlines; dashed lines, isotachs in knots. (After Yanai, 1964.)

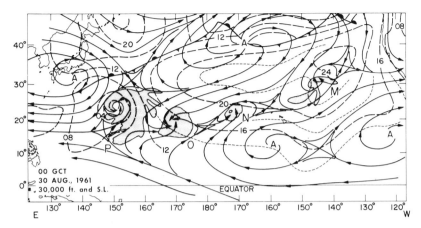

FIG. 14.21 Sea-level isobars (dashed) and 30,000-ft streamlines at 00 GCT Aug. 30, 1961. Shaded areas show organized cloud systems, extracted from TIROS III nephanalyses. (Redrawn, after Sadler, 1963.)

spirals in low- and middle-cloud systems. In vortex *N*, which did not extend to low levels, reconnaissance aircraft found extensive middle and high cloud, but little disturbance from the normal trade cumulus regime in low levels. The same was apparently true of system *O*. Sadler suggests that the large sea-level vortex *P* might have originated as a cold-core vortex developed downward from high levels.[11] Vortices of the kind shown in Fig. 14.21 are presumably the "distal" cyclones discussed by Palmer. They regularly form in the Pacific trough and, according to Sadler, move southwestward along it with an average speed of 12 to 15 knots. In the Atlantic analyses by Aspliden *et al.* (1966), the movements of upper-level cyclonic vortices were irregular and often eastward.

14.10 SEVERE THUNDERSTORM SYSTEMS

In certain land regions of the tropics, organized thunderstorm systems are observed which are evidently similar in nature to the middle-latitude squall-line thunderstorms described in Chapter 13. Two regions where these are especially notable are northeast India-Pakistan, and northwest Africa. Ramaswamy (1956) provides a thorough discussion of the severe

---

[11] Sadler cautions that although vortices of this kind may develop into typhoons, the appearance of such a cloud system or of a cyclonic circulation in the middle troposphere is inadequate evidence of the presence of a typhoon, since the system may have a cold core.

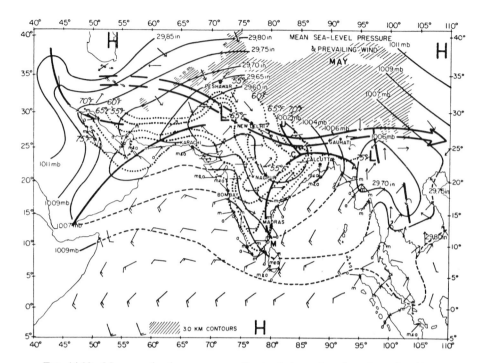

FIG. 14.22 Mean sea-level pressure, resultant winds, and surface dewpoint (dotted lines) for May. (After Ramaswamy, 1956.) Approximate axis of strongest resultant winds at 300 mb (double arrow) has been added. The strongest resultant winds are 35 to 40 knots over northwest Indo-Pakistan and 25 to 30 knots over the east, but the winds on individual days may be much stronger.

thunderstorms of northern India. These occur mostly in the three months preceding the onset of the summer monsoon, being most common in May. Most daily sea-level maps in May strongly resemble the mean chart in Fig. 14.22, which thus typifies the low-level environmental conditions on a storm day.

In the premonsoon season, severe thunderstorms occur in the region between about 20°N and the Himalayas; those farther south are of a milder variety. East of about 83°E, the common type of severe storm is the "nor'-wester," characterized by strong surface winds (generally over 40 and up to 100 mph), heavy rain, and often destructive hail. Farther west the "andhi," with similarly strong winds and dust storms, but little surface rain, is characteristic. "Nor'westers" occur in the region occupied by southerly and easterly winds in the low-level, marine tropical air, while "andhis" occur in the region of low-level westerly or northwesterly winds.

The air masses involved have much the same thermodynamic structure as those discussed in Section 13.2. In low levels, fresh marine air invades the coastal region east of trough MA (Fig. 14.22), part of this turning westward toward trough UC. Thus the southerlies and easterlies contain relatively cool air, which, however, has a high vapor content and high θ_w. Over the central and northwestern plains where the surface is strongly heated in spring, the structure is like Fig. 13.4b (solid curves), with steep lapse rates and low relative humidity near the ground. East of trough MCU, the marine air is overrun at a modest elevation by the warmer and drier westerly current, and the structure is qualitatively similar to Fig. 13.4a (solid curves). The dry overrunning air from the west has an appreciably lower θ_w than does the low-level marine air over the Gangetic plain, and marked convective instability is present.

In neither type of air mass is significant deep convection likely to result from surface heating alone. In the one case, this is inhibited by the presence of a stable layer over the moist air; and in the other, by the condition of low humidity. In both cases, *modification of the air masses by organized lifting favors the development of deep convection*, as in temperate latitudes (Section 13.2).

On the basis of a large number of synoptic cases, Ramaswamy gives convincing evidence that low-level synoptic patterns provide no conclusive indication as to the likelihood of occurrence of severe thunderstorms; therefore clues must be sought in the upper levels. He found that large outbreaks of thunderstorms and of severe squalls occurred when the upper-tropospheric flow pattern indicated significant divergence aloft. The most widespread outbreaks occurred east of troughs in the westerlies, another common location (particularly in the western dry region) being just upstream from pronounced upper ridges where there was a marked decrease of vorticity downstream.

As brought out in earlier chapters, it can be generally expected that the upper-level divergence is most pronounced, not only where the variation of vorticity along the current is appreciable but also where the wind speed is great. Thus Ramaswamy concluded that *the presence of the subtropical jet stream and its perturbations over northern Indo-Pakistan in the premonsoon season are related in an essential way to the occurrence of large convective systems.* From an examination of other regions he observes: "The subtropical jet stream...seems to produce similar convection in the subtropics all over the world, wherever it overruns on its equatorward side, moist air with pronounced latent instability."

For reasons discussed in Section 13.4, the condition of dry air with low θ_w overrunning the marine air favors development of the cold downdrafts and intense wind squalls characteristic of "nor'westers." The physical significance of this stratification was apparently recognized first by workers in India (Desai and Mal, 1938; Normand, 1946). In the "andhis" of the western plains, evaporation of rain into a deep and dry subcloud layer is important in generating squalls. It is noteworthy that strong squalls are generally absent (except in the drier region of northwestern Indo-Pakistan) after the southwest monsoon becomes fully established, despite an overall increase in convection. This can be evidently attributed to the absence of very dry air in the middle troposphere.

The "disturbance line" (DL) is a line of thunderstorms, frequently observed over the central and western areas of north Africa, generally moving with a westward component. Despite the sparse observations in this region, Hamilton and Archbold (1945) provided a remarkably clear and detailed description of the phenomenon. This has been elaborated upon by Eldridge (1957), who gives examples and statistics on the dimensions of the lines and other aspects.

The DL's occur in all months, but are much more frequent and most highly developed in summer when the southwest monsoon extends farthest northward (Fig. 14.1b). The lower-level humid westerlies are then overlain by dry and hot winds from the Sahara, with easterlies in the high troposphere (Fig. 3.13b). Near their northern bounds, the moist westerlies are normally too shallow for significant convection to occur, while farther south (e.g., on the Guinea coast) the moist layer is deep and marked instability is not present. The DL occurs between these two regions, where moderately deep moist air is available and (with the dry overrunning current) pronounced convective instability is present.

While the origin of a DL is difficult to establish, development from afternoon convection is common; some persist through the night, and a few evidently start at night, according to Eldridge. Regula (1936) related the thunderstorm systems to westward-moving pressure waves (see also Johnson, 1964). Hamilton and Archbold, and Eldrige, identify several regions of elevated terrain as likely places of origin.

Hamilton and Archbold (1945) were apparently the first to suggest that this type of disturbance propagates as a gravity wave, with a speed (averaging 25 to 30 knots) depending on the height and intensity of the interface between the moist layer and the overlying drier and warmer air. They deduced a circulation in a vertical plane which is very like the patterns in organized convective systems that were arrived at later by investigators working in other regions. Their description, and that by Eldridge, indicate that the DL is in all essentials similar to the squall line of temperate latitudes. Whereas the latter forms in an environment with an upward increase of west wind, the DL environment is characterized by vertical shear of the opposite sort. Correspondingly, following the discussion in Section 13.4, thunderstorm downdrafts transfer momentum and air with low θ_w from the upper-tropospheric easterlies. This causes the air within the thunderstorms to converge with the low-level moist westerlies, resulting in a continuous regeneration of convection on the west side toward which the DL moves. Owing to the presence of moist low-level air and convective instability over a broad range of longitude, such a disturbance is capable of lasting for a long time, and may move uninterruptedly from its place of inception inland to the west coast of Africa and past it.

14.11   CHARACTER OF THE TRADE-CONFLUENCE ZONE

Alpert (1945, 1946) published extensive statistics derived from numerous aircraft traverses and which revealed the low-level wind structure of the trade-confluence zone in the easternmost Pacific. The resultant wind patterns agree generally with McDonald's analyses (1938) in this region (Fig. 14.1). In particular, Alpert's data showed that in late winter, when the TC is closest to the Equator, the mean winds are from an easterly direction on both sides, whereas in late summer, when the TC is farthest north, the winds on the south side are from southwest. In a zone several hundred kilometers wide, the resultant winds were relatively weak, mainly as a result of variable wind direction in the well-known "doldrums."

Although the mean streamlines in Fig. 14.1 suggest a simple meeting of the northeast and southeast trades, numerous studies have indicated that the actual structure is far more complex. Fletcher (1945) and R. H. Simpson (1947) concluded that there are two zones of convergence, at least in parts of the tropics. From a study of observations over Central America and South America, Simpson indicated that the region intermediate to these zones has a lower temperature and θ_w than the air on either side, and that the zones of disturbed weather "may result from horizontal convergence within a stream of Northern Hemisphere or Southern Hemisphere air rather than at the intersection of Northern and Southern Hemisphere streams."

This view is supported by satellite cloud pictures and other data. Kornfield *et al.* (1967) present photographic integrations of satellite observations. These indicate comparatively very little cloudiness over the geographical Equator from about 160°E eastward to 80°W and over the Atlantic during the periods illustrated.[12] During the period Mar. 16–31, 1967, bands of cloudiness were present over the central and eastern Pacific, north and south of the Equator; the southern band was virtually nonexistent in January and June. Kornfield *et al.* note that "these bands are relatively stationary and vary in intensity rather than migrate across the equator."

With regard to the region south of Asia, Ramage (1968) states that "an east-west oriented pressure trough exists throughout the year in the tropics of each hemisphere." He indicates also that each of the troughs migrates (or dissolves and re-forms elsewhere) within the hemisphere, being located farthest poleward over the continents in summer when the trough is best developed. The double-trough system and the migration within the hemispheres were evident during much of the IGY period, as shown by daily tropical analyses (Deutscher Wetterdienst, 1965). Thus the earlier concept of a single TC migrating between hemispheres is evidently oversimplified.

Utilizing satellite observations, Sadler (1964) has made analyses that support the views of Fletcher and of Simpson. Over the eastern Pacific, Sadler generalizes a structure (in summer) wherein the trades do not meet head-on; rather, there is a zone of variable winds on each side of which there is a confluence at the edges of the northeast trades and of the turned westerly trades from the Southern Hemisphere.

Persistent cloudiness is present mainly in the bordering regions, while in the central part large cloud masses appear with clear regions between them. It appears reasonable to suppose that these cloud systems are groupings of cumulonimbus clouds, with large anvil sheets that account for their massive appearance, and that a large part of the vertical energy transport in the tropical zone takes place within these agglomerations. According to Sadler, many of them apparently develop into hurricanes. There are also indications that in some cases these large cloud masses have a limited lifetime (of the order of 1 or 2 days), and that one such mass may collapse, to be replaced by another in a different location. Too little is known at present to attempt a description of their circulations and behavior.

The foregoing comments on these and other systems of the trade-confluence region are admittedly sketchy, and are intended only to suggest the complex nature of its structure. This complex structure, and its dependence on longitude and seasons, should also be considered in the interpretation of the mean meridional mass circulations presented in Chapters 1 and 2.

---

[12] The same was not true in other longitudes. Kornfield *et al.* also state: "Cumulus activity, over the continental areas, is much more intense on the 'summer side' of the equatorial cloud band."

REFERENCES

Alpert, L. (1945, 1946). Intertropical convergence zone of the Eastern Pacific Ocean. *Bull. Am. Meteorol. Soc.* **26**, 426–432; **27**, 15–29.

Aspliden, C. I., Dean, G. A., and Landers, H. (1966). Satellite study, tropical North Atlantic, 1963, Part II. Rept. No. 66–4, 16 pp. plus charts. Dept, Meteorol., Florida State Univ.

Brooks, C. E. P., and Braby, H. W. (1921). The clash of the trades in the Pacific. *Quart. J. Roy. Meteorol. Soc.* **47**, 1–13.

Colón, J. A. (1964). On interactions between the southwest monsoon current and the sea surface over the Arabian Sea. *In* "Proceedings of the Symposium on Tropical Meteorology" (J. W. Hutchings, ed.), pp. 216–229. New Zealand Meteorol. Serv., Wellington.

Cressman, G. P. (1948). Relations between high- and low-latitude circulations. *Misc. Rept.* No. 24, pp. 68–101. Dept. Meteorol., Univ. Chicago.

Dallas, W. L. (1887). Memoir on the winds and monsoons of the Arabian Sea and north Indian Ocean. Supt. Govt. Printing, Calcutta, India.

Desai, B. N., and Mal, S. (1938). Thundersqualls of Bengal. *Beitr. Geophys.* **53**, 285–304.

Deutscher Wetterdienst, Seewetteramt, Hamburg. (1965). "IGY World Weather Maps, Tropical Zone." Deut. Wetterdienst, Offenbach.

Dunn, G. E. (1940). Cyclogenesis in the tropical Atlantic. *Bull. Am. Meteorol. Soc.* **21**, 215–229.

Eldridge, R. H. (1957). A synoptic study of West African disturbance lines. *Quart. J. Roy. Meteorol. Soc.* **83**, 303–314.

Ficker, H. (1936). Die Passatinversion. *Veroeffentl. Meteorol. Inst. Univ. Berlin* **1**, No. 4.

Fletcher, R. D. (1945). The general circulation of the tropical and equatorial atmosphere. *J. Meteorol.* **2**, 167–174.

Forsdyke, A. G. (1960). Problems of synoptic meteorology in the tropics; Synoptic models of the tropics; Tropical synoptic techniques. *In* "Tropical Meteorology in Africa" (D. J. Bargman, ed.), pp. 1–6, 14–23, and 35–46. Munitalp Found., Nairobi.

Freeman, J. C. (1948). An analogy between the equatorial easterlies and supersonic gas flows. *J. Meteorol.* **5**, 138–146.

Frost, R., and Stephenson, P. M. (1964). Mean stream lines for standard pressure levels over the Indian Ocean and adjacent land areas. *In* "Proceedings of the Symposium on Tropical Meteorology" (J. W. Hutchings, ed.), pp. 96–106. New Zealand Meteorol. Serv., Wellington.

Garstang, M. (1967). Sensible and latent heat exchange in low latitude synoptic scale systems. *Tellus* **19**, 492–508.

Hamilton, R. A., and Archbold, J. W. (1945). Meteorology of Nigeria and adjacent territory. *Quart. J. Roy. Meteorol. Soc.* **71**, 231–264.

Hubert, L. F. (1949). High tropospheric westerlies of the equatorial west Pacific Ocean. *J. Meteorol.* **6**, 216–224.

Johnson, D. H. (1964). Weather systems of West and Central Africa. *In* "Proceedings of the Symposium on Tropical Meteorology" (J. W. Hutchings, ed.), pp. 339–346. New Zealand Meteorol. Serv., Wellington.

Johnson, D. H., and Mörth, H. T. (1960). Forecasting research in East Africa. *In* "Tropical Meteorology in Africa" (D. J. Bargman, ed.), pp. 56–137. Munitalp Found., Nairobi.

Kornfield, J., Hasler, A. F., Hanson, K. J., and Suomi, V. E. (1967). Photographic cloud climatology from ESSA III and V computer produced mosaics. *Bull. Am. Meteorol. Soc.* **48**, 878–883.

Koteswaram, P. (1958). Asian summer monsoon and general circulation. *In* "Monsoons of the World" (S. Basu, P. R. Pisharoty, K. R. Ramanathan, U. K. Bose, eds.), pp. 105–110. India Meteorol. Dept., New Delhi.

Koteswaram, P., and George, C. A. (1958a). On the formation of monsoon depressions in the Bay of Bengal. *Indian J. Meteorol. Geophys.* **9**, 9–22.

Koteswaram, P., and George, C. A. (1958b). A monsoon depression in Bay of Bengal. *In* "Monsoons of the World" (S. Basu, P. R. Pisharoty, K. R. Ramanathan, U. K. Bose, eds.), pp. 145–156. India Meteorol. Dept., New Delhi.

LaSeur, N. E. (1960). Synoptic analysis in the tropics: The general problem; Methods of tropical synoptic analysis; Tropical synoptic models. *In* "Tropical Meteorology in Africa" (D. J. Bargman, ed.), pp. 7–13, 24–34, and 47–55. Munitalp Found., Nairobi.

McDonald, W. F. (1938). "Atlas of Climatic Charts of the Oceans," 130 charts. W. B. No. 1247. U.S. Govt. Printing Office, Washington, D. C.

Malkus, J. S. (1956). On the maintenance of the trade winds. *Tellus* **8**, 335–350.

Malkus, J. S., and Riehl, H. (1964). "Cloud Structure and Distributions over the Tropical Pacific Oceans," 229 pp. Univ. of California Press, Berkeley, California.

Malurkar, S. L., and Desai, B. N. (1943). Notes on forecasting weather in India. Tech. Note No. 1, 21 pp. Govt. Central Press, Bombay.

Merritt, E. S. (1964). Easterly waves and perturbations, a reappraisal. *J. Appl. Meteorol.* **3**, 367–382.

Mintz, Y., and Dean, G. A. (1952). The observed mean field of motion of the atmosphere. *Geophys. Res. Papers* **17**, 1–65.

Neiburger, M., Beer, C. G. P., and Leopold, L. B. (1945). The California stratus investigation of 1944. U. S. Weather Bur., Washington, D. C.

Neiburger, M., Johnson, D. S., and Chien, C. W. (1961). "Studies of the Structure of the Atmosphere over the Eastern Pacific Ocean in Summer. I. The Inversion over the Eastern North Pacific Ocean," 94 pp. Univ. of California Press, Berkeley, California.

Normand, C. W. B. (1946). Energy in the atmosphere. *Quart. J. Roy. Meteorol. Soc.* **72**, 145–167.

Normand, C. W. B., and Rao, K. N. (1946). Distribution of wet bulb potential temperature in latitude and altitude. *Nature* **158**, 128.

Palmer, C. E. (1951). On high-level cyclones originating in the tropics. *Trans. Am. Geophys. Union* **32**, 683–696.

Palmer, C. E. (1952). Tropical meteorology. *Quart. J. Roy. Meteorol. Soc.* **78**, 126–164.

Palmer, C. E., Wise, C. W., Stempson, L. J., and Duncan, G. H. (1955). The practical aspect of tropical meteorology. Air Force Surveys in Geophysics, AFCRCTN-55-220, No. 76, 195 pp. Air Force Cambridge Res. Lab., Bedford, Massachusetts.

Palmer, C. E., Nicholson, J. R., and Shimaura, R. M. (1956). An indirect aerology of the tropical Pacific. Final Rept., Contr. AF 19(604)-546, 131 pp. Oahu Res. Center, Inst. Geophys., Univ. Calif.

Peixoto, J. P., and Crisi, A. R. (1965). Hemispheric humidity conditions during the IGY, 166 pp. Sci. Rept. No. 6, Planetary Circ. Proj. Dept. Meteorol., M.I.T., Cambridge, Massachusetts.

Pisharoty, P. R. (1964). Monsoon pulses. *In* "Proceedings of the Symposium on Tropical Meteorology" (J. W. Hutchings, ed.), pp. 373–379. New Zealand Meteorol. Serv., Wellington.

Ramage, C. S. (1962). The subtropical cyclone. *J. Geophys. Res.* **67**, 1401–1411.

Ramage, C. S. (1964). Some preliminary research results from the International Meteorological Centre. *In* "Proceedings of the Symposium on Tropical Meteorology" (J. W. Hutchings, ed.), pp. 403–408. New Zealand Meteoro'. Serv., Wellington.

Ramage, C. S. (1968). Problems of a monsoon ocean. *Weather* **23**, 28–37.

Raman, C. R. V., and Dixit, C. M. (1964). Analyses of monthly mean resultant winds for standard pressure levels over the Indian Ocean and adjoining continental areas. *In* "Proceedings of the Symposium on Tropical Meteorology" (J. W. Hutchings, ed.), pp. 107–118. New Zealand Meteorol. Serv., Wellington.

Ramaswamy, C. (1956). On the sub-tropical jet stream and its role in the development of large-scale convection. *Tellus* **8**, 26–60.

Regula, H. (1936). Barometrische Schwankungen und Tornados an der Westküste von Afrika. *Ann. Hydrograph.* (*Berlin*) **64**, 107.

Riehl, H. (1945). Waves in the easterlies and the polar front in the tropics. *Misc. Rept.* No. 17, 79 pp. Dept. Meteorol., Univ. Chicago.

Riehl, H. (1948). On the formation of typhoons. *J. Meteorol.* **5**, 247–264.

Riehl, H. (1950). On the role of the tropics in the general circulation of the atmosphere. *Tellus* **2**, 1–17.

Riehl, H. (1954). "Tropical Meteorology," pp. 1, 93–97, 220, and 374–375. McGraw-Hill, New York.

Riehl, H. (1965). Varying structure of waves in the easterlies. *Proc. Intern. Sympo., Dynamics Large-Scale Atmospheric Processes, Moscow, 1965* pp. 411–416.

Riehl, H., and Malkus, J. S. (1958). On the heat balance in the equatorial trough zone. *Geophysica* (*Helsinki*) **6**, 503–537.

Riehl, H., Malkus, J. S., Yeh, T.-C., and LaSeur, N. E. (1951). The north-east trade of the Pacific Ocean. *Quart. J. Roy. Meteorol. Soc.* **77**, 598–626.

Rosenthal, S. L. (1960). A simplified linear theory of equatorial easterly waves. *J. Meteorol.* **17**, 484–488.

Sadler, J. C. (1963). Utilization of meteorological satellite cloud data in tropical meteorology. *In* "Rocket and Satellite Meteorology" (H. Wexler and J. E. Caskey, Jr., eds.), pp. 333–356. North-Holland Publ., Amsterdam.

Sadler, J. C. (1964). Tiros observations of the summer circulation and weather patterns of the Eastern North Pacific. *In* "Proceedings of the Symposium on Tropical Meteorology" (J. W. Hutchings, ed.), pp. 553–571. New Zealand Meteorol. Serv., Wellington.

Sadler, J. C. (1965). The feasibility of global tropical analysis. *Bull. Amer. Meteorol. Soc.* **46**, 118–130.

Saha, K. R. (1966). A contribution to the study of convection patterns in the equatorial trough zone using TIROS-IV radiation data. Tech. Rept. No. 74, 29 pp. Dept. Atmospheric Sci., Colorado State Univ., Fort Collins, Colorado.

Schmidt, F. H. (1951). Streamline patterns in equatorial regions. *J. Meteorol.* **8**, 300–306.

Simpson, J., Garstang, M., Zipser, E. J., and Dean, G. A. (1967). A study of a non-deepening tropical disturbance. *J. Appl. Meteorol.* **6**, 237–254.

Simpson, R. H. (1947). Synoptic aspects of the intertropical convergence near Central and South America. *Bull. Am. Meteorol. Soc.* **28**, 335–346.

Simpson, R. H. (1952). Evolution of the Kona storm, a subtropical cyclone. *J. Meteorol.* **9**, 24–35.

Starr, V. P., and Peixoto, J. P. (1964). The hemispheric eddy flux of water vapor and its implications for the mechanics of the general circulation. *Arch. Meteorol., Geophys. Bioklimatol.* **A14**, 111–130.

Thompson, B. W. (1965). "The Climate of Africa," 132 pp. Oxford Univ. Press, Nairobi.

van de Boogaard, H. M. E. (1964). A preliminary investigation of the daily meridional transfer of atmospheric water vapour between the equator and 40°N. *Tellus* **16**, 43–54.

Watts, I. E. M. (1955). "Equatorial Weather," 224 pp. Oxford Univ. Press (Univ. London), London and New York.

Yanai, M. (1964). Formation of tropical cyclones. *Rev. Geophys.* **2**, 367–414.

Yanai, M., and Nitta, T. (1967). Computation of vertical motion and vorticity budget in a Caribbean easterly wave. *J. Meteorol. Soc. Japan* **45**, 444–466.

Zipser, E. J. (1966). The distribution and depth of convective clouds over the tropical Atlantic Ocean, as determined from meteorological satellite and other data. Ph. D. dissertation, Chapter 4. Florida State Univ., Tallahassee.

15

TROPICAL CYCLONES, HURRICANES, AND TYPHOONS

Tropical disturbances occasionally develop into intense cyclones known as hurricanes in the Atlantic and east Pacific regions and as typhoons in the western area of the north Pacific. In contrast to cyclones in middle and high latitudes, severe tropical storms develop only over oceans and in regions where the baroclinity of the basic current is weak. Hence they are not manifestations of conversion between preexisting available potential energy and kinetic energy, as are extratropical cyclones. Their origin and maintenance instead depend on the ability of the atmosphere to produce available potential energy through the agency of interior heat sources resulting from the disturbed motion field, depending on the potential or convective instability of the tropical atmosphere.

Extensive recent explorations by the National Hurricane Research Project, and by Air Force and Navy reconnaissance aircraft, have greatly improved our knowledge of the structure of tropical cyclones. The conditions required for the often rapid transformation of a weaker disturbance into a severe storm are, however, inadequately known. While broadly favorable conditions exist over large parts of the tropics, and lesser disturbances abound, there is a global average of only about 50 hurricanes per year. Although several hypotheses have been proposed to explain their genesis, arrival at a final solution has been hampered by the difficulty of attaining adequate observations in the early stages.

In this chapter, although we discuss certain characteristics in some depth, we cannot deal with all. For more complete discussions, we refer to the chapter in Riehl (1954) and to the book by Dunn and Miller (1960), which is a valuable source of material relevant to all aspects of hurricanes. We cannot attempt to summarize the results of numerical experiments. The proceedings of a Mexico City conference (Adem, ed., 1963, 1964) contain several papers in this area, in which significant advances are now being made. The proceedings of a conference in Tokyo (Japan Meteorological Agency, 1963) are an especially valuable source of papers on climatological and behavioral aspects of tropical cyclones.

15.1   Necessary Conditions for Formation of Tropical Cyclones

A mature tropical cyclone is a disturbance whose central part is considerably warmer than the surroundings, up to very high levels. The essential dynamical features were first enunciated by Ferrel (1856) who, following Espy, indicated that the violence and duration of the hurricane "depend upon the quantity of vapor supplied by the currents flowing in below." Ferrel introduced the principle of conservation of angular momentum to account for the intense "gyratory motion" achieved by the air as it moves inward, "which it still, in some measure, retains after ascending to the regions above, where the [strong radial pressure gradient] does not prevail," causing the air to flow outward in upper levels.

Shaw (1922) described the tropical storm as a depression endemic only to areas with high sea temperature. He visualized the storm as having a core composed of air that rises moist adiabatically from the heated surface layer, extending to a probable height of 15 km, and calculated the possible pressure reduction that could result. Despite this estimate, many held to the view that tropical cyclones were shallow disturbances, until Haurwitz (1935) demonstrated conclusively that the sea-level pressure distribution could be accounted for only if the central air columns were appreciably warmer than the surroundings up to at least 10 to 11 km. The first aerological evidence that the warm core extends through an even greater depth was presented by Simpson (1947). He also indicated that the tropopause is elevated over the core, where in the uppermost troposphere and lower stratosphere the air is colder than the surroundings. Koteswaram (1967) shows that in some cases the overlying cold core extends to at least 27 km, and suggests that it may be caused by forced ascent above the intense convection at lower levels.

A warm-core structure does not seem typical of tropical disturbances in general (Chapter 14). Hence *the formation of a warm core is the first decisive sign of tropical cyclone formation*. Its genesis, due to liberation of latent heat in the inner cloud region, requires a vertical stratification such that

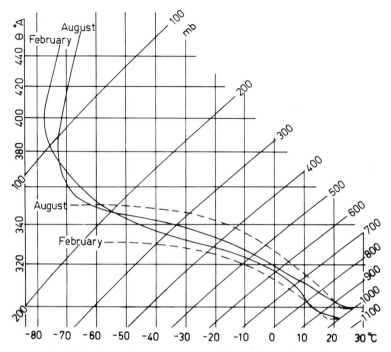

FIG. 15.1   Mean-temperature soundings in the Caribbean region in August and February (after C. L. Jordan, 1958a) and the corresponding temperatures of an air parcel lifted pseudo-adiabatically from the 1000-mb level.

ascending air becomes warmer than the undisturbed surroundings. C. L. Jordan's mean soundings over the Caribbean (1958a), in Fig. 15.1, show that this is possible in August, whereas in February the lifted air would generally be colder. The "positive area" on this diagram provides a measure of the overall baroclinity that may be attained between the inner rain area and the undisturbed surroundings. With moist-adiabatic ascent and a level of no disturbance at 100 mb, Malkus and Riehl (1960) indicate that the defect of sea-level pressure follows the rule

$$- \delta p_0 \text{ (mb)} \approx 2.5 \delta \theta_e \text{ (°K)} \qquad (15.1)$$

in the tropical range of θ_e. With typical saturation deficits at the surface, θ_e changes by about 4.4°K for each 1°C change of surface temperature; thus $- \delta p_0$ (mb) $\approx 11 \delta T_0$ (°C). These expressions neglect the contribution of the warmer air within the eye. Hence the sensitivity of the lowest possible pressure to the surface temperature is even greater.

While the temperature and moisture content of the lowest tropospheric layer depend closely on the sea-surface temperature, the temperature in the middle and upper troposphere does not. Hence the potential instability of the atmosphere, upon which hinges the possibility of developing a warm core, is strongly influenced by the sea temperature, according to Eq. (15.1) and Fig. 15.1. Intense tropical cyclones are therefore confined to areas with warm surface waters, excluding much of the tropical ocean (Palmén, 1948). This is demonstrated by Fig. 15.2, after Bergeron (1954).

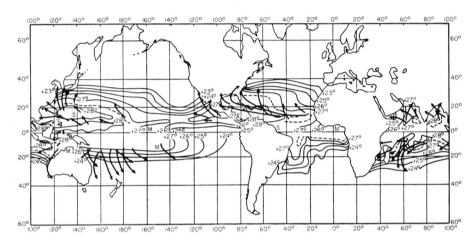

FIG. 15.2   Principal tracks of tropical cyclones in relation to mean sea-surface isotherms, in September for the Northern Hemisphere, and in March for the Southern Hemisphere. (After Bergeron, 1954).

The frequency of hurricanes over the Atlantic (Fig. 15.3) is well correlated with the seasonal variation of sea-surface temperature, shown for the Caribbean Sea according to Colón (1960). The potential to form a warm core may be characterized by an "instability index," showing the temperature difference between ascending surface air and the undisturbed atmosphere at a fixed level such as 300 mb. This index, for the West Indies area (Hebert and Jordan, 1959), is also shown in Fig. 15.3. Its incomplete accord with the trend of hurricane frequency indicates that although potential instability is a factor, other influences are involved.

In the western Pacific, the typhoon frequency (crosses in Fig. 15.3) follows a somewhat similar annual course, which Kasahara (1954) found to be correlated with the potential instability[1] and the water temperature.

[1] This is also demonstrated by McRae (1956) for the region near Australia.

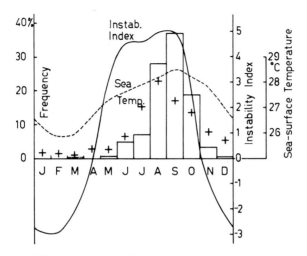

FIG. 15.3   Instability index in the West Indies area (after Hebert and Jordan, 1959) and mean sea-surface temperature over the Caribbean (Colón, 1960) compared with monthly frequency of hurricanes over Atlantic Ocean. Crosses indicate monthly percentage of typhoons over western Pacific.

Table 15.1 shows that at Guam the mean temperature below about 800 mb is lower in February than in August, but above that level it is actually higher than in August. Both circumstances contribute to the marked seasonal

TABLE 15.1

MEAN TEMPERATURE OVER GUAM (13.4°N, 144.6°E) IN AUGUST AND FEBRUARY COMPARED WITH THE TEMPERATURE OF AN AIR PARCEL LIFTED PSEUDO-ADIABATICALLY FROM THE SEA SURFACE[a]

Level	August			February		
	Mean temp. (°C)	Temp. of lifted air	Diff.	Mean temp. (°C)	Temp. of lifted air	Diff.
Surface	27.8			26.4		
850 mb	18.5	19.5	1.0	17.1	18.3	1.2
700 mb	9.5	12.8	3.3	12.1	10.5	−1.6
500 mb	− 6.2	− 0.5	5.7	− 3.1	− 3.5	−0.4
400 mb	−16.5	−10.3	6.2	−14.9	−14.0	0.9
300 mb	−31.8	−25.0	6.8	−30.2	−29.0	1.2
200 mb	−53.7	−49.0	4.7	−51.9	−53.0	−1.1

[a] After Kasahara (1954).

variation of stability shown by the difference values in Table 15.1. The few cool-season typhoons are generally confined to low latitudes, where the instability may on occasion be considerable.

The preceding examples show that ascending (unmixed) air becomes appreciably warmer than the undisturbed atmosphere only in suitable regions and seasons. Palmén (1948) found that *the lowest water temperature for which warm-core cyclones are likely to form is about 26 to 27°C*. Also, since strong rotation can, according to V. Bjerknes' circulation theorem, be generated readily only in those regions where the coriolis parameter exceeds a certain minimum value, *a zone close to the Equator has to be excluded even where the water temperature is high*.[2] Palmer (1952) states that "the great majority of the cyclones of the tropical north Pacific form in latitudes south of 6°N but intensify rapidly only in the zone 6°N to 15°N."

The establishment of a warm core around a quasi-vertical axis, characteristic of a fully developed tropical cyclone, is difficult to achieve in an atmosphere with appreciable vertical shear. To the necessary conditions mentioned above, we may therefore add the rule that *the basic current should have only a weak vertical shear*. This rule accounts for the exceptional seasonal distribution of tropical cyclones over the north Indian Ocean. During the height of the summer Asiatic monsoon, this region is occupied by a steady-wind system with pronounced vertical shear (Section 8.7). Tropical cyclones tend to be absent then, but form in the pre- and postmonsoon months when the basic circulation is nearly barotropic and also (Section 14.5) the water temperatures are highest in the Arabian Sea and Bay of Bengal.[3]

In summary, the necessary (but not sufficient) climatological-geographical conditions for formation of intense tropical cyclones (Palmén, 1948, 1956; Bergeron, 1954) are:

1. Sufficiently large sea areas with a water temperature so high that moist air lifted from the lowest layers of the atmosphere (with almost the temperature of the sea surface), expanding pseudo-adiabatically, remains considerably warmer than the surrounding undisturbed atmosphere at least up to a level of about 12 km.

---

[2] In regions where the trade confluence is characterized by cyclonic shear, this may augment the circulation due to the earth's rotation as an initial condition. However, this shear is usually appreciable only when the TC is a considerable distance from the Equator (Section 14.11).

[3] Most tabulations for the north Indian Ocean show a continuance of "tropical cyclones" throughout the summer monsoon season. However, as indicated in Section 14.8, these mid-monsoon cyclones are more closely akin to middle-latitude baroclinic disturbances than to warm-core tropical cyclones.

2. A coriolis parameter larger than a certain minimum value, thus excluding a belt of about 5 to 8° latitude on either side of the Equator.

3. Weak vertical wind shear and, correspondingly, weak baroclinity in the basic current in a deep tropospheric layer.

15.2 STRUCTURE OF MATURE TROPICAL CYCLONES

The characteristic structure of mature tropical cyclones follows from the principles listed above, and may be discussed in reference to the schematic Fig. 15.4. In this figure, the left side shows the profiles of isobaric surfaces,

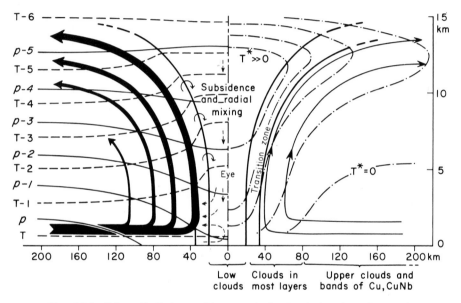

FIG. 15.4   Schematic features of inner part of a tropical cyclone (see text).

whose vertical spacings are consonant with the inward increase of temperature shown by the isotherms. The circulation in a vertical plane along the radial direction (generally much weaker than the tangential wind component) is in a direct solenoidal sense and produces kinetic energy (Section 16.3).

Outside the eye the baroclinic structure is due to intense release of latent heat in the rising branch of the circulation. The temperature achieved here is determined by the θ_e of the air rising from the lowest atmospheric layer, if entrainment from the colder and dryer environment of the cyclone can be disregarded in the inner cloud region. In the inflow layer, the air is subjected to cooling (and lowering of its dew point) by decompression as it moves toward the central low-pressure region. As a consequence (Byers, 1944), the transfer of sensible and latent heat from the underlying sea surface is enhanced, and may be so intense that it effectively counteracts the influence of expansion.[4] Hence the process is nearly isothermal, the energy transfer increasing the θ_e by an amount depending on the pressure decrease between the outer undisturbed moist layer and the innermost part of the rain zone. Also, because the air moving inward and partially rising at different distances from the center will have been subjected to varying pressure drops and increases of θ_e, this accounts for the intense temperature gradients along isobaric surfaces above the surface layer.

The resulting thermal field of the hurricane thus depends not only upon the temperature and specific humidity of the inflow layer at the outskirts, but also upon the heat and moisture flux from the sea within the interior. Although the energy added is a minor part of the total energy budget (Section 15.6), its effect is very great. This was stressed by Riehl (1951), who noted that the overall baroclinity of the cyclone outside the eye could be doubled as a result of the process (see Fig. 15.7).

A systematic description of the structure, given by Deppermann (1947), has been broadly confirmed but improved upon as a result of later aerological explorations. A mature tropical cyclone can be divided into four parts: (1) an outer region with inward-increasing cyclonic wind speeds and limited convection; (2) a belt, in the inner portion of which the wind reaches hurricane strength, characterized by squall lines and heavy convection; (3) a more or less ring-formed inner rain region with very heavy rain and squalls and maximum wind strength; and (4) the eye, inside a "transition zone" through which there is a rapid inward decrease of wind speed.

The general cloud distribution is indicated on the right of Fig. 15.4. Radial inflow is essentially confined to a layer 1 to 2 km deep, being strongest in the subcloud layer. Convergence in this layer is most pronounced near the outer boundary of the transition zone, where ascending motions

---

[4] This is suggested by observations cited by Haurwitz (1935), which show that the surface temperature often did not fall in typhoons entering the Philippines. During the passage of an 898-mb typhoon over a Japanese naval fleet (Arakawa, 1954), the wet-bulb temperature remained almost constant (indicating an increase of the surface θ_e from 360°K in the outskirts to 385°K in the center). The most thorough documentation of a hurricane modifying the sea-surface temperature, and thereby possibly its heat source, is that by Leipper (1967). The process is upwelling, following divergence of the warm surface layer caused by wind stresses. A theoretical analysis is given by O'Brien and Reid (1967).

are strongest and mixed convective and stratiform types occupy a "cloud ring" (not always completely closed) through most of the troposphere. The spreading upper parts congregate to form the extensive and stable layer-cloud mass carried outward by the upper-level outflow. Although this is partly nourished by updrafts from bands of cumulus and cumulonimbus outside the main inner rain ring, observations have essentially borne out the conclusion from the first hurricane reconnaissance flight, by Wexler (1945), that "the immense hurricane cloud system [is] maintained by ascending motion through a relatively narrow throat near the storm's center."

The characteristic isopleths of temperature departure from the mean standard atmosphere, shown on the r.h.s. of Fig. 15.4, reflect the result of strong heat production in the central part of the cyclone and the spreading outward of warm air in upper levels. The slightly negative temperature departures in the lower troposphere at the outer radii indicate an absence of large amounts of moist-adiabatic ascent of surface air (LaSeur and Hawkins, 1963).

Most cyclones show some asymmetry, largely associated with their forward motion. However, their main aspects may first be described in terms of quasi-symmetric toroidal circulations. Despite the great variety in individual cyclones, the typical structure may be best illustrated by a real synoptic case, somewhat modified to give the general features. For this purpose we select hurricane "Helene" on Sept. 26, 1958, which Krishnamurti (1962), Miller (1962), Schauss (1962), Colón (1964), and Gentry (1964) have thoroughly investigated.[5]

From flights at about 560 mb in different quadrants of this hurricane, the smoothed mean profiles of tangential wind velocity v_θ, temperature T, and dewpoint temperature T_d are shown in Fig. 15.5. The lines at radii 15 and 30 km enclose a zone with especially strong inward increase of T and decrease of T_d and strong cyclonic wind shear. Outside this zone, anticyclonic shear prevails. The innermost band of strong radar echoes, shown in Colón's analysis (1964), had a mean width of about 15 km between 23- and 38-km radius, and was centered on the radius of maximum tangential wind speed (48 m/sec) at which the highest T_d values were also observed. *Hence the "cloud wall," or ring of heaviest precipitation, where low-level*

---

[5] We should note that Krishnamurti's study of this and other hurricanes dealt specifically with their nonsymmetrical character. He demonstrated by numerical experiments that the vertical motions connected with the asymmetrical wind field exhibit a spiral form, similar to that of the principal convective bands actually observed. Averaging around the circumference would give the "symmetrical" field discussed here.

*convergence and ascent are greatest, is evidently identified with the outer
boundary of the thermal "eye wall," or transition zone, where the strongest
wind is observed.*

Values of θ_e are plotted at selected radii along the T_d curve. An increase
from 340°K to 353°K (or 354°K if the air is saturated) is observed between
100 and 30 km, while both within the eye wall and in the eye, θ_e has the
nearly constant value 354°K. From this it may be inferred that the air in
this region was ultimately derived from the inflow layer that feeds into the

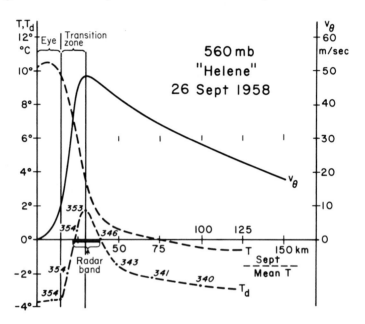

Fig. 15.5   Mean radial distribution of temperature (*T*), dewpoint temperature (T_d),
and wind velocity (v_θ) at the 560-mb level in hurricane Helene.

ring of maximum ascending motion. The inward increase of θ_e, especially
pronounced between 50- and 30-km radius, is largely a result of transfer
of sensible and latent heat from the sea surface in the region outside the
eye wall and of its transport upward to the 560-mb level.

With the aid of data presented by Colón for this and other levels, the
idealized Fig. 15.6 has been constructed. This scheme, and much of the
following discussion, is largely influenced by the model presented by Kleinschmidt (1951). On the right are given the inner and outer boundaries of
the eye wall, the temperature, and tangential wind field; on the left are
the eye wall, the field of θ_e, and the cloud and rain distribution. As a coun-

terpart to the θ_e isopleths, on the right side are shown selected lines of constant absolute angular momentum about the central axis of the hurricane, represented by

$$M_r = v_\theta r + \frac{f}{2} r^2 \qquad (15.2)$$

Under the assumption that both θ_e and M_r are conserved in the free atmosphere, these lines should be parallel and correspond to two-dimensional streamlines in the plane of the vertical section. As discussed in Section 15.5, the air in the lower layers loses angular momentum through friction as it moves inward. This accounts for the large inward decrease of M_r in the "free atmosphere" above, which draws its M_r from the surface layers.

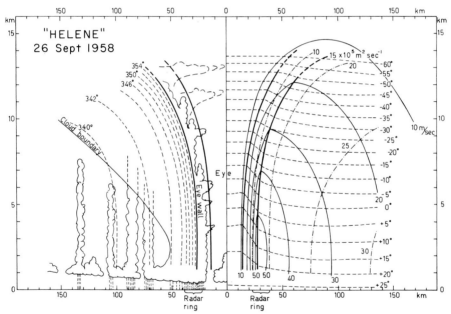

FIG. 15.6 Mean structure of a mature hurricane (Helene). To the right are given the boundaries of the eye-transition zone, temperature at 5°C interval, tangential wind velocity at 10 m/sec interval, and dash-dotted lines showing the angular momentum (for intervals of 5×10^5 m^2 sec^{-1}). To the left are given the same eye-wall boundaries, clouds and rain, and lines of constant equivalent potential temperature (2°K interval).

The pattern of temperature departure from the mean tropical atmosphere was similar to that in Fig. 15.4, with an excess temperature of 11 to 13°C in the eye between 5- and 12-km height. At the outer boundary of the eye wall, the temperature was 5 to 7°C lower (Fig. 15.5), with a more gradual decrease farther out. At this time the central pressure was about 950 mb; lowest pressure of 932 mb occurred about 12 hr later.

There are exceptions to the idealized structures discussed above. For example, Simpson (1956) reports that in some storms, while the strongest winds are observed in low levels near the center, farther out they are sometimes found in the middle troposphere. Also, Simpson and Riehl (1958) show examples of "ventilation" in which there is a systematic transverse flow of mid-tropospheric air relative to the storm. This brings air with low θ_e into proximity with the core. They suggest that this may account for the observation in some hurricanes that the temperature in the rain area is lower than that given by moist-adiabatic ascent from the surface layers, and thus constitutes a brake on the storm intensity. Despite these and other exceptions, the discussion above is valid with regard to the fundamental processes.

15.3   Characteristics of the Eye

The eye deserves special attention because it governs the lowest central pressure in the cyclone (Haurwitz, 1935; Miller, 1958). In Fig. 15.7, curve A represents undilute ascent for the undisturbed tropical atmosphere; curve B, a parcel rising in the inner rain region where the surface pressure is assumed

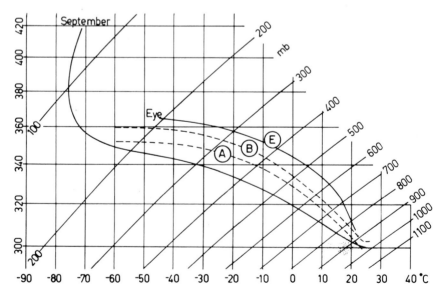

Fig. 15.7 September mean tropical sounding. Curves: (A) temperature in moist-adiabatic ascent from surface layer ($p_0 = 1000$ mb); (B) same for air with $p_0 = 960$ mb and same surface temperature; (E) typical eye temperature in moderate hurricane.

to be 50 mb lower. While the possible hydrostatic reduction of surface pressure due to sea-air transfer is considerably increased in this case, the typically observed eye temperatures (curve E) indicate a markedly greater reduction. A satisfactory theory for the eye must account for, in addition to its high temperature, the observed humidity and angular momentum distribution.

C. L. Jordan (1958b) has investigated the mean eye temperature, dividing dropsonde observations into six groups according to the central surface pressure (Table 15.2). The surface temperature varies by only a small

TABLE 15.2

MEAN TEMPERATURE IN THE EYE OF TROPICAL CYCLONES[a]

Group:	I	II	III	IV	V	VI	Mean tropical atmosphere
Surface pressure (mb):	998–980	979–965	964–950	949–939	925–906	901–883	
No. of cases:	14	15	13	4	17	10	
Surface	25.2	25.7	25.0	26.5	26.7	27.1	24.7
900 mb	21.7	22.9	23.2	24.5	26.2	—	18.3
800 mb	17.6	19.3	19.3	20.7	22.9	24.5	13.1
700 mb	12.7	14.7	15.8	18.2	18.9	22.1	7.9
600 mb	7.1	8.8	11.0	14.0	...	...	1.0
500 mb	− 2.4	0.7	3.4	6.0	...	...	− 7.6

[a] After C. L. Jordan (1958b).

amount, the somewhat higher temperature in Groups IV through VI indicating that the most intense cyclones may form only if the surface waters are exceedingly warm. Higher up, the eye temperature increases with decreasing surface pressure, in accord with hydrostatic considerations. Hurricane Helene ($p_0 \approx 950$ mb) was intermediate to Groups III and IV; the eye temperature of 10°C at 560 mb in Fig. 15.5 agrees well with Jordan's mean soundings.

The observed eye temperature at 560 mb in Helene could be explained by subsidence to this level from about 370 mb, initially with the mean tropical temperature. However, even if the air were initially saturated, its relative humidity at 560 mb would be only 12% compared with an observed humidity of about 40%. If, instead, we assume that the air in the eye of a mature cyclone essentially originates in the low-level moist inflow, its high temperature and moderate relative humidity could be considered the result of some kind of "föhn" process, with descent in the eye after ascent with latent heat release outside it. Eye soundings in the extremely deep cyclones Marge and Ione (Fig. 15.8) illustrate this process. The θ_e, with almost constant values 373°K and 354°K at different levels, corresponds to pseudo-adiabats starting for Marge at 920 mb (assumed surface pressure at the eye

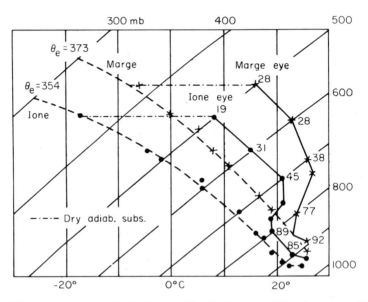

FIG. 15.8 Temperature and relative humidity in the eyes of hurricane Ione (C. L. Jordan, 1957) and typhoon Marge (Simpson, 1952), and corresponding temperatures of the "cloud wall," approximately on the moist adiabats with the θ_e-values 354°K and 373°K, respectively. Dash-dotted lines indicate dry-adiabatic descent that would bring the "cloud air" to the temperatures and humidities observed in the eyes of the storms.

wall) with a temperature of 26°C and near-saturation, and for Ione at 950 mb with a temperature of 25.5°C and a relative humidity of 85%. The dash-dotted lines show dry adiabats which, if the air were saturated in the eye wall at the start of descent (at 375 and 360 mb), account closely for the observed humidities at 500 mb. A similar agreement is obtained at lower levels.

Thus *the thermodynamic properties indicate that the air up to at least 500 mb mostly originates in the innermost ring of moist air, which has risen from low levels, mixed inward in the upper troposphere, and descended within the eye.* These combined processes account for the almost constant θ_e of the eye equal to that in the innermost cloud wall (Fig. 15.5). As observed by Colón (1964), lateral mixing alone would chill the eye, since the air outside it is cooler and evaporation would further reduce the temperature. Hence subsidence must be a continuing process with downward heat transport through upper levels of the eye, to balance its heat budget.

The necessity for lateral mixing of moisture into the eye was pointed out by Simpson (1952) and C. L. Jordan (1952). Miller (1958) devised a quantitative model in which the air begins descent at tropopause level with a temperature "assumed to be that which would be produced by moist adiabatic ascent from the surface." At successively lower levels, increasingly larger proportions of saturated air from the wall cloud are mixed into the descending air. This diminishes its temperature and increases its humidity, partially countering the effect of subsidence. Using empirically determined mixing rates, along with the θ_e of the wall cloud prescribed by the sea-surface temperature, Miller estimates the lowest sea-level pressure that could result from the process if all other conditions were favorable. This ranges from 987 mb for $T_w = 26°C$ to 894 mb for $T_w = 31°C$.

The inmixture discussed above would also augment the angular momentum in the eye, where the momentum is actually low. Malkus (1958) considered the eye as a region of gentle downdraft in which air originating at the top has no initial rotation, angular momentum being imparted to it from the surrounding cloud ring as the air descends. Entrainment consistent with the thermodynamic properties, together with the influence of surface friction in the lower troposphere, provides a satisfactory structure in which the angular momentum inside the eye is considerably less than that outside. Malkus concluded (as did Durst and Sutcliffe, 1938) that the net result of the angular-momentum transfer processes is to create supergradient winds in the lower levels, where air is flung outward. Such an expulsion, at some distance above the surface, is (as Simpson notes) necessary to balance both the downward mass flux from above and the influx in low levels (Fig. 15.4). The existence of low-level frictional inflow is demonstrated by Simpson's (1952) observation that the pressure gradient and wind are appreciable even in the eye, which is partly filled with low cloud with an elevated "hub" in its central part.

Malkus also concluded that significant evaporation occurs in the eyes of some hurricanes; this evaporation results from water originating largely from the melting of ice particles that fall from cumulonimbus tops spreading into the eye. Radar observations of this process are described by Kessler (1958). Since evaporation chills the air, Malkus considers this as a brake on hurricane development and points out that in some extremely deep cyclones, the eye is free of upper cloud[6].

The simple process earlier described in connection with Fig. 15.8 is not inconsistent with the more complicated processes outlined later. This is because air entrained from the cloud region dominates the thermodynamic properties of the lower troposphere within the eye (Malkus estimates that 97% of the eye air at 700 mb has originated from the surrounding clouds).[7] Since the θ_e of the entrained air is constant, as was also the assumption in Fig. 15.8, essentially the same relative humidities are obtained by subsidence from a given level as by the more complex mixing processes.

15.4   The Eye Wall and Contiguous Regions

According to the discussion above, differential vertical motion takes place between the ascending air outside and the subsiding air within the eye. This circulation concentrates the gradients of temperature and of tangential momentum, and in accord with the principles discussed in Section 9.3, an eye boundary with "frontal" characteristics is maintained. This corresponds to the "transition zone" in Figs. 15.4 and 15.6.

Ferrel (1856) indicated that inflowing air could not enter the central part of a hurricane because, with the tangential velocity achieved by close approach to the center, the centrifugal force would be so great that the air could not be forced farther inward by the radial pressure gradient. A dynamical analysis proving that the eye is a necessary feature of a hurricane was given by Syōno (1951), and extended in somewhat different terms by Kuo (1959) and by Riehl (1963). With conservation of angular momentum, v_θ would become infinite as a low-level streamline approaches the center. Kuo observes that "the total kinetic energy that can be gained by the current is limited [by the available pressure drop], and therefore this cannot happen. In other words, the converging current cannot penetrate a certain minimum radius r_m and must turn upward and eventually outward at higher levels, since the radial pressure gradient usually decreases with elevation. The surface of revolution defined by this streamline is identified with the wall of the eye."

---

[6] As in typhoon Marge, described by Simpson (1952). C. L. Jordan (1952) reports that in 20 typhoon penetrations, a middle or high overcast was present in the eye on 13 occasions, while the upper eye was clear on only 4. In no case was the eye clear at low levels.

[7] On the basis of tritium measurements, Östlund (1967) suggests that the air within the eye at 800 mb may be made up of 5% stratospheric air, the remainder being drawn from the cloud wall.

As an example, with a pressure drop of 50 mb along the streamline a maximum wind speed of 94 m/sec would be possible. An air parcel starting at $r = 500$ km in latitude $15°$ would, with conservation of M_r, achieve this tangential velocity at 50-km radius. Inward of this radius it would be impossible to satisfy both the laws of conservation of angular momentum and energy. Since friction decreases the angular momentum, it makes possible a further inward movement, but this is still limited.

In the inflow layer where interactions with the surface are strong, a horizontal streamline cuts the isolines of both M_r and θ_e (Fig. 15.6). Above the inflow layer, these interactions become less important, and the θ_e and M_r isopleths approximate the movement of air ascending in the inner rain region of a mature hurricane. With increasing radius, this statement becomes less valid as the systematic toroidal motions of the inner region give way to predominantly convective clouds, which, in a less organized fashion, transfer properties between lower and upper levels by eddy processes.

With the assumption that absolute angular momentum is conserved in the free atmosphere, the slope of a streamline in the inner region of ascent can be determined from

$$\tan \psi = \frac{\delta z}{\delta r} = -\frac{\partial M_r / \partial r}{\partial M_r / \partial z} \tag{15.3}$$

Using Eq. (15.2), we get

$$\tan \psi = -\frac{v_\theta + r(\partial v_\theta / \partial r) + fr}{r(\partial v_\theta / \partial z)} \tag{15.4}$$

At the radius of maximum v_θ, the second term of the numerator vanishes, while the third term is small compared with the first. Hence, at this radius,

$$\tan \psi_m \approx -\frac{v_\theta}{r \, \partial v_\theta / \partial z} \tag{15.5}$$

which represents the slope of the outer boundary of the eye wall. If we also substitute for $\partial v_\theta / \partial z$ from Eq. (8.12), neglecting f against $2v_\theta / r$ at small r, this becomes

$$\tan \psi_m \approx -\frac{2v_\theta^2 T}{gr^2 \, \partial T / \partial r} \tag{15.6}$$

Typically, the temperature contrast between the eye and the rain region is greatest in the upper-middle troposphere, as in Fig. 15.4. Considering this and the smaller values of v_θ in upper levels, Eq. (15.6) is consistent with

Fig. 15.6, in which the eye boundary is almost vertical in low levels and gradually becomes quasi-horizontal in the upper troposphere.

It is of special interest to evaluate the radius r_{ca}, where the circulation about the center changes from cyclonic to anticyclonic. This can be found from Eq. (15.2) by putting $v_\theta = 0$, giving (neglecting a minor term)

$$r_{ca} = \sqrt{\frac{2(v_\theta r)_m}{f}} \qquad (15.7)$$

where $(v_\theta r)_m$ represents the angular momentum at the sloping axis of maximum wind, just at the outer boundary of the eye wall. From Fig. 15.6, in the Helene case, $(v_\theta r)_m = 15 \times 10^5$ m² sec^{-1} and $f = 0.74 \times 10^{-4}$ sec^{-1}. This gives $r_{ca} = 200$ km as the outer limit of the upper cyclonic circulation (cf. Syōno, 1951), in approximate agreement with observations. Riehl (1963) found that Eq. (15.7), in combination with the law $v_\theta r^{0.5} = $ const. (which typifies the frictional inflow layer), prescribes an eye radius that agrees well with observations.[8]

The slope of the outer boundary of the eye wall may also be computed from the formula for a discontinuity surface of first order, Eq. (7.12). For flow with a small radius of curvature (i.e., $2kV \gg f$),

$$\tan \psi_m \approx -\frac{2v_\theta T}{gr} \left\{ \frac{(\partial v_\theta/\partial r)_1 - (\partial v_\theta/\partial r)_2}{(\partial T/\partial r)_1 - (\partial T/\partial r)_2} \right\} \qquad (15.8)$$

where subscripts 1 and 2 denote the air at the outer and inner sides of the surface. Applying this formula at 5 km in Fig. 15.6, the slope is determined to be 0.67, whereas Eq. (15.5) gives a slope of 0.55. The agreement is fair, considering uncertainties of measurement and also that the streamline radius r in Eq. (15.5) may differ somewhat from the trajectory radius that should be used in Eq. (15.8).

The dynamical necessity for the general arrangement of the angular-momentum field shown in Fig. 15.6 was brought out by Durst and Sutcliffe (1938). They expound the principle that with a warm-core storm in which the radial pressure gradient decreases upward, rising rings of air are brought to levels where the centrifugal and coriolis forces (corresponding to their initial angular momentum) overbalance the weaker pressure gradients

---

[8] The eye radius varies between storms and with time in the same storm. In most, the eye becomes smaller as the cyclone deepens, and larger as it weakens, but the opposite is often true (C. L. Jordan, 1961; Ito, 1963). Based on 46 aircraft penetration reports, C. L. Jordan (1952) finds an eye diameter variation between 13 and 140 km, with no clear relation to central pressure.

aloft, and must move outward. Before aerological observations became available, Haurwitz (1935) demonstrated that the eye must be funnel-shaped. On the assumption of a vanishing pressure gradient at a high level (e.g., 10 km), he considered the sea-level pressure as being due to the added weight of the column below, this being made up of a layer of warm "eye" air and a layer of cooler air in the surrounding rain area. This analysis indicated that the strong pressure gradient in a typhoon is largely due to the slope of the eye boundary, with a mean temperature increase toward the center due to the contribution of an increasingly deep layer of high-temperature eye air.

Malkus and Riehl (1960) have explained the sea-level pressure gradient in a "moderate" hurricane outside the eye wall as being the result of the intense heat flux from the sea surface. They assumed the radius of the eye wall to be about 30 km, which corresponds to the outer boundary of the transition zone in Fig. 15.6. The surface pressure of about 966 mb at this radius would be hydrostatically explained by the temperature distribution of the vertically ascending low-level air that, during its radial inflow, has gained energy from the sea. The corresponding θ_e should, according to Malkus and Riehl, be 362°K, about 8° more than that computed for the Helene case.

The difference is partly a result of the assumption made by Malkus and Riehl that the air at this distance ascends vertically and that hence any outward slope of the eye wall may be disregarded. With a structure as in Fig. 15.6, however, this assumption would not be strictly valid, and the surface-pressure gradient outside the eye would be influenced by the out-ward-sloping eye wall. In this case a more moderate interior heat flux from the ocean suffices to account for the sea-level pressure gradient. The inward increase of θ_e from 100 to 30 km in Fig. 15.5 shows clearly the contribution of heat transfer from the sea, but the limit of θ_e is about 354°K rather than 362°K.

The outward spreading of the angular momentum lines with height, as in Fig. 15.6, must be a general characteristic of warm-core storms. Evidently, however, expression of this varies from storm to storm. In hurricane Daisy (C. L. Jordan et al., 1960) the clouds in the "radar wall" were in some pictures nearly vertical up to 15 to 18 km, but in typhoon Marge, Simpson (1952) describes a "coliseum" of clouds, with walls rising to about 11 km where the upper rim was smoothly rounded off. Jordan et al. point out that Daisy was a small intense storm, unusual in that the velocity profile approximated a vr-vortex in the inner part outside the ring of maximum winds. Based on a very thorough analysis of radar and photographic data,

Malkus *et al.* (1961) concluded that when Daisy was near maximum intensity, "hot towers" penetrating to the upper troposphere occupied only about 4% of the rain region ($r < 200$ mi).[9] Ackerman (1963) found that at this stage, "convective" liquid water content was observed in about 30 to 40% of the inner region ($r < 60$ mi) in the middle troposphere (but suggests that, especially in immature cyclones, many of the convective clouds do not penetrate to the high troposphere).

It seems clear from these observations and from descriptions of turbulence encounters that, in some cyclones, ascent in the inner region is dominated by cumulonimbus, as emphasized by LaSeur (1962), while in others a broader-scale and more uniform ascent takes place. Gherzi (1956) stresses that thunderstorm activity is absent in the inner parts of fully developed typhoons and that electrical activity is a sign of their decay. Evidently, in different cases, the vertical circulation in the cyclone interior may be of the types sketched in Fig. 15.9. Figure 15.9a corresponds to a weakly con-

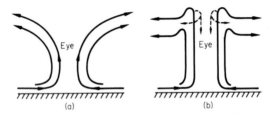

FIG. 15.9   Schematic air flow in the inner parts of cyclones not dominated by strong convection (a), and with intense convection in the cloud wall (b).

vective core, and Fig. 15.9b to a cyclone in which strong convection dominates the inner rain region. In the latter case, vertical accelerations due to buoyancy would considerably exceed the radial accelerations resulting from the solenoid field. Vigorous convective elements thus tend to conserve v_θ in their more nearly vertical movement, in which case a more upright eye wall would be indicated by Eq. (15.5). Upper-level expulsion of air from the inner region may in that case be viewed in terms of the process described by Malkus (1958). While active clouds rise almost vertically, after decay has set in a convective column will acquire an outward tilt. Since precipitation is generated mainly in the vigorous updrafts, the active convection detected by radar does not indicate the outward movement of air

---

[9] Asai and Kasahara (1967) suggest that this observation is connected with their theoretical finding that convective clouds carry heat upward most efficiently when they cover only a few percent of a given area.

in upper levels, which takes place in the decayed clouds. The substantial decrease of wind with height, observed even in the strongly convective hurricane Daisy, shows the influence of an average outward movement in the eye wall.

It may be stressed that our example of the structure of a typical medium hurricane is simplified in some respects. The hurricane has been treated as an axially symmetric vortex. Usually, the wind speed is strongest to the right of the track; this asymmetry is reduced when the motion of the storm is subtracted, to give the air motions relative to a coordinate system moving with the storm. In hurricane Cleo, LaSeur and Hawkins (1958) found that the *geostrophic* wind field was nearly symmetrical. They demonstrated that, as a result of the movement of the cyclone, the trajectory curvature is very much sharper on its left than on its right side; correspondingly, according to Eq. (8.2), the wind speed must be appreciably stronger on the right side.

15.5   Angular-Momentum Balance

The fundamental processes involved in maintaining a quasi-steady vortex may be clarified by studying the balance of angular momentum about the cyclone axis. If we consider a vertical wall at a given radius, the distribution in Fig. 15.6 implies a net import of absolute angular momentum, since the largest values of M_r are in the inflow layer. This condition is necessary to balance the loss due to tangential surface stresses inside the wall considered. Since these stresses destroy angular momentum everywhere within the surface cyclonic circulation, the influx of M_r must increase outward to the limit of this circulation.

The radial influx $(M_r)_r$ through a cylindrical surface at radius r and between the pressure surfaces p_1 and p_2 is given by

$$(M_r)_r = - \frac{2\pi r^2}{g} \left\{ \int_{p_2}^{p_1} \bar{v}_\theta \bar{v}_r \, dp + \int_{p_2}^{p_1} \overline{v'_\theta v'_r} \, dp + \frac{fr}{2} \int_{p_2}^{p_1} \bar{v}_r \, dp \right\} \qquad (15.9)$$

if f is assumed constant over the area occupied by the cyclone. Similarly, the upward flux between radii r_1 and r_2 is

$$(M_r)_z = - \frac{2\pi}{g} \left\{ \int_{r_1}^{r_2} \bar{v}_\theta \bar{\omega} r^2 \, dr + \int_{r_1}^{r_2} \overline{v_\theta' \omega' } r^2 \, dr + \frac{f}{2} \int_{r_1}^{r_2} \bar{\omega} r^3 \, dr \right\} \qquad (15.10)$$

where a bar denotes the mean value at a given radius and the primes are deviations from this mean. The first r.h.s. terms in Eqs. (15.9) and (15.10) give the radial and vertical fluxes of angular momentum due to the axially symmetric part of the wind field, whereas the second terms represent the eddy fluxes arising from asymmetry and from small-scale disturbances in the vertical velocity field. The last terms express the fluxes of earth angular momentum.

The angular-momentum balance has been investigated in individual cyclones, but mainly for the region close to the center. In view of this and the considerable variations between cyclones and between calculations for different times in the same cyclone, the main principles of this balance are best clarified by using the data for a "mean cyclone." The computations reported below were based upon wind data given for upper levels by E. S. Jordan (1952) and for sea level by Hughes (1952), with some adjustments to assure mass balance.

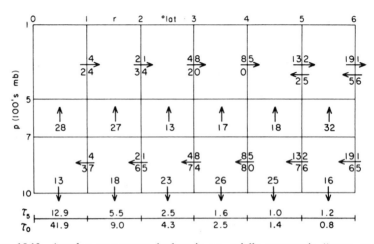

FIG. 15.10 Angular momentum budget in an axially symmetric "mean tropical cyclone." Horizontal arrows denote fluxes across radii (unit: 10^{22} g cm^2 sec^{-2}). Downward arrows at bottom show momentum transport to sea, and arrows between 700 and 500 mb show upward momentum transport through this layer. Upper figures at the horizontal arrows give transport of earth angular momentum; lower figures, of relative momentum. Below, tangential stresses at 500 mb and at surface. (After Palmén and Riehl, 1957.)

Figure 15.10 gives the budget for an axially symmetric cylone (Palmén and Riehl, 1957). In this figure, inflow is confined to the layer between 1000 and 700 mb and outflow to the layer between 500 and 100 mb. Assuming no vertical flux at 100 mb, the difference between the total fluxes of M_r across the cylindrical walls at two radii, between 100 and 1000 mb, must represent the flux of angular momentum to the earth within the ring enclosed, due to surface friction. The transfer $(M_r)_0$ to the earth is expressed by

$$(M_r)_0 = 2\pi \int_{r_1}^{r_2} \tau_{\theta,0} r^2 \, dr \tag{15.11}$$

where $\tau_{\theta,0}$ is the tangential surface stress. Average values for radial rings are shown at the bottom of Fig. 15.10. More detailed calculations for the innermost regions of mature hurricanes by Riehl and Malkus (1961) and by Miller (1962) give comparable stresses, with a strong inward increase inside the 60- to 80-nmi radii they considered. The budget shown here is not complete because the eddy term in Eq. (15.9) had to be disregarded. It gives, however, a general scheme for the momentum balance, at least in the more central parts of a mature cyclone where the asymmetry often seems to be of minor importance (Riehl and Malkus, 1961).[10]

At all radii, $\bar{v}_r$ was assumed zero in the 500 to 700 mb layer. Correspondingly, the vertical flux through the middle troposphere is determined as a residual, giving balance in the lower and upper blocks. A part of the upward flux through the intermediate layer is due to the vertical-mass circulation; another part results from the eddy stresses. In the original determination of the $\bar{v}_r$ profiles, it was assumed that $v_r r =$ const. inward to the 2° radius, in accord with Hughes' finding that the surface divergence is very small outside this radius. This condition prescribes no net mass transport vertically through the middle troposphere between 2° and 6° radius; hence the $(M_r)_z$ values shown are entirely due to eddy stresses represented by the middle r.h.s. term of Eq. (15.10). Inside radius 2°, the transfer by the mass circulation, expressed by the other two terms, accounted for 76% of the whole; thus the eddy stresses were of comparatively minor importance in the inner part.

Considering the lower layer, it is seen that a large amount of earth angular momentum is brought in at the outermost radius. The inward decrease of this import expresses a conversion of earth angular momentum to relative momentum, according to the equation for a steady symmetrical vortex

---

[10] Riehl (1961) concluded from calculations based on surface and low-level winds in a large sample of cyclones that "the momentum flow due to asymmetries...may be omitted as a small term, at least within the 600-km radius and in the low troposphere." Except for the innermost region, a similar conclusion for the upper outflow layers has not been established. Both Pfeffer (1958) and Palmén and Riehl (1957) have computed the "eddy transfer" of M_r from Jordan's data. Although these averaged data do not adequately represent the true eddy fluxes, the results suggest that the eddy transfer increases in importance outward from the center.

(see, e.g., Haurwitz, 1941; Miller, 1962):

$$\frac{d\bar{v}_\theta}{dt} = \bar{v}_r \frac{\partial \bar{v}_\theta}{\partial r} = - \frac{\bar{v}_\theta \bar{v}_r}{r} - f\bar{v}_r + \frac{1}{\varrho}\frac{\overline{\partial \tau_{\theta z}}}{\partial z} \tag{15.12}$$

Differentiation of Eq. (15.2) shows that this expression corresponds to constant absolute angular momentum with radius, in the absence of friction. In that case, according to Eq. (15.9), the inward decrease of the radial transfer of ($fr^2/2$) would be matched by an equal inward increase of the radial transfer of ($v_\theta r$).

Examination of the lower blocks of Fig. 15.10 shows that this is not the case, the transfer of relative momentum increasing only slowly to the 4° radius and decreasing at smaller radii. The interpretation has been mentioned earlier, namely, that momentum is removed from the lower layers by frictional stresses at both the surface and in the middle troposphere. Similarly, in the upper outflow layer the relative angular momentum is reconverted to earth angular momentum, the source of the relative momentum being partly that horizontally transferred outward and partly that imparted to the upper layer by eddy stresses in the middle troposphere. Outside 4° radius, an inward transfer of relative angular momentum is indicated in the upper levels. This is expressed formally in the first r.h.s. term of Eq. (15.9) by an association between positive v_r and negative v_θ, i.e., by the outflow extending to radii where there is anticyclonic circulation. Palmén and Riehl noted that in a symmetrical cyclone, the upper anticyclonic circulation must extend beyond the outermost limit of the lower-level cyclonic circulation, to provide the net import of angular momentum required to balance the loss by surface stresses in the interior.

The transfer from the lower to the upper layer is in the normal fashion down the angular-momentum gradient. Since the upward transfer by the mass circulation is significant only in the interior part, the principal mechanism at greater radii must be eddies. Gray (1966) has analyzed the $v_\theta' w'$ correlation, expressed by the middle r.h.s. term of Eq. (15.10), deduced from aircraft data in hurricanes. He evaluated equivalent stresses, due essentially to the updrafts and downdrafts in convective clouds, which correspond approximately to the magnitudes indicated in Fig. 15.10.

Gray (1967) further discusses the implications of these stresses for the mechanics of hurricanes. In a warm-core disturbance, the result of low-level inflow and high-level outflow is to augment the shear (v_θ decreasing upward) in a manner that, in the absence of internal stresses, would eventually lead to a balance between the solenoidal acceleration and the circulation acceleration (coriolis and centripetal) associated with the vertical wind shear. The direct circulation would then cease. Convective clouds, transferring tangential momentum between lower and upper levels, continually degrade the vertical shear so that this balance is not achieved and the direct solenoidal circulation required to regenerate it is maintained.

The outer cloud bands, which represent a total condensation heat source that is small compared to that in the inner part of the tropical cyclone, are especially important in this respect. Being located between the inner heat source and the outer undisturbed tropical atmosphere, they are in the region where the vertical shear is strong and they can draw upon the large momentum differences between lower and upper levels. Moreover, in this outer region the potential instability of the lower troposphere is greatest, favoring cumulus convection.

Fujita *et al.* (1967) demonstrate that the outer rainbands of a typhoon, by augmenting the angular momentum in upper levels, enable movement of air outward to large radii where M_r is large. The right side of Fig. 15.11a shows a symmetrical wind field on the assumption that all outflow is derived from an inner rainband surrounding the eye, with conservation of absolute angular momentum in the outflow layer. If there is an outer rain band (left side), both mass and cyclonic angular momentum from lower levels are added to the upper outflow layer. With conservation of M_r outside this convective zone, a weaker anticyclonic circulation is present in the outer part. The effect is consistent with observations, which show that the anticyclonically circulating winds in the upper troposphere are weaker than would be expected under conservation of angular momentum over the whole region. Although this computation corresponds (Fig. 15.11b) to a single rain band such as the huge squall lines sometimes observed outside the main cirrus canopy of a typhoon, the effects of the spiral rain bands closer to the center are obviously the same. For discussions of different characteristics of these spiral rain bands, we refer to Ushijima (1958), Senn and Hiser (1959), and Gentry (1964).

In summary, the scheme in Fig. 15.10 shows that *a tropical cyclone represents a sink region of angular momentum about the cyclone axis, brought in from the outside atmosphere.* The scheme is analogous with that in·Chapter 1, where the poleward parts of the atmosphere were considered as sink regions of zonal angular momentum transferred from the tropical source regions. The conversions between earth angular momentum and relative angular momentum in the Hadley cells are very similar to the processes in tropical cyclones. In both cases the generation of relative momentum is mainly determined by a symmetrical circulation, although the outward transfer in

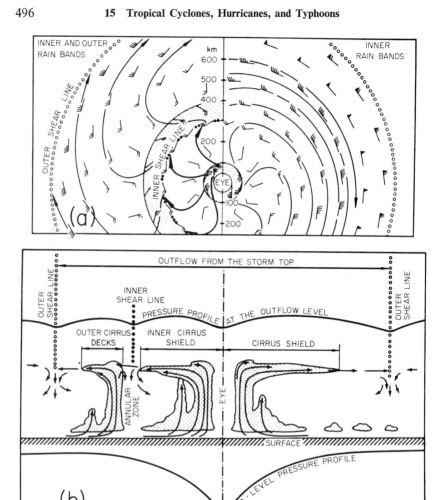

FIG. 15.11   (a) Two-dimensional distribution of outflow winds for a model typhoon with an inner rain band between 50 and 100 km from the center (right half). The left half represents a model with an outer rain band in addition to the inner. (b) Vertical section corresponding to these two models. Note that the radial dimensions do not correspond in the two figures. (After Fujita *et al.*, 1967.)

upper levels (in one case, away from the cyclone; in the other case, away from the tropics) is influenced by horizontal eddies. Also, in both cases the vertical exchange of momentum necessary to maintain a direct circulation depends on frictional stresses (surface and internal) as well as upon the heat sources and sinks (Section 17.1).

15.6 THE ENERGY BUDGET

A deeper knowledge of the rules governing the formation and maintenance of tropical cyclones can be achieved only by studies of the energy balance. In Chapter 2 it was pointed out that the kinetic energy of the atmosphere is a small quantity compared with the total energy. Hence it is most meaningful to treat these forms of energy separately. In Chapter 16 the energy conversions will be discussed more generally; these can probably be visualized more readily in the tropical cyclone than in any other large-scale disturbance.

The most direct method of studying the energy processes is to investigate the outflow of total energy at an arbitrary radius from the center of the cyclone. To simplify the problem, we again treat an axially symmetric vortex. If the heat source in a region limited by radius r_0 and isobaric surfaces p_0, p_H is denoted by dQ/dt, the energy equation for a steady-state cyclone can be expressed by

$$\frac{dQ}{dt} = \frac{2\pi r_0}{g} \int_{p_H}^{p_0} (c_p T + \Phi + k) v_r \, dp \qquad (15.13)$$

where all quantities on the r.h.s. are understood to be circumferential averages when a symmetrical cyclone is considered. Here dQ/dt is analogous to the heating function used in Chapter 2 for computation of the divergence of the meridional energy flux. Equation (15.13) is a form of Eq. (16.4), omitting boundary stresses and assuming no energy flux through p_H, taken at the "top" of the storm. Using the same symbols as in Chapter 2, Eq. (15.13) takes the form

$$\pi r_0^2 [Q_s + R_a + LE] - \frac{2\pi r_0}{g} \int_{p_H}^{p_0} Lq v_r \, dp = \frac{2\pi r_0}{g} \int_{p_H}^{p_0} (c_p T + \Phi + k) v_r \, dp \qquad (15.14)$$

The last l.h.s. term represents the water-vapor energy flux through a vertical wall at r_0, and together with the evaporation must (for steady state) equal the energy released by condensation and precipitation. Thus the entire l.h.s. may also be expressed by $\pi r_0^2 (Q_s + LP + R_a)$, where P is the average precipitation rate inside r_0. The release of latent heat could therefore be estimated from measured precipitation, observations of which, however, are generally lacking.[11]

---

[11] Brooks (1946) made comparisons between the water-vapor influx and observed rainfall in a landed hurricane; see also Bergeron (1954).

For the "mean tropical cyclone" discussed in Section 15.5, Palmén and Riehl (1957) computed the net outflow of energy at a radius of 2° latitude and the corresponding net inflow of latent heat. According to Eq. (15.14), the difference between these gives the sum of the three first l.h.s. terms. The results are shown in Table 15.3. Since radiation represents a heat sink, the "additional heat source" implies that the flux of latent and sensible heat from the underlying surface was larger than 10^{10} kJ/sec.

TABLE 15.3

RADIAL OUTFLOW OF REALIZED ENERGY $(c_p T + \Phi)$ AND INFLOW OF LATENT HEAT[a]

Outflow of energy	56.4
Inflow of latent heat	55.4
Additional heat source	1.0
Total interior heat source	56.4

[a] At radius 2° Latitude in a "Mean Hurricane," with additional Heat Source $(Q_s + LE + R_a)$ Estimated as a Residual. Units are 10^{10} kJ/sec. After Palmén and Riehl (1957). Here the radial flux of kinetic energy is omitted, as a very small quantity at this radius.

Riehl and Malkus (1961) and Miller (1962) have computed energy budgets for hurricanes Daisy and Helene. Referring the reader to these detailed studies, which deal with several other aspects of hurricane dynamics, we limit our discussion here to the energy budgets of these storms inside a radius of 1° latitude, where the heat sources are most active. In these studies not only the lateral fluxes across the outer boundary but also the flux of latent and sensible heat from the sea inside the boundary were estimated, whereas the effect of radiation was derived as a residual quantity. The results in Table 15.4 show that the kinetic energy is insignificant for the total budget, whereas the flux of latent and sensible heat from the sea surface accounts for only 9 to 16% of the total heat source. The estimate of radiation is quite uncertain because it represents the small difference between larger quantities that may contain errors.

The computations reviewed above show clearly that *the essential source of total energy is derived from the lateral influx of water vapor in the moist surface layer, but that also the additional flux of latent and sensible heat from the sea in the core region represents a heat source not to be neglected.* As long as only the core region of a hurricane is considered, the influence of radiation seems to be of minor importance.

TABLE 15.4

RADIAL OUTFLOW OF ENERGY AND INFLOW OF LATENT HEAT[a]

Hurricane	Daisy[b]	Helene[c]
Outflow of $(c_p T + \Phi)$	36.9	37.4
Outflow of k	− 0.4	− 0.2
Total interior heat source (I)	36.5	37.2
Inflow of latent heat	34.1	31.9
Flux of latent and sensible heat from sea	3.4	6.2
Heat sources without radiation (II)	37.5	38.1
Radiation (I minus II)	− 1.0	− 0.9

[a] At Radius 1° Latitude, and the Corresponding Interior Heat Sources, for Hurricanes Daisy and Helene. Units are 10^{10} kJ/sec.
[b] After Riehl and Malkus (1961).
[c] After Miller (1962).

As observed earlier, the total heat flux from the sea surface depends largely on the inward-decreasing surface pressure. The strong winds and air-sea temperature difference due to cooling by decompression favor rapid transfer of latent and sensible heat from the sea. Despite the relative smallness of this additional heat source, it has a preferred influence on the generation of kinetic energy. This is because (Section 15.2) the increased θ_e results in a higher temperature of the rising air in the inner precipitation region (Fig. 15.7) and contributes substantially to the baroclinity of the cyclone and thus to its capacity to generate kinetic energy.

The intensity of rainfall in the central regions of a tropical cyclone can be computed from the influx of water vapor and the additional evaporation from the sea surface. For hurricanes Daisy and Helene (Table 15.4), the total latent heat source inside a 1° radius was about 37×10^{10} kJ/sec, equivalent to 14.4×10^7 kg/sec rainfall within this area if all condensed vapor precipitates. This corresponds to a mean rainfall rate of about 13.4 mm/hr, which is probably representative of the inner parts of tropical cyclones. Concerning this and the radial distribution of rain intensity, the reader is referred to the paper by Riehl and Malkus (1961).

The kinetic-energy budget of a tropical cyclone can be computed if the distributions of radial velocity and pressure are known. The change of total kinetic energy (K) inside a fixed radius r_0 is determined (Section 16.1) by

$$\frac{\partial K}{\partial t} = -\frac{2\pi r_0}{g}\int_{p_H}^{p_0} kv_r\, dp - \frac{2\pi}{g}\int_{p_H}^{p_0}\int_0^{r_0} v_r\frac{\partial \Phi}{\partial r} r\, dr\, dp - D \quad (15.15)$$

if the upper isobaric surface p_H is taken at a level where the vertical transfer of kinetic energy may be neglected. The first r.h.s. term gives the net influx of kinetic energy at r_0; the second term, the work done by the radial pressure gradient (production of kinetic energy); and D symbolizes the rate of frictional dissipation. If the integration is performed for different radii, a budget of kinetic energy for concentric rings is achieved. The formula was applied by Riehl and Malkus to the inner part of hurricane Daisy. The result is presented in Table 15.5, where the dissipation represents the residual necessary to give balance.[12]

TABLE 15.5

Kinetic-Energy Budget of the Inner Region of Hurricane Daisy, Aug. 27, 1958[a]
(Units are 10^{10} kJ/sec)

Kinetic energy process	Radial band (nmi)				
	0–20	20–40	40–60	60–80	0–80
Convergence	0.18	0.12	0.08	−0.04	0.34
Production	0.12	0.23	0.17	0.16	0.68
Time change	0.01	0.01	0.01	0.01	0.04
Total dissipation	0.29	0.34	0.24	0.11	0.98

[a] After Riehl and Malkus (1961).

The total dissipation comprises the effects of both ground friction and internal friction in the atmosphere, the latter being larger, according to Riehl and Malkus. The last column of Table 15.5 shows that the kinetic energy was partly maintained against frictional losses by a considerable net influx of kinetic energy across the outer boundary. In the whole region, there was a positive generation of kinetic energy which, per unit area, increased inward. Inside a radius of $1°$ latitude the total energy production amounted to 0.52×10^{10} kJ/sec. In Table 15.4 the heat source in the same

[12] Gentry (1964) estimated from the $\omega'T'$ correlation (Section 16.1) that an additional kinetic-energy production of 0.32×10^{10} kJ/sec (1/3 of the total) was provided by convective-cloud eddies in the region $r < 60$ nmi.

region was given as 36.5×10^{10} kJ/sec; hence, only about 1.4% of this was converted to mechanical energy in this inner part.

This efficiency as a heat engine is quite low and comparable with that of the whole atmosphere (Section 16.2). However, kinetic energy is also produced in the outer portion, where the heat source is relatively small, so

TABLE 15.6

KINETIC-ENERGY BUDGET OF A "MEAN HURRICANE"[a] (UNITS ARE 10^{10} kJ/sec)

Kinetic energy process	Radial Band (deg lat.)			
	0.25–2	2–4	4–6	0.25–6
Convergence	0.37	−0.08	−0.29	0.00
Production	0.48	0.58	0.44	1.50
Total dissipation	0.85	0.50	0.15	1.50

[a] After Palmén and Jordan (1955).

that the overall efficiency is larger than that given above. This has been estimated by Palmén and Jordan (1955) on the basis of an energy budget for the "mean hurricane." From Table 15.6, the production of kinetic energy inside a 6° latitude radius was 1.5×10^{10} kJ/sec. Hughes (1952) indicates that about 90% of the total rainfall occurred within 2° latitude of the center of the mean cyclone. Adjusting the values in Table 15.3 accordingly, the heat source in the whole cyclone would be about 63×10^{10} kJ/sec. The corresponding efficiency is 2.4%, somewhat larger than the mean for the whole atmosphere. The efficiency will vary between cyclones; for example, both generation and dissipation of kinetic energy per unit area were much larger in Daisy (Table 15.5), which was more intense than the "mean hurricane."

It has earlier been stressed that the ability to convert heat to kinetic energy depends on the difference in temperature between the ascending air and the surrounding undisturbed atmosphere. The export of a large amount of realized energy from the core region of a hurricane (Tables 15.3 and 15.4) is essentially confined to the uppermost troposphere and does not very much influence the layers underneath. If a cyclone could be treated as a mechanically closed system (Section 16.3) with no heat sinks, its outer part would gradually be heated adiabatically, owing to the descent required by mass continuity, and the baroclinity would diminish. This would lead to reduced kinetic-energy production and a gradual weakening of the cyclone. Only an effective heat sink in the outer parts of the cyclone, due to radiation, could prevent this.

If we assumed the radiative heat sink arrived at in Table 15.4 (10^{10} kJ/sec in a 1° latitude radius) to be representative, a radiating area of radius 8° latitude would be required to dispose of the 63×10^{10} kJ/sec heat source estimated above for the "mean hurricane". This radiation rate is possibly excessive, since the average summer radiative loss at appropriate latitudes (Fig. 2.9b) amounts to only about 0.4×10^{10} kJ/sec for an area 1° latitude in radius. Hence a much larger area would be required, much in excess of the dimensions of a hurricane, showing that *the hurricane circulation must be considered as an open system.*[13]

It was indicated in Section 15.5 that to account for the angular momentum budget, the system must also be considered as open in the sense that the outflow aloft must extend beyond the bounds of the surface cyclone, or there must be an eddy import, or both. The implications that eddy-exchange processes play an important role not only for the energy budget, but also for the angular momentum budget in the outer part, appear to be supported by synoptic evidence. Most tropical cyclones are characterized by pronounced asymmetry, at least outside the core region, which should be considered in any complete energy and momentum budget.

15.7 FORMATION AND GROWTH OF SEVERE TROPICAL CYCLONES

The most difficult problem in the theory of tropical cyclones is to account for the initial formation. Many writers (e.g., Riehl, 1951) have pointed out that the development is not spontaneous; rather, intense storms always evolve from some disturbance of lesser intensity, which may have been in existence for a long time.

The character of the initial disturbance is varied. While early investigators often took the view that tropical cyclones form predominantly in the "doldrums," Dunn and Miller (1960) indicate that most Atlantic hurricanes form within the northeast trades. Palmer (1952) stated that in the central area of the north Pacific, "all tropical storms whose origin can be traced began as the result of instability in a wave in the easterlies; either as an equatorial

---

[13] We do not intend to imply that radiative influences are of small consequence. Some authors have suggested, for example, that the "clear ring" just outside the edge of the cirrus canopy of the hurricane, which Fett (1964) takes as evidence of subsidence extending down to the surface, is importantly influenced by radiative cooling.

wave [Sec. 14.7] or the easterly wave of the Caribbean [type]. ..." It has long been recognized that many mid-season Atlantic hurricanes come from the vicinity of Cape Verde Islands. Erickson (1963) gives a complete synoptic analysis of a tropical storm that formed just off the west coast of Africa, originating from an easterly wave that could be readily traced from central Africa. On the other hand, according to Sadler (1964), hurricanes of the eastern Pacific form from preexisting weak vortices in the zone between the northeast trades and the westerly branch of the turned southeast trades north of the Equator. Gabites (1956) indicates that in the south Pacific, easterly waves are not commonly observed, and hurricanes generally start in the intertropical convergence zone.

According to Dunn and Miller (1960), about one of six Atlantic hurricanes originate as perturbations moving away from the trade-confluence zone in the Panama region, and most of the remainder form from easterly waves. During the warmer months, at least one easterly wave is present almost every day over the Atlantic. In that region an average of only eight disturbances per year reach tropical storm intensity (maximum wind at least 17 m/sec), and about 60% of these achieve hurricane force (33 m/sec). Thus *a weak disturbance has a poor chance of becoming a tropical storm, but one that has achieved tropical-storm intensity has an excellent prospect of becoming a full hurricane.*

Only in the past two decades has emphasis concerning the formative processes shifted from lower levels to the higher troposphere, although it was earlier recognized that the real problem is that of evicting aloft the air that converges in lower levels. Riehl (1948) drew attention to the presence in the upper tropical troposphere of circulations whose scale and behavior are largely independent of the lower-level disturbances. He concluded that since the eviction of air aloft is favored by the warm-core structure, "it should assist the growth of the storms if they were located in a region where high pressure exists in a broadscale sense, so that the general pressure field is added to the pressure distribution produced by the storm itself." While hurricane development appears to be favored by certain patterns in the upper-contour field (Miller, 1958), there is much variation from case to case. In a study of 13 cases, Zipser (1964) found that practically all hurricane formations took place in the general regions of greatest 500- to 200-mb thickness, where this significantly exceeded that of the mean tropical atmosphere. However, nondeveloping disturbances were observed with similar, although generally weaker, patterns.

The incipient cyclone often forms beneath an equatorward extension of a trough in the mid-latitude westerlies, in which event intensification most commonly occurs during long-wave progression or discontinuous retrogression (Riehl, 1948; Riehl and Burgner, 1950), which brings an upper ridge over the surface cyclone. Northeast of Australia, McRae (1956) finds that tropical cyclogenesis ordinarily occurs where the anticyclonic curvature and shear increase downstream on the equatorward side of the subtropical jet stream. Although the vertical shear of the basic current is weak over the locale of cyclogenesis, most tropical cyclones forming in this region soon move poleward beneath the westerlies and assume a broad (nonfrontal) baroclinic structure. The region is unique in that (even in summer) strong westerlies aloft are found in fairly low latitudes.

Hurricane Ella (October 1962) formed beneath the southwest current in advance of an upper-tropospheric trough (Dunn and Staff, 1963), originating from an easterly wave that moved into this region (Alaka and Rubsam, 1964). Alaka (1962) gives another example wherein hurricane intensity was achieved beneath a strong southwesterly subtropical jet stream on its anticyclonic flank, where he indicated there was negative vorticity. In both cases, at the 200-mb level there was a change from cyclonic curvature upstream to anticyclonic downstream. This suggests development in response to organized upper divergence in the manner typical of extratropical cyclones (Section 11.2). The sequence of events in the Ella case indicated clearly that the upper-tropospheric anticyclone developed during or following surface intensification. This was observed first in the middle rather than the high troposphere (Alaka and Rubsam), possibly a consequence of the appreciable vertical shear between the northeast trades and the southwest winds at high levels.

The examples cited above illustrate some of the varied flow patterns that can exist at the time of initial cyclogenesis. Nevertheless, a feature that appears commonly (see, e.g., Yanai, 1961a; Fett, 1964; Koteswaram, 1967) is that the upper anticyclone surrounding a well-developed hurricane tends to be in phase with an upper-tropospheric ridge connected with the broad-scale, middle-latitude, westerly waves. On the outer limit of the hurricane outflow, a moderately strong anticyclonic jet stream appears, indicating that the outflow takes place in a region of relatively low dynamic stability on its anticyclonic flank.

At least in the case of tropical storms that develop from easterly waves, it appears (Yanai, 1961a) that the rising motions take place in a troposphere initially colder than the surroundings. This condition is favorable for convection, since with relatively uniform low-level θ_e, it implies greater than normal potential instability. Most investigators concur that this "cold core" is then gradually transformed to a warm core through release of latent heat and its transfer upward in the convective clouds. The most instructive study of the early stages is probably that of typhoon Doris by Yanai. Badly needed are several similar studies to show whether the formation of a warm-core vortex in general follows a similar pattern.

Although it is generally agreed that a "convective" hypothesis alone is inadequate to account for hurricane formation, it is instructive to examine this process in a qualitative way. A general scheme is illustrated in Fig. 15.12, in which the development is represented in four stages (which represent a continuous evolution of a symmetrical storm). These may be described as follows:

Stage I. Enhanced convection is present in some region where there is initially a feeble low-pressure system at sea level. As a cumulative effect of latent heat release and its eddy flux upward in towering cumulonimbus clouds, the temperature rises slowly in the upper troposphere and the isobaric surfaces above are elevated. This increasingly favors outflow from the region aloft, lowering the central sea-level pressure and enhancing low-level inflow of moist air to nourish the convective clouds.

Stage II. Through this inflow, the cyclonic circulation increases in low levels. This is communicated upward by the convective clouds, initiating a cyclonic vortex in the central part in the upper troposphere. A weak central low develops, surrounded by a divergent ring of high pressure from which radial outflow continues. The release of latent heat and its upward flux increase, and the temperature rise spreads downward and outward. As the cyclonic circulation increases, the rising branches begin to diverge outward from the warm core in upper levels, under conservation of angular momentum. Inside the diverging branches, the direct effect of condensation heating is absent and subsidence begins in the interior upper part.

Stage III. As the solenoid field caused by interior heating becomes more intense, the vertical mass circulation becomes more orderly and concentrated in rings around the central axis. Air rising from lower levels has an increasingly larger circulation and, according to Eq. (15.7), the limit of the cyclonic circulation in the upper outflow (connected with the ring of maximum tangential winds in low levels) expands outward. Inside this region, subsidence with adiabatic warming develops to lower levels, as required to maintain dynamic and static equilibrium, and the eye works downward. This process, in connection with the increased mass convergence in low levels and a greater mass divergence in upper levels, results in a rapid pressure fall in the central region. Turbulent mass exchange starts between the surrounding upward flow of moist cloud air and the eye.

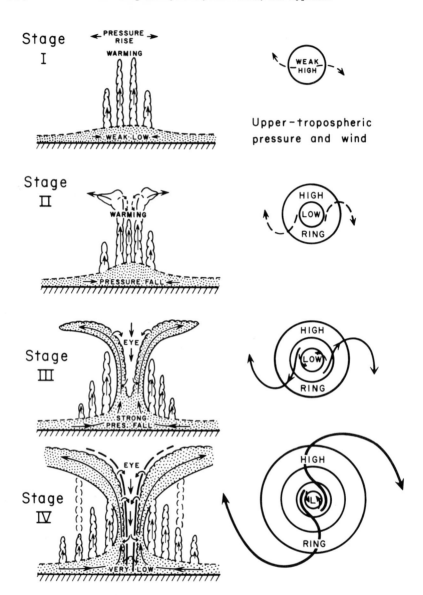

Fig. 15.12   Schematic stages in formation of a symmetrical hurricane (see text).

Stage IV. The vertical mass circulation and the connected release of latent heat and kinetic energy reach their final values. Rings of rapidly upward-moving air, in which the mass and heat transport is partly turbulent, surround the eye. Into the eye, air from the surrounding saturated region is introduced by turbulent mixing, and sinking takes place, compensated by upper convergence and lower divergence. The final quasi-steady state is determined by a balance between production and dissipation of kinetic energy and by balance of the total energy and angular-momentum budgets.

Sawyer (1947) pointed out that the effect of the process described above depends not only on the necessary potential instability in the vertical, but also upon a condition that would allow the air to flow freely outward in the upper levels. Without this condition, Sawyer argues that although the processes in Stages I and II would operate initially, a final stage would be reached in which there would be a balanced anticyclone aloft, uniform pressure at sea level, and a cessation of vertical circulation. His "dynamic instability" theory for the removal of air aloft has formed the basis for a number of later investigations. A concise and lucid critique of the various instability theories, and of numerical computation experiments related to hurricanes, has been given by Spar (1964).

Yanai has applied the theory of the stability of an axially symmetric vortex in which, at the initial stage (identified as an incipient cyclone), the balance is expressed by the gradient wind relation and the hydrostatic equation. Hence, if these initial conditions are marked by a bar, we have (Yanai, 1961b, 1964)

$$\left(\frac{\partial \bar{\Phi}}{\partial r}\right)_p = \bar{v}_\theta \left(f + \frac{\bar{v}_\theta}{r}\right) \tag{15.16a}$$

$$\frac{\partial \bar{\Phi}}{\partial p} = -\bar{\alpha} = -\frac{R\bar{T}}{p} \tag{15.16b}$$

$$\left(f + \frac{2\bar{v}_\theta}{r}\right) \frac{\partial \bar{v}_\theta}{\partial p} = -\left(\frac{\partial \bar{\alpha}}{\partial r}\right)_p \tag{15.16c}$$

where the last equation expresses the thermal wind relationship.

If the equilibrium expressed by Eqs. (15.16) is disturbed, tangential and radial accelerations will ensue. Following Yanai, these are expressed by the approximate equations (wherein we consider only the friction F_θ acting upon the tangential motion)

$$\frac{\partial v_\theta'}{\partial t} + \eta v_r' + \frac{\partial \bar{v}_\theta}{\partial p} \omega' = F_\theta \tag{15.17a}$$

$$\frac{\partial v_r'}{\partial t} - \xi v_\theta' + \frac{\partial \Phi'}{\partial r} = 0 \tag{15.17b}$$

$$\frac{\partial \Phi'}{\partial p} = -\alpha' \qquad\qquad (15.17c)$$

$$\frac{\partial v_r'}{\partial r} + \frac{v_r'}{r} = -\frac{\partial \omega'}{\partial p} \qquad\qquad (15.17d)$$

Here $\eta = f + (\bar{v}_\theta/r) + (\partial \bar{v}_\theta/\partial r)$ is the absolute vorticity of the initial field and $\xi = f + (2v_\theta/r) \equiv 2M_r/r^2$ is related to its angular momentum. Equation (15.17d) represents the continuity equation. Yanai also considered an equation expressing the local change of specific volume or temperature caused by a heat source and by the disturbed motion. The resulting circulation is very complicated, and an exact solution of the equation system cannot be readily achieved. The final equation derived by Yanai contains three stability coefficients that depend on the initial distribution of the tangential wind field, the baroclinity, and the static stability. We discuss here only the first stability criterion.

By time-differentiating Eq. (15.17b), substituting for $\partial v_\theta/\partial t$ from Eq. (15.17a) and for $\partial \bar{v}_\theta/\partial p$ from Eq. (15.16c), we obtain

$$\frac{\partial^2 v_r'}{\partial t^2} = -\xi \eta v_r' - \frac{\partial}{\partial t}\left(\frac{\partial \Phi'}{\partial r}\right) + \omega' \frac{\partial \bar{\alpha}}{\partial r} + \xi F_\theta \qquad\qquad (15.18)$$

This expression was first derived by Sawyer (1947), who used it to study the rapid growth of radial velocity in an incipient cyclone. Sawyer was especially interested in the increase of the outward movement of concentric air rings in the upper troposphere, caused by a disturbance in the radial pressure gradient ($\partial \Phi'/\partial r < 0$) that results from heating of the air beneath. At the level concerned, the vertical velocity ω' and the friction F_θ may be disregarded. At the moment of the initial-pressure disturbance, $v_r' = 0$ and $\partial^2 v_r'/\partial t^2 > 0$ if $(\partial/\partial t)(\partial \Phi'/\partial r) < 0$. Somewhat later, when $v_r' > 0$, the first r.h.s. term gives a positive contribution if $\xi \eta < 0$. Hence, Sawyer concluded that a rapid growth of the outward velocity occurs if either

$$(1) \quad \xi < 0 \qquad \text{and} \qquad \eta > 0$$

or

$$(2) \quad \xi > 0 \qquad \text{and} \qquad \eta < 0$$

Case 1 represents an anticyclonic vortex whose rotation exceeds the component of the earth's rotation about the vertical, but whose vertical component of absolute vorticity is positive. Case 2 represents either a cyclonic or anticyclonic vortex whose rotation in space is cyclonic but whose absolute vorticity is negative.

If we consider an incipient cyclonic vortex with a cold core or a barotropic cyclone, it is obvious that $\xi > 0$. In this case it would be difficult to see how the corollary condition $\eta < 0$ for case 2 could be realized, since this would require a rather large anticyclonic shear. However, if heating due to condensation occurs in a deep layer of the atmosphere, we could visualize a stage of the development when an upper anticyclone could become superposed on a low-level cyclonic vortex. If in this special case it were somehow possible for condition 1 to be fulfilled (requiring $\xi < 0$), this condition would evidently be annihilated quickly. In this case a rapid increase of the outflow would be favored by the combined influence of the first and second r.h.s. terms in Eq. (15.18), leading to rapid increase of radial inflow in low levels, supported here by F_θ, and to a pressure drop at the surface caused by the upper mass divergence. Through this radial circulation, moist air with high values of θ_e would be transported upward, but also cyclonic circulation would be simultaneously brought up to the upper troposphere, resulting in a gradual change of ξ from negative to positive. As a result, the contribution of the first r.h.s. term in Eq. (15.18) would become negative in the core region. Thus, only during a short time interval might an upper anticyclonic vortex with a rotation exceeding the component of the earth's rotation about the vertical be essential for the development.

On other grounds, Sawyer dismissed case 1 as a factor in development after the initial stages, and concluded that a required condition is "a distribution of circulation at some level aloft such that anticyclonic [absolute] circulation increases outward...." Since the circulation of a ring of air is $2\pi M_r$, differentiation of Eq. (15.2) shows that $\eta = r^{-1} \partial M_r/\partial r$. Hence Sawyer's criterion specifies $\eta < 0$, case 2. The interpretation is: "If the air in the expanding ring [conserving its circulation] finds itself with a greater angular velocity in space...than that of the air which previously occupied its new position, the pressure gradient will be inadequate to balance the centrifugal force on the air and it will be thrown outwards." It is difficult to show synoptically that either negative vorticity or negative absolute angular momentum is present in the sense required for dynamic instability, although Alaka (1962) contends that both are significant.[14]

----

[14] In hurricane Daisy, Alaka shows that both negative vorticity and negative angular momentum were present at 240 mb in regions that partly overlapped. The interpretation is rendered difficult because ξ and η in Eq. (15.18) apply to a symmetrical vortex and thus involve only the v_θ wind component, whereas consideration of the v_r component and large curvatures not concentric with the cyclone center result in the negative values computed by Alaka. When averaged at a radius, the absolute vorticity was positive (Alaka, 1962), as indicated by the M_r distribution in Fig. 15.6.

At any rate, the term involving inertial instability appears to be of small consequence. If we consider a time when the development has gotten well under way, v_r' might be assigned a value of, say, 3 m/sec. Supposing that $\xi\eta = -f^2$, where $f = 0.5 \times 10^{-4}$ sec^{-1}; the first r.h.s. term of Eq. (15.18) would, with these values, be $-\xi\eta v_r' = 0.75 \times 10^{-8}$ m sec^{-3}. By comparison, if the isobaric surfaces are being elevated by heating near the center and the outward isobaric height gradient is changing at the rate of 25 m/50 km in 10 hr, the term $-(\partial/\partial t)(\partial\Phi'/\partial r) = 14 \times 10^{-8}$ m sec^{-3} is 20 times greater. Considering that if ξ or η is negative, its value is likely to be small (most likely $|\xi\eta| \ll f^2$) so that $-\xi\eta v_r'$ is even smaller and is negligible compared with the other terms. A similar conclusion was reached by Yanai (1964) by a more elegant analysis.

Kleinschmidt (1951), in a theory further developed by Yanai, considered inertial instability ($\eta < 0$) as an essential mechanism during the initial development, but in his model the instability was taken along surfaces of equal equivalent potential temperature. He considered a stage when a warm core has been established. In this case, surfaces of constant angular momentum M_r have an outward slope. For instability, it is required that the slope of the θ_e-surfaces be larger than the slope of the M_r-surfaces. It is very difficult to show that this really is the case in the initial stage. During the mature stage, it was suggested that both systems of surfaces should approximately coincide as in Fig. 15.6, which then would represent neutral stability.

In a developing cyclone the radial velocity changes according to the schematic Fig. 15.13. During the time interval $(t_0 - t_1)$, $\partial^2 v_r'/\partial t^2 > 0$, and during interval $(t_1 - t_2)$, $\partial^2 v_r'/\partial t^2 < 0$, if t_2 represents the time when the hurricane has reached maturity. Considering the uppermost level at time

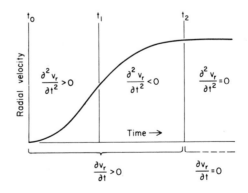

FIG. 15.13   Time changes of radial velocity (see text).

t_0 when $v_r' = 0$, the increase of radial velocity is, according to Eq. (15.18), entirely determined by $(\partial/\partial t)(\partial \Phi'/\partial r)$, which is negative in the upper troposphere when an upper warm core starts to develop. Later, when at this level $v_r' > 0$, the first r.h.s. term becomes significant, its contribution depending on the signs of ξ and η. As indicated earlier, even if $\xi < 0$ at some time early in the development, it is likely that this condition would soon be eradicated.[15] Also, as indicated in Section 15.5, since the air in the inflow layer loses M_r toward the center and this air feeds the outflow layer, most likely $\partial M_r/\partial r > 0$ so that $\eta > 0$. Thus, after the development has gotten under way, $(\xi \eta) > 0$. Hence the first r.h.s. term of Eq. (15.18) generally represents a brake on the acceleration of the outward flow, which therefore must depend upon the other terms.

To appraise the significance of these terms in a qualitative way, we may consider the times $t = t_1$ and $t > t_2$ when (Fig. 15.13) $\partial^2 v_r/\partial t^2 = 0$. In these cases, Eq. (15.18) may be written (Palmén, 1956) as

$$v_r(t = t_1) = \frac{\omega(\partial \alpha/\partial r) - (\partial/\partial t)(\partial \Phi/\partial r)}{\xi \eta} + \frac{F_\theta}{\eta} \qquad (15.19a)$$

$$v_r(t > t_2) = \frac{\omega(\partial \alpha/\partial r)}{\xi \eta} + \frac{F_\theta}{\eta} \qquad (15.19b)$$

The bars and primes have been dropped in these equations, which correspond to special conditions at times t_1 and $t > t_2$. They are equivalent to an expression discussed by Durst and Sutcliffe (1938), derived directly from the equation of motion for v_θ; in this form use has been made of the thermal and gradient wind equations.

The processes at time t_1 are essentially the same as at time $t > t_2$ except that the magnitudes of $\omega \partial \alpha/\partial r$ and F_θ are weaker because the cyclone is not yet fully developed, and there is a contribution (of uncertain sign) from $(\partial/\partial t)(\partial \Phi/\partial r)$. We shall discuss only the steady-state mature hurricane in which the radial velocity is governed by Eq. (15.19b) and the vertical velocity by Eq. (15.17d).

At a fixed radius the vertical distributions of v_θ and v_r may be given by the curves in Fig. 15.14, in which it is assumed that the velocity components vanish at 100 mb. The schematic figure gives the magnitudes of all quantities $\omega \, \partial \alpha/\partial r$, ξ, η, and F_θ for the friction layer (1000 to 850 mb), the lower part of the free atmosphere (850 to 550 mb), the upper part of the free atmosphere (550 to 250 mb), and the uppermost troposphere (250 to 100 mb).

---

[15] A further difficulty is that in their early stages many tropical storms are situated beneath high-tropospheric troughs in the larger-scale flow.

In the inner part, η is very large in lower levels (Hughes, 1952) and decreases slowly upward through the middle troposphere, but rapidly upward in the upper troposphere (E. S. Jordan, 1952). The same is true of ξ, which is proportional to M_r (Fig. 15.6). The term $\omega \, \partial \alpha / \partial r$, which represents an

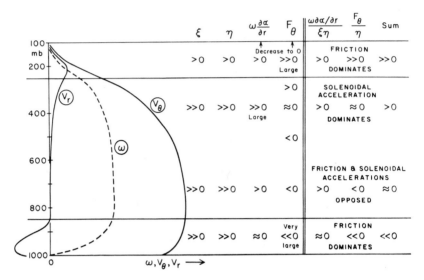

FIG. 15.14   General magnitudes of factors influencing radial inflow or outflow (see text).

outward movement of air as a result of its ascent through levels in which the radial pressure gradient weakens upward, is small near the earth's surface and greatest in the upper troposphere. The friction F_θ is very large in the surface layer, where it removes tangential momentum, and again large in the uppermost troposphere, to which momentum is transferred by convective clouds from lower levels; the effect being nil somewhere around 300 to 400 mb, in line with Gray's estimate (1967) based on a statistical population of convective clouds and their properties.

In the lower boundary layer, F_θ has a very strong negative value. The radial inflow in this layer is thus conditioned mainly by the requirement for the frictional torque to be balanced by an inward movement of rings of air that have a large absolute circulation about the center. This is expressed by the angular momentum budget in the lower part of Fig. 15.10. Near the top of the boundary layer, the frictional influence becomes less important and is counteracted by the effect of upward motion. Through the layer 850 to 250 mb, the first r.h.s. term of Eq. (15.19b) increases gradually upward, whereas the frictional term remains small and negative through most of the layer, and then increases rapidly to large positive values in the upper troposphere. Here the effects of the two terms reinforce each other to produce a large outflow component in this layer, where η is small and the dynamic stability is weak. Above about 200 mb, the organized vertical motion, the baroclinity, and the frictional influence, all decay so that v_r declines.

It should be stressed that the foregoing reasoning is very qualitative. It is, however, in agreement with the scheme in Sections 15.5 and 15.6 concerning the angular momentum and energy balance. In the inflow layer there is divergence of the vertical angular-momentum flux and a convergence in the uppermost outflow layer. The kinetic energy of the system is maintained against the dissipating forces by the positive work done by the radial pressure gradient in the lowest layer, which is much larger than the negative work in the outflow layer.

Summarizing the essential results of the present theories, it can be stated that the role of a possible dynamic or inertial instability in the initial transformation of a cold-core cyclone into a warm-core vortex is not yet fully clarified. Possibly this type of instability, computed on an isentropic surface or more likely on a surface of constant equivalent potential temperature, may play some role for the radial outward mass transport in the upper troposphere. However, since most incipient cyclones appear as asymmetric disturbances, it seems unlikely that any theory starting from the assumption of a symmetric vortex in the sense of Sawyer or Kleinschmidt could lead to a general solution of the problem.

The possible importance of any inertial instability seems to be associated with the occurrence of heating in the middle and upper troposphere as a result of the release of latent heat. Only this effect seems to be able to produce the proper type of instability. However, since this type of heat source in a deep layer already presupposes the existence of an orderly mass inflow in low layers, it seems quite questionable whether the upper outflow or the lower inflow should be considered the initial triggering effect. Obviously, they support each other. In this respect the question of the possible triggering effects necessary for the eventual release of the potential instability in the tropical atmosphere is not yet fully solved. Also, it is not yet clear whether at any time period of the development any type of real instability is needed. Perhaps it is sufficient to reduce the inertial stability to a certain low value that would suffice to permit the proper increasing mass outflow in the upper layers under the influence of a gradual increase of the internal heat source.

Yanai (1964) has given a description of this gradual heating of the upper troposphere in an easterly wave that transformed into a typhoon. He suggested that the initial warming, greatest at levels of 300 to 400 mb, is due to upward heat transfer in tall convective clouds. This process can take place while there is still adiabatic cooling due to organized ascent at lower levels. Hence the warm core develops first in upper levels and gradually works downward as heat is accumulated; organized outflow starts at high levels, and cyclone development ensues, basically following the scheme described earlier in connection with Fig. 15.12.

The earlier discussion was, for the purpose of exposing certain aspects relevant to development theories, couched in terms of a radially symmetrical structure. Actually, many disturbances show a pronounced asymmetry. For example, Malkus et al. (1961) considered the cloud patterns of hurricane Daisy in its early stages to be more typical of an easterly wave, with greater symmetry in the mature stage.

The characteristic cloud structure revealed by satellites has been illustrated by Fritz et al. (1966) and synthesized into a synoptic model by Fett (1964). In the earliest stages, an irregular and extensive cloud mass is evident. Later, but still fairly early in the development, spiraling cumulus bands appear, but the most solid cloud masses, often becoming organized in a "comma" shape, remain east of the circulation center. At this stage, when the depth of the depression is only a few millibars, Fritz et al. suggest that unless a pronounced cirrus outflow occurs, "presumably such disturbances will not develop rapidly." In a cyclone that develops further, generation of cirrus takes place, and the circulation center becomes surrounded by the upper cloud. An eye may be visible long before hurricane intensity is reached. While in some mature storms the central shield is quite circular, in others a considerable asymmetry is observed. Merritt and Wexler (1967) show that this is in accord with the upper wind field, which generally exhibits asymmetry of outflow. They indicate that, with cirrus production in the central core, it would be possible in a typical case to produce in 12 to 18 hr time a cirrus canopy comparable to those observed by the satellites. The diameter achieved by this canopy is highly correlated with the maximum wind speed in the hurricane (Fritz et al., 1966).

The cloud sequence described above is consistent with Yanai's analysis of the formation of typhoon Doris. Relatively high temperatures were observed over the convective region east of the wave trough. As the development progressed, this warm region approached the trough from the east. The original asymmetrical structure of the easterly wave trough, tilting toward the cooler air on its east side, gradually disappears as this air is heated by the convective processes, and the vertical warm-core system eventually dominates.

As noted earlier, significant advances have recently been made in numerical model studies of hurricane circulations, by a number of investigators. We could not, however, do justice to these important researches in a brief review, and have chosen to limit ourselves to a presentation of empirical results.

15.8 DECAY OR TRANSFORMATION OF TROPICAL CYCLONES

According to Dunn and Miller (1960), very few tropical cyclones dissipate while remaining over tropical or subtropical waters of the Atlantic. However most cyclones decay rapidly when they more in over land (Malkin, 1959) or move outside the warm waters of the principal cyclone regions shown in Fig. 15.2. This is in accord with the previous results concerning the energy balance, which strongly depends on the θ_e of the air in the core of the storm and thus on the sea temperature. Illustrations of the filling of hurricanes after recurvature poleward over cooler water are given by Colón (1962) for the Atlantic and by McRae (1956) for the region east of Australia. The cause of filling may, however, be a combination of this and other factors. Sadler (1964) indicates that movement over cooler waters is likely to cause dissipation for any cyclone moving out of the restricted warm-water region of the eastern Pacific. However, those moving in a direction north of west also come under strong upper-tropospheric westerlies, with appreciable vertical shear of the basic current, which is an adverse condition (Section 15.1).

When a tropical storm moves inland, the increased friction might be expected to upset the previous balance between production and dissipation of kinetic energy, leading to a rapid decrease of kinetic energy at least until the rate of dissipation slows down as a result of the decreasing wind velocity. From this one may conclude that tropical cyclones over land would survive, but at a lower-energy level, if the energy production determined by the vertical circulation and the baroclinity remained unchanged. It is to be expected that the immediate consequence of the increased ground friction over land would be an intensification of the inflow, and there are sometimes suggestions of intensified rainfall resulting from this. Hubert (1955) investigated the influence of enhanced inflow due to increased surface roughness and concluded that this is inadequate to account for rapid filling, which should not be expected unless there is a decrease in the energy supply. He observed that filling is mainly evident in the innermost part of the vortex, with weak effects at a moderate distance from the center, resulting in a decreased pressure gradient near the ring of strongest winds.

This observation is consistent with the proposition that *filling results because the heat flux from the earth's surface becomes negligibly small when a storm moves inland, resulting in a reduction of the temperature excess of the core.* From the discussion in Sections 15.1 and 15.2 it is evident that such a reduction is very effective in weakening the solenoid field responsible for the production of kinetic energy. The marked lowering of the surface θ_e in a filling hurricane entering the southeastern United States has been illustrated by Bergeron (1954). His conclusion that filling comes mainly from reduction of the θ_e of the central rising air is confirmed by Miller's study (1963) of hurricane Donna when it moved in over Florida. In line with Hubert's findings, the radial mass flow changed very little. However, the production of kinetic energy was sharply reduced in the core region, primarily as a result of the decreased pressure gradient over land, which was in turn caused by the rapid removal of the oceanic heat source. Miller therefore concluded that the increased dissipation by stronger surface drag was of minor importance compared to the removal of the oceanic heat source. When hurricane Donna again moved out into the Atlantic over the very warm waters of the Gulf Stream region, renewed deepening ensued.

The mechanism of the filling process may be described as follows: Owing to removal of the oceanic heat source in the inner region, the baroclinity is reduced, since the air ascending in the inner cloud wall now has a somewhat lower θ_e. As a result, the outward radial wind component in upper levels is reduced. The previous balance between the mass inflow in low levels and mass outflow in upper levels is thus temporarily disturbed, leading to an integrated net mass convergence and pressure rise. During this phase the cyclone tends to approach a depth around 1000 mb, according to Malkus and Riehl (1960), determined only by the release of latent heat intrinsic to the moist surface layer in its outer parts.

A comparison with extratropical cyclones in this respect is instructive. These storms often form or intensify over land because they derive most of their kinetic energy from the preexisting available potential energy of the baroclinic atmosphere. By contrast, tropical cyclones *produce and maintain* their own available potential energy, this being derived from the liberation of latent heat of the ascending low-level moist air augmented by the additional flux of latent and sensible heat from the sea.

Decay upon moving over cooler water is caused by an intrusion of potentially colder low-level air into the core of the cyclone. This is often a relatively slow process governed by the poleward decrease of the sea-surface temperature. If, however, a rapid intrusion of a deep cold-air mass occurs when a tropical cyclone approaches a strong baroclinic zone of the temperate west-wind belt, a new energy source becomes active and results often in a simultaneous regeneration and transformation into an extratropical cyclone. This type of regeneration is common in the vicinity of the North American east coast and in the region of Japan, preferentially late in the hurricane or typhoon season when outbreaks of polar-air masses start to penetrate into lower latitudes. Such rejuvenated tropical cyclones may survive as intense storms for several days, occasionally reaching the European west coast or even as far north as Iceland, with some residual characteristics of tropical disturbances. Some of the most severe storms occurring in September over these regions, far from the tropics, originate in this way.[16]

An example of regeneration of a tropical hurricane and its simultaneous transformation into a very severe extratropical cyclone over land was hurricane Hazel (Palmén, 1958). This destructive hurricane moved in across the east coast of the United States on Oct. 15, 1954, simultaneous with a very deep cold-air outbreak over the central and eastern parts of the continent. As shown by Fig. 15.15, the expected weakening of the hurricane over land never took place. Following incorporation of an intense cold front (Fig. 8.2) into its circulation and the ensuing intense conversion of available potential energy, the storm remained very destructive when moving far northward into Canada, eventually becoming a regular occluded cyclone. The 300-mb chart in Fig. 16.6, corresponding to the middle time in Fig. 15.15, shows that the hurricane had already come under the characteristic upper-level divergence region ahead of the middle-latitude trough, which aids in the eviction of air in upper levels from the hurricane circulation. An instructive account of the development of this trough, and its role in maintaining the severe cyclone, was given by Hughes *et al.* (1955). These authors describe how it was possible to predict the strong increase in intensity of the upper-level trough, which followed the general pattern outlined in Section 11.5, on the basis of prior events observed upstream.

Sekioka (1956, 1957, 1959) has examined the structure of 21 typhoons moving over Japan and the Sea of Japan. Of these storms, 18 had the complex structure, with a more or less distinct frontal system characteristic of the transition stage to extratropical storms, and only 3 could still at this relatively high latitude be considered pure tropical typhoons. It is significant that these three typhoons all occurred in July when the Asian continent is at its warmest and only weak cold outbreaks could be expected

---

[16] As an example of these "quasi-tropical" storms may be mentioned the cyclone north of the British Isles on Sept. 6, 1966, which started as hurricane Faith in the Caribbean region about ten days before. The 500-mb temperatures of -6 to $-8°C$ over England indicate the real tropical origin of the warm-air mass of this cyclone.

from this region. This result shows that the survival of tropical storms is especially favored in extratropical regions where energy conversions are supported by interaction with middle-latitude disturbances of the regular

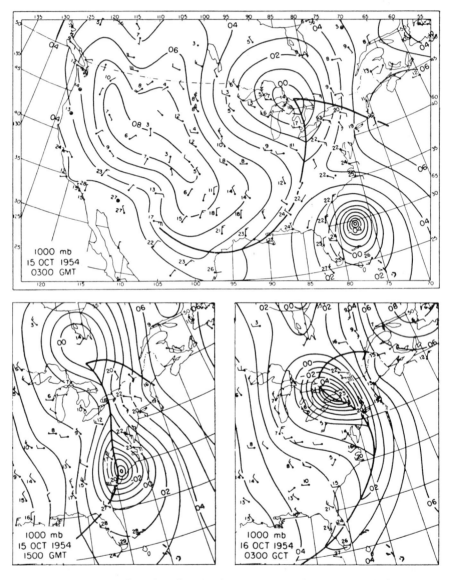

FIG. 15.15   Fronts (heavy) and 1000-mb contours (hundreds of feet) during transformation of hurricane Hazel into extratropical cyclone (after Palmén, 1958). Top, 03 GCT, Oct. 15, 1954; bottom, charts 12 and 24 hr later.

baroclinic type. Tropical cyclones recurving and reaching latitudes poleward of 30° in either hemisphere tend to undergo this transformation into extratropical storms if the baroclinic structure of the atmosphere is favorable. If this is not the case, a more rapid filling may be expected, owing to a gradual intrusion of cooler air in the surface layer.

REFERENCES

Ackerman, B. (1963). Some observations of water content in hurricanes. *J. Atmospheric Sci.* **20**, 288–298.

Adem, J., ed. (1963, 1964). Proceedings of the Third Technical Conference on Hurricanes and Tropical Meteorology. *Geofis. Intern.* **3**, 132–150; **4**, 169–230.

Alaka, M. A. (1962). On the occurrence of dynamic instability in incipient and developing hurricanes. *Monthly Weather Rev.* **90**, 49–58.

Alaka, M. A., and Rubsam, D. T. (1964). Some notes on the formation of hurricanes with particular reference to Hurricane Ella, 1962. *In* "Proceedings of the Symposium on Tropical Meteorology" (J. W. Hutchings, ed.), pp. 641–649. New Zealand Meteorol. Serv., Wellington.

Arakawa, H. (1954). On the pyramidal, mountainous and confused sea in the right or dangerous semi-circle of typhoons. *Papers Meteorol. Geophys. (Tokyo)* **5**, 114–123.

Asai, T., and Kasahara, A. (1967). A theoretical study of the compensating downward motions associated with cumulus clouds. *J. Atmospheric Sci.* **24**, 487–496.

Bergeron, T. (1954). The problem of tropical hurricanes. *Quart. J. Roy. Meteorol. Soc.* **80**, 131–164.

Brooks, E. M. (1946). An analysis of an unusual rainfall distribution in a hurricane. *Bull. Am. Meteorol. Soc.* **27**, 9–14.

Byers, H. R. (1944). "General Meteorology," p. 432. McGraw-Hill, New York.

Colón, J. A. (1960). On the heat balance of the troposphere and the water body of the Caribbean Sea. Natl. Hurricane Res. Proj. Rept. No. 41, 65 pp.

Colón, J. A. (1962). Changes in the eye properties during the life cycle of tropical hurricanes. Natl. Hurricane Res. Proj. Rept. No. 50, Part II, pp. 341–354.

Colón, J. A. (1964). On the structure of hurricane Helene (1958). Natl. Hurricane Res. Proj. Rept. No, 72, 56 pp.

Deppermann, C. E. (1947). Notes on the origin and structure of Philippine typhoons. *Bull. Am. Meteorol. Soc.* **28**, 399–404.

Dunn, G. E., and Miller, B. I. (1960). "Atlantic Hurricanes," 326 pp. Louisiana State Univ. Press, Baton Rouge, Louisiana.

Dunn, G. E. *et al.* (1963). The hurricane season of 1962. *Monthly Weather Rev.* **91**, 199–207.

Durst, C. S., and Sutcliffe, R. C. (1938). The importance of vertical motion in the development of tropical revolving storms. *Quart. J. Roy. Meteorol. Soc.* **64**, 75–84.

Erickson, C. O. (1963). An incipient hurricane near the West African coast. *Monthly Weather Rev.* **91**, 61–68.

Ferrel, W. (1856). An essay on the winds and the currents of the ocean. *Nashville J. Med. Surg.* **11**, 375–389.

Fett, R. W. (1964). Aspects of hurricane structure: New model considerations suggested by TIROS and Project Mercury observations. *Monthly Weather Rev.* **92**, 43–60.

Fritz, S., Hubert, L. F., and Timchalk, A. (1966). Some inferences from satellite pictures of tropical disturbances. *Monthly Weather Rev.* **94**, 231–236.

Fujita, T., Izawa, T., Watanabe, K., and Imai, I. (1967). A model of typhoons accompanied by inner and outer rainbands. *J. Appl. Meteorol.* **6**, 3–19.

Gabites, J. F. (1956). A survey of tropical cyclones in the South Pacific. *In* "Proceedings of the Tropical Cyclone Symposium, Brisbane," pp. 19–24. Australian Bur. Meteorol., Melbourne.

Gentry, R. C. (1964). A study of hurricane rainbands. Natl. Hurricane Res. Proj. Rept. No. 69, 85 pp.

Gherzi, E. (1956). Methodology in the study of the origin and motion of tropical cyclones. *In* "Final Report of the Caribbean Hurricane Seminar," pp. 19–33. Govt. Dominican Republic, Ciudad Trujillo.

Gray, W. M. (1966). On the scales of motion and internal stress characteristics of the hurricane. *J. Atmospheric Sci.* **23**, 278–288.

Gray, W. M. (1967). The mutual variation of wind, shear and baroclinicity in the cumulus convective atmosphere of the hurricane. *Monthly Weather Rev.* **95**, 55–73.

Haurwitz, B. (1935). The height of tropical cyclones and the eye of the storm. *Monthly Weather Rev.* **63**, 45–49.

Haurwitz, B. (1941). "Dynamic Meteorology," p. 151. McGraw-Hill, New York.

Hebert, P. J., and Jordan, C. L. (1959). Mean soundings for the Gulf of Mexico area. *Monthly Weather Rev.* **87**, 217–221.

Hubert, L. F. (1955). Frictional filling of hurricanes. *Bull. Am. Meteorol. Soc.* **36**, 440–445.

Hughes, L. A. (1952). On the low-level wind structure of tropical cyclones. *J. Meteorol.* **9**, 422–428.

Hughes, L. A., Baer, F., Birchfield, G. E., and Kaylor, R. E. (1955). Hurricane Hazel and a long-wave outlook. *Bull. Am. Meteorol. Soc.* **36**, 528–533.

Ito, H. (1963). Aspects of typhoon development as viewed from observational data in the lower troposphere. *In* "Proceedings of the Inter-Regional Seminar on Tropical Cyclones in Tokyo," pp. 103–119. Japan Meteorol. Agency, Tokyo.

Japan Meteorological Agency. (1963). "Proceedings of the Inter-Regional Seminar on Tropical Cyclones in Tokyo," 315 pp. Japan Meteorol. Agency, Tokyo.

Jordan, C. L. (1952). On the low-level structure of the typhooneye. *J. Meteorol.* **9**, 285–290.

Jordan, C. L. (1957). Mean soundings for the hurricane eye. Natl. Hurricane Res. Proj. Rept. No. 13, 10 pp.

Jordan, C. L. (1958a). Mean soundings for the West Indies area. *J. Meteorol.* **15**, 91–97.

Jordan, C. L. (1958b). The thermal structure of the core of tropical cyclones. *Geophysica (Helsinki)* **6**, 281–297.

Jordan, C. L. (1961). Marked changes in the characteristics of the eye of intense typhoons between the deepening and filling stages. *J. Meteorol.* **18**, 779–789.

Jordan, C. L., Hurt, D. A., Jr., and Lowrey, C. A. (1960). On the structure of Hurricane Daisy on August 27, 1958. *J. Meteorol.* **17**, 337–348.

Jordan, E. S. (1952). An observational study of the upper wind-circulation around tropical storms. *J. Meteorol.* **9**, 340–346.

Kasahara, A. (1954). Supplementary notes on the formation and the schematic structure of typhoons. *J. Meteorol. Soc. Japan* **32**, 31–52.

Kessler, E., III (1958). Eye region of hurricane Edna, 1954. *J. Meteorol.* **15**, 264–270.

Kleinschmidt, E., Jr. (1951). Grundlagen einer Theorie der tropischen Zyklonen. *Arch. Meteorol., Geophys. Bioklimatol.* **A4**, 53–72.

Koteswaram, P. (1967). On the structure of hurricanes in the upper troposphere and lower stratosphere. *Monthly Weather Rev.* **95**, 541–564.

Krishnamurti, T. N. (1962). Some numerical calculations of the vertical velocity field in hurricanes. *Tellus* **14**, 195–211.

Kuo, H. L. (1959). Dynamics of convective vortices and eye formation. *In* "The Atmosphere and the Sea in Motion" (B. Bolin, ed.), pp. 413–424. Rockefeller Inst. Press, New York.

LaSeur, N. E. (1962). On the role of convection in hurricanes. Natl. Hurricane Res. Proj. Rept. No. 50, Part II, pp. 323–334.

LaSeur, N. E., and Hawkins, H. F. (1963). An analysis of hurricane Cleo (1958) based on data from research reconnaissance aircraft. *Monthly Weather Rev.* **91**, 694–709.

Leipper, D. F. (1967). Observed ocean conditions and hurricane Hilda, 1964. *J. Atmospheric Sci.* **24**, 182–196.

McRae, J. N. (1956). The formation and development of tropical cyclones during the 1955–56 season in Australia. *In* "Proceedings of Tropical Cyclone Symposium, Brisbane," pp. 233–261. Australian Bur. Meteorol., Melbourne.

Malkin, W. (1959). Filling and intensity changes in hurricanes over land. Natl. Hurricane Res. Proj. Rept. No. 34, 17 pp.

Malkus, J. S. (1958). On the structure and maintenance of the mature hurricane eye. *J. Meteorol.* **15**, 337–349.

Malkus, J. S., and Riehl, H. (1960). On the dynamics and energy transformations in steady-state hurricanes. *Tellus* **12**, 1–20.

Malkus, J. S., Ronne, C., and Chaffee, M. (1961). Cloud patterns in hurricane Daisy. *Tellus* **13**, 8–30.

Merritt, E. S., and Wexler, R. (1967). Cirrus canopies in tropical storms. *Monthly Weather Rev.* **95**, 111–120.

Miller, B. I. (1958). On the maximum intensity of hurricanes. *J. Meteorol.* **15**, 184–195.

Miller, B. I. (1962). On the momentum and energy balance of hurricane Helene (1958). Natl. Hurricane Res. Proj. Rept. No. 53, 19 pp.

Miller, B. I. (1963). On the filling of tropical cyclones over land. Natl. Hurricane Res. Proj. Rept. No. 66, 82 pp.

O'Brien, J. J., and Reid, R. O. (1967). The non-linear response of a two-layer, baroclinic ocean to a stationary, axially-symmetric hurricane: Part I. Upwelling induced by momentum transfer. *J. Atmospheric Sci.* **24**, 197–207.

Östlund, H. G. (1967). Hurricane tritium II: Air-sea exchange of water in Betsy 1965. Mimeograph, 23 pp., Inst. Marine Sci., Univ. of Miami, Florida.

Palmén, E. (1948). On the formation and structure of tropical hurricanes. *Geophysica* (*Helsinki*) **3**, 26–38.

Palmén, E. (1956). Formation and development of tropical cyclones. *In* "Proceedings of Tropical Cyclone Symposium, Brisbane," pp. 213–231. Australian Bur. Meteorol., Melbourne.

Palmén, E. (1958). Vertical circulation and release of kinetic energy during the development of hurricane Hazel into an extratropical storm. *Tellus* **10**, 1–23.

Palmén, E., and Jordan, C. L. (1955). Note on the release of kinetic energy in tropical cyclones. *Tellus* **7**, 186–188.

Palmén, E., and Riehl, H. (1957). Budget of angular momentum and energy in tropical cyclones. *J. Meteorol.* **14**, 150–159.

Palmer, C. E. (1952). Tropical meteorology. *Quart. J. Roy. Meteorol. Soc.* **78**, 126–164.

Pfeffer, R. L. (1958). Concerning the mechanism of hurricanes. *J. Meteorol.* **15**, 113–120.

Riehl, H. (1948). On the formation of typhoons. *J. Meteorol.* **5**, 247–264.

Riehl, H. (1951). Aerology of tropical storms. *In* "Compendium of Meteorology" (T. F. Malone, ed.), pp. 902–913. Am. Meteorol. Soc., Boston, Massachusetts.

Riehl, H. (1954). "Tropical Meteorology," Chapter 11. McGraw-Hill, New York.

Riehl, H. (1961). On the mechanisms of angular momentum transports in hurricanes. *J. Meteorol.* **18**, 113–115.

Riehl, H. (1963). Some relations between wind and thermal structure of steady state hurricanes. *J. Atmospheric Sci.* **20**, 276–287.

Riehl, H., and Burgner, N. M. (1950). Further studies of the movement and formation of hurricanes and their forecasting. *Bull. Am. Meteorol. Soc.* **31**, 244–253.

Riehl, H., and Malkus, J. S. (1961). Some aspects of hurricane Daisy, 1958. *Tellus* **13**, 181–213.

Sadler, J. C. (1964). Tropical cyclones of the eastern North Pacific as revealed by TIROS observations. *J. Appl. Meteorol.* **3**, 347–366.

Sawyer, J. S. (1947). Notes on the theory of tropical cyclones. *Quart. J. Roy. Meteorol. Soc.* **73**, 101–126.

Schauss, C. E. (1962). Reconstruction of the surface pressure and wind field of hurricane Helene. Natl. Hurricane Res. Proj. Rept. No. 59, 45 pp.

Sekioka, M. (1956, 1957, 1959). A hypothesis on complex of tropical and extratropical cyclones for typhoon in the middle latitudes. *J. Meteorol. Soc. Japan* **34**, 276–287 (Part I); **34**, 336–345 (Part II); **35**, 170–173 (Part III); **37**, 111–114 (Part IV).

Senn, H. V., and Hiser, H. W. (1959). On the origin of hurricane spiral rain bands. *J. Meteorol.* **16**, 419–426.

Shaw, W. N. (1922). The birth and death of cyclones. *Geophys. Mem.* **2**, No. 19, 213–227.

Simpson, R. H. (1947). A note on the movement and structure of the Florida hurricane of October 1946. *Monthly Weather Rev.* **75**, 53–58.

Simpson, R. H. (1952). Exploring eye of typhoon "Marge", 1951. *Bull. Am. Meteorol. Soc.* **33**, 286–298.

Simpson, R. H. (1956). Some aspects of tropical cyclone structure. *In* "Proceedings of Tropical Cyclone Symposium, Brisbane," pp. 139–157. Australian Bur. Meteorol., Melbourne.

Simpson, R. H., and Riehl, H. (1958). Mid-tropospheric ventilation as a constraint on hurricane development and maintenance. *Proc. Tech. Conf. Hurricanes, Miami, Fla.,* 1958 D4, pp. 1–10. Am. Meteorol. Soc., Boston, Massachusetts.

Spar, J. (1964). A survey of hurricane development. *Geofis. Intern.* **4**, 169–178.

Syōno, S. (1951). On the structure of atmospheric vortices. *J. Meteorol.* **8**, 103–110.

Ushijima, T. (1958). Outer rain bands of typhoons. *J. Meteorol. Soc. Japan* **36**, 1–10.

Wexler, H. (1945). The structure of the September, 1944 hurricane when off Cape Henry, Virginia. *Bull. Am. Meteorol. Soc.* **26**, 156–159.

Yanai, M. (1961a). A detailed analysis of typhoon formation. *J. Meteorol. Soc. Japan* **39**, 187–214.

Yanai, M. (1961b). Dynamical aspects of typhoon formation. *J. Meteorol. Soc. Japan* **39**, 282–309.

Yanai, M. (1964). Formation of tropical cyclones. *Rev. Geophys.* **2**, 367–414.

Zipser, E. J. (1964). On the thermal structure of developing tropical cyclones. Natl. Hurricane Res. Proj. Rept. No. 67, 23 pp.

16

ENERGY CONVERSIONS

IN ATMOSPHERIC CIRCULATION SYSTEMS

Studies of energy conversions and transfers are, as brought out in Chapters 1 and 2, essential to the whole problem of understanding the general circulation and the roles of disturbances of different kinds. Evaluations of energy transformations provide a useful check on the plausibility of an analysis or hypothesis. For example, such evaluations are important in comparing the results of numerical models of the general circulation, against the partitions of energy transformations determined from observations in the real atmosphere.

Depending on the problem being considered, general expressions for the conversion may be formulated either in terms of particle behavior or of the integrated effects of processes occurring within a specified region. Examples of the former were given in Sections 8.2 and 8.3, dealing with the trajectories of air particles in the jet stream, and in Chapter 13 concerning the behavior of selected elements in convective storms. We are concerned here with the overall energy conversions in entire circulation systems. For reasons that will become evident, it is necessary in this case to consider the integrated effects of processes that take place in a volume comprising as much of the whole system as is feasible.

The difficulty of studying energy conversions, particularly in extratropical disturbances, arises from three principal sources: (1) ambiguities in defining the boundary of a synoptic system; (2) uncertainties in evaluating the transfer of energy through the boundary selected; and (3) the multiplicity of forms in which the energy may be expressed. The first and second of these are of consequence because the transport of energy across the boundary may be of the same order as the energy source within the included volume unless the region selected is so large as to contain several disturbances. Concerning the third, we may note that, with particular regard to kinetic energy, its small magnitude compared with the total internal and potential energy in the atmosphere makes its changes difficult to evaluate from the simultaneous changes of other energy forms. This problem is magnified by the impossibility to consider any system as mechanically and thermodynamically closed.

A very extensive literature exists on the atmospheric-energy processes, and we make no attempt to summarize all aspects here. Present concepts have evolved largely from the principles laid down by Reynolds (1895) and Margules (1905), as well as from the classical thermodynamic formulations. General expressions governing the production, redistribution, and dissipation of energy are summarized by Haurwitz (1941), Starr (1951), Miller (1951), Van Mieghem (1951), and Saltzman (1957). Our object in this chapter is to show how these relationships may be formulated in order to make their use practicable and to apply them in ways that elucidate the mechanics of certain circulation systems.

16.1   BASIC ENERGY EQUATIONS

The energy processes in the atmosphere are governed by two fundamental relationships: the equation of mechanical energy and the first law of thermodynamics. The mechanical-energy equation (the first equation below) can be derived directly from the equation of motion combined with the equation of mass continuity. It expresses the process of conversion between kinetic energy and potential energy, since the pressure field is related to the distribution of potential energy. The second equation is an identity, in which the left side is obtained by expanding $(g\varrho \, dz/dt)$ and applying the continuity equation, and expresses the change of potential energy associated with vertical movement. The third equation is obtained from the first law of thermodynamics and the continuity equation.

Following Van Mieghem (1958) and Eliassen and Kleinschmidt (1957), the local changes of kinetic, potential, and internal energy per unit volume can be expressed by

$$\frac{\partial \varrho k}{\partial t} + \nabla_3 \cdot (\varrho k \mathbf{V} + p\mathbf{V} - \mathscr{F} \cdot \mathbf{V}) = p\nabla_3 \cdot \mathbf{V} - \mathscr{F} \cdot \nabla_3 \mathbf{V} - g\varrho w$$

$$\frac{\partial \varrho \varPhi}{\partial t} + \nabla_3 \cdot \varrho \varPhi \mathbf{V} = g\varrho w \qquad (16.1)$$

$$\frac{\partial \varrho c_v T}{\partial t} + \nabla_3 \cdot \varrho c_v T \mathbf{V} - \varrho \frac{dh}{dt} = -p\nabla_3 \cdot \mathbf{V} + \mathscr{F} \cdot \nabla_3 \mathbf{V}$$

where k, $\varPhi$, and $c_v T$ represent kinetic, potential, and internal energy per unit mass, $\mathscr{F}$ the tensor of the Navier-Stokes viscosity stresses, $\mathbf{V}$ the three-dimensional wind velocity, and dh/dt the heat added per unit mass and time by the combined processes of radiation, conduction, and condensation of water vapor. When expressed in the forms above, the changes of energy within a fixed volume of air are determined by the internal sources combined with the energy fluxes through its boundary and the effect of work done by pressure and viscous stresses at the boundary.

On the right sides of these equations all terms appear twice with opposite signs. Hence *these terms can be interpreted as the rate of conversion between the three energy forms.* Such conversions are of prime importance in meteorology (Miller, 1950).

By addition of the three equations, the local change of total energy is

$$\frac{\partial \varrho(k + \varPhi + c_v T)}{\partial t} = -\nabla_3 \cdot (k + \varPhi + c_v T)\varrho \mathbf{V} - \nabla_3 \cdot p\mathbf{V} + \nabla_3 \cdot \mathscr{F} \cdot \mathbf{V} + \varrho \frac{dh}{dt}$$
$$(16.2)$$

Alternatively, since $p\mathbf{V} = (c_p - c_v)\varrho T \mathbf{V}$,

$$\frac{\partial \varrho(k + \varPhi + c_v T)}{\partial t} = -\nabla_3 \cdot (k + \varPhi + c_p T)\varrho \mathbf{V} + \nabla_3 \cdot \mathscr{F} \cdot \mathbf{V} + \varrho \frac{dh}{dt} \qquad (16.3)$$

The second r.h.s. term represents the work done by friction per unit volume; the last term, the heat added by conduction, radiation, and liberation of latent heat.

If Eq. (16.3) is integrated over a fixed volume bounded by a surface σ, the change of the sum of total kinetic energy K, total potential energy P, and total internal energy I of the volume can be written (by use of Gauss' theorem) in the form

$$\frac{\partial (K + P + I)}{\partial t} = -\int_\sigma (k + \varPhi + c_p T)\varrho v_n \, d\sigma + \int_\sigma \mathbf{V} \cdot \mathscr{F} \cdot d\boldsymbol{\sigma} + \frac{dQ}{dt} \qquad (16.4)$$

where the last term represents the heating function, $d\boldsymbol{\sigma}$ is an area element (whose unit normal is defined positive outward from the volume), $d\sigma$ its scalar value, and v_n the outward normal wind component.

Equation (16.4) is the well-known energy equation according to which the total energy change in a fixed volume of the atmosphere is equal to the total heat added, diminished by the outflux of kinetic energy, potential energy, and enthalpy across the boundary of the volume. The second integral can be essentially considered as an expression of kinetic-energy flux. For example, when a shear exists across the boundary with a greater tangential velocity on the outer side, the action of shearing stress would be to abstract momentum from the environment and impart it to the air within the volume. When one boundary of the volume is the earth's surface, the surface stress diminishes the momentum of the atmosphere, kinetic energy being largely converted to heat.

Equation (16.4) was used in Chapter 2 for a computation of the mean meridional energy flux from the meridional distribution of the heating function, disregarding the comparatively small local change of total energy within a season. As already pointed out in that particular application, the kinetic-energy flux and the heating due to frictional dissipation could, as small quantities, be ignored in a first approximation.

In principle, Eqs. (16.1) could be used for an estimate of the conversion between total potential and internal energy on the one hand and total kinetic energy on the other, but for this purpose the equations are not practical. This depends on the well-known facts (Lorenz, 1955) that the kinetic energy of an atmospheric system is a very small quantity compared with the total potential and internal energy, and that the heating function generally has a very high numerical value compared with the rate of change of kinetic energy. Hence it is difficult to estimate the rate of conversion in this respect because the very large terms cannot be evaluated with sufficient accuracy.

More useful formulas can, however, be achieved if only the equation of mechanical energy in simplified form is used. For this purpose we assume that in any large-scale circulation system, the kinetic energy is entirely determined by the kinetic energy of horizontal motion. This is permissible for "synoptic-scale" systems wherein the vertical wind component is small, owing to the shallowness of the atmosphere (w^2/V_h^2 being typically of the order 10^{-5}), but not for small-scale disturbances. By neglecting the change of vertical kinetic energy, hydrostatic balance is assumed. Then the individual change of kinetic energy per unit mass is determined by

$$\frac{dk}{dt} = - \mathbf{V} \cdot \nabla \Phi + \alpha \mathbf{V} \cdot \frac{\partial \boldsymbol{\tau}}{\partial z} \tag{16.5}$$

Here and in the ensuing discussion $\mathbf{V}$ denotes the *horizontal* wind vector and $\nabla \Phi$ the horizontal gradient of geopotential along an isobaric surface, and $\boldsymbol{\tau}$ is the horizontal component of the eddy stress. In synoptic-scale systems the latter may be expressed by

$$\boldsymbol{\tau} = \mu \frac{\partial \mathbf{V}}{\partial z}$$

where μ is the eddy viscosity coefficient.[1]

If the total derivative in Eq. (16.5) is expanded into the local change plus the horizontal and vertical advective changes, the local change of kinetic energy per unit mass may be written as

$$\frac{\partial k}{\partial t} = -\nabla \cdot k\mathbf{V} - \frac{\partial k\omega}{\partial p} - \nabla \cdot \Phi\mathbf{V} + \Phi\nabla \cdot \mathbf{V} + \alpha\mathbf{V} \cdot \frac{\partial \boldsymbol{\tau}}{\partial z} \qquad (16.6)$$

Further, substituting $-\Phi \, \partial\omega/\partial p$ for the next-to-last term and observing that $\partial\Phi/\partial p = -\alpha$, we get the expression derived by White and Saltzman (1956):

$$\frac{\partial k}{\partial t} = -\nabla \cdot (k + \Phi)\mathbf{V} - \frac{\partial (k + \Phi)\omega}{\partial p} - \alpha\omega + \alpha\mathbf{V} \cdot \frac{\partial \boldsymbol{\tau}}{\partial z} \qquad (16.7)$$

Since in Eq. (16.7) Φ is (except for the layer close to the earth's surface) generally much larger than k, one might conclude that the fluxes of k could be disregarded in comparison with the much larger fluxes of Φ. That such a conclusion is not permissible will be shown in the following discussion.

Let A be the area limited by a boundary whose total length is L in a fixed isobaric surface. We denote the *boundary* mean values by circumflex accents, deviations from the boundary mean by primes, and the normal outward wind component across L by v_n. Then by use of Gauss' theorem, the third r.h.s. term of Eq. (16.6), when integrated over the area, may be written as

$$-\int_A \nabla \cdot \Phi\mathbf{V} \, dA = -\widehat{\Phi v_n}L = -L(\hat{\Phi}\hat{v}_n + \widehat{\Phi'v'_n})$$

Similarly, if brackets denote the *areal* mean values and double primes the deviations therefrom, the next-to-last term of Eq. (16.6) may be trans-

[1] The use of this simple expression for the frictional stress is again justified by the shallowness of the atmosphere, the vertical shear being in general very much larger than the horizontal shear, and the horizontal variations of the vertical velocity negligible compared with either. This assumption is, of course, not valid in systems (such as convective clouds) with very strong local variations of all wind components.

formed by the sequence

$$\int_A \Phi \nabla \cdot \mathbf{V} \, dA = A([\Phi][\nabla \cdot \mathbf{V}] + [\Phi'' \nabla \cdot \mathbf{V}''])$$

$$= L[\Phi]\hat{v}_n - A\left[\Phi'' \frac{\partial \omega''}{\partial p}\right]$$

$$= L[\Phi]\hat{v}_n - A\left[\frac{\partial \Phi'' \omega''}{\partial p} - \omega'' \frac{\partial \Phi''}{\partial p}\right]$$

$$= L[\Phi]\hat{v}_n - A\left[\frac{\partial \Phi'' \omega''}{\partial p}\right] - A[\omega'' \alpha'']$$

Upon substitution of the final expressions above for the appropriate terms in the area-integrated Eq. (16.6), we get

$$\int_A \frac{\partial k}{\partial t} \, dA = -\, \widehat{Lkv_n} - A\left[\frac{\partial k \omega}{\partial p}\right] + L([\Phi] - \hat{\Phi})\hat{v}_n - \widehat{L\Phi' v_n'} - A\left[\frac{\partial \Phi'' \omega''}{\partial p}\right]$$

$$- A[\omega'' \alpha''] + \int_A \alpha \mathbf{V} \cdot \frac{\partial \boldsymbol{\tau}}{\partial z} \, dA \qquad (16.6')$$

This expression may now be integrated between the isobaric surfaces p_0 (approximately at the earth's surface) and p_H corresponding to an upper level, observing that $(p_0 - p_H)/g$ is mass per unit area and substituting RT/p for α. Then, if ω is assumed to vanish at the earth's surface, we get the final expression (Palmén, 1960):

$$\frac{\partial K}{\partial t} = -\underbrace{\frac{L}{g} \int_{p_H}^{p_0} \widehat{kv_n} \, dp}_{(1)} + \underbrace{\frac{A}{g}[k\omega]_{p_H}}_{(2)} - \underbrace{\frac{L}{g} \int_{p_H}^{p_0} (\hat{\Phi} - [\Phi])\hat{v}_n \, dp}_{(3)}$$

$$- \underbrace{\frac{L}{g} \int_{p_H}^{p_0} \widehat{\Phi' v_n'} \, dp}_{(4)} + \underbrace{\frac{A}{g} [\Phi'' \omega'']_{p_H}}_{(5)} - \underbrace{\frac{RA}{g} \int_{p_H}^{p_0} \frac{[T'' \omega'']}{p} \, dp}_{(6)} + \underbrace{D}_{(7)}$$

$$(16.8)$$

where D represents the frictional dissipation of horizontal kinetic energy. Alternatively, from Eq. (16.5),

$$\frac{\partial K}{\partial t} = -\frac{L}{g} \int_{p_H}^{p_0} \widehat{kv_n} \, dp + \frac{A}{g}[k\omega]_{p_H} - \frac{A}{g} \int_{p_H}^{p_0} [\mathbf{V} \cdot \nabla \Phi] \, dp + D \quad (16.9)$$

where K again denotes the total kinetic energy in the prescribed volume in which the total mass remains constant, owing to the assumptions of a fixed lateral boundary and fixed values of p_0 and p_H. In Eq. (16.8) the numbered terms (1) to (5) represent the net horizontal and vertical outflux of kinetic and potential energy through the boundary of the region, whereas term (6) represents the conversion between available potential energy and kinetic energy inside the boundary. The quantities $(\hat{\Phi} - [\Phi])$, Φ' and Φ'' are not very large and are fully comparable with k, demonstrating that the flux of k cannot be neglected against the flux of Φ even though generally $\Phi \gg k$.

The sixth (kinetic-energy generating) term gives a positive contribution if T'' and ω'' are negatively correlated. That is, *kinetic energy is generated if the ascending air is in general warmer than the descending air in the region.* For a closed system, this is equivalent to a decrease of the available potential energy. For a dynamically open system, the sum of the terms (3) through (6) in Eq. (16.8) is equal to the work done upon or by the air in moving with a component in the direction of or against the horizontal pressure-gradient force. This work, expressed by the term in $- \mathbf{V} \cdot \nabla \Phi$ of Eq. (16.9) for an isobaric coordinate system, is seen to be equivalent by a comparison of the two equations. In both equations the two first r.h.s. terms represent the outfluxes of existing kinetic energy through the vertical walls and the upper-level isobaric surface bounding the volume.

For the dissipation D, a simple approximation can be derived. Integrating the last term of Eq. (16.5) over the area A and from the surface to the height H corresponding to pressure p_H, we get

$$\int_0^H \int_A \left(\alpha \mathbf{V} \cdot \frac{\partial \boldsymbol{\tau}}{\partial z} \right) \varrho \, dA \, dz = \int_0^H \int_A \frac{\partial \mathbf{V} \cdot \boldsymbol{\tau}}{\partial z} \, dA \, dz - \int_0^H \int_A \boldsymbol{\tau} \cdot \frac{\partial \mathbf{V}}{\partial z} \, dA \, dz$$

or upon introduction of the coefficient of eddy viscosity,

$$D = A \left[\mu \mathbf{V} \cdot \frac{\partial \mathbf{V}}{\partial z} \right]_H - A [\mathbf{V}_0 \cdot \boldsymbol{\tau}_0] - A \int_0^H \left[\mu \left(\frac{\partial \mathbf{V}}{\partial z} \right)^2 \right] dz \qquad (16.10)$$

The first r.h.s. term represents the work done by eddy stresses at the upper isobaric surface, and it vanishes if the integration is either performed through the entire atmosphere or (for practical purposes) up to the level of maximum wind speed. If it does vanish, the remaining two terms express the dissipation simply as the sum of the dissipation by friction at the earth's surface and as the internal dissipation by shearing stresses within the atmosphere. A considerable part of this latter dissipation occurs in the "friction layer" (Holopainen, 1965; Kung, 1966). The ratio between the rates of dissipation in the friction layer (including the surface friction) and in the "free atmosphere" is not very well known. According to an estimate by Brunt (1944), the total mean rate of dissipation amounts to about 5 W/m², of which he apportioned about 3 W/m² to the friction layer between the ground and 1 km and about 2 W/m² to dissipation in the free atmosphere (taken as 1 to 10 km). This still very crude estimate of the total dissipation is in satisfactory agreement with the generation of kinetic energy computed from the vertical energy flux (Fig. 2.13).

16.2　APPLICATION TO THE GENERAL CIRCULATION

Equations (16.8) and (16.9) may be applied to the general circulation in order to achieve a mean budget of kinetic energy. If we consider only the zonal mean values for selected time periods, as in Chapter 1, we can write the equations for kinetic energy of the mean zonal motion (k_λ) and of the mean meridional motion (k_φ) in the form

$$\left.\frac{dk_\lambda}{dt}\right|_F = \bar{u}\bar{v}\left(f + \frac{\bar{u}}{a}\tan\varphi\right) \tag{16.11a}$$

$$\left.\frac{dk_\varphi}{dt}\right|_F = -\bar{u}\bar{v}\left(f + \frac{\bar{u}}{a}\tan\varphi\right) - \bar{v}\frac{\partial\bar{\Phi}}{a\,\partial\varphi} \tag{16.11b}$$

if friction is neglected (the assumption of frictionless motion being denoted by subscript F). In Eq. (16.11b) the last term corresponds to $f\bar{v}\bar{u}_g$; since the zonal wind above the friction layer is predominantly almost in gradient balance, this term is nearly equal and opposite to the first term, and the generation of mean meridional kinetic energy is generally small. A corresponding term $(-f\bar{u}\bar{v}_g)$ does not appear in Eq. (16.11a) because, when averaged all around the hemisphere, $\partial\Phi/\partial\lambda = 0$. Thus the generation of kinetic energy of the zonally averaged motions, $k_m \equiv (\bar{u}^2 + \bar{v}^2)/2$, appears almost entirely in the zonal wind component, although it is derived from the mean meridional motion and corresponds to the work done by the mean meridional pressure forces. This is simply an expression of the fact that a change of kinetic energy (in this case, of the vector mean wind) can be effected only by the component of force along the wind direction.

Stated in another way, the appearance of identical terms with opposite signs in Eqs. (16.11a) and (16.11b) indicates a transformation between mean zonal and mean meridional kinetic energy. Thus, if integrated over a latitude belt and through the whole depth of the atmosphere, the total generation of kinetic energy of mean motion is given by

$$\left.\frac{\partial K_m}{\partial t}\right|_{CF} = -\frac{2\pi a^2}{g}\int_{p_H}^{p_0}\int_{\varphi_1}^{\varphi_2}\bar{v}\frac{\partial\bar{\Phi}}{a\,\partial\varphi}\cos\varphi\,d\varphi\,dp \tag{16.12a}$$

or, by interpretation of the next-to-last term of Eq. (16.8),

$$\left.\frac{\partial K_m}{\partial t}\right|_{CF} = -\frac{2\pi a^2 R}{g} \int_{p_H}^{p_0} \int_{\varphi_1}^{\varphi_2} \frac{\bar{T}\bar{\omega}}{p} \cos \varphi \, d\varphi \, dp \qquad (16.12b)$$

if the flux terms are neglected (subscript *CF* denoting that the partial derivative pertains to a closed system and does not include the effect of friction). Neglect of the flux terms is permissible if the integration is extended over the total regions of the mean Hadley or Ferrel cells, at the boundaries of which $\bar{v}$ is assumed to vanish, and if the "eddy terms" are neglected. Under this assumption, the mean frictional dissipation in the same cells would on the average counterbalance the generation of K_m. However, it will be recalled (Section 1.5) that most of the kinetic energy produced in the Hadley circulation is actually exported (by eddies) across its poleward bounds, and that within the latitudes of the Ferrel circulation there is a large conversion from eddy to zonal kinetic energy.

TABLE 16.1

PRODUCTION OF KINETIC ENERGY BY THE MEAN MERIDIONAL CIRCULATIONS IN THE NORTHERN HEMISPHERE[a]

Lat. belt	0–30° N	30–90° N	Whole hemisphere
Winter	30	−11	19
Summer	3	− 2	1

[a] After Holopainen (1965). Units are 10^{10} kW.

Independent of this limitation, it is of interest to determine the values of the integrals in Eqs. (16.12). Holopainen has made (1965) separate evaluations for the winter (December–February) and summer (June–August) seasons and for the regions of the hemisphere equatorward of and poleward of 30°N.[2] From evaluations of the r.h.s. of Eq. (16.12a), Holopainen obtained the generations of mean kinetic energy presented in Table 16.1.

[2] Note that by choice of fixed geographical boundaries, the generations in the two regions do not (especially in summer) correspond strictly to those in the Hadley cell and outside it, since the boundaries of the mean meridional cells show a considerable seasonal variation.

The rate of production of kinetic energy of mean motion is, as would be expected, large in the southern region and dominated by the strong Hadley circulation in winter. Considering the whole hemisphere, the negative contribution by the Ferrel circulation reduces the total rate to the more moderate value 19×10^{10} kW. In summer, the rate of kinetic-energy production by the mean meridional circulations is small in both belts, with a positive value in the southern part and a negative value in the northern part.

In Section 2.6 an attempt was made to evaluate the average rate of generation of kinetic energy during the winter season in the extratropical region. Here this total rate, including the negative rate of the Ferrel cell, was estimated from the vertical large-scale heat flux to be 5.3 W/m². For the whole region considered, this amounts to a total generation rate of somewhat more than 60×10^{10} kW. No similar estimate was made for the tropical region (roughly between the Equator and 30°N). Since here the Hadley cell contributes about 30×10^{10} kW and tropical disturbances of different types in all probability give a positive contribution, *the total rate of conversion between available potential energy and kinetic energy in the whole Northern Hemisphere in winter may be estimated at about* 100×10^{10} kW, a value that is reconverted to heat (or internal and potential energy) as a result of frictional dissipation according to Eq. (16.1). We have made no corresponding estimate in this manner for summer. However, owing to the considerably weaker circulation and disturbances, a much smaller kinetic energy production would be expected. This is borne out by the available calculations (Section 16.5). The seasonal variation of kinetic-energy production is likely much smaller in the Southern Hemisphere, where the synoptic disturbances as well as the mean meridional circulation of the tropics are also pronounced in summer.

On a global scale, the generation rate must on the average equal the dissipation rate. Hence an estimate of the latter would give an independent value of the rate of generation of kinetic energy. Such estimates are necessarily very crude, but appraisals by Sverdrup (1917) and Brunt (1944) are in satisfactory agreement with the hemispheric value of 100×10^{10} kW given above.

Owing to lack of adequate data, it is not yet possible to present more accurate values for the mean rate of conversion between available potential energy and kinetic energy on a global scale. It may, however, be of interest to use the given values for an estimate of the efficiency of the atmosphere as a thermodynamic engine. If the solar constant is taken as 1.94 ly/min and this is reduced by the estimated average albedo of about 39%, the radiation absorbed amounts to 118×10^{12} kW. Compared with this, a mean global rate of energy conversion of about 200×10^{10} kW gives an efficiency of roughly 1.7%.

It is somewhat questionable how the concept of efficiency should be defined. In our definition we have considered only the conversion associated with large-scale atmospheric processes. The problem has been treated in different ways by several authors (see, e.g., Haurwitz, 1941; Brunt, 1944; Defant and Defant, 1958; Hess, 1959), and about 2% is generally considered an upper limit.[3] The small value of the rate of conversion compared with the total incoming solar radiation, and the small value of the kinetic energy compared with potential and internal energy, motivated the exclusion of kinetic energy from the discussion of the total atmospheric-energy budget in Chapter 2.

16.3   GENERATION OF KINETIC ENERGY IN EXTRATROPICAL DISTURBANCES

Any application of Eqs. (16.8) or (16.9) to extratropical disturbances is rather complicated because these are mostly very far from closed systems and all "flux terms" must be considered. In the simple case of an axially symmetric circular vortex, wherein the radial velocity v_r disappears at the distance r_1 from the central axis and the vertical velocity vanishes at pressures p_0 and p_H, the production of kinetic energy (without considering dissipation) within the volume is given by the next-to-last term of Eq. (16.6). Under the assumptions above, all other terms disappear and $\partial K/\partial t = dK/dt$ when integrated over the volume. The production can then be written as

$$\left.\frac{dK}{dt}\right|_{CF} = \frac{2\pi}{g} \int_{p_H}^{p_0} \int_0^{r_1} (\Phi \nabla \cdot \mathbf{V}) r \, dr \, dp = \frac{2\pi}{g} \int_{p_H}^{p_0} \int_0^{r_1} \Phi \frac{\partial r v_r}{\partial r} dr \, dp \quad (16.13a)$$

Since the system is considered closed at r_1 and p_H, the right-hand term may also be written as

$$\left.\frac{dK}{dt}\right|_{CF} = -\frac{2\pi}{g} \int_{p_H}^{p_0} \int_0^{r_1} v_r \frac{\partial \Phi}{\partial r} r \, dr \, dp \quad (16.13b)$$

[3] Based on the conception by Carnot, the efficiency of a reversible heat engine is $(Q_1 - Q_2)/Q_1$, where Q_1 is the quantity of heat taken in and Q_2 the quantity rejected. Since the heat per unit mass is proportional to the temperature, this may be written as $(T_1 - T_2)/T_1$. In the atmosphere, the temperatures of the ascending branches of the circulation (in which heat is mainly realized) are on the average not greatly different from the effective radiation temperature at which heat is rejected. Hence the portion of the heat converted to mechanical energy is small compared with the total incoming radiation.

which could also be obtained directly from Eq. (16.5). Also, the l.h.s. of Eq. (16.13a) may be written as

$$\frac{dK}{dt}\Big|_{CF} = -\frac{2\pi R}{g} \int_{p_H}^{p_0} \int_0^{r_1} \frac{T\omega}{p} r\, dr\, dp \qquad (16.13c)$$

from Eq. (16.8), in which (with a closed system) all other terms drop out if friction is neglected. In the expressions above, Φ, v_r, T, and ω are functions of r and p only.

It is useful to consider the kinetic-energy production in the three different forms in Eqs. (16.13a–c) because each of these has a special interpretation with regard to the properties of a cyclone or anticyclone. The interrelations among these properties can be readily seen because all the expressions are completely equivalent.

The processes are illustrated schematically in Fig. 16.1. In the case of a mass circulation with inflow at high pressure (i.e., in lower levels) and outflow at lower pressure, which is connected with the characteristic vertical circulation in a cyclone, it follows from Eq. (16.13c) that *kinetic energy is produced if the rising core of the region is warmer than the outer parts.* This characteristic is illustrated by Fig. 16.1a, in which profiles of the isobaric surfaces are drawn with vertical separations that correspond to the temperature distribution and which consider the hydrostatic relationship. In terms of Eq. (16.13b) it can be seen that the horizontal branches connected with the vertical circulation are toward lower pressure (or toward lower geopotential on an isobaric surface) in both upper and lower levels, an arrangement suitable for generating kinetic energy according to Eq. (16.5). The interpretation in terms of Eq. (16.13a) is also evident, since the divergence associated with these horizontal branches is in regions of large geopotential, and convergence is in regions of smaller geopotential. The importance of this correlation between Φ and $\nabla \cdot \mathbf{V}$ was emphasized by Starr (1948, 1951).[4] In the opposite case of a cold core with ascent in it (Fig. 16.1b), it follows

---

[4] Starr's clear and concise discussion of energy principles is recommended to those not familiar with it. A minor flaw is his interpretation that "the sources of kinetic energy are to be found in the regions of horizontal divergence" and that "cyclonic areas...act as sinks." This is true in one sense that he mentions, namely, that part of the kinetic energy produced in anticyclonic regions flows out of them into cyclonic regions. However, kinetic energy is also largely produced in the latter. In the example considered here, at a given isobaric surface the area-averaged divergence disappears and $[\Phi \nabla \cdot \mathbf{V}]$ in Eq. (16.13a) may be replaced by $[\Phi'' \nabla \cdot \mathbf{V}'']$. The association of convergence with low geopotential (negative Φ'') as well as that of divergence with high geopotential express a generation of kinetic energy due to the flow of air between regions of high and low geopotential.

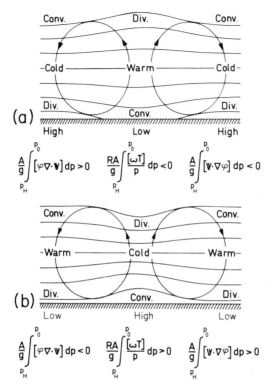

$$\frac{A}{g}\int_{P_H}^{P_0}[\varphi\nabla\cdot\mathbf{v}]\,dp > 0 \qquad \frac{RA}{g}\int_{P_H}^{P_0}\frac{[\omega T]}{p}\,dp < 0 \qquad \frac{A}{g}\int_{P_H}^{P_0}[\mathbf{v}\cdot\nabla\varphi]\,dp < 0$$

$$\frac{A}{g}\int_{P_H}^{P_0}[\varphi\nabla\cdot\mathbf{v}]\,dp < 0 \qquad \frac{RA}{g}\int_{P_H}^{P_0}\frac{[\omega T]}{p}\,dp > 0 \qquad \frac{A}{g}\int_{P_H}^{P_0}[\mathbf{v}\cdot\nabla\varphi]\,dp > 0$$

Fig. 16.1   Examples of circulations in the vertical plane which (a) produce and (b) destroy kinetic energy. The ordinate is height, and the thin solid lines are profiles of isobaric surfaces. In a stable atmosphere, the vertical motions would diminish the available potential energy in case (a) and increase it in case (b). In either case, a steady potential energy field could be maintained if heat were added in the ascending branches and abstracted from the descending branches.

that kinetic energy is consumed or reconverted into available potential energy.

It also follows that *with a cold-core system, kinetic energy is produced in the case of upper radial inflow and lower radial outflow, with sinking in the cold core.* The latter case represents the energy production if a low-level anticyclone weakens with height or changes to a high-level cyclone (as is characteristic in the cold-air tongues west of sea-level cyclones of extratropical latitudes). Raethjen (1964) has described the cyclone in terms of a couplet, consisting of a system of this type adjoining and connected to a system of the kind in Fig. 16.1a, the whole being superposed on a basic current with vertical shear. A sloping "cyclogenetic axis," corresponding to the tilt observed in synoptic systems, connects the convergent lower region beneath the rising branch with the convergent upper region over-lying the descending branch. Raethjen also stresses the connection between the kinetic-energy generation by the cyclone and the vertical heat transfer implied by this arrangement.

In a system of the type in Fig. 16.1a, the essential part of the energy generated appears as kinetic energy of the circulation in surfaces normal to the radial plane, as a result of the transformation between radial (k_r) and tangential (k_θ) kinetic energy. The first r.h.s. terms in

$$\frac{dk_\theta}{dt}\bigg|_F = v_r v_\theta \left(f + \frac{v_\theta}{r} \right) \quad \text{and} \quad \frac{dk_r}{dt}\bigg|_F = - v_r v_\theta \left(f + \frac{v_\theta}{r} \right) - v_r \frac{\partial \Phi}{\partial r}$$

(16.14)

determine this transformation.[5]

It should be stressed that the result in Eq. (16.13c), concerning the cor-relation between T and ω and its influence on energy production, follows entirely from application of the continuity equation. It is therefore an expression of the work done by the horizontal-pressure forces expressed by the integral term in Eq. (16.13b). The scheme in Fig. 16.1a may be applied to tropical cyclones (Section 15.6). In some cases the scheme in Fig. 16.1b may be applied to cold extratropical anticyclones or even to upper cold lows with a weak surface low (Section 10.2).

To elucidate how the generation term (6) in Eq. (16.8) acts in extra-tropical cyclones, we shall use the synoptic case investigated by Danard (1964). Figure 16.2b shows a typical wave cyclone over the eastern United States, and Fig. 16.2a shows the corresponding 500-mb chart with the prin-cipal frontal zone at that level. Kinematically computed vertical motions in the middle troposphere (Fig. 16.2c) indicate maximum ascent nearly over the surface cyclone and its frontal system. Comparison with Fig. 16.2a suggests a close correspondence of the general region of ascent with the warm air east of the trough and of sinking air with the cold-air tongue.

Figure 16.3 illustrates the frontal structure, temperature field, and observ-ed winds in a west-east section along approximately 40°N. Figure 16.4 repeats the frontal layer in the same section and also gives the ω-distri-bution based on Danard's computations from vertical integration of the

---

[5] The interpretation here is the same as that following Eqs. (16.11), and is valid to the extent that the circulation is nearly in gradient-wind equilibrium. In a synoptic-scale system in which the divergence is small and $v_\theta \gg v_r$ (for a symmetrical vortex), this trans-formation predominates. In a small-scale system such as a cumulus or cumulonimbus cloud, such a generalization obviously does not hold.

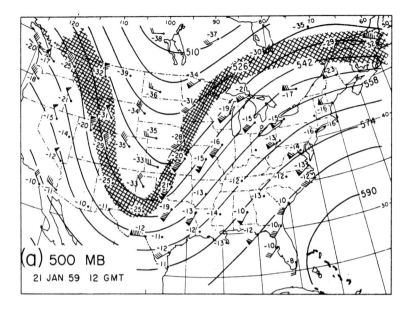

FIG. 16.2a

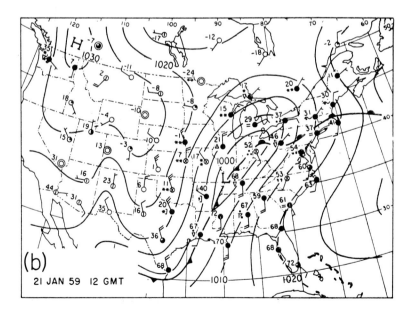

FIG. 16.2b

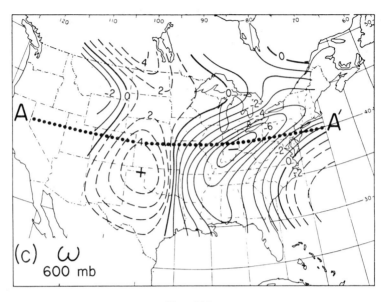

FIG. 16.2c

FIG. 16.2   Analyses at 12 GCT Jan. 21, 1959, by Danard (1964). (a) 500-mb contours (decameters) and front; (b) surface front and sea-level isobars; (c) vertical motion (μb/sec) at 600 mb.

isobaric divergence field. In the upper part of Fig. 16.4 is the temperature profile at the 500-mb surface and the observed (somewhat smoothed) rate of precipitation at the ground, both along the line of the cross section. As can be seen, the precipitation rate is closely correlated with the intensity of ascending motion, strongly supporting the reality of the computed ω-field. These figures also show very clearly the negative correlation between ω and T, indicating a strong rate of conversion from available potential energy to kinetic energy. This synoptic case may be considered characteristic of polar-front cyclones during their periods of strongest development, when they are typified by a pronounced space lag between the pressure and thermal troughs.

In his investigation of the energy development in the preceding synoptic case, Danard did not use the ω-field computed from divergence, but rather the ageostrophic winds achieved by solving the "diagnostic ω-equation" with addition of the latent heat released. Essentially, this means a computation of the work done by horizontal pressure forces according to the next-to-last term of Eq. (16.9). Since the system considered was an open system, the work term in this computation represents the sum of terms (3) through

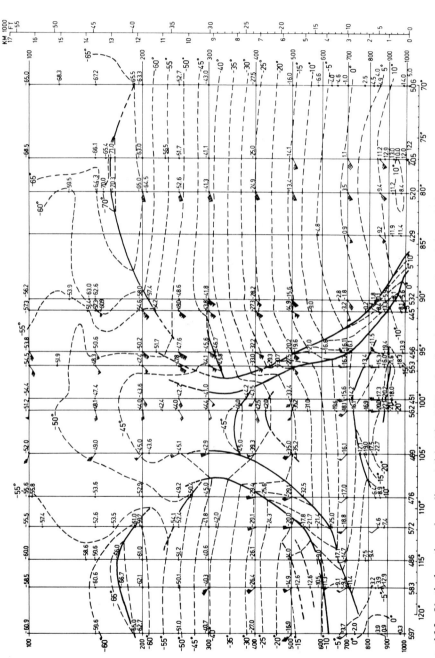

FIG. 16.3 Vertical section along line AA' in Fig. 16.2c, showing isotherms, frontal boundaries and tropopauses, and winds plotted relative to north at top of figure.

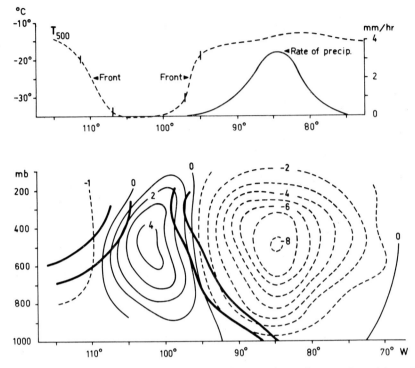

FIG. 16.4   Bottom diagram corresponds to Fig. 16.3; heavy lines are frontal boundaries and thin lines are isopleths of vertical velocity dp/dt (μb/sec). At top, profile of 500-mb temperature, and rate of precipitation (mm/hr).

(6) in Eq. (16.8). Generally, however, the last of these terms is the dominating one if the computation is extended over a sufficiently large region.

In a similar synoptic case discussed in Section 12.6, Palmén and Holopainen (1962) computed all the individual terms in Eq. (16.8). Also in this case, the relationship between ω and T was similar to that in Danard's case, showing clearly the process of energy conversion inside the synoptic system. Although the sources of error involved in such a computation must be acknowledged, the estimated values of the different terms (except D) in Eq. (16.8) are presented in Table 16.2 to show their magnitudes in a real synoptic case. The area used in the computation (34 to 49°N and 79 to 119°W in Fig. 12.17a) was 5.6×10^6 km², and the 300-mb surface was used as the upper boundary p_H. Because of this selection of the upper boundary, the numerical values of the vertical flux and work done upon that boundary were relatively large. In other cases of similar developing cyclones, many of the terms could be very different. This is especially true

TABLE 16.2

VALUES OF THE SIX FIRST RIGHT-HAND TERMS IN EQ. (16.8), COMPUTED FOR A SYNOPTIC
CASE REPRESENTING A DEVELOPING CYCLONE[a]

Term	Character of term	Value[b]
1	Kinetic energy flux through lateral boundary	8.8
2	Kinetic energy flux through 300-mb surface	−9.8
3	Mean flux of potential energy through lateral boundary[c]	4.3
4	Eddy flux of potential energy through lateral boundary[c]	1.1
5	Eddy flux of potential energy through 300-mb surface[c]	− 5.9
6	Internal generation of kinetic energy	11.7
1–6	Sum ($\partial K/\partial t$, not considering friction)	10.2

[a] After Palmén and Holopainen (1962). Units are 10^{10} kW.

[b] In the values for the flux, a negative sign means that energy is removed through the bounding walls or top surface, and a positive sign indicates that the flux contributes to increase the energy within the volume.

[c] These fluxes, although expressed in terms of potential energy, express the work done by the surrounding atmosphere in changing the energy within the volume. In the basic expression for kinetic energy generation, $dk/dt = -\mathbf{V}\cdot\nabla\Phi$, this can also be interpreted as an "advection" of potential energy.

concerning the fluxes of kinetic energy, terms (1) and (2) in Eq. (16.8), and the work done by the surrounding atmosphere, terms (3) through (5), whereas the sixth term, representing the rate of energy conversion inside the region in developing cyclones, mostly has a large positive value if the area is sufficiently large.

Other studies of the same problem have been made by Palmén (1958) for the Hazel cyclone and by Väisänen (1961) for a cyclone over England. Table 16.3 gives a summary of the kinetic-energy production per unit time in these four synoptic situations.[6] The first of these cases was exceptional,

[6] Phillips (1949) has discussed the work done by subsiding cold-air masses upon their surroundings, expressed by the term $-V_3\cdot p\mathbf{V}$ in Eq. (16.2). A subsiding cold dome expands in lower levels (at higher pressure) and contracts at lower pressure, thus losing energy in its interior and transferring it into the environment by work against the pressure forces. For a situation involving a strong cyclone with subsiding cold air, Phillips evaluated the mean work across the polar-front surface at 30×10^{10} kW. For reasons discussed by him, this value is not equivalent to the kinetic-energy production, being rather an overestimate. The preference shown by cold-air domes to move equatorward, while at the same time subsiding and releasing potential energy, has been discussed theoretically by Rossby (1949).

TABLE 16.3

Generation of Kinetic Energy in Selected Cyclones

Case	Area (m^2)	Production (kW)	Production (W/m^2)
Palmén (1958)	3.7×10^{12}	19×10^{10}	51
Väisänen (1961)	1.8	3	17
Palmén and Holopainen (1962)	5.6	12	21
Danard (1964)	10.4	17	16

being a relatively rare combination of a tropical and extratropical disturbance. In the three other cases the production rate varied between 16 and 21 W/m^2, which may be considered characteristic of polar-front cyclones in winter during their time of strong development. In Väisänen's example, the cyclone was comparable in size to the last two listed, but the area of computation was restricted essentially to the central portion. It seems proper to consider a cyclone in the broad sense as including the neighboring anticyclonic regions and thus as a couplet including both the rising and descending branches, as in Fig. 16.2. Then the size may be taken as about 10×10^{12} m^2 and the generation as about 18 W/m^2, giving an energy production by one cyclone of 18×10^{10} kW.

From the earlier estimate of 5.3 W/m^2 for the overall winter energy conversion rate north of 32°N, the overall production rate for this whole region is 64×10^{10} kW. Allowing for the negative production by the Ferrel cell (Table 16.1), the production rate due to eddies within this cap would be about 75×10^{10} kW. Comparing with the preceding estimate for one cyclone, we conclude that *only about four or five developing cyclones of typical size and intensity are required to account for the entire kinetic energy generated in the extratropical cap of the hemisphere.* This conclusion is harmonious with that in Section 10.6, where it was estimated that a small number of well-developed polar-front disturbances could accomplish the required meridional and vertical heat exchange in middle latitudes. While the examples in Table 16.3 may be more vigorous than average cyclones, this comparison does not take account of the contributions by weaker disturbances in the roughly two-thirds of the polar cap outside the strong disturbances.

The kinetic energy generated in a cyclonic region does not necessarily appear as an increase of kinetic energy in the same region, but often largely as a strong net export of energy from the region. This is true also if the local frictional dissipation is considered. In the Hazel case mentioned above, the total horizontal export of kinetic energy from the region was almost equal to the generation of energy inside the same region. Vertical profiles of the generation and export of kinetic energy for this case are shown in Fig. 16.5. These have been subjectively modified below 850 mb,

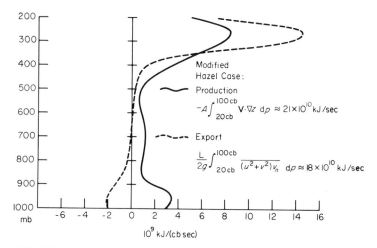

Fig. 16.5   Distribution with pressure of the production (solid curve) and export (dashed curve) of kinetic energy within or from the area of Fig. 16.6 between 30° and 45° N, and from 100° W to the east coast. Below 850 mb, Palmén's curves (1958) have been subjectively modified. In this case, the production (and export) were dominant in the eastern part of the area, which partly encompassed a vigorous surface cyclone; the numerical values will vary from case to case, depending upon the synoptic case and the computation area chosen.

where it is likely that both generation and import of kinetic energy were underestimated in the original computations. The curves are characteristic, although the magnitudes exceed the usual values for a cyclone. The main features are generation in both the lower inflow and upper outflow layers[7] together with an import of kinetic energy in the lower troposphere and an export in the upper troposphere. These features would be in accord with the central portion of the schematic Fig. 16.1a if the system were considered

[7] Kung (1966) and others have shown by computations over a large area and time that this typifies the overall atmospheric behavior, and that the mid-tropospheric generation is small. Kung found that the dissipation below 700 mb was nearly equal to the generation, suggesting that the large kinetic energy generated in higher levels is also essentially annihilated by processes in the upper levels.

as open, with the descending branches of the circulation removed a considerable distance from the region of the cyclone.

A large part of the kinetic-energy generation in the Hazel case took place in upper levels, where the export of kinetic energy was also concentrated. The cross-contour flow at 300 mb (Fig. 16.6) was exceptionally

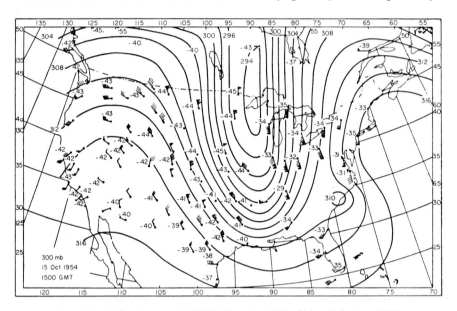

FIG. 16.6   300-mb chart, 15 GCT Oct. 15, 1954. (After Palmén, 1958.)

pronounced in this case. As noted in Section 9.7, the air in the corresponding region east of the upper trough could be identified with that rising from lower levels in the cyclone farther southeast. This example demonstrates graphically how the kinetic energy produced by a disturbance and exported from its vicinity is fed into the jet stream, and suggests how the kinetic energy of the jet stream can be maintained against dissipative influences by the cumulative contributions of such disturbances around the globe.

In the surroundings of cyclonic disturbances, obviously only a part of the kinetic energy generated by them is dissipated by frictional forces, whereas another part is reconverted into potential and internal energy as a result of "indirect" circulations with descending warmer air and ascending colder air (positive correlation between ω and T). A part of this process probably occurs in the subtropical high-pressure belt south of the main polar front. It may also be noted that the kinetic-energy budgets of disturbances must vary considerably during their lifetimes. In the early stages of some cyclones, there is evidently an import of kinetic energy into the locality of development, although considering the whole lifetime there must be an export.

Some interesting features concerning the generation of kinetic energy in polar-front disturbances may be illustrated as follows: Figure 16.7 shows

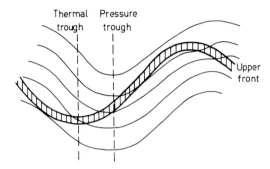

FIG. 16.7 Schematic picture of phase lag between pressure and thermal troughs in a developing perturbation associated with upward and poleward flux of heat and generation of kinetic energy.

schematically the characteristic phase lag between a pressure trough and a thermal trough in the middle troposphere, the thermal trough being identified with a southward extension of the polar front from its mean position. The strongest ascent ahead of the trough often occurs in the warm air at a somewhat higher latitude than the strongest descent behind it, which typically takes place in the southern parts of the cold air (Chapter 10). This type of three-dimensional air flow gives: (1) northward flux of heat due to positive correlation between v and T; (2) upward flux of heat due to negative correlation between ω and T; and (3) positive conversion between available potential energy and kinetic energy because of the same negative correlation. This latter energy production is partly a result of direct circulations in a *zonal* plane.

Let us now consider the processes in a *mean meridional* plane. In Fig. 16.8 the circle represents a fixed parallel at latitude φ; and the sinusoidal curve, a "polar front" with five thermal troughs and ridges. The plus and minus signs mark the regions of strongest ascending and descending motion, respectively. As marked in the figure, the average position of regions with strong ascent is north of the fixed latitude and the average position of regions with strongest descent is south of the same latitude. Depending on the method of averaging the vertical motions around the hemisphere,

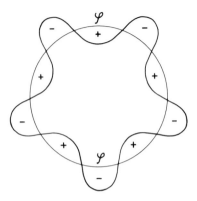

FIG. 16.8   Scheme of the regions of strongest ascending (+) and descending(−) motions in relation to a circumpolar front with five disturbances.

the mean circulation in a vertical section across the flow will appear in one of the opposite forms in Fig. 16.9. *Whereas averaging with respect to a fixed latitude results in the Ferrel circulation, which is indirect and kinetic-energy consuming because on the average the temperature decreases northward, averaging relative to the polar-front zone would result in a direct circulation that converts available potential energy to kinetic energy.* This proposition has been demonstrated by Riehl and Fultz (1957) by means of corresponding computations of the motions in a "rotating dishpan," which simulated the extratropical atmospheric circulation.

   The schemes in Fig. 16.9 show how, at the same time, the kinetic energy of the mean *zonal* motion is reduced by the reversed mean meridional

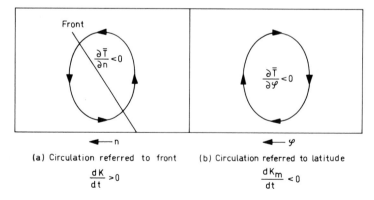

(a) Circulation referred to front   (b) Circulation referred to latitude

FIG. 16.9   The sense of the "meridional" circulation when referred to a meandering frontal zone (a) and to a fixed latitude (b).

circulation, whereas the *total* kinetic energy may increase. The difference lies, of course, in the facts that the generation of kinetic energy in extra-tropical storms occurs largely through vertical solenoidal circulations in planes oriented more or less zonally (see Figs. 16.2 and 16.4), and that this component of the circulation is taken into account by the method of averaging for Fig. 16.9a, but not for Fig. 16.9b. The energy generated is largely in the form of meridional kinetic energy, but in its total effect the mean toroidal motion about the polar-front system described by Fig. 16.9a contributes to maintain the circumpolar jet stream. As noted in Section 1.5, there is also a significant transfer of zonal kinetic energy from the tropics, especially in winter, which helps to counter the dissipation in extratropical latitudes.

As Thompson (1959, 1961) has demonstrated theoretically, the generation of kinetic energy in disturbances depends fundamentally upon a phase lag between the pressure trough and the thermal trough, as in Fig. 16.7. Alternatively, Thompson writes the condition for wave instability in a form showing that the eddy kinetic-energy generation is proportional to the poleward heat transfer, indicating that these effects are a maximum when the thermal trough lags by 90° in a sinusoidal disturbance. In connection with the occlusion process, this phase lag decreases; hence the conversion rate between available potential energy and kinetic energy also decreases. If the troughs and ridges of the isotherms and pressure contours become more or less in phase, little kinetic energy is produced because rising and descending motions (ahead of and behind the pressure trough) occur at nearly the same temperature and the correlation between ω and T is almost nil.

Such a change in the phase lag also implies that the axis of the cyclone becomes more vertical, as in the example of Figs. 11.4 through 11.6. Frictionally forced convergence in low levels then partially underlies cold air, and the generation may become negative. Since, furthermore, the frictional dissipation always reduces the kinetic energy, any such cyclone becomes a sink region of kinetic energy and the filling stage of the cyclone has started. Often the filling is a quite slow process because, during filling, the vertical circulation forced by surface friction declines gradually as the frictional inflow itself weakens along with the diminishing horizontal-vortex circulation; also, the release of latent heat prolongs the process if the air is not too stable.

However, the local change of total kinetic energy always depends so much on the flux terms of Eq. (16.8) that no simple rules can be given concerning the actual change of kinetic energy in such an open system. As another instance, many "upper cold lows" persist over long periods, as illustrated in Section 10.2. The energetics of quasi-symmetrical cold vortices, in which both frictional dissipation and character of the vertical circulation appear sometimes to favor the destruction of the vortex, is not yet fully solved and may depend on the existence of an energy flux from the surroundings into the disturbance.

16.4   INFLUENCE OF HEAT SOURCES AND SINKS

In the preceding discussion, only a limited attempt has been made to consider the influence of heat sources and sinks on the energy conversions in the atmosphere. Certain fields of potential-energy distribution were assumed for illustration, but their origin and maintenance were not discussed. However, the question of maintenance, especially in the disturbances, becomes important if the conversion processes over extended time periods (for instance, during the whole life cycle of a cyclone) are to be treated. Raethjen (1966) has stressed that even over periods not longer than 24 to 48 hr, the influence of radiation on the maintenance of the potential energy might be considerable.

Because of the vertical stability of the atmosphere, any solenoidal circulation associated with ascending warm air and descending cold air would after some time effectively reduce the available potential energy and hence also the rate of conversion. Only if there are processes acting in the opposite direction, so that the potential-energy distribution in the disturbance can be maintained or reduced at a slower rate, could the conversion continue to produce kinetic energy for a longer time.

The possibilities in this respect may be illustrated by a simple example (see Petterssen *et al.*, 1962, for a more elaborate discussion). Consider a sinusoidal wave in which the isotherms lag the streamlines by 90° phase. The local temperature change at a given level is determined by

$$\frac{\partial T}{\partial t} = - w(\gamma_a - \gamma) - \mathbf{V} \cdot \nabla T + \frac{1}{c_p} \frac{dh}{dt} \qquad (16.15)$$

where γ_a denotes the adiabatic vertical-lapse rate of temperature; γ, the real lapse rate; and dh/dt, the heat added per unit time and mass. The last term represents the diabatic influence on the local temperature change. We shall disregard here the radiative effect,[8] but consider the release of latent heat and the turbulent flux of sensible heat.

---

[8] Suomi and Shen (1963) computed, from 13 days of satellite data over a portion of the middle latitudes, a mean generation of eddy available potential energy due to dif-

If the wave moves eastward with speed c (the eastward component of the wind relative to the moving system being $u - c$), and there is no local temperature change in a coordinate system fixed to the moving wave, from Eq. (16.15) we get

$$w = \frac{(1/c_p)(dh/dt)}{\gamma_a - \gamma} - \frac{(u - c)(\partial T/\partial x) + v(\partial T/\partial y)}{\gamma_a - \gamma} \qquad (16.16)$$

This equation determines the vertical velocity that is necessary for the maintenance of the temperature field of the moving system.

The value of w is, of course, a complicated function of space. However, let us consider only a mean value of w in a deep tropospheric layer and separately in the warm and cold air. The first r.h.s. term in Eq. (16.16) represents the effect of diabatic heating; and the second term, the effect of advection. To simplify the discussion, we assume $(u - c)$ to be small as a mean value for the layer (considering, say, the middle levels of a short wave). As the mean value of $(\gamma_a - \gamma)$, we select 4°C/km. Let us further assume that $\partial T/\partial y = - 8°C/1000$ km in both masses and $v = + 20$ m/sec in the warm air and $- 20$ m/sec in the cold air. With these crude assumptions, the value of the last term becomes ± 4 cm/sec in the warm and cold air, respectively.

The effect of diabatic heating can be estimated in the following way: Let the heating be 1 ly/min (equivalent to a rainfall rate of 1 mm/hr). For the depth of the troposphere (roughly between 1000 and 300 mb) this corresponds to a heat source of 0.017 cal/sec for a total mass of 700 g/cm². According to Eq. (16.16), the corresponding mean vertical velocity induced by heating would then be about 2.5 cm/sec. If only the latent heat is considered, this latter effect will be essentially confined to the warm air, in which most of the condensation occurs during the developing phase of a polar-front cyclone. Under all these assumptions, the vertical velocity necessary to maintain the local temperature in the moving system would be $+ 6.5$ cm/sec in the warm air and $- 4$ cm/sec in the cold air.

In individual cyclones the diabatic heating and the effect of advection could, of course, vary greatly in space. As a consequence, strong local variations occur in the generation or destruction of available potential energy. The estimates given above, which are intended only for illustration, suggest

---

ferential radiation in warm and cold air, amounting to 0.6 W/m². For a longer winter period, using data at about 40°N, Corcoran and Horn (1965) found an average value of 0.055 W/m². This is less than 1% of the kinetic-energy production rates in the last column of Table 16.3.

that under the influence of latent heat release, the maximum ascending motion in the warm air is likely to be stronger than the descending motion in the cold air (also, if heat were added to the cold air, its descending motion would be diminished). On the other hand, most rain areas are relatively limited, so that this difference is probably compensated with regard to vertical mass flux by a correspondingly larger extent of the areas with descending motion.

In the preceding discussion it was assumed for the sake of simplicity that the thermal structure of the schematic disturbance was conserved. In such a case, the horizontal advection could be viewed as a mechanism that continuously transforms mean available potential energy (expressed by the mean meridional temperature gradient) into "eddy available potential energy." This offsets the tendency for the latter to be destroyed by the vertical motions, which at the same time express a transformation into kinetic energy.

Thus the meridional advection by disturbances acts to create or maintain eddy available potential energy, expressed largely by temperature variations in zonal planes. Furthermore, the conversion between eddy available potential energy and eddy kinetic energy largely occurs through solenoidal circulations in zonal planes according to the processes described above. On the other hand, the condensation phenomena, which represent an intense heat source in the warm air of cyclonic disturbances and only moderately affect the cold air, act to restore the eddy available potential energy.

The effect of heat flux from the earth's surface further modifies the picture, since this heat source largely acts in the cold air and hence tends to reduce the temperature differences. This influence is especially strong during the colder season when cold polar- or continental-air masses are being advected over warm-water surfaces. However, it may be observed that the heat received at the earth's surface is generally transferred upward through a relatively shallow layer (wherein it essentially counteracts the influence of advective cooling), whereas the latent heat released in the warm air is more likely to be communicated to the upper troposphere.

Petterssen et al. (1962) have computed the transfer of sensible and latent heat from the north Atlantic Ocean in different parts of cyclones. Similar investigations have been made by Laevastu (1965) for the north Pacific Ocean, showing comparable results. The strongest flux of sensible and latent heat occurred in the regions of active cold outbreaks (Fig. 12.5), whereas both heat fluxes were weak or even negative in the warm air. In cold outbreaks, the transfer of sensible heat reached values of 0.2 to 0.5 ly/min, with somewhat higher values of 0.2 to 0.8 ly/min for latent heat. Of the latter, only a portion heats the atmosphere locally through condensation, whereas another portion increases the humidity content of the air. Except in the relatively limited regions of strong shower activity where the cold air flows cyclonically, indicating lower-level convergence and ascending motion, the actual heating of the cold air could be estimated at an average value of about 0.5 ly/min or less (see Section 12.2).

In the absence of precipitation measurements in these oceanic cyclones, the release of latent heat is not known. However, a mean rate of 0.5 mm/hr would not be unreasonable. This would correspond to a latent heat release (essentially in the warm air above the frontal surfaces) of 0.5 ly/min, which is equivalent to the maximum heat source in the cold air. In such typical winter cyclones the influence of the distribution of heat sources upon the available potential energy therefore depends on the precipitation rate (essentially a heat source in the warm air) and the flux of sensible heat from the earth (essentially a heat source in the cold air).

From this it can be seen that the heat sources in cyclone development cannot be neglected. Maintenance of the next-to-last term in Eq. (16.8), expressing the conversion to kinetic energy, depends upon preservation of the available potential energy of the eddies. The eddy potential energy converted is partly restored by horizontal advective processes, drawing upon the zonal available potential energy. This process is, however, importantly modified by the heat sources. In an extratropical cyclone, the overall effect depends upon which of the opposing influences discussed above is dominant.

In tropical cyclones (Section 15.6), the primary heat source (mostly release of latent heat, but partially also transfer of sensible heat from the oceans) is located in the central parts of the storm where the temperature also is higher than in the outer parts. Hence the two heat sources cooperate rather than oppose. The available potential energy in this type of disturbance is therefore a result of the air circulation, and the same can be said about the heat source. In this case, on the average, the increase of available potential energy due to the heat source may easily compensate the decrease of available potential energy due to transformation into kinetic energy. In extratropical cyclones, however, the initially available potential energy is gradually converted into kinetic energy, and the existing heat sources may or may not contribute to the maintenance of available potential energy. However, also in these cases, the latent heat is mainly released in the warmer air mass and gives a positive contribution to the available potential energy. From this we can conclude that *the heat of condensation in extratropical disturbances also gives an important contribution to the total energy develop-*

ment. Without this additional heat source, polar-front cyclones could not reach the intensities observed.[9]

It follows that the development of a cyclone is strongly influenced by its geographical position, since the heat sources show a similar dependence. This may be illustrated by the following examples. In a continental cyclone approaching the American east coast in winter, very little sensible heat is added to the cold air as long as the cold air remains over the continent. At the same time, a large release of latent heat is favored by the high moisture content in the warm air east of the frontal system over the Atlantic seaboard. This position therefore strongly favors the maintenance of the temperature contrast or the available potential energy. If the cyclone moves farther east and the cold air overruns the warm surface waters east of the Gulf Stream, more heat could be added to the cold air than is released in the precipitation region, resulting in a decrease of available potential energy and hence also in the possible rate of energy conversion.[10]

On the American west coast the conditions are quite different. For a cyclone approaching this coast in winter, the heat source in the cold air (still over the ocean) is considerable, whereas the moisture content of the warm air to the east is smaller than it is over the ocean off the east coast. Hence the distribution of heat sources is not so favorable for energy conversion. This agrees with the observation that the eastern coasts of continents during the cold season are the more favorable regions for strong cyclogenesis.

The importance of latent heat for cyclone development can also be seen when a cyclonic disturbance moves eastward over the American continent. When moist air from the Gulf region becomes involved in the cyclonic circulation east of the Rocky Mountains, a pronounced deepening often follows, as is well known among American forecasters. A similar situation confronts meteorologists of the Eastern Hemisphere. Here, pronounced cyclone developments occur as a result of the moisture and sensible heat

---

[9] See the discussion in Section 3 of Danard's paper, concerning the production of vorticity by the vertical-motion field associated with latent heat release, and its influence in counteracting the dissipation by friction.

[10] In the examples given by Petterssen *et al.*, the release of latent heat computed from diagnostic equations was smaller than the sensible heat transfer from the surface into the cold air. However, they point out that, for practical reasons, the cyclones selected for study were well out in the Atlantic, east of the average cyclone track. Most cyclones forming near the east coast of North America move northeastward toward Iceland. In such a case, the western part of the cyclone would be over land or relatively cold water, and the sensible heat transfer into the cold air would be appreciably smaller than if the cyclone followed a more eastward course.

sources in the air from the Mediterranean and Black Sea regions, whereas most cyclone development is quite moderate over Siberia despite the often very strong initial baroclinity of the air in this region.

In the simple example discussed earlier, we considered a hypothetical wave whose amplitude did not change with time. More generally, the effect of horizontal advective processes in a disturbance can be seen in an actual growth in amplitude of its thermal pattern. In such a case the isotherms move slower than the winds (in the regions where the wind is strong), and the general pattern of vertical motions is similar to that described above. The growth process was illustrated schematically in Fig. 11.3, and actual examples of large increases in the amplitudes of large-scale disturbances were given in Chapters 10 and 11. Although the end product of an amplifying wave may take very different forms (Fig. 10.4), the final destruction of its eddy potential energy is accomplished largely by subsidence of the cold air accompanied by diabatic heating from the surface (in the lower-latitude parts) and by ascent together with radiative cooling (in the initially warm higher-latitude parts, as in an occlusion). The generation and decay of individual disturbances that express eddy available potential energy is a sporadic process, the overall eddy energy being maintained at something like a mean state through the decay of some disturbances accompanied by the growth of others.

16.5 GENERAL SCHEME OF ATMOSPHERIC-ENERGY CONVERSIONS

Overall schemes of the atmospheric-energy processes have been presented by several authors. These have been applied in two general classes of computations, namely, diagnosis from real data, and numerical experiments in which the boundary conditions providing energy sources and sinks are specified and the model atmosphere adjusts to an operating condition. The first experiment in the latter category was accomplished by Phillips (1956), who gives a lucid description of the conversions between the various energy forms. Although the circulation systems that developed in Phillips' experiment were not quantitatively comparable to those in the real atmosphere, remarkable success was achieved in reproducing the most essential physical processes, with respect to both eddy motions and mean meridional circulations. More recently, such experiments on a hemispheric or global scale by Y. Mintz and A. Arakawa, J. Smagorinsky, S. Manabe and their collaborators, C. Leith, and A. Kasahara and W. Washington have closely reproduced the essential climatological features and, with somewhat less fidelity, the mobile cyclones and anticyclones. Here we are concerned only with the results based on real data diagnosis, useful summaries of which have been given by Oort (1964) and Wiin-Nielsen (1965).

Common to all schemes of energy transformation is the impossibility of presenting more-or-less final quantitative results, but qualitatively they provide a rather satisfactory idea of the atmospheric-energy processes. The general framework of the computations is founded on the formulations by Van Mieghem (1952) and Lorenz (1955) of the energy forms: kinetic energy of the mean motion and of the eddy motion, and available potential energy of the mean and eddy fields. Wiin-Nielsen (1962) has further distinguished between the "barotropic" portion of the kinetic energy (that of the vertically averaged flow) and the kinetic energy associated with the vertical shear. Further divisions are possible, but here we discuss only the simplest overall aspects.

In most schemes the principal energy processes are the following: Since the latitudinal distribution of net radiation in combination with the energy exchange between earth and atmosphere result in an average surplus of heat in low latitudes and a deficit in higher latitudes, a continuous generation of "mean available potential energy" (A_m) occurs. This may be directly converted into "kinetic energy of mean motion" (K_m). Such a conversion $(A_m \rightarrow K_m)$ takes place in the Hadley cells of low latitudes, whereas a reversed conversion $(K_m \rightarrow A_m)$ occurs in the Ferrel cells. Whether the overall positive or negative rate of conversion is the dominating one depends on the differential heating of the atmosphere in the meridional direction and upon the importance of the eddy processes. In the winter hemispheres of the globe, the mean conversion has the direction $A_m \rightarrow K_m$, owing to the dominating influence of the Hadley circulation, whereas in the summer hemispheres this type of conversion is of decreased importance for the whole budget of kinetic energy (Holopainen, 1965).

Owing to the presence of disturbances superposed on the zonal flow, A_m is continually being transformed into "eddy available potential energy" (A_e). This transformation is determined by the product of the eddy transport of heat and the gradient of mean temperature in the meridional direction. A_e is partly converted into "eddy kinetic energy" (K_e) via the processes described previously. A_e can, however, also be produced or destroyed directly as a result of release of latent heat or through turbulent heat flux. As an example of this generation we may consider an initial state determined by the longitudinally averaged mean meridional temperature distribution. In such a case, oceanic regions act as heat sources and continental regions as relative heat sinks during winter, whereas the opposite distribution of relative sources and sinks characterize the summer season. Hence, this would generate A_e. On the other hand, already existing A_e may be either reduced as a result of turbulent flux of sensible heat (maximum flux into cold air) or increased by release of latent heat (maximum release in warm air). Since the generation of K_e from eddy conversion processes is on the average positive, the total increase of A_e as a result of all the foregoing processes should be commensurate with the rate of generation of K_e. This rate was previously estimated at about 80×10^{10} kW for the Northern Hemisphere in winter. The rate of transformation between K_e and K_m is determined by the product of the eddy transport of momentum and the gradient of zonal momentum, both taken in the meridional direction. On the average, this transformation is positive ($K_e \rightarrow K_m$).

Based on an appraisal of the computations from different sources, Oort (1964) has arrived at a tentative energy cycle for the Northern Hemisphere, averaged over the year, as shown in Fig. 16.10. Estimates of the total existing amounts are indicated in the four boxes for A_m, A_e, K_m, and K_e, the estimated uncertainties being given in parentheses. Arrows between boxes indicate the directions and rates of energy transformation; external arrows, the generation or destruction of the corresponding energies through diabatic effects and frictional dissipation.

According to this scheme, *the energy cycle generally proceeds from mean available potential energy through eddy available potential energy to eddy kinetic energy and finally to mean kinetic energy.* On an annual basis, only a negligible part of the mean available potential energy is (according to Fig. 16.10) directly converted into mean kinetic energy as a result of mean meridional circulations, the sign of this conversion being questionable. Although in this scheme the conversions are intended to represent the hemi-

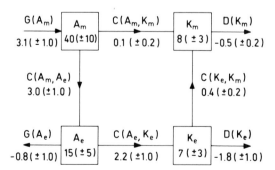

FIG. 16.10   Flow diagram of the atmospheric energy averaged over the whole year in the Northern Hemisphere, according to Oort (1964). Units for existing energy amounts (in boxes) are 10^5 J/m². For energy transformations (arrows), units are J/(m² sec) or W/m². G denotes generation, C conversion, and D dissipation.

spheric integral, it was emphasized in Chapters 1 and 2 that the dominant energy-conversion processes differ in high and low latitudes. Zonal kinetic energy is, in lower latitudes, generated mainly by the mean meridional circulations, and most of the zonal kinetic energy generated by the Hadley cells is exported by eddy processes to the higher latitudes (Section 1.5). Thus, if only the tropics were considered, $C(A_m, K_m)$ would have a much larger value than that shown in Fig. 16.10, especially in winter, and the other conversions would be smaller.

It is of further interest to note the conversions between waves of different scales. On the whole, as noted above, $K_e \rightarrow K_m$. Based on 500-mb data for the Northern Hemisphere, Saltzman and Teweles (1964) find that this holds also for all individual harmonic wave numbers 1 to 15. The aggregate of waves, $n = 2$ to 15, transfer energy to wave numbers 0 and 1, thus maintaining the asymmetric circumpolar vortex. Wave numbers 2, and 5 to 10, are indicated as sources that transfer kinetic energy both to mean motion and to other wave numbers. This reflects respectively "a large forced conversion of potential energy on this scale [$n = 2$] associated with the continent-ocean structure" and ($n = 5$ to 10) the conversion by cyclone-scale disturbances (Section 16.3). Wave numbers 3 and 4, characteristically the long waves, evidently have kinetic energy fed into them by the super-posed cyclone-scale waves. While this analysis indicates that, for the most part, the shorter waves feed kinetic energy into the longer ones (including wave number 0, which represents the mean zonal flow), the reverse is true for the conversion of available potential energy. Fjørtoft (1959) has shown theoretically that this must be so. In accord with Fig. 16.10, such a con-version from the large-scale potential-energy field to the smaller synoptic-scale eddies is necessary, since the latter generate kinetic energy through processes that dissipate the variances of the temperature field.

From our earlier discussion of the energy processes, it is evident that, in winter, most of the numerical values would be larger than those in Fig. 16.10. For instance, the value of $C(A_m, K_m)$ would probably be about 0.7 W/m² in winter and about 0.0 W/m² in summer (for the whole Northern Hemisphere), according to Table 16.1. Considering these values, it seems likely that the annual value in Fig. 16.10 is an underestimate. For the con-version rate $C(A_e, K_e)$, a more probable value would be about 4 W/m² in winter and 2 W/m² or somewhat less in summer. The higher winter value is in better accord with the estimate of the energy conversion in Section 2.6 for the extratropical part of the Northern Hemisphere.[11] The rates of fric-

---

[11] The conversion rate $C(A_e, K_e)$ in winter was estimated at about 6 W/m², if the

tional dissipation must obviously show a similar variation. Seasonal diagrams of the type in Fig. 16.10 (Wiin-Nielsen, 1965) indicate, in line with the comments above, that $C(A_m, A_e)$ is about twice as great in winter as in summer, the same being true of $C(A_e, K_e)$. An even larger seasonal variation was computed by Krueger *et al.* (1965). While it seems to be firmly established that the directions of energy conversion shown in Fig. 16.10 are typical for winter, Wiin-Nielsen points to the month of January 1963, studied by Wiin-Nielsen *et al.* (1964), as an exception. In this month there were unusually vigorous eddies in wave numbers 2 and 3. $C(A_m, A_e)$ was in the usual direction, but much stronger than usual, and the conversions $C(A_m, K_m)$ and $C(K_e, K_m)$ were opposite to those in Fig. 16.10.

It may be stressed that the energy transformations $C(A_m, K_m)$ and $C(A_e, K_e)$ generally are somewhat underestimated, owing to the methods used. Since most computations of the vertical velocity are not made directly from wind data, but from models in which heat sources are not properly considered, this results in too small values of the production of eddy kinetic energy.[12] Also, in the models and area coverage used, the general upward motion in the lower-latitude part of the hemisphere could not be properly considered. As a result of this, the contribution of the Hadley cell to the conversion of A_m into K_m is considerably underestimated, especially during the cold season. This is partly a consequence of the special difficulty in applying most numerical computation models to the special conditions in the equatorial belt. However, it also emphasizes the need for improvement of the observational data in this region which represents the most important heat source in the whole atmosphere.

---

negative rate of the Ferrel cell (-0.9 W/m²) according to Table 16.1 is considered. A conversion rate of only 2 W/m² in the tropical region (roughly south of 30°N) would give the mean value of 4 W/m² in the whole Northern Hemisphere.

[12] An exception is the study by Kung (1966), who used wind and height data to evaluate the cross-contour flow over North America, daily for an 11-month period and up to the 50-mb level. He computed a mean annual generation rate of 9.5 W/m² and a winter rate of 15.4 W/m². These values and the dissipation rates he obtained are three to four times the energy conversion rates in Fig. 16.10 and in our winter estimate based on requirements for vertical heat transfer. We incline to the opinion that they may be too large, although it appears plausible that North America is an area of considerably stronger kinetic-energy production than most other regions of the Northern Hemisphere. Dutton and Johnson (1967) also estimate a value for $G(A_m)$ which is almost twice that in Fig. 16.10, and discuss in detail why the energy conversions are generally underestimated. Their acceptance of Kung's estimate of dissipation for North America, as representative of the hemisphere, does not seem completely justified.

REFERENCES

Brunt, D. (1944). "Physical and Dynamical Meteorology," 2nd ed., pp. 75–76 and 286–287. Cambridge Univ. Press, London and New York.

Corcoran, J. L., and Horn, L. H. (1965). The role of synoptic scale variations of infrared radiation in the generation of available potential energy. *J. Geophys. Res.* **70**, 4521–4527.

Danard, M. B. (1964). On the influence of released latent heat on cyclone development. *J. Appl. Meteorol.* **3**, 27–37.

Defant, A., and Defant, F. (1958). "Physikalische Dynamik der Atmosphäre," Chapter 12. Akad. Verlagsges., Frankfurt.

Dutton, J. A., and Johnson, D. R. (1967). The theory of available potential energy and a variational approach to atmospheric energetics. *Advan. Geophys.* **12**, 333–436.

Eliassen, A., and Kleinschmidt, E. (1957). Dynamic meteorology. *In* "Handbuch der Physik" (S. Flügge, ed.), Vol. 48, pp. 1–154. Springer, Berlin.

Fjørtoft, R. (1959). Some results concerning the distribution and total amount of kinetic energy in the atmosphere as a function of external heat sources and ground friction. *In* "The Atmosphere and the Sea in Motion" (B. Bolin, ed.), pp. 194–211. Rockefeller Inst. Press, New York.

Haurwitz, B. (1941). "Dynamic Meteorology," Chapters 4 and 12. McGraw-Hill, New York.

Hess, S. L. (1959). "Introduction to Theoretical Meteorology," Chapter 19. Holt, New York.

Holopainen, E. O. (1965). On the role of mean meridional circulations in the energy balance of the atmosphere. *Tellus* **17**, 285–294.

Krueger, A. F., Winston, J. S., and Haines, D. A. (1965). Computation of atmospheric energy and its transformation for the Northern Hemisphere for a recent five-year period. *Monthly Weather Rev.* **93**, 227–238.

Kung, E. C. (1966). Large-scale balance of kinetic energy in the atmosphere. *Monthly Weather Rev.* **94**, 627–640.

Laevastu, T. (1965). Daily heat exchange in the North Pacific, its relations to weather and its oceanographic consequences. *Soc. Sci. Fennica, Commentationes Phys.-Math.* **31**, No. 2, 1–53.

Lorenz, E. (1955). Available potential energy and the maintenance of the general circulation. *Tellus* **7**, 157–167.

Margules, M. (1905). Über die Energie der Stürme. *Jahrb. Zentralanst. Meteorol. Geodyn., Wien.* [English transl., "The Mechanics of the Earth's Atmosphere. A Collection of Translations by Cleveland Abbe." *Smithsonian Inst. Misc. Collections* **51**, No. 4, 533–595 (1910)].

Miller, J. E. (1950). Energy transformation functions. *J. Meteorol.* **7**, 152–159.

Miller, J. E. (1951). Energy equations. *In* "Compendium of Meteorology" (T. F. Malone, ed.), pp. 483–491. Am. Meteorol. Soc., Boston, Massachusetts.

Oort, A. H. (1964). On estimates of the atmospheric energy cycle. *Monthly Weather Rev.* **92**, 483–493.

Palmén, E. (1958). Vertical circulation and release of kinetic energy during the development of hurricane Hazel into an extratropical storm. *Tellus* **10**, 1–23.

Palmén, E. (1960). On generation and frictional dissipation of kinetic energy in the atmosphere. *Soc. Sci. Fennica, Commentationes Phys.-Math.* **24**, No. 11, 1–15.

Palmén, E., and Holopainen, E. O. (1962). Divergence, vertical velocity and conversion between potential and kinetic energy in an extratropical disturbance. *Geophysica* (*Helsinki*) **8**, 89–113.

Petterssen, S., Bradbury, D. L., and Pedersen, K. (1962). The Norwegian cyclone models in relation to heat and cold sources. *Geofys. Publikasjoner, Norske Videnskaps-Akad. Oslo* **24**, 243–280.

Phillips, N. A. (1949). The work done on the surrounding atmosphere by subsiding cold air masses. *J. Meteorol.* **6**, 193–199.

Phillips, N. A. (1956). The general circulation of the atmosphere: A numerical experiment. *Quart. J. Roy. Meteorol. Soc.* **82**, 123–164.

Raethjen, P. (1964). Zur Energetik und Dynamik kräftiger Zyklogenesis. *Beitr. Physik Atmosphäre* **37**, 197–211.

Raethjen, P. (1966). Zur Energetik kräftiger Frontogenesis. *Beitr. Physik Atmosphäre* **39**, 182–198.

Reynolds, O. (1895). On the dynamical theory of incompressible viscous fluids and the determination of the criterion. *Phil. Trans. Roy. Soc. London* **A186**, 123–164.

Riehl, H., and Fultz, D. (1957). Jet stream and long waves in a steady rotating-dishpan experiment. *Quart. J. Roy. Meteorol. Soc.* **83**, 215–231.

Rossby, C.-G. (1949). On a mechanism for the release of potential energy in the atmosphere. *J. Meteorol.* **6**, 163–180.

Saltzman, B. (1957). Equations governing the energetics of the larger scales of atmospheric turbulence in the domain of wave number. *J. Meteorol.* **14**, 513–523.

Saltzman, B., and Teweles, S. (1964). Further statistics on the exchange of kinetic energy between harmonic components of the zonal flow. *Tellus* **16**, 432–435.

Starr, V. P. (1948). On the production of kinetic energy in the atmosphere. *J. Meteorol.* **5**, 193–196.

Starr, V. P. (1951). Applications of energy principles to the general circulation. *In* "Compendium of Meteorology" (T. F. Malone, ed.), pp. 568–574. Am. Meteorol. Soc., Boston, Massachusetts.

Suomi, V. E., and Shen, W. C. (1963). Horizontal variation of infrared cooling and the generation of eddy available potential energy. *J. Atmospheric Sci.* **20**, 62–65.

Sverdrup, H. U. (1917). Über den Energieverbrauch der Atmosphäre. *Veroeffentl. Geophys. Inst. Univ. Leipzig* **2**, 173–196.

Thompson, P. D. (1959). Some statistical aspects of the dynamical processes of growth and occlusion in simple baroclinic models. *In* "The Atmosphere and the Sea in Motion" (B. Bolin, ed.), pp. 350–358. Rockefeller Inst. Press, New York.

Thompson, P. D. (1961). "Numerical Weather Analysis and Prediction," Chapter 10. Macmillan, New York.

Väisänen, A. (1961). Investigation of the vertical air movement and related phenomena in selected synoptic situations. *Soc. Sci. Fennica, Commentationes Phys.-Math.* **26**, No. 7, 1–72.

Van Mieghem, J. (1951). Application of the thermodynamics of open systems to meteorology. *In* "Compendium of Meteorology" (T. F. Malone, ed.), pp. 531–538. Am. Meteorol. Soc., Boston, Massachusetts.

Van Mieghem, J. (1952). Energy conversions in the atmosphere on the scale of the general circulation. *Tellus* **4**, 334–351.

Van Mieghem, J. (1958). On the interpretation of the energy equations in dynamic meteorology. *Geophysica* (*Helsinki*) **6**, 559–576.

White, R. M., and Saltzman, B. (1956). On conversion between potential and kinetic energy. *Tellus* **8**, 357–363.

Wiin-Nielsen, A. (1962). On transformation of kinetic energy between the vertical shear flow and the vertical mean flow of the atmosphere. *Monthly Weather Rev.* **90**, 311–323.

Wiin-Nielsen, A. (1965). Some new observational studies of energy and energy transformations in the atmosphere. *World Meteorol. Organ., Tech. Note* 66, 177–202.

Wiin-Nielsen, A., Brown, J. A., and Drake, M. (1964). Further studies of energy exchange between the zonal flow and the eddies. *Tellus* **16**, 168–180.

17

SUMMARY OF THE
ATMOSPHERIC CIRCULATION PROCESSES

The atmospheric balance of angular momentum and energy, as well as the roles of different types of circulation systems in effecting this balance, have been discussed in various chapters. As a conclusion, it is appropriate to summarize briefly the principal aspects in relation to the general circulation. In this discussion, we shall confine ourselves mainly to the Northern Hemisphere in winter, when the energy generation and exchange processes are best developed. With the exception of the Asiatic monsoon and a greater amount of convection, the circulations in summer are essentially similar except for changes in latitudes and intensities of the circulation systems.

17.1 GENERAL ASPECTS

The broad function performed by the atmosphere-ocean system is that of storing and redistributing the energy received from the sun. Considering the differences between the seasons in a hemisphere, the change in atmospheric-energy storage is relatively small because of the small mass of the atmosphere and the very large heating excess or deficit involved. Of the radiative excess or deficit of the atmosphere-earth system over a whole hemisphere in summer or winter, only a minor part is compensated by transequatorial energy flux. The predominant part goes into storage, principally in the oceans, during spring-summer, and is given up to the atmosphere during autumn-winter. A reasonably modest seasonal change in temperature of the upper oceanic layer suffices to account for the change in energy content (Section 2.1).

This storage explains why the trade-confluence zone (TC), averaged around the globe, undergoes only a weak seasonal variation in latitude despite the large radiative imbalance within a season. This small displacement is observed particularly over oceanic regions. By contrast, especially in the Asia-Africa and Australia regions, the TC shows a considerable latitudinal change in response to the continental influence. Intense insolation in the summer hemisphere along with small heat storage in the soil results in a large thermal contrast between a continent and the neighboring ocean whose temperature is fairly constant. This heating contrast, augmented by latent heat release and radiation, drives the monsoon circulations. In effect, the displacement of the TC corresponds to an encroachment of the rising branch of the Hadley cell of one hemisphere into the other hemisphere. A heated continental region is then "ventilated," with relatively cool oceanic trade-wind air entering it in lower levels and leaving it in the opposing branch in upper levels, with a higher energy content. This represents a net energy export from one hemisphere to the other, whose intensity depends on the extent of displacement of the TC from the Equator, which is in turn obviously dependent on the land-ocean contrast or the difference in heat storage. In the Asia-Australia sector, the large migration is of course influenced by the simultaneous net gain of radiation by a continent on one side of the Equator, and net cooling by the continent on the other.

Except in the lower latitudes (Fig. 2.4), the poleward transport of heat in ocean currents is a small fraction of the total meridional transfer needed to offset the effect of heat sources and sinks. Consequently, the atmosphere must accomplish most of this transfer. At the same time, the continual radiative loss of heat by the atmosphere (Fig. 17.1) requires a compensating upward transfer from the earth's surface, where there is a radiative surplus (in winter, up to about 55°N, and in summer at all latitudes, according to London, 1957). Thus, in general, there is required a poleward and upward energy transfer. The different components of these transfers were discussed in Chapter 2 and are elaborated further in Section 17.3.

With regard to the vertical transfer of energy, it was brought out in Chapters 2 and 16 that, except in the "boundary layer," this is accomplished by vertical circulations of different scales ranging from cumulus clouds to the mean meridional circulations. These are important partly because they transfer realized energy and partly because in the ascending branches latent heat is released, representing a heat gain by the atmosphere corresponding to the heat loss at the earth's surface through the process of evaporation. In extratropical latitudes, the atmospheric radiative cooling is most intense somewhat above the 800-mb level (Fig. 17.1). With temperatures character-

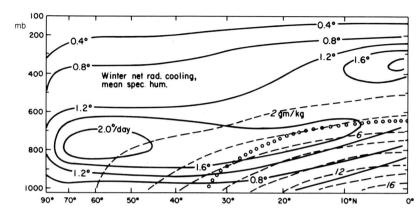

FIG. 17.1 Solid lines show net radiative cooling of the atmosphere (°C/day) in winter. (After London, 1957.) Dashed lines are mean specific humidity isopleths for the cooler six months of the IGY. (After Peixoto and Crisi, 1965.) Line of circles corresponds to envelope A in Fig. 17.4.

istic of these latitudes, about 60% of the latent heat release by a rising parcel takes place below 700 mb (Fig. 2.11); the actual proportion may be somewhat greater, since many cloud systems of these latitudes do not extend into the upper troposphere.

Consequently, in the lower levels where the radiative loss is greatest (this also being enhanced, as pointed out by London, by the predominance of low clouds), this loss is almost balanced by the combined influences of latent heat release and sensible heat import from the tropics (Fig. 2.11), which is also greatest in lower levels. In the higher troposphere, although the radiative loss is smaller, this exceeds the two sources mentioned above. To effect a balance, the upward heat transfer by eddies must diminish rapidly with increasing height, with a vertical convergence of heat transfer much larger in the middle and upper than in the lower troposphere. Considering the probable influence of "small eddies," including those in the friction layer as well as cumulus convection, the additional upward transport required to be accomplished by organized circulation systems is greatest in the middle troposphere (Fig. 2.12), where vertical motions in these systems are characteristically largest (Chapters 10 and 11).

Since the heat transfer by the Ferrel circulation is downward, the needed upward transport has to be effected by eddies in which ascending motions are on the whole associated with higher temperature and water-vapor content, and descending motions with lower temperature and less water vapor. Through mass-continuity requirements, the vertical branches of the circulation are connected with horizontal branches in such a way that (because the atmosphere is nearly hydrostatic) there is on the whole a horizontal flow from higher to lower pressure at different levels. Thus the sinking of cold and rising of warm air implies a generation of kinetic energy through the action of pressure forces with components along the air motion. The decrease of potential energy implied by the systematic vertical motions is countered by suitably distributed heat sources and sinks. In extratropical latitudes the effect of these is expressed largely by conversion to eddy potential energy from the zonal available potential energy derived from the basic meridional distribution of net radiation; i.e., by repeated distortions of zonally oriented isotherms into warm and cold tongues that successively degrade. In addition, direct generation of eddy potential energy arises from the ocean-continent heating contrasts and from the release of latent heat in the disturbances.

A continual degradation of kinetic energy takes place through surface friction and viscous stresses within the atmosphere. Thus the atmospheric circulation can be maintained only through a corresponding release of kinetic energy in the form of the direct solenoidal circulations described above, which at the same time transfer heat upward.

The frictional influences, which cause a loss of kinetic energy, are at the same time largely responsible for the maintenance of the kind of solenoidal circulations that regenerate kinetic energy. Considering a circular vortex symmetrical about the pole, obviously no meridional circulation would be generated if the circulation acceleration due to the solenoids in a meridional plane were in balance with the sum of the coriolis and centrifugal accelerations (i.e., if the actual vertical shear equaled the thermal shear).

The closest approach to such a hemispheric-scale symmetrical vortex is the wind system of the Hadley-cell latitudes. In the surface easterlies, an eastward-acting stress is exerted, while at the level of strongest westerlies, internal viscous stresses (transferring momentum upward and downward) act to diminish the west-wind speed. Consequently, as a result of surface and internal friction, there is a continual tendency to diminish the actual vertical shear (and the corresponding circulation acceleration associated with coriolis and centrifugal forces) below the value required to balance the solenoidal circulation acceleration. If the zonal-wind system is to be maintained at an essentially constant state, this means that a solenoidally direct circulation must be present, which generates kinetic energy (or vertical shear). Thus, in a near-symmetrical vortex, the atmospheric motion cannot be maintained without systematic deviations from the gradient-wind law, of the kind expressed by the direct Hadley cells.

In the absence of heat sources and sinks, any direct circulation of this sort would tend to destroy the solenoid field, since in a thermally stable atmosphere ($\partial\theta/\partial z > 0$) there would be local cooling where the air rises and warming where it descends. This tendency toward destruction is countered by the distribution of incoming and outgoing radiation together with the influences of condensation and evaporation (considering also earth storage of heat and its transfer to the atmosphere). Thus, on the average, there must be an equilibrium between the destruction of the solenoid field and its regeneration by heat sources and sinks, parallel with the destruction of kinetic energy by friction and its regeneration by the solenoidal circulation. The same principles apply to the maintenance of cyclones and anticyclones, although (as stressed in Chapter 16) their application is much more complicated, owing to the asymmetry of these systems and the import or export of kinetic and other forms of energy.

Eliassen (1952) has given a complete theoretical discussion of the principles qualitatively stated above. He concludes that "heat sources in the outer, warmer part of the vortex, and heat sinks in the inner, colder part will give direct circulations. Likewise, sources of angular momentum in the lower part of the vortex, where the vortex motion is slow, and sinks of angular momentum in the upper part, where the vortex motion is fast, will give direct meridional circulations." Also, "the speed of the meridional circulation, for given sources [and sinks] of heat or angular momentum, will increase with decreasing [dynamic] stability of the vortex."

The partial dependence on dynamic stability, which has been employed in several contexts in earlier chapters, is evident from the simplified relationship (neglecting eddies, vertical motion, and curvature)

$$\frac{d\bar{u}}{dt} \approx \bar{v}\frac{\partial\bar{u}}{\partial y} \approx f\bar{v} - D$$

applied in upper levels, D being the dissipation of westerly momentum. Rewriting,

$$\bar{v}\left(f - \frac{\partial\bar{u}}{\partial y}\right) \approx D$$

For a given dissipation it is clear that if the shear is cyclonic (expressing large dynamic stability), $\bar{v}$ will be smaller than if it is anticyclonic. The local dissipation is countered both by the generation (expressed by $f\bar{v}$) and the local increase of $\bar{u}$ due to importation from regions of higher momentum, the latter effect being negative with anticyclonic shear. The two main kinds of more-or-less symmetrical cyclonic vortices in the atmosphere (namely, the Hadley cells and tropical cyclones) have (in the main) weak dynamic stability in upper levels, which favors strong direct circulations or appreciable departures from gradient wind.

The principles treated by Eliassen for symmetrical vortices have been extended by Kuo (1956), who further considers the influence of eddies upon the balance requirements. As brought out in Chapter 1, asymmetrical eddies in the upper troposphere transport large amounts of relative angular momentum poleward across the bounds of the Hadley cells. This corresponds to a removal of westerly kinetic energy from the upper levels of the tropical latitudes, contributing in the same sense as the viscous dissipation toward a requirement for a direct solenoidal circulation to sustain the mean zonal circulation. At the same time, as demonstrated by Mintz, Kuo, and others, the eddy convergence of westerly momentum in extratropical latitudes is so large as to require an indirect mean meridional circulation there in order to maintain a balanced state.

The mean meridional circulation in winter is shown in Fig. 17.2, redrawn from Fig. 1.5 on a sine latitude scale to portray the extent of the circulations with reference to areas of the earth between latitude belts.[1] In addition to indicating the areal significance of the Hadley cell, this latitude scale relegates the third (direct) cell that probably exists in high latitudes to its proper geographical dimensions, suggesting its unimportance for the hemispheric energy and angular-momentum budget.

As a measure of the comparative influence of disturbances, the standard deviation of the meridional wind component is also shown in Fig. 17.2 (the standard deviation of the west-wind component is quite similar in distribution and magnitude). The square of the vector standard deviation, which gives the mean eddy kinetic energy, has been derived from Crutcher's data (1961) by Kao and Taylor (1964). When mapped around the hemisphere, the eddy kinetic-energy maximum lies on the average a short distance poleward of the axis of the mean strongest winds. This close association is, of course, related to the presence of traveling synoptic disturbances in this barocline region. The effect of the eddy components in transferring heat and angular momentum is large because the $v'T'$ and $v'u'$ correlations in these disturbances are significant. Figure 17.2 illustrates that although the Hadley mean circulation dominates in low latitudes, the influence of eddies becomes increasingly important in its poleward parts, toward which at the same time the mean meridional circulation diminishes in strength.

---

[1] As pointed out in Section 2.2, the indicated mass circulation in the Hadley cell may be too great, by perhaps as much as 30%. For consistency with the earlier chapters, we retain the mean circulations shown in Fig. 1.5. This also affects the energy transfers shown in Fig. 17.8, but does not influence the general discussion.

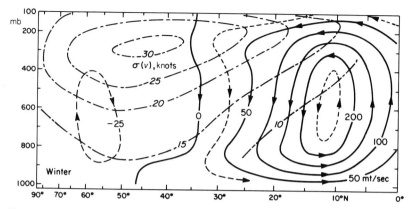

FIG. 17.2 Streamlines of the mean meridional circulation in winter, from Fig. 1.5, in millions of tons per second. Dash-dotted lines show standard deviation (knots) of meridional wind component, averaged around the hemisphere. (From Crutcher's data, 1961, for winter.)

The relative influences of these two characteristics of the atmospheric motions were discussed in Chapter 2 with regard to energy (Figs. 2.7 and 2.8) and in Chapter 1 in relation to the poleward transfer of absolute angular momentum. This transfer is the second principal requirement placed upon the atmosphere, in addition to the transfer of energy outlined above. The requirement exists because of the presence of (1) surface easterlies in the tropics where the earth's angular momentum is large (implying a frictional transfer of absolute angular momentum from the earth's surface), and of (2) surface westerlies in extratropical latitudes where, in order for the total mean atmospheric momentum to remain unchanging, a corresponding loss is necessary.

As indicated in Chapter 1, the vertically integrated poleward transfer of angular momentum by the mean meridional circulation, although significant in low latitudes, is increasingly dominated toward the poles by the transport by eddies. The eddies account entirely for this transfer where it reaches a maximum at the latitudes of the subtropical highs and the overlying subtropical jet streams. This transfer, as stressed by Jeffreys, Bjerknes, and Starr, is associated with a departure of the wave disturbances from a sinusoidal shape, giving a positive $u'v'$ correlation.[2]

Thus the significance of the mean meridional circulation of the Hadley

---

[2] On an annual average, according to Holopainen (1966), the meridional transfer of westerly momentum is strongly dominated by the *transient* eddies, whose maximum effect (at the 200-mb level, 35°N) is about eight times as great as the maximum transfer by standing eddies (at 300 mb, 50°N).

cells lies not in its effectiveness in transferring angular momentum but in the conversion of earth (Ω) angular momentum into relative ($\bar{u}$) angular momentum. This results in maintenance of the subtropical jet stream, or of $\bar{u}$-momentum, which can (through the agency of the Jeffreys-Reynolds stresses) be exported to sustain partially the westerlies of extratropical latitudes.

In addition, however, the vertical branches of the mean meridional circulations are responsible for the greatest part of the vertical transfer of absolute angular momentum (Fig. 1.11), both within the tropics (upward) and in extratropical latitudes (downward). In this respect it may be noted that the Ω-angular momentum (which dominates in this transfer) does not differ greatly between the latitudes of the ascending and descending branches of the Hadley cell, but this difference is several times as large between the latitudes of the ascending and descending branches of the Ferrel cell. Correspondingly, the vertical transports of Ω-momentum accomplished by these cells are more nearly comparable than would be suggested by the very large difference between their mass circulations. It may be observed further that if the mean circulation cells did not exist, the required vertical transfers by eddies would be extremely large and difficult to account for.

Consequently, the overwhelming dominance of eddies in the meridional transfer of angular momentum does not imply that the mean meridional cells are of minor significance for an understanding of the mechanism of the general circulation. To summarize, a well-developed Hadley circulation is essential because (1) frictional influences continually degrade the vertical shear, which can be restored only through such a circulation with poleward drift in upper levels and equatorward drift in lower levels; (2) conversion from Ω-angular momentum to $\bar{u}$-momentum by such a circulation is necessary in order to furnish the momentum exported from the tropics by the eddies; and (3) since the absolute angular momentum gained at the surface in the tropics is exported to higher latitudes predominantly in the upper levels, a large upward transfer through the middle troposphere is required. In the absence of a strong direct meridional-mass circulation, this would have to depend on an eddy transfer from levels where the winds are easterly to levels where they are westerly, or up the gradient of momentum.

17.2 Dominant Features of Circulation

The main features of the winter hemispheric circulation are summarized schematically in Fig. 17.3. Included are the principal air masses: polar air (PA), middle-latitude air (MLA), and tropical air (TA), and their corre-

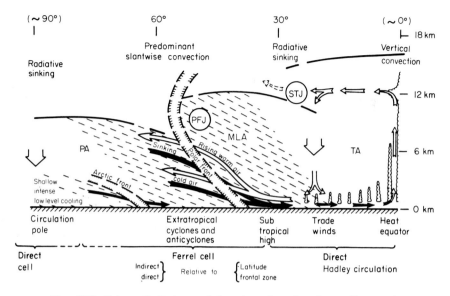

Fig. 17.3 Schematic features of the atmosphere in winter. (See text.)

sponding tropopauses at different characteristic levels. As discussed in Chapter 4, these are separated by barocline zones of varying intensities. Between TA and MLA the baroclinity is greatest in the upper troposphere, beneath the subtropical jet stream (STJ). Between MLA and PA, the polar-front zone is culminated by the polar-front jet stream (PFJ). The frontal zone is in places characterized by a concentrated but indistinct barocline belt, but in others by a well-defined frontal layer that may extend through the whole troposphere or through only the upper or lower troposphere.[3] The tropospheric frontal zone has a more diffuse counterpart in the lower stratosphere, where the slope and sense of the baroclinity are reversed.

In the Hadley cell that dominates the tropics, there are maximum mean meridional motions of 2 to 3 m/sec at 12°N, equatorward at about 950 mb and poleward near 200 mb, during the coldest season (December–February). The connected vertical branches have maximum intensities of about 0.7 cm/sec upward over the thermal equator, and 0.6 cm/sec downward near 15 to

---

[3] The arrows denoting ascending warm air and descending cold air in Fig. 17.3 cut the polar front that is represented in its approximate mean position. This should not be interpreted as suggesting that the air currents penetrate the frontal layer. Rather, the front undergoes changes both in position and slope. Partly considering the movements of fronts, the slopes of air trajectories are typically smaller than the slope of the polar front (Sections 12.3 and 12.4), although the *relative* trajectories may have steeper slopes.

20°N, at the 500-mb level. The upward mass transport takes place essentially in convective (cumulus and cumulonimbus) clouds, wherein ascent of the order of several meters per second in clouds covering a small portion of the equatorial belt accounts for the small area-averaged vertical motion noted above. In the descending branch in the poleward part of the tropics, gentle broad-scale descent takes place mainly as a result of radiative cooling, which allows the air to settle through the isentropic surfaces.

A similar gentle sinking probably prevails in the polar region of the Northern Hemisphere, as it does in the Southern Hemisphere over Antarctica. In the subtropical and polar regions of dominant descending motions, radiation-induced sinking prevails because, in the absence of significant latent heat release, there is no mechanism (except in the limited number of synoptic disturbances) whereby ascent of appreciable amounts of air can take place to compensate the sinking. Although it may be that a similar radiative sinking occurs between the cloud systems of the tropical belt, this is overcompensated by the large upward mass transport in the convective clouds.

Convection in the tropical clouds is locally intense, owing to the thermodynamical instability, and takes place essentially in the vertical. In middle latitudes, by contrast, the dominant process may be termed "slantwise convection" in which (since the air masses generally have a stable stratification) the broad-scale vertical motions in disturbances are a few centimeters per second compared with horizontal winds having speeds of tens of meters per second. Thus, although air parcels may undergo vertical displacements of several kilometers in a typical disturbance while essentially preserving their thermodynamic properties, they are at the same time subjected to horizontal displacements of hundreds or thousands of kilometers. An instructive discussion of the air movements in terms of "isentropic relative-flow trajectories" has been given by Green *et al.* (1966). As discussed in Chapter 4, the bulk of the upper-tropospheric air in middle latitudes circulates round and round the hemisphere, meandering successively northward and southward following the broad-scale upper-level waves, and this MLA has typical properties intermediate between those of PA and TA.

As also emphasized in Chapter 4, a meandering of this sort does not properly describe the air motions where there is significant sustained vertical movement in synoptic disturbances. Rather, when the motions relative to the upper-level wave patterns are considered, equatorward movement takes place in the descending polar-air branches, with poleward movement in the ascending tropical-air branches (Fig. 4.3). As discussed in Chapters 10 and 16, these exchanges, in which large quantities of cold and warm air are irreversibly transferred both vertically and meridionally, represent energy transfers in these directions, with at the same time a generation of kinetic energy.

Of the cold air that sinks and moves equatorward in this manner, a part is modified and returns poleward within subtropical latitudes; however, a very large part penetrates into the tropics on the east sides of both migratory and semipermanent subtropical anticyclones. There the shallow cold-air masses are strongly modified by addition of sensible heat and water vapor from the warmer surface, the latent heat addition predominating if longitudinal mean values are considered. On reaching low latitudes, the thermodynamic-energy content is sufficient to cause the air to rise, when subjected to convergence and the release of latent heat, into the middle and upper troposphere in cumulus convection. The energy released, mostly near the "thermal equator", is then transported poleward in the upper branch of the Hadley circulation.[4] This mean circulation dominates the transfer process in low latitudes, but its intensity diminishes poleward of about 10 to 15° latitude, and the role of meridional energy transfer is increasingly taken over by the eddies. As is the case with angular momentum, these accomplish essentially the entire poleward energy transfer near the latitudes of the subtropical anticyclones where the mean meridional circulation becomes nil (Fig. 2.7).

In the poleward halves of the Hadley cells, the air settling through the influence of radiation aloft with divergence near the surface is in large part reincorporated into the mixed marine layer through the process discussed in Section 14.3. The subsided air of subtropical latitudes then partly recirculates equatorward in the trade winds and partly moves poleward as a moist layer of restricted depth around the west sides of the subtropical highs. On moving poleward, this real tropical air together with the air that is modified within extratropical latitudes feeds the warm ascending branches of extratropical cyclones.

Essentially, the same features of the broad-scale circulation pattern are present in summer, although in the Northern Hemisphere these are very much weaker because of the lesser requirement for meridional exchange of energy. During this season, both the thermal equator and the regions occupied by the principal air masses are shifted poleward from their posi-

---

[4] Here it should not be imagined that the latitude near 30° (in winter) indicated as the poleward bound of the Hadley cell is inflexible. As mentioned in Section 8.4, on occasion a large part of the upper-tropospheric tropical air of the Hadley cell penetrates far poleward in connection with intense disturbances of the subtropical jet or branches thereof.

tions during the coldest season. The heat source in the thermal equator, situated entirely in the Northern Hemisphere, is dominated by the release of latent heat transferred across the geographical Equator from the Southern Hemisphere, whereas the accumulation of moisture due to the much weaker trade-wind branch of the Northern Hemisphere Hadley cell is of much less importance. The upward mass and energy transport has generally the same character as during the winter season, although a somewhat larger amount of the vertical energy exchange is taken over by disturbances in the easterlies. Occasionally and locally, tropical cyclones are very effective in releasing latent heat and transporting it upward, but their overall importance for the global processes is small compared with other smaller-scale but more numerous convective systems.

Consonant with the decreased need for meridional exchange of energy in summer, the "slantwise convection" in middle latitudes is less well developed than in winter. This can be seen in a weakening of the polar-front activity and the corresponding jet stream, whereas the intensified solar radiation causes locally strong convection and heat flux in the vertical, preferentially over continental regions and largely in connection with synoptic disturbances.

During the transition seasons (March–May and September–November) the circulation pattern undergoes a transition from the winter type to the summer type, or vice versa. Because the change occurs rather rapidly at times, especially in certain geographical regions (Fig. 3.14), it is somewhat meaningless to present average values for the circulations in these transition periods.

The very different mechanisms employed by the atmosphere in transferring properties, in different latitudes, may be viewed as follows: One important consideration is the difference between the rates of rotation of the earth about the vertical in low and higher latitudes. Both theory and experiments in the laboratory (Fultz *et al.*, 1959) have demonstrated that, for a given differential heating, the Hadley type of meridional circulation tends to dominate when the rate of rotation about the vertical is slow. However, when this rate is fast, the currents tend to be more in gradient balance, and eddies that have the character of middle- and high-latitude waves with cyclones and anticyclones are dominant.

A second consideration resides in the relative requirements for vertical and horizontal transfer of energy. In the zone of the thermal equator, where the heat source is strongest in the central portion, the meridional energy flux is weak but combined with strong divergence. However, because of the large energy gain by the trade-wind air in contact with the earth's surface, in combination with radiative cooling which in these latitudes is quite intense in the higher troposphere (Fig. 17.1), there is a strong requirement for vertical transfer. It follows that vertical convection is the dominant process in this region. By contrast, in middle latitudes where both meridional and vertical transfers of large amounts of energy are required, and where the air masses are typically stable even for moist-adiabatic ascent except in restricted parts of disturbances, slantwise convection is the most effective means of accomplishing both transfers simultaneously.

As already mentioned, there is a strong need to transfer energy away meridionally just from the central tropical belt, which is characterized by an intense source of energy mainly due to condensation. Since the upward transfer in the central region involves a net movement of large quantities of mass as well as energy from the lower to the upper troposphere (from which, because of the vertical stratification, the air cannot then descend except by the slow process of radiative energy loss), the removal of mass from this zone must take place by mean circulations poleward in upper levels. At the same time, since the mean meridional circulation is very effective in transporting energy, there is no need for a significant transfer by eddies. This need arises farther poleward, where the mean meridional circulation slackens, owing to the stabilizing effect of the earth's rotation (greater geostrophic control), although the requirement for meridional energy transfer increases up to middle latitudes (see Figs. 2.4 and 2.7).

The preference for different kinds of circulations at high and low latitudes may be considered in the context of the mean distribution of static energy, $E = (c_p T + gz + Lq)$, shown in Fig. 17.4. The distribution will differ greatly from this mean picture at different longitudes in synoptic systems, and correspondingly there are large variations with time at a given location, especially in higher latitudes (see, e.g., Fig. 12.15b). Nevertheless, the vertical and horizontal gradients of E are for the most part qualitatively similar, both in the tropics and in extratropical latitudes, to that seen in the mean section.

In the tropics, following the presentation for the Asiatic monsoon longitudes by Normand and Rao in Fig. 14.12, we have entered the dashed lines A and B, which are significant for the vertical transfer by convective clouds (Section 14.5). Line A represents the level of minimum E; thus air parcels rising from below this level carry energy upward through the layer.[5]

---

[5] The extent to which the total energy distribution is dominated by the water-vapor distribution in the lower tropical troposphere, may be appreciated by comparing Fig. 17.4 with the specific humidity isopleths in Fig. 17.1. Note that $(c_p T + gz)$ increases upward if $\partial\theta/\partial z > 0$, which is the normal condition except in the boundary layer. Also,

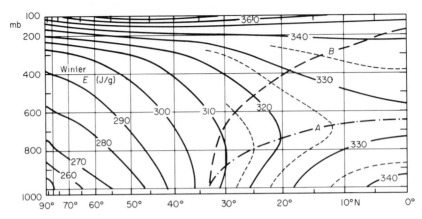

FIG. 17.4   Static energy distribution ($c_p T + gz + Lq$) averaged over Northern Hemisphere in winter. Temperatures are those used in Fig. 1.1a; isobaric heights taken from Heastie and Stephenson (1960), and specific humidities from Peixoto and Crisi (1965).

The distribution is equivalent to that of θ_w in Fig. 14.12. Free convection can most readily take place below level A, but most clouds will suffer a loss of buoyancy due to the entrainment of air with low energy content, and only a small proportion will rise higher.

Evidently any air escaping upward from the near-surface layers in the zone of mean convergence near the Equator must carry with it the high energy that has been acquired by trade-wind air approaching low latitudes, through transfer of sensible and latent heat from the surface. As discussed in Sections 14.4 and 14.5, the only obvious way for this high-energy air to rise to high levels where it can move poleward (Fig. 17.2), is for it to ascend in large-diameter cumulonimbus clouds. It is possible for air in the cores of such clouds (protected from entrainment) to approach level B, corresponding to the value of E (or θ_w) near the earth's surface.

The processes of the vertical energy transport in the equatorial branch of the Hadley cell may be illustrated in the following way: Figure 17.5a shows (cf. Fig. 1.5) the maximum mean vertical velocity in the belt 0 to 5°N in winter. The vertical distribution of the observed mean static energy (E_{obs})

---

a comparison with the radiative cooling isopleths in Fig. 17.1 is instructive. As trade-wind air moves equatorward, within the layer below envelope A of Fig. 17.4, an air column suffers much more rapid radiative cooling at the top than at the bottom; at the same time it gains both sensible heat and latent energy from the surface. These processes, acting in concert, account for an increase of potential instability toward the equator in the layer below A; however, the generation of extreme instability is prevented by overturning in the trade-wind cumuli.

is given by the solid curve in Fig. 17.5b, which agrees closely with the curve used by Riehl and Malkus for the equatorial trough. If the air moved uniformly and slowly upward at the rates shown in Fig. 17.5a, it would lose energy as it rose according to the radiation rates in Fig. 17.1, and its static energy would conform to the dashed curve E_s in Fig. 17.5b. This demonstrates that such a slow uniform ascent would be impossible, since in that case $E_s > E_{obs}$ below 480 mb and $E_s < E_{obs}$ above that level. If, on the other hand, ascent took place in protected-core cumulonimbi as indicated by Riehl and Malkus, the rising air would follow curve E_F (approximating to a moist adiabat), arriving in the upper troposphere with an energy content corresponding to that at low levels.

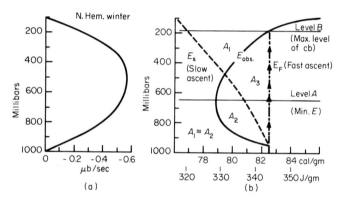

FIG. 17.5 Maximum mean vertical motion (a) and specific energy E_{obs} (b) in latitude belt 0 to 5°N, Northern Hemisphere winter. (See text.)

We may now compare the vertical distributions of energy in the mean equatorial atmosphere, corresponding to these different processes. Under the assumption that the radiative cooling rate would not be influenced by the mechanism of vertical heat flux, the total integrated energy between the surface and the layer to which the ascending branch reaches would be the same. If we assume that the air ascends from 950 to 180 mb, the integrated energy loss during slow ascent would be given by

$$\frac{1}{g} \int_{180\ mb}^{950\ mb} (E_F - E_s)\ dp = \frac{1}{g} (A_1 + A_3)$$

where A_1 and A_3 are the areas in Fig. 17.5b. Similarly, $(A_2 + A_3)$ represents the radiative loss by the mean equatorial atmosphere, in the case wherein the net mass ascent to level B takes place in fast upward currents in cumulonimbi occupying only a small part of the total area. A_1 and A_2 are nearly equal, consistent with the assumption above.

Comparing these two mechanisms, A_2 represents a deficit of the atmospheric energy, below that which would be realized in the case of uniform ascent, while A_1 represents an excess. This shows that the upper troposphere is heated more by convection and the lower troposphere less, but that the total energy content of the atmospheric column is the same as it would be if the air had moved upward at the mean rate; i.e., $\int E_s \, dp = \int E_{\mathrm{obs}} \, dp$. The height to which the convection can penetrate, 180 mb in Fig. 17.5b, is governed by the energy content in the surface layers. A certain part of the mass ascending in the equatorial belt reaches a higher level, but this part obviously ascends with a somewhat greater value (83 to 83.5 cal/g) of E from the warmer belt 0 to 5°S is this season. Likewise, in parts of the tropics where the surface energy is smaller, the maximum height reached by convection will be lower; and in all regions, most convective clouds will fail to attain level B because of erosion due to entrainment.

Outside the tropics, the situation is entirely different, since (in the mean) the static energy E increases with height in the lower as well as the upper troposphere. Widespread deep vertical convection being generally impossible, the different mechanism of *slantwise* convection in extratropical cyclones most be called upon. Air cannot rise vertically as in the tropics, but ascending motion with energy transport to the middle or upper troposphere can take place if there is poleward movement at the same time, and descent can take place with equatorward movement. In the case of ascending motion, as stressed in Chapter 10, the release of latent heat is very important, just as it is in the tropics. For example, the contribution of Lq at the surface near 30°N is about 20 J/g. From Fig. 17.4 it is evident that air with the corresponding total energy content could, in an extratropical disturbance, rise to the tropopause level; without the energy provided by latent heat release, ascent would be possible only to the middle troposphere. In individual disturbances, the energy isopleths of Fig. 17.4 would be displaced poleward on the east sides and equatorward on the west sides of cyclones, in harmony with the distortion of the temperature and moisture field at a given level. Thus a simultaneous upward and poleward transfer of E is implied by the correlations between v, $-\omega$, and E at a given latitude in the disturbances, as discussed for sensible heat in Section 16.3.

It should be noted that there are local exceptions to the preceding generalization in which vertical convection occurs in extratropical latitudes. This takes place, for example, in parts of deep cold outbreaks over warm ocean surfaces in winter (Figs. 12.4 through 12.6), and over land regions with strong heating of the surface layers in the warm season. Also, vertical instability develops in the warm sectors of some cyclones, even in winter, through the process of differential advection discussed in Section 13.2. In these cases the static energy distribution is in restricted localities strongly distorted from the pattern in Fig. 17.4, and the air can move vertically in the same fashion as in the tropics. At the same time it should be emphasized that Fig. 17.4 illustrates longitudinal averages in the tropics; although convection is the dominant form of cloudiness, the stable type of cloud and precipitation may be observed even in quite low latitudes, especially over the eastern parts of the oceans.

The meridional distribution of static energy in summer is similar; however, the potentially unstable regions limited by lines *A* and *B* extend poleward to about 45°N, when averaged around the hemisphere. There are, of course, marked longitudinal deviations in the seasonal mean conditions, as in winter.

17.3 Contributions of Energy Transfer in Different Forms

In Chapter 2 the energy transfers across latitudes were presented as vertically integrated quantities. As an aid in visualizing them in relation to the circulation features shown in Fig. 17.2, we illustrate here the principal transfers in a different way.

The mean flux of water vapor, based on Peixoto and Crisi's computations (1965) for the six cooler months of 1958, is shown in Fig. 17.6. In

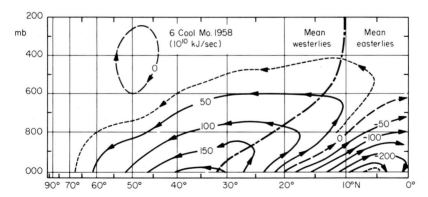

Fig. 17.6 Total flux of water vapor in Northern Hemisphere, six cooler months of 1958. (From data of Peixoto and Crisi, 1965.) Each channel carries 50×10^{10} kJ/sec latent heat or 2×10^{11} g/sec water vapor. Dash-dotted line is southern boundary of mean westerly winds.

this figure, each channel between arrows transports 50×10^{10} kJ/sec of latent energy, or 2×10^{11} g/sec of water vapor (integrated amounts around the hemisphere). The flux lines are meaningful with regard to the vapor transports at different heights across latitude circles and to the water flux through the earth's surface. Thus a flux channel leaving the earth's surface indicates that, within the latitude belt embraced, the evaporation exceeds the precipitation returned to the surface by 2×10^{11} g/sec.

As seen from Fig. 17.6, almost 60% of the excess vapor acquired by the atmosphere in the belt 5 to 35°N is transferred southward in winter, partly into the Southern Hemisphere and partly supplying the vapor for rainfall just north of the Equator. Equatorward of about 20°N, the meridional flux is dominated by transport in the lower or trade-wind branch of the Hadley cell, where (Fig. 17.1) the mean specific humidity is large. North of this latitude, the flux by eddies predominates (cf. Fig. 14.4). This general scheme conforms with what would be expected from the flow characteristics described in Fig. 17.2.

While the greatest poleward transfer of latent heat takes place near 35°N, the corresponding *geostrophic eddy flux* of sensible heat (Fig. 17.7) reaches a maximum near 50°N. In this figure, only the meridional fluxes are meaningful because of the heat sources and sinks within the atmosphere and the interchange between sensible heat and various other forms of energy.

The partition between different mechanisms of energy transfer can be most conveniently illustrated by the vertical profiles in Fig. 17.8. These are shown for three latitudes: 50°N, where eddy disturbances predominate

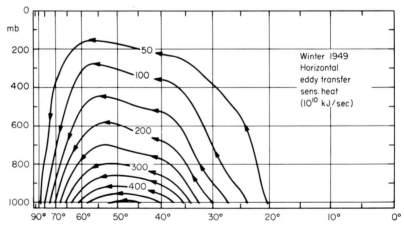

Fig. 17.7 Meridional geostrophic eddy flux of sensible heat; winter, 1949. (From data of Mintz, 1955.) Channels carry 50×10^{10} kJ/sec. Only the horizontal component of flux is meaningful. (See text.)

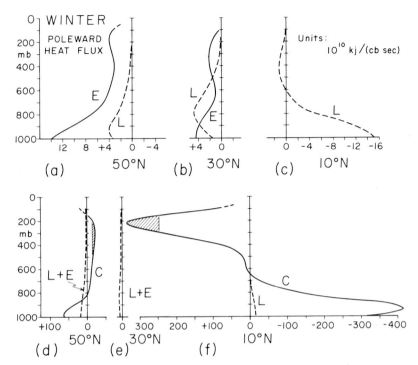

FIG. 17.8 Vertical profiles of meridional energy fluxes at 50°N, 30°N, and 10°N in winter. (a) to (c) show eddy flux of sensible heat (*E*) and total flux of latent heat (*L*), as in Figs. 17.6 and 17.7; (d) to (f) show mean circulation flux (*C*) of energy ($c_pT + gz$) and, for comparison, the sum of the latent and eddy heat fluxes. (For remarks on compatibility of the data, see text.)

and the Ferrel circulation is also significant; 30°N, where eddies are important but the mean meridional circulation is essentially absent; and 10°N, where the Hadley circulation is strong but the eddy flux is weak.

It should be stressed that the data used are not completely compatible. The circulation flux is for mean winter (December–February) conditions; the eddy heat flux, for a particular winter; and the latent heat flux, for the six cooler months. Considering the strong seasonal change of the Hadley circulation, the latent heat flux at 10°N may, for example, be much too low in comparison with the other fluxes. However, our purpose here is only to illustrate the main features, which would not be much changed if compatible seasons were used.

Because of its great size in the low and high troposphere, the circulation flux of realized energy $\bar{v}(c_pT + gz)$ is presented in Fig. 17.8d-f on a scale different from that of the eddy sensible heat flux and total latent heat flux in Fig. 17.8a-c. Since $(c_pT + gz)$ increases with height in a stable atmosphere, opposing poleward and equatorward mass transports in upper and lower levels imply that in the Hadley cell, there is a net poleward transfer of realized energy, whereas in the Ferrel cell, there is a net equatorward transfer. The net circulation flux of realized energy is represented by shaded areas in Figs. 17.8d and f for comparison with the other two forms of energy flux $(L + E)$.

At 50°N (Fig. 17.8a), the required poleward flux of energy is accomplished mainly by eddy heat transfer, although the poleward transfer of latent heat is significant. The opposing effect of the Ferrel circulation is minor, amounting to about 15% of the eddy sensible heat flux. At 30°N (Fig. 17.8b), the eddy fluxes of sensible and latent heat are almost equal; the circulation flux, small and uncertain, is not shown. At 10°N (Fig. 17.8c), the eddy flux of sensible heat is small and not shown; a large equatorward latent heat flux takes place, but this is more than compensated by the poleward flux of realized energy due to the mean Hadley circulation. From the standpoint of the energy budget in the 0 to 10° latitude belt, an overall poleward energy flux at 10°N is possible, since heat is added from the surface within this belt; in addition (Table 2.5), a significant northward flux of $(c_pT + gz)$ takes place across the Equator, exceeding the southward flux of latent heat there.

In the central latitudes of both Ferrel and Hadley cells, the circulation flux of energy at upper- and lower-tropospheric levels is much greater than the combined eddy sensible heat and latent heat transfers at corresponding levels (Figs. 17.8d, f). The net circulation flux is in both cases very small compared with the flux in the upper or lower branch. Comparing Figs. 17.8e and 17.8f, it is seen that the poleward energy transfer in the upper troposphere at 30°N is very small relative to that at 10°N. The interpretation is, of course, that practically all energy transported poleward in the upper branch of the Hadley cell is carried downward in its descending branch and thence equatorward in the lower branch.

To illustrate the disposition of energy, we consider the ring of air above 650 mb, and between 10°N and the poleward boundary of the Hadley cell. In the mean circulation, the same amount of air leaves this ring through the 650-mb level as enters it across the vertical wall at 10°N, but it leaves with a somewhat lower energy content (Fig. 17.4). The meridional convergence of energy is thus greater than the vertical divergence of energy. The difference is partly expressed by a radiative-energy loss, and partly by an export across the poleward boundary by eddy heat transfer. For the most part, the air in the Hadley cell is simply recirculated, huge amounts of energy being carried in its branches, slightly more in a poleward than in an equatorward direction to provide the necessary net meridional energy transfer.

The heat budgets in different latitudes were discussed in detail in Chapter 2. At this point it may be useful to give a broad summary of the fluxes that balance the heat budget of the extratropical regions. These are shown, considering *annual averages*, in Table 17.1, which is based on Tables 2.3 and 2.4. Here, the various contributions are expressed in terms of their effect upon changing the mean temperature of the entire atmosphere over the regions concerned.

TABLE 17.1

COMPONENTS OF ATMOSPHERIC HEATING AVERAGED OVER THE EXTRATROPICAL REGIONS
OF NORTHERN AND SOUTHERN HEMISPHERES

Region:	30–90° N		30–90° S	
Influence of net radiation:	-0.87°C/day	100%	-0.93°C/day	100%
Components of heating:				
Meridional flux of $(c_p T + gz)$	$+0.15$°C/day	17%	$+0.18$°C/day	19%
Meridional flux of latent heat	$+0.08$°C/day	9%	$+0.10$°C/day	11%
Sensible heat flux from earth	$+0.19$°C/day	22%	$+0.12$°C/day	13%
Latent heat flux from earth	$+0.45$°C/day	52%	$+0.53$°C/day	57%
Total heating from fluxes	$+0.87$°C/day	100%	$+0.93$°C/day	100%

Considering the whole regions poleward of 30° latitude, the average net radiative loss by the atmosphere is about 0.9°C/day, being slightly greater in the Southern than in the Northern Hemisphere. About 26 to 30% of this deficit is compensated by energy flux from the tropics, two-thirds of this being in the form of sensible heat and geopotential, and the remainder being latent heat flux. The remaining 70 to 74% of the radiative deficit is compensated by heat transfer from the earth's surface within the extratropical regions concerned, about three-quarters of this being in the form of latent heat of evaporation. As would be expected from its mainly oceanic nature, the portion of this transfer expressed by evaporation is relatively greater in the Southern Hemisphere. Of the total heat supply for the extratropical regions, 61% is derived from latent heat released through condensation and precipitation in the Northern Hemisphere, and 68% originates in this way in the Southern Hemisphere.

17.4 WAVELENGTH AND GEOGRAPHICAL CONSIDERATIONS

In connection with the preceding summary and that in Section 16.5, it is appropriate to give some indications of how the existing energy of the disturbances is distributed according to wave number, and also how this is related to heat and angular-momentum transport and to energy conversions. The data cited below all pertain to the Northern Hemisphere and may be expected to differ considerably in the Southern Hemisphere because of its unlike physiography.

Eliasen (1958) gives a complete harmonic analysis of daily maps at 1000 and 500 mb from Oct. 21 to Nov. 30, 1950. At latitudes south of about 50°N, the kinetic energy of the zonally averaged flow (at 500 mb) was larger than that of the eddies, whereas at higher latitudes there was considerably more energy in the eddies. Wave numbers 1 to 4 collectively contained roughly the same amount of kinetic energy as the group in wave numbers 5 to 10. In the longest waves, most of the kinetic energy was in the zonal component of motion, while in wave numbers 4 to 10 most of the energy was in the meridional component during this period.

Based on harmonic analyses of 500-mb charts at 50 to 55°N for January–February 1951 and 1953, Van Mieghem *et al.* (1959) found a pronounced peak in the mean meridional kinetic energy ($\bar{K}_y$) at hemispheric wave numbers 3 and 4, with a broader secondary maximum centered on wave number 7. A further analysis of different aspects was carried out for January–April 1953. This disclosed that 80% of the high $\bar{K}_y$ in wave number 3 was contributed by the quasi-stationary waves, which dominate the winter mean circulation (Fig. 3.1). At wave number 4, which commonly characterizes the long waves, the contribution from stationary and moving waves was about equal. For higher wave numbers, $\bar{K}_y$ was predominantly associated with moving waves.

The meridional eddy flux of sensible heat, and also of westerly momentum, was strongest for wave number 3, with a secondary maximum for wave number 1 (which expresses the off-center circulation with respect to the geographic pole). The eddy heat flux by the moving shorter waves was appreciably smaller. Since the computations were made for the 500-mb level and most of the flux of sensible as well as latent heat takes place below that level, this cannot be taken as a conclusive comparison for the disturbances through all depths. However, Shaw (1966) computed the heat advection from the surface to 100 mb for four midseason months. For January, he found that this was strongly dominated by the mean waves, particularly the two troughs centered over the continental east coasts;

while in agreement with the data of Van Mieghem *et al.*, there was a lesser contribution by the shorter mobile waves. In interpreting this comparison it is necessary to bear in mind that the mean waves are themselves to a large extent a statistical manifestation of the repeated formation of strong cyclones and cold outbreaks in these localities.[6]

Haines and Winston (1963) provide data on the seasonal and longitudinal variations of heat transport, based on daily computations for $3\frac{1}{2}$ yr at 850 and 500 mb (which Winston has demonstrated to be well correlated with the transport through the whole troposphere). These bring out the strong seasonal variation, the maximum poleward transfer of sensible heat in midsummer being only about one-sixth of that in midwinter (see Figs. 2.7 and 2.8). In agreement with other computations (Chapter 2 and Section 17.3), the geostrophic eddy heat flux is strongest at 45 to 50°N, with maximum heat flux convergence in latitudes 55 to 65°N and maximum flux divergence at 30 to 40°N. Computations based on the $v'T'$ correlation at 45°N disclose three principal regions where the contribution to poleward heat transport is greatest during the colder half-year. The sharpest maximum, anchored near 120°E, is associated with northerly flow of cold air from Asia. The other maxima are less pronounced and showed a greater variation from year to year. These are, according to Haines and Winston, mainly contributed by southerly flow of warm air to the rear of the ridge over western North America (strongest transport near 140°W) and the poleward flow of warm air over the Atlantic (strongest transport near 40°W). Although the effect of equatorward flow of cold air off the east coast of North America is evident, this is much less pronounced than the effect of the corresponding flow west of the East Asiatic trough. Somewhat surprisingly, the mean northerly flow over west and central North America was associated with a minimum (sometimes negative values) of poleward heat flux, a consequence of the fact that air moving southward to the east of a well-developed ridge is often relatively warm in this region (and also over western Europe).

Wiin-Nielsen (1959) has evaluated the generation of kinetic energy for January and April 1959, based on daily computations of the vertical motion

---

[6] In the Southern Hemisphere with its smaller influence of continents, the relative contribution of migratory disturbances must be considerably greater. Comparison of the 500-mb mean charts for the two hemispheres in winter (Figs. 3.1 and 3.4) or in summer (Figs. 3.2 and 3.3) indicates the smaller average influences of geographically fixed waves in the Southern Hemisphere. On a daily basis, the meridional flow associated with waves is slightly weaker in southern winter than in northern winter, but appreciably stronger in southern summer than in northern summer (van Loon, 1965).

field by the Joint Numerical Weather Prediction Unit in Washington. Although the vertical motions are generally underestimated, with a resulting underestimate in the energy conversion, Wiin-Nielsen's computations of the partition according to wave number are of interest. These indicate a distinct maximum of generation in wave numbers 6 and 7 (which, as noted above, also contain the greatest existing meridional kinetic energy). Wiin-Nielsen points out that this, equivalent to a wavelength of 4000 to 4700 km, "coincides well with the most unstable waves as predicted from the linear perturbation theory." Furthermore, his finding that little conversion takes place at wave numbers 10 or greater ($L \leq 2800$ km) corresponds to the abrupt cutoff in wave instability predicted by the theory for short waves.

In broad agreement with the findings of Van Mieghem *et al.*, cited above, Wiin-Nielsen *et al.* (1963) determined for January 1962 that about half of the poleward transport of both heat and westerly momentum is carried out by wave numbers 1 to 4, and somewhat more than a quarter of the total by wave numbers 5 to 8 (the relative amounts varying with latitude).

In the illustrations given above and in earlier chapters, the basic features have generally been illustrated as longitudinal averages. For maps of heat sources and sinks over the Northern Hemisphere, reference may be made to Clapp (1961), Adem (1964), and Asakura and Katayama (1964). Clapp provides a synthesis of the results of other workers which, while indicating the validity of the broad aspects, at the same time emphasizes the great uncertainties in quantitative values derived by different methods.

The principal deviations from the longitudinal average, shown by such maps, are the large net heating rates in belts extending northeastward from the low-latitude east coasts of Asia and North America. These are identified with the release of latent heat in cyclones moving over these favored tracks and also with the large ocean-to-air sensible heat transfer off the continental east coasts in winter. As Clapp illustrates, when mean conditions are considered, "the field of heating is almost 90° out of phase with that of temperature," but somewhat less; a requirement that for energy production (i.e., restoration of potential energy consumed in producing kinetic energy) the heat sources be positively correlated with the temperature field.

17.5 CONCLUSION

Although there are many gaps in our present knowledge of atmospheric circulation systems, when one compares the present state with that of two to three decades ago, he is forced to acknowledge that remarkable progress has taken place in this short time. As in many spheres of endeavor, a firm physical groundwork had earlier been laid for studying the behavior of the atmosphere. On examining the contributions from times when only rudimentary observations were available, one is struck with the brilliance of some of the earlier deductions. Recent advances have been largely stimulated by the circumstance of vastly improved observations, which for the first time have ensured a fairly confident analysis of circulation features that earlier could be only speculated. The ability to utilize the staggering amount of observational information has, of course, been possible in some types of studies only because of the parallel development of data processing and electronic numerical computation techniques.

There still remain many basic features of the atmosphere whose nature is incompletely understood. Among the most important of these are the processes of energy transfer at the earth's surface and in the adjacent "boundary layer" of the atmosphere, and the mechanisms of transfer of properties by convective clouds and their interactions with the large-scale circulations, especially those of the tropics. Although some of the principal mechanisms governing cyclogenesis have been exposed, it cannot be said that the process is completely understood. For example, it has been possible to produce replicas of cyclones in numerical prediction experiments without taking into account the existence of fronts (although in some cases these develop in rudimentary form). Since well-developed fronts are characteristic of the real atmosphere and much of the available potential energy and kinetic energy are concentrated in their vicinity, there is still to a certain extent an open question about the fundamental physical relationships between frontogenesis and cyclogenesis. Such questions as these, and many others, suggest that there is still abundant scope for investigating the workings of the atmosphere.

References

Adem, J. (1964). On the physical basis for the numerical prediction of monthly and seasonal temperatures in the troposphere—ocean—continent system. *Monthly Weather Rev.* **92**, 91–104.

Asakura, T., and Katayama, A. (1964). On the normal distribution of heat sources and sinks in the lower troposphere over the Northern Hemisphere. *J. Meteorol. Soc. Japan* **42**, 209–244.

Clapp, P. F. (1961). Normal heat sources and sinks in the lower troposphere in winter. *Monthly Weather Rev.* **89**, 147–162.

Crutcher, H. L. (1961). Meridional cross sections. Upper winds over the Northern Hemisphere. *Tech. Paper* No. 41, 307 pp. U. S. Weather Bur., Washington, D. C.

Eliasen, E. (1958). A study of the long atmospheric waves on the basis of zonal harmonic analysis. *Tellus* **10**, 206–215.

Eliassen, A. (1952). Slow thermally or frictionally controlled meridional circulation in a circular vortex. *Astrophys. Norvegica* **5**, 19–59.

Fultz, D., Long, R. R., Owens, G. V., Bohan, W., Kaylor, R., and Weil, J. (1959). Studies of thermal convection in a rotating cylinder with some implications for large-scale atmospheric motions. *Meteorol. Monographs* **4**, No. 21, 1–104.

Green, J. S. A., Ludlam, F. H., and McIlveen, J. F. R. (1966). Isentropic relative-flow analysis and the parcel theory. *Quart. J. Roy. Meteorol. Soc.* **92**, 210–219.

Haines, D. A., and Winston, J. S. (1963). Monthly mean values and spatial distribution of meridional transport of sensible heat. *Monthly Weather Rev.* **91**, 319–328.

Heastie, H., and Stephenson, P. M. (1960). Upper winds over the world. *Geophys. Mem.* No. 103, 1–217.

Holopainen, E. O. (1966). Some dynamic applications of upper-wind statistics. Finnish Meteorol. Off. Publ. No. 62, 22 pp.

Kao, S.-K., and Taylor, V. R. (1964). Mean kinetic energies of eddy and mean currents in the atmosphere. *J. Geophys. Res.* **69**, 1037–1049.

Kuo, H. L. (1956). Forced and free meridional circulations in the atmosphere. *J. Meteorol.* **13**, 561–568.

London, J. (1957). A study of the atmospheric heat balance. Final Rept., Contr. AF 19 (122)-165, 99 pp. Dept. Meteorol. Oceanog., New York University.

Mintz, Y. (1955). Final computation of the mean geostrophic poleward flux of angular momentum and of sensible heat in the winter and summer of 1949 (Article V). Final Rept., Gen. Circ. Proj., Contr. AF 19(122)-48. Dept. Meteorol., Univ. California, Los Angeles.

Peixoto, J. P., and Crisi, A. R. (1965). Hemispheric humidity conditions during the IGY. Sci. Rept. No. 6, 166 pp. Planetary Circ. Proj. Dept. Meteorol., M.I.T., Cambridge, Massachusetts.

Shaw, D. B. (1966). Note on the computation of heat sources and sinks in the atmosphere. *Quart. J. Roy. Meteorol. Soc.* **92**, 55–66.

van Loon, H. (1965). A climatological study of the atmospheric circulation in the Southern Hemisphere, Part I: 1 July 1957–31 March 1958. *J. Appl. Meteorol.* **4**, 479–491.

Van Mieghem, J., Defrise, P., and Van Isacker, J. (1959). On the selective rôle of the motion systems in the atmospheric general circulation. *In* "The Atmosphere and the Sea in Motion" (B. Bolin, ed.), pp. 230–239. Rockefeller Inst. Press, New York.

Wiin-Nielsen, A. (1959). A study of energy conversion and meridional circulation for the large-scale motion in the atmosphere. *Monthly Weather Rev.* **87**, 319–331.

Wiin-Nielsen, A., Brown, J. A., and Drake, M. (1963). On atmospheric energy conversions between the zonal flow and the eddies. *Tellus* **15**, 261–279.

AUTHOR INDEX

Numbers in italics refer to the pages on which the complete references are listed.

A

Ackerman, B., 490, *519*
Adem, J., 164, *165*, 472, *519*, 584, *585*
Alaka, M. A., 229, *233*, 504, 509, *519*
Alpert, L., 465, *467*
Anderson, C. E., 412, *422*
Anderson, R., 108, *115*
Andre, M. J., 394, *422*
Angell, J. K., 207, *233*
Arakawa, A., 77, *99*, 114, *116*, 225, *234*
Arakawa, H., 212, *233*, 478, *519*
Archbold, J. W., 464, *467*
Árnason, G., 316, *350*, 376, *387*
Asai, T., 55, *64*, 418, *423*, 490, *519*
Asakura, T., 584, *585*
Aspliden, C. I., 447, 459, 461, *467*
Assmann, R., 120, *136*
Austin, E. E., 3, *23*, 67, *98*, 178, *193*
Austin, J. M., 83, *97*, 147, *166*, 241, *271*
Austin, P. M., 419, *422*

B

Baer, F., 334, *350*, 517, *520*
Ball, F. K., 192, *193*
Bannon, J. K., 91, *101*, 178, *193*, 369, 379, *387*
Barrett, E. W., 93, *97*
Bates, F. C., 398, 410, 413, 420, *422*
Batten, E. S., 78, 79, *97*
Baur, F., 27, 45, *64*
Beckwith, W. B., 395, 396, *422*
Beebe, R. G., 398, *422*
Beer, C. G. P., 435, *468*
Bellamy, J. C., 220, *233*

Belmont, A. D., 82, *97*
Benton, G. S., 27, *64*, 384, *387*
Bergeron, T., 1, *23*, 27, 29, *65*, 102, 103, 104, 108, *115*, *116*, 118, 119, 120, 122, 125, 130, *136*, *137*, *138*, 164, *166*, 190, 191, 192, *193*, 238, 239, *270*, 352, 353, 354, *387*, *388*, 401, 402, *422*, 474, 476, 497, 516, *519*
Berggren, R., 109, 115, *116*, 176, 177, *193*, 199, *233*, 274, 276, *313*
Berkofsky, L., 68, *97*
Berson, F. A. 262, *270*
Bertoni, E. A., 68, *97*
Bigelow, F. H., 119, *137*
Birchfield, G. E., 334, *350*, 517, *520*
Bjerknes, J., 1, 9, 16, *23*, 27, 29, 43, *64*, *65*, 73, *97*, 102, 109, *116*, 123, 124, 125, 127, 128, 129, 131, 133, 134, 135, *137*, *138*, 141, 145, 146, 147, *165*, 171, 182, 190, *193*, 201, *233*, 262, *270*, 335, 340, *350*, 352, 353, *388*, 417, *422*
Bjerknes, V., 1, *23*, 102, *116*, 122, 124, *137*, 205, *233*
Blackadar, A. K., 232, *233*, 398, *422*
Blasius, W., 119, *137*
Bleck, R., 288, *313*
Bleeker, W., 267, *270*, 394, *422*
Bohan, W., 572, *586*
Bolin, B., 68, 71, *98*, 115, *116*, 274, 276, *313*
Bonner, W. D., 230, *233*, 345, *350*
Borovikov, A. M., 353, 369, *387*
Boucher, R. J., 353, *387*
Boville, B. W., 83, *98*, *100*, 108, *115*
Boyden, C. J., 357, *387*

587

SUBJECT INDEX

A

Aerology
  early studies, 131-133
  indirect, air mass properties, 130
Ageostrophic motion
  curved flow, 201, 220-224
  dynamically unstable ridge, 335-336
  frontogenesis, 243-245, 251-253, 262, 267-268
  Hadley cell, 564
  jet streaks, 201-204, 209-210
  kinetic energy production, 21, 534-536, 544, 564-565
  low-level jet stream, 230-232
Air mass, 102-115
  depth change in meridional movement, 289-292
  relation to main jet streams, 107-108, 113-115, 569
    to meridional cells, 104-105
  simplified classification, 103-104, 568-569
  structure in tropics, 438, 473, 475
Advection, horizontal
  development of thermal field, 322-331
  differential, stability changes, 397-399
  thermal, and vertical motion, 548-549
  vorticity, related to divergence, 317-319
Air mass exchange, see also Disturbances
  cutoff lows and highs, 274-278
  cyclone families, relation to, 128-129
  related to evolution of disturbances, 341-344
  three-dimensional trajectories in, 106, 156-157, 308-312
Air mass modification, 156, 361-365, 436-437, see also Evaporation, Heat transfer
Angular momentum

budget, northern winter, 15-16
conversion, earth- to relative-, 17, 493-494, 568
maintenance of zonal circulation, 17-20
Angular momentum flux, 8-23, 566-568, see also Hurricane, Torque
  convective storms, 421-422
  eddy, 9-13
  mean circulation, 15
  vertical, 14-16, 22, 286-288, 568
Anticyclone, see also Blocking, Disturbances
  latitudinal frequency, 92-93
  subtropical, boundary of tropics, 9-10, 103-106, 426
  upper, low latitude, 459
  warm, high latitude, 274-278
Atmospheric structure, hemispheric scale
  disturbance frequency and tracks, 92-96
  mean isobaric contours, 67-73, 428
  mean temperature field, 3, 79
  mean wind field, 4, 74, 77, 79, 88, 94, 428, 567
  meridional wind component, 5
  schematic illustrations, 105, 114, 151, 155, 158, 569

B

Baroclinity, see also Cyclone, Front, Jet stream, Potential energy, Vertical shear
  basic current, related to disturbance type, 144-145, 444-445
Blocking, 157, 274-278

C

Circulation, meridional, see Ferrel cell, Hadley cell